## COMPACT
# FIELD GUIDE TO AUSTRALIAN BIRDS

## TEXT & ILLUSTRATIONS BY
## MICHAEL MORCOMBE AM

Woodslane Press Pty Ltd
10 Apollo Street
Warriewood, NSW 2102
Email: info@woodslane.com.au
Tel: 02 8445 2300  Website: www.woodslanepress.com.au

Originally published in Australia in 2004 by Steve Parish Publishing
2nd edition published in Australia in 2016 by Pascal Press
This 3rd edition published in Australia in 2024 by Woodslane Press

Text, maps & images © 2004 and 2024 Michael Morcombe
© 2024 Woodslane Press

The right of Michael Morcombe to be identified as moral rights author and as illustrator is asserted in accordance with the *Copyright Amendment (Moral Rights) Act 2000*.

This work is copyright. All rights reserved. Apart from any fair dealing for the purposes of study, research or review, as permitted under Australian copyright law, no part of this publication may be reproduced, distributed, or transmitted in any other form or by any means, including photocopying, recording, or other electronic or mechanical methods, without the prior written permission of the publisher. For permission requests, write to the publisher, addressed "Attention: Permissions Coordinator", at the address above.

The information in this publication is based upon the current state of commercial and industry practice and the general circumstances as at the date of publication. Every effort has been made to obtain permissions relating to information reproduced in this publication. The publisher makes no representations as to the accuracy, reliability or completeness of the information contained in this publication. To the extent permitted by law, the publisher excludes all conditions, warranties and other obligations in relation to the supply of this publication and otherwise limits its liability to the recommended retail price. In no circumstances will the publisher be liable to any third party for any consequential loss or damage suffered by any person resulting in any way from the use or reliance on this publication or any part of it. Any opinions and advice contained in the publication are offered solely in pursuance of the author's and publisher's intention to provide information and have not been specifically sought.

A catalogue record for this book is available from the National Library of Australia

Printed in China through Asia Pacific Offset Ltd
Original book design by Leanne Nobilio
Revisions in this edition by Luke Harris
Editor: Wynne Webber
Proofreaders: Debra Hudson & Karin Cox
Publisher: Andrew Swaffer
Base map of Australia supplied by MAPgraphics, Brisbane, Australia
Front cover, top to bottom: Varied Honeyeater, Azure Kingfisher, Red-backed Fairy-wren
Back cover: Eastern Rosellas
Title page: Rainbow Bee-eater
Page 2: top, Superb Fairy-wren;
right: Australian Owlet-nightjar
Page 3: Large-tailed Nightjar
Page 4: top, Gouldian Finches;
bottom, Rufous-crowned Emu-wrens
Page 5: top, Azure Kingfisher;
bottom, Long-tailed Finches (yellow-billed form)

MIX
Paper | Supporting responsible forestry
FSC® C136333

# GUIDE TO FAMILY GROUPS

See the back endpapers for colour-coded Quick Index and Guide to Family Groups, or refer to colour tags on page edges.

| # | Family Groups | Pages |
|---|---|---|
| 1 | Emu, Cassowary, Megapodes, Quails | 10–13 |
| 2 | Geese, Swans, Ducks, Grebes, Penguins | 14–27 |
| 3 | Petrels, Shearwaters, Prions, Albatrosses, Storm-Petrels, Diving Petrels | 28–59 |
| 4 | Tropicbirds, Boobies, Darter, Cormorants, Pelican, Gannets, Frigatebirds | 60–69 |
| 5 | Egrets, Herons, Bitterns, Spoonbills, Ibises, Stork, Cranes | 70–81 |
| 6 | Osprey, Kites, Eagles, Buzzard, Harriers, Baza, Sparrowhawk, Goshawks, Kestrel, Hobby, Falcons | 82–101 |
| 7 | Moorhen, Swamphen, Native-hens, Coot, Bush-hen, Crakes, Rails, Bustard | 102–109 |
| 8 | Button-quails, Snipes, Dowitchers, Godwits, Curlews, Shanks, Yellowlegs, Sandpipers, Knots, Tattlers, Sanderling, Stints | 110–135 |
| 9 | Phalaropes, Jacanas, Stone-curlews, Stilts, Pratincoles, Oystercatchers, Plovers, Dotterels, Lapwings | 136–151 |
| 10 | Skuas, Jaegers, Gulls, Terns, Noddies | 152–169 |
| 11 | Doves, Pigeons | 170–179 |
| 12 | Cockatoos, Lorikeets, Parrots, Rosellas | 180–205 |
| 13 | Cuckoos, Coucal, Owls, Frogmouths, Nightjars | 206–221 |
| 14 | Swifts, Kingfishers, Bee-eater, Dollarbird | 222–231 |
| 15 | Pittas, Lyrebirds, Scrub-birds, Bristlebirds, Treecreepers | 232–237 |
| 16 | Fairy-wrens, Emu-wrens, Grasswrens | 238–249 |
| 17 | Pardalotes, Scrubwrens, Heathwrens, Fieldwrens, Redthroat, Gerygones, Thornbills, Whitefaces | 250–269 |
| 18 | Wattlebirds, Friarbirds, Honeyeaters, Miners, Spinebills, Chats, Gibberbird | 270–293 |
| 19 | Scrub-robins, Robins, Flycatchers, Logrunner, Chowchilla, Babblers | 294–303 |
| 20 | Whipbirds, Wedgebills, Quail-thrushes, Sittella, Bellbird, Shrike-tit, Whistlers, Shrike-thrushes | 304–313 |
| 21 | Monarch Flycatchers, Boatbill, Fantails, Willie Wagtail, Drongo, Magpie-lark | 314–321 |
| 22 | Trillers, Cuckoo-shrikes, Cicadabird, Orioles, Figbird, Woodswallows, Butcherbirds, Magpie, Currawongs | 322–333 |
| 23 | Riflebirds, Manucode, Ravens, Crows, Chough, Apostlebird, Catbirds, Bowerbirds | 334–341 |
| 24 | Thrushes, Flycatchers, Starlings, Swallows, Martins, White-eyes, Mistletoebird, Sunbird | 342–349 |
| 25 | Warblers, Grassbirds, Spinifexbird, Cisticolas, Sparrows, Wagtails, Pipits, Feral Finches | 350–357 |
| 26 | Finches, Firetails, Mannikins, Munia | 358–365 |

Michael Morcombe took his first photograph of a bird when he was in high school. It was a black-and-white shot of a Grey Fantail on its nest.

From that time, Michael was rarely out of hides in locations from swamps to the very tree tops, watching and following birds, finding their nests, learning their calls and behaviour – all in pursuit of the perfect photo of each species. Along the way, he found that his interest had moved a little from photographing birds to the birds themselves: he had gained vast up-close and personal experience of birds, their calls, habits, habitats, nest site preferences and behaviour around nests. And he began to spend time sketching birds as well as photographing them.

In 1965, Michael published his own large-format bird calendar, which he followed with another in 1966. His career as a photographer–author then took off: his first book was *Wild Australia* in 1968, quickly followed by *Australia's National Parks* in 1969. He is author of some forty titles, many of them coauthored by his wife, Irene, who tends to specialise in flora. Michael's books span the range of natural history and landscape, but Australian birdlife remains his chief passion: *Birds of Australia* appeared in 1970; *Australian Bush Birds* in 1974; *The Great Australian Bird Finder* in 1984; and, in 1990, *Australian Birds in the Wilderness*, which won the NSW Royal Zoological Society's 1991 Whitley Award for Best Photography Book. He has won numerous awards in international photography exhibitions, and the 1969 C.J. Dennis award, given by the Fellowship of Australian Writers for the best work on an Australian subject.

But throughout this time, Michael experienced the frustration of being unable to get that perfect shot of a particular species, or of taking the perfect shot to illustrate a bird's behaviour only to find that the photograph was ruined by blurring or distance. Over time, he turned to painting and sketching birds, drawing on his detailed knowledge of birds and their habits. Over the years, he spent much time refining what he would find most useful in a bird guide, and so he wrote, illustrated and assembled the *Field Guide to Australian Birds*, an authoritative, easy-to-use handbook that quickly became a favourite. Michael felt that a more portable book was needed in the field, and here it is, the *Compact Field Guide to Australian Birds*.

# CONTENTS

| | |
|---|---|
| Dedication & Acknowledgements | 4 |
| Introduction | 5 |
| How to use this guide | 6 |
| Parts of the plumage | 8 |
| Native bird orders; families; family groups | 8 |
| Field marks & illustrated glossary | 9 |
| 1  Emu, cassowary, megapodes & quail | 10 |
| 2  Geese, ducks, swans, grebes & penguins | 14 |
| 3  Albatrosses, petrels & shearwaters | 28 |
| 4  Pelican & tropicbirds to frigatebirds | 60 |
| 5  Herons & egrets to cranes | 70 |
| 6  Raptors | 82 |
| 7  Native-hens, crakes, rails, coot & bustard | 102 |
| 8  Button-quail & migratory waders | 110 |
| 9  Phalaropes to plovers & pratincoles | 136 |
| 10  Skuas, jaegers, gulls, terns & noddies | 152 |
| 11  Doves & pigeons | 170 |
| 12  Cockatoos, lorikeets & parrots | 180 |
| 13  Cuckoos, owls, frogmouths & nightjars | 206 |
| 14  Swifts, kingfishers, bee-eater & dollarbird | 222 |
| 15  Pittas, lyrebirds | 232 |
| 16  Fairy-wrens, emu-wrens, grass-wrens | 238 |
| 17  Pardalotes & bristlebirds to scrubwrens | 250 |
| 18  Honeyeaters & chats | 270 |
| 19  Scrub-robins & robins to babblers | 294 |
| 20  Whipbirds & quail-thrushes to bellbirds | 304 |
| 21  Monarch flycatchers & drongo to fantails | 314 |
| 22  Trillers to currawongs | 322 |
| 23  Riflebirds to bowerbirds | 334 |
| 24  Thrushes to sunbird | 342 |
| 25  Reed-warblers to pipits | 350 |
| 26  Grassfinches | 358 |
| BirdLife Australia; Australian Wildlife Conservancy; | 366 |
| Author's note on photography & birdwatching | 366 |
| Glossary; Bibliography | 367 |
| Index of scientific names | 368 |
| Index of common names | 376 |

# DEDICATION & ACKNOWLEDGEMENTS

## DEDICATION

This book is dedicated to my wife Irene for her support and assistance throughout this project, a contribution most valued and appreciated.

## ACKNOWLEDGEMENTS

The assistance I have received from a great many people extends beyond the time taken for this book or its larger-format predecessor. This book benefits from my lifetime's experience in observing and photographing birds, during which time many people have given encouragement, assistance, advice, information or pleasant company on my travels. This help spans the years, and some 40 subsequent titles, since my first book was published. I wish to express my appreciation to all those who helped make this possible and enjoyable over the years.

For help with this book and others before it, my thanks to Ron Johnstone, Curator of Ornithology at the Western Australian Museum, for access to the Museum's collection of Australian and many South-East Asian birds. On a great many occasions, he gave his time generously, answering my queries and giving his thoughts on the taxonomy of species or race, or on details of a plumage.

With each project – books on birds, wildflowers, national parks – I have tried to acknowledge my thanks for assistance. Lack of space prevents repeating the full list from those past projects, except to acknowledge that a book like this draws on experience from all previous work; thus my thanks include the earliest books. All those whose help and expertise were acknowledged in the large handbook edition have, of course, by that assistance, had a hand in the completion of this compact edition. In addition, I want to thank Bob Forsythe, Noel and Dudley Gross, Briony Fremlin and David James, Eric and Margaret McCrum, and Judy MacKinnon for suggestions, contributions, observations, loan of photographs and other assistance.

I wish to express my appreciation for Steve Parish's support in encouraging me to follow up the original, handbook-sized field guide with this portable compact edition. I am especially appreciative, as any author would be, of the freedom given to me to create this book, including choice of page size and shape, compilation of contents, and most other details. I wish also to give my most sincere thanks to the publishing teams at Steve Parish Publishing, and later Pascal Press and now Woodslane Press, who have supported me through several editions of this title. They include Kate Lovett, Phil Jackson, Wynne Webber, Debra Hudson, Leanne Nobilio, Cristina Pecetta, Ann Wright, Lynne Dickinson, Luke Harris, Karin Cox and Andrew Swaffer. My sincere thanks to all, and to others I may have omitted, for their contribution.

# INTRODUCTION

In any compact or pocket guide, text space is limited. In this guide, the main text is for information that cannot be illustrated, especially behaviour, song and habitat, and, within these, only material that is relevant to species identification is included. Detailed distribution notes are rendered superfluous by the distribution maps. Maps are more informative for many species because the boundaries of range are not clear-cut – they fade away through a fringing zone where occasional vagrants have been observed. In such situations, the palest tint on the graduated maps shows the fringing 'unlikely but possible' part of the range, and the deeper gradations of colour show the reliable extent of a species' range.

All identification information is in the captions surrounding the illustrations, rather like notes on field sketches, with short pointers indicating the parts of relevance. The overall effect can be 'busy', but this guide's purpose is to give the best possible information to aid identification.

Illustrations in this Compact Guide are reproduced at a size that allows an optimum number of species to each page for comparison across species. The paintings are not intended as art; that is best left to large display volumes. This guide aims to give as much relevant content as possible in the most practical size and weight for taking into the field, where other references are likely to be too big, too heavy or out of reach.

The maps, while showing races and where best to find each species, do not include breeding range. Unlike Europe and North America, information for most Australian species is far too sparse to allow accurate maps that show breeding range. Maps based on sparse data can be quite misleading.

Some nests are included in this edition, usually where distinctive nests or eggs can help in species identification. Nowadays, few observers can find and identify nests, particularly those that are small and well-hidden. The most skilled nest finders were egg collectors – and there are many fewer egg collectors today than in the past – and bird photographers. Until the relatively recent advent of powerful telephoto lenses, most birds were photographed at their nests, so photographers became skilled at finding carefully hidden nests of species such as the Rufous-crowned Emu-wren, and at spotting the beautifully camouflaged nests of birds such as the Varied Sittella, the Lemon-bellied Flycatcher, the Southern Scrub-Robin and nightjars, to name but a few.

The demise of the nineteenth-century practice of egg collecting and the decline of bird-at-nest photography may have resulted in less disturbance of birds at their nests. But this was never significant, except in the case of very rare species, compared with predation by cats, foxes, raptors and goannas, or the take-over of nest hollows by introduced birds and bees. Even one pair of Square-tailed Kite, a native species of raptor that preys on small birds' nestlings, would each year account for the loss of more birds than the disturbance caused by all the egg collectors and photographers of yore.

As a result of this gradual loss of observers skilled in finding nests, fewer nest sites and breeding ranges are recorded. Consequently, early nest records and photographs that have added to the overall bank of bird knowledge, along with collections of bird skins in museums, are likely to increase in significance with time.

# HOW TO USE THIS GUIDE

Each of the family groups or chapters is introduced with a box of small bird illustrations across one or occasionally two pages. This introduction gives a concise display of the species in that group. Usually there is one illustration per species, typically the most recognisable plumage or that most commonly seen in Australia, and usually of the male.

These introductions follow a taxonomic sequence that puts together birds most closely related, and is similar to the sequence used by Birds Australia in the multi-volume reference work *Handbook of Australian, New Zealand and Antarctic Birds* (HANZAB). In the text pages, I have used convenient groupings of look-alike birds rather than following the more formal taxonomic order.

In the introductions, the birds are usually to the same scale. At the top right-hand corner of the page, a common, widespread species appears for comparison. Within the text of this and most other field guides, the scale on any page is constant but differs from page to page, so that, for instance, the Emu and Southern Cassowary on one page appear to be a similar size to a robin or pardalote on another page. While length measurements are given for all species, and wingspan is given for many, the difference in bulk is not so readily visualised from numbers, especially when length is measured from tip of beak to tip of tail with the bird stretched out. The introductory pages to each chapter should prove valuable to novice birdwatchers, as they show that small kingfishers like the Little and Azure are tiny beside the Sacred Kingfisher, let alone the Blue-winged Kookaburra. Similarly huge variation in size, concealed by scale adjustment from page to page, occurs in practically every other family group.

## USING THE INTRODUCTION TO FAMILY GROUPS

- Most species in the group to same scale, with a common bird as a size comparison; for this group, a Scarlet Robin.
- Sequence similar to that in the earlier publication *Handbook of Australian, New Zealand and Antarctic Birds*.
- Families are separated by thick blue lines; the page number for the first text entry for each genus is listed.

Scale: one of several very common birds

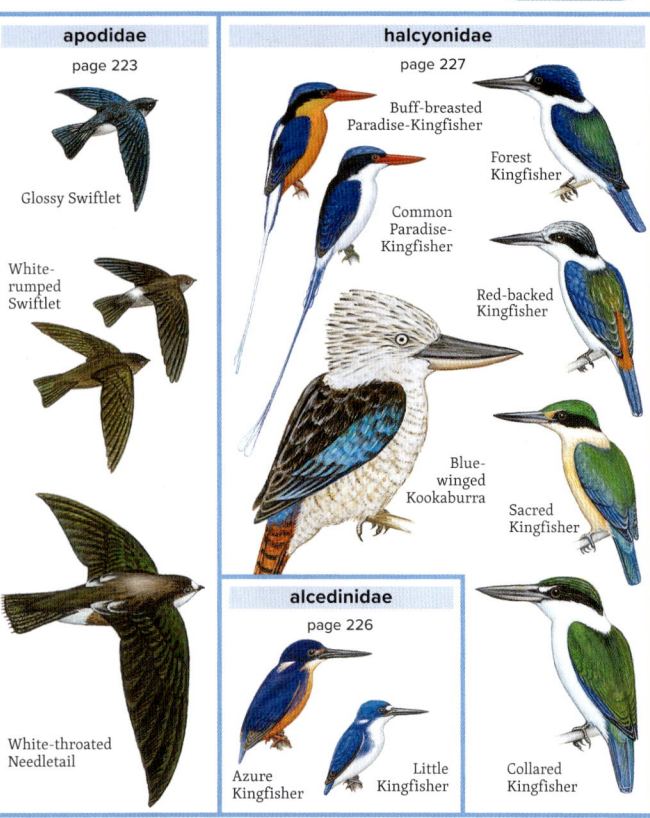

# HOW TO USE THIS GUIDE

This complete compact edition retains the earlier features that make it so user-friendly for new and inexperienced birdwatchers. Each chapter or family group is identified by its own colour tab on the edge of each page. An abbreviated list of species and groups to be found in the chapter is reproduced with the colour code and page numbers on the front and back flaps of the cover. The complete colour code is printed on the front endpapers beside an explanation of how to use the guide to family groups and the Quick Index at the back of the book.

The Quick Index on the back endpapers lists main bird names such as 'Albatross' and 'Heron' as well as singular birds such as 'Malleefowl' and gives the page number for the first of that group of species or for the single species. The colour code for the family group is also given. The Quick Index is an easy way to find a bird when the user has a general but imprecise idea of the species under observation. One page back from the Quick Index are sample illustrations of species in every family group, which will help new users of the guide to locate groups quickly.

Species are fully indexed by scientific name, including races or subspecies listed in the text, and by common name, including many widely used names as well as all the official common names adopted by BirdLife Australia. For instance, 'Jabiru' is cross-referenced to the Black-necked Stork, and 'Peewee' leads users to the entry for the Magpie-lark. Once the family group of a species has been located, the layout of text and illustrations offers several unique features.

Unlike most guides, this book has colour plates on both left- and right-hand pages so that each species has a more compact, squarish presentation. Sometimes, two closely related species, one of which was formerly a subspecies, share a map or a profile split with a vertical line. This layout reduces the distance the user's eyes must travel from the illustration to the distribution map and text entry, thus minimising the chance of losing one's place. The layout has also enabled me to place similar species together, so they can be compared horizontally as well as vertically, and to put groups of four to six species with marked similarities on a single spread. The graduated distribution maps help pinpoint the most likely sites for locating each species.

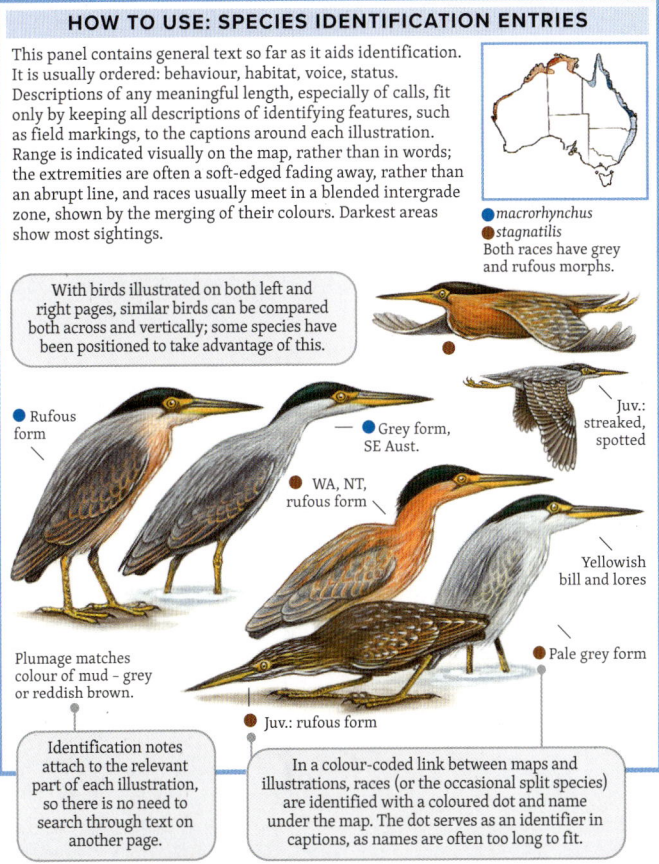

## HOW TO USE: SPECIES IDENTIFICATION ENTRIES

This panel contains general text so far as it aids identification. It is usually ordered: behaviour, habitat, voice, status. Descriptions of any meaningful length, especially of calls, fit only by keeping all descriptions of identifying features, such as field markings, to the captions around each illustration. Range is indicated visually on the map, rather than in words; the extremities are often a soft-edged fading away, rather than an abrupt line, and races usually meet in a blended intergrade zone, shown by the merging of their colours. Darkest areas show most sightings.

● *macrorhynchus*
● *stagnatilis*
Both races have grey and rufous morphs.

With birds illustrated on both left and right pages, similar birds can be compared both across and vertically; some species have been positioned to take advantage of this.

● Rufous form

● Grey form, SE Aust.

● WA, NT, rufous form

Juv.: streaked, spotted

Yellowish bill and lores

● Pale grey form

Plumage matches colour of mud – grey or reddish brown.

● Juv.: rufous form

Identification notes attach to the relevant part of each illustration, so there is no need to search through text on another page.

In a colour-coded link between maps and illustrations, races (or the occasional split species) are identified with a coloured dot and name under the map. The dot serves as an identifier in captions, as names are often too long to fit.

## PARTS OF THE PLUMAGE

PLUMAGE FEATURES

The exact number of feathers in tails, primaries, secondaries and other tracts varies between families, but in the field there is rarely an opportunity to count feathers. The relative proportions vary greatly, as is shown in the example where a swift's wing is compared with an albatross's wing.

## NATIVE BIRD ORDERS; FAMILIES; FAMILY GROUPS

Struthioniiformes: Casuariidae: ch. 1, p. 10
Galliformes: Megapodiidae and Phasianidae: ch. 1, p. 10
Anseriformes: Anatidae: ch. 2, p. 14
Podicipediformes: Podicipedidae: ch. 2, p. 14
Sphenisciformes: Spheniscidae: ch. 2, p. 14
Procellariiformes: Procellariidae to Hydrobatidae and Oceanitidae: ch.3, p. 28
Pelecaniformes: Phaethonidae to Fregatidae: ch. 4, p. 60
Ciconiiformes: Ardeidae to Ciconiidae: ch. 5, p. 70
Falconiformes: Accipitridae and Falconidae: ch. 6, p. 82
Gruiformes: Gruidae: ch. 5, p. 70; Rallidae and Otididae: ch. 7, p. 102
Turniciformes: Turnicidae: ch. 8, p. 110
Charadriiformes: Pedionomidae, Rostratulidae and Scolopacidae: ch. 8, p. 110; Jacanidae to Glareolidae: ch. 9, p. 136; Laridae to ch. 10, p. 152
Columbiformes: Columbidae: ch. 11, p. 170
Psittaciformes: Cacatuidae & Psittacidae: ch. 12, p. 180
Cuculiformes: Cuculidae & Centropodidae: ch. 13, p. 206
Strigiformes: Strigidae & Tytonidae: ch. 13, p. 206
Caprimulgiformes: Podargidae to Aegothelidae: ch. 13, p. 206
Apodiformes: Apodidae: ch. 14, p. 222
Coraciiformes: Alcedinidae to Coraciidae: ch. 14, p. 222
Passeriformes: Pittidae: ch. 15, p. 232, to Passeridae: ch. 26, p. 358

# FIELD MARKS & ILLUSTRATED GLOSSARY

Whether using the introduction to each chapter, or going direct to the main entry for a species, identification is often helped by a bird's distinctive 'jiz'. This is the combination of shape, behaviour and character that can make it immediately recognisable, even in instances when the bird is but a distant silhouette against the sky. Often, the differences between species are slight and a number of features must be compared: plumage markings, shape and colour. Some examples of field marks are set out below, attached to illustrations as text labels. Also, in italics, are glossary items, or brief notes on ornithological terms, that are placed near illustrations to which they apply.

*Plumages: The first true plumage, usually acquired in the nest after the down of the hatchlings, is the juvenile, often followed by one or more immature stages. Then, finally, comes adult plumage, in which the sexes may be the same or different. There may be seasonal changes, breeding plumage (usually summer), being brighter than nonbreeding (winter).*

*Supercilium: Brow line, dark or light*

Black line, widest through eye

*Speculum: A colourful, iridescent, glossy panel, often greenish and purple, in the wings*

White 'armpits' (axillaries)

*Carpal joint or 'wrist': Forms the bend of the wing. Especially prominent on all frigatebirds, which hold wings with this joint thrust far forward, a distinctive silhouette*

Underwing linings white, in contrast with dark surrounds

Frigatebirds are instantly recognisable by their jiz or character, including way of moving.

Wings far forward and wide at carpal joint, giving unique flight silhouette or jiz

Long straight bill

Crest becomes much smaller in the winter nonbreeding season.

*Casque: Helmet or crest-like tall hard ridge on head. Example: Cassowary*

*Eye-line: from bill through eye*

Neck frill

*Pelagic: Describing the oceanic habitat of birds such as the storm-petrels, i.e. often on and near the ocean surface far from land*

Broad black band

*Collar: A band around the neck*

Bright rufous-buff band across wing

*Tarsus: The shank or lower part of the leg*

*Soaring: Usually in circles, rising in updrafts or thermals; wings outstretched, held in a steep or flattish dihedral, but different from that used in gliding*

Broad white band across outer tips

In flight, dark blackish flight feathers emphasise rufous wing band.

Brown Goshawk's glide: rounded, backswept

Australian Hobby's glide: wings pointed, backswept

*Gliding: Wings rather rigid, straight, outstretched, wing tips slightly or strongly backswept, held in a steep upwards to shallowly downwards dihedral. Without flapping, the bird usually follows a fairly straight path, descending gradually.*

Tail folded in glide; long, rounded tip

*Graduated: Usually referring to tail, gradually narrower towards tip*

*Morph or Phase: Distinctive form within a population. The Brown Falcon has colour morphs from very dark to light, and distinctly different morphs may be seen at the same nest.*

*Fingered: Usually used in: describing the widely spread primaries of raptors*

Distinctive facial markings help identify the Brown Falcon.

Spotted Harrier with wings held in deep V dihedral

Brown morph

*Cere: Bare fleshy area at the base of bill; contains nostrils.*

Dark morph

Dark patch on all morphs

Brown Falcon gliding with wings in a shallow dihedral

Black Falcon gliding on drooped wings

Downward dihedral or angle of wings

# EMU, CASSOWARY, MEGAPODES & QUAIL

The family Struthionidae is represented in Australia by the feral Ostrich. The other very large flightless birds, the Cassowary and Emu, are both within one family, Casuariidae. The family Megapodiidae has three Australian species: Malleefowl, Australian Brush-turkey and Orange-footed Scrubfowl; these are builders of large incubation mounds that are warmed by fermenting vegetation or the summer sun's heat. The family Phasianidae includes the native *Coturnix* quails and a miscellaneous collection of feral birds. The introduced California Quail is sole member of the family Odontophoridae.

Magpie = 42 cm

## struthionidae
### page 11

Ostrich, Cassowary and Emu are shown at about 70% of average size; this is also about equal to the smallest adults within the size range for each species.

Ostrich

## megapodiidae
### page 12

Malleefowl

Orange-footed Scrubfowl

Australian Brush-turkey

## phasianidae
### page 12

Stubble Quail

King Quail

## odontophoridae
### page 13

California Quail

## casuariidae
### page 11

Southern Cassowary

Wild Turkey

Brown Quail

Emu

Red Junglefowl

Common Pheasant

Indian Peafowl

# OSTRICH, EMU, CASSOWARY

**Ostrich** *Struthio camelus* 1.7–2.7 m
Feral species. Very large, much taller than Emu or Cassowary. Introduced; now rare in wild. Typically in semi-arid grassland, shrubland or very open dry woodland. Usually silent, except males in breeding season. Booming sounds: in distance 'oo-oo-ooom'; at close range a loud, booming roar.

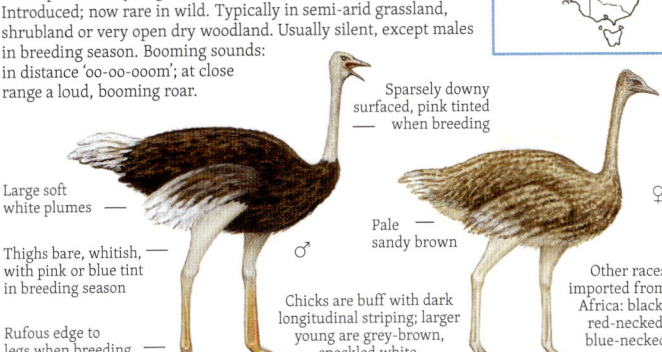

- Sparsely downy surfaced, pink tinted when breeding
- Large soft white plumes
- Thighs bare, whitish, with pink or blue tint in breeding season
- Rufous edge to legs when breeding
- Pale sandy brown
- Chicks are buff with dark longitudinal striping; larger young are grey-brown, speckled white.
- Other races imported from Africa: black; red-necked; blue-necked

**Emu** *Dromaius novaehollandiae* 1.5–2 m
Largest of Australian native birds, widespread and familiar. Feathers hair-like and loose, giving untidy floppy or shaggy appearance. Semi-arid grassland, scrub, open woodland; less commonly heath, alpine areas, tall dense forest. Extremely well adapted to semi-arid regions: migratory or nomadic, wandering in response to seasonal conditions; common. Usually silent, occasionally low intensity booming. In breeding season, deep thudding booming and drumming by female.

● *novaehollandiae*
● *rothschildii*: darker, no ruff; WA except the Kimberley

- Plumage long, loose and shaggy; bounces as the bird runs.
- Bare skin, blue to white
- Wings vestigial
- Individual feathers double-shafted
- White ruff when breeding
- Females average larger and darker than male
- Striping of downy young is gradually replaced by dark grey-brown juvenile plumage and is no longer visible by age 4 to 5 months.
- Downy young
- Juv.

**Southern Cassowary** *Casuarius casuarius* 1.5–1.75 m
Flightless, almost as tall as emu, but more solidly built. Massive legs, coarse plumage, tall casque on crown. Rainforest, especially clearings, margins, vicinity of streams; occasionally in adjoining melaleuca swamp, eucalypt forest. Feeds on fruits of rainforest trees. Threatened by loss of habitat, illegal hunting, attacks by dogs. Gives deep, thudding, resonant boomings, abrupt rough grunts, hissing and roaring sounds. Shy, usually solitary, but always potentially dangerous.

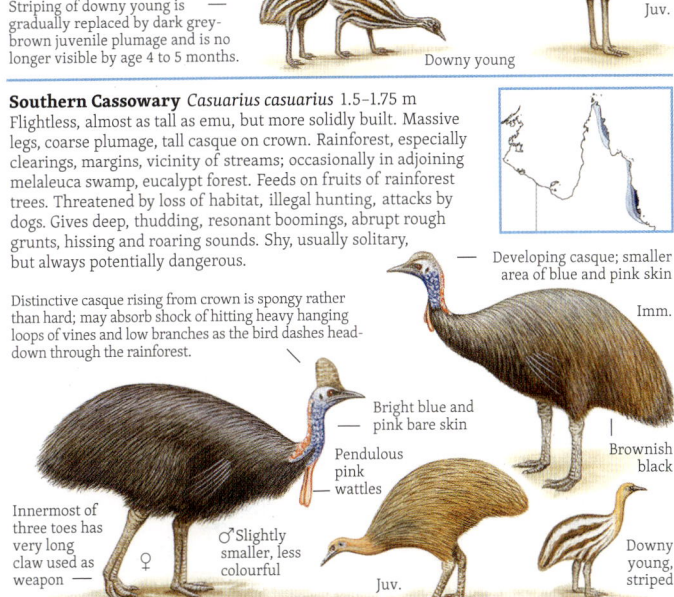

- Distinctive casque rising from crown is spongy rather than hard; may absorb shock of hitting heavy hanging loops of vines and low branches as the bird dashes head-down through the rainforest.
- Developing casque; smaller area of blue and pink skin
- Imm.
- Bright blue and pink bare skin
- Pendulous pink wattles
- Brownish black
- Innermost of three toes has very long claw used as weapon
- ♂ Slightly smaller, less colourful
- Juv.
- Downy young, striped

# MEGAPODES; FERAL FOWL

**Malleefowl** *Leipoa ocellata* 55–60 cm
In unburnt mallee or woodland with abundant litter and low scrub; forages in leaf litter. Quiet, shy, wary; rarely put to flight. Male has a deep, double noted, booming territorial call; female a high crowing. Also various clucks, soft chuckles, grunts. Scarce, reduced range; endangered by loss and deterioration of habitat.

Plumage coloured and intricately patterned to blend with the browns, fawns, black and white of bark, leaf and twig litter of the mallee habitat

Tail long, barred

Heavily built, powerful

Black line

Wings broad; rounded tips

Barred, streaked, blotched rufous, grey and brown, fringed and banded buff and white

**Orange-footed Scrubfowl** *Megapodius reinwardt* 40–50 cm
Active fossicker of ground litter in rainforest, riverine gallery forest, monsoon forest, vine thickets. Noisy, calling both day and night. Often many birds together, deep and powerful, gurgling 'ok-ok-owk-owwwwk-ok-ok-ok-ok'; various chuckles and screams. Sedentary; common. Many races outside Australia.

Dark blue-grey
Appears darker in gloomy, shadowy rainforest.
Orange legs, feet

● *tumulus*: NT and WA
● *yorki*: Cape York
● *castanotus*: Qld S of Cape York

Only slight differences among races

**Australian Brush-turkey** *Alectura lathami* 60–70 cm
Rainforest, from temperate to tropical; densely vegetated gullies in wet eucalypt forest; monsoon and gallery forest; some drier scrub. Scratches about the forest floor litter in search of fallen fruits, seeds and invertebrates. Common, widespread. Gives vibrating low grunts, loud 'gyok-gyok', probably only from male. Tame at tourist sites, gardens, very timid in bush.

Large, turkey-like, blackish body, bare red skin on neck and head

Long, flat-sided tail

Ridge of short black bristles

Red, bare; ♂ nonbr. reduced and dull. ♀ Dull red face only

Yellow wattle at base of neck ●

♂ Br.

● *lathami*
● *purpureicollis*: Cape York, Qld

Collar violet, tinted white

Imm.: similar, but less red on neck, colours dull ♂

**Wild Turkey** *Meleagris gallapavo* 90–125 cm
Feral: wild Turkey of N America and Mexico. Wild and semi-domesticated birds on farmland, grassland, forest edges. Usually in small flocks, foraging on ground. Common on King and Flinders Is., Bass Str.

♂  Feral species: scale = about 1/2

**Red Junglefowl** *Gallus gallus* 40–70 cm
Feral, similar to Asian Red Junglefowl; released in parts of Aust., established on Heron and North-West Is. Rainforest and other dense closed forest, open understorey areas. Gives various clucks, cackles. Male's territorial crows are like domestic rooster, at higher pitch, noisiest at dawn. Small colonies, declining.

♂   ♀

**Common Pheasant** *Phasianus colchicus* 55–100 cm
Feral, King Is. in Bass Str. and Rottnest Is., WA; varied habitat. Terrestrial and diurnal, usually in small to large parties or flocks. Explodes from cover with clapping of wings, strong swift flight. Loud, hoarse crowing, clucks, hisses and other noises. Small populations where introduced; locally rare to common.

♀   ♂

**Indian Peafowl** *Pavo cristatus* ♂ 2–2.3 m; ♀ to 1 m
Feral; conspicuous, noisy. Forages on ground; flight strong despite length of tail; roosts high in trees, seeking high perch if disturbed. Calls from male an extremely loud, carrying, raucous, drawn out 'heirr-elp'. From both sexes, cluckings, shrieks and chuckles.

♀   ♂

## CALIFORNIA QUAIL; TRUE QUAIL

**California Quail** *Callipepla californica* 24–27 cm
Feral; native to W parts of N America; survives on King Is. in Bass Str. Terrestrial, seeking seeds and edible vegetation. If disturbed, all run very fast, then explode upward into flight. Contact calls given when regrouping. In breeding season, territorial crowings from males; many other calls ranging from soft and plaintive to loud.

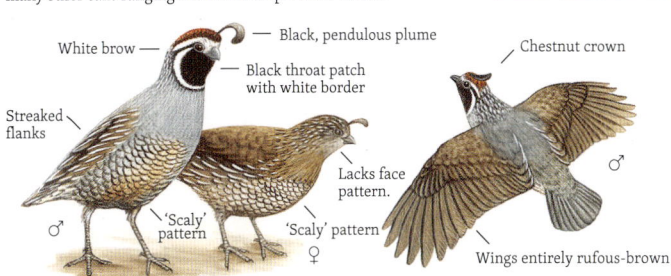

**Stubble Quail** *Coturnix pectoralis* 17–20 cm
Named for its liking for short grasses; often found in paddocks where low but dense cover of wheat-stalk stubble remains. Otherwise in short and tall grasses: spinifex, saltbush and blue-bush. Bursts up on whirring wings, a whistling, metallic sound, fast, low, then dropping abruptly. In pairs when breeding; otherwise single birds or coveys up to 20. Gives a clear, ringing 'cheery-wit' and 'too-too-weep', last note higher, louder, given repeatedly.

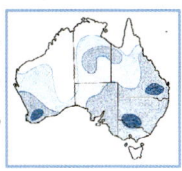

**King Quail** *Excalfactoria chinensis* 13–15 cm
In pairs or family groups in dense wet vegetation of swampy grassland, sedge and heath; also in drier crops and grassland. Common in N Aust., uncommon to rare in S. If disturbed, runs; looks small, dark. More often seen when flushed, rising with little wing noise; soon drops to cover. Gives drawn out whistle, two or three notes, descending, husky and plaintive, 'kee-er, kee-er...' or 'kee-eee-er, kee-ee-er...'. Also quiet peeps, low husky growling.

- *victoriae*
- *colletti*: slightly smaller; NT and WA

**Brown Quail** *Coturnix ypsilophora* 17–20 cm
Typically in grass, crops, heaths, rainforest edges, grassy and spinifex woodland; prefers damp, rank vegetation. Singly, in pairs or in coveys of 10–30. Flight low, fast, usually brief before plunging to cover again. Call is a double noted, ascending whistle: 'pi-pieer', or 'tu-wieep'. Gives a sharp chirp in alarm. Common.

- *ypsilophora*: Tas.
- *australis*: mainland

# 2 GEESE, DUCKS, SWANS, GREBES & PENGUINS

The Magpie Goose (Anseranatidae), sole member of its family, has a long convoluted windpipe enabling the strong honking calls. Swans, geese and ducks (Anatidae), include some species well-adapted to a largely dry continent, nomadic and quick to breed on temporary floodwaters. Grebes (Podicipedidae) are secure in their almost exclusively aquatic habitat, rarely on land. Penguins (Spheniscidae) are but infrequent visitors to southern Australian seas, except for the Little Penguin, a common resident along southern coasts.

Magpie = 42 cm

## anseranatidae
page 15

Magpie Goose

Cape Barren Goose

Mute (White) Swan (not illustrated)

Black Swan

Plumed Whistling-Duck

Wandering Whistling-Duck

Spotted Whistling-Duck

Blue-billed Duck

Musk Duck

## anatidae
page 15

Australian Wood Duck

Pink-eared Duck

Grey Teal

Chestnut Teal

Garganey

Northern Pintail

Pacific Black Duck

Australasian Shoveler

Northern Shoveler

Northern Mallard

Australian Shelduck

Radjah Shelduck

Freckled Duck

Hardhead

Green Pygmy-Goose

Cotton Pygmy-Goose

## podicipedidae
page 24

Australasian Grebe

Little Grebe: vagrant; has more red on face and fore neck.

Hoary-headed Grebe

Great Crested Grebe

## spheniscidae
page 25

Penguins: Gentoo

Adelie

Chinstrap

Magellanic

Little Penguin

King Penguin

Macaroni ('Royal')

Race — schlegeli

Southern Rockhopper

Fiordland

Erect-crested

Snares

## MAGPIE GOOSE, CAPE BARREN GOOSE, SWANS

**Magpie Goose** *Anseranas semipalmata* 75–90 cm
Conspicuous, well-known water bird of tropical floodplains, at times in flocks of thousands. When disturbed, takes to air rather than to deeper water. Feeds in shallows of dams, irrigated crops and swampy margins of deeper waterways. Has a loud honking, that of male louder and higher. A flock makes a cacophony of nasal trumpeting. Common to locally abundant across N Aust.

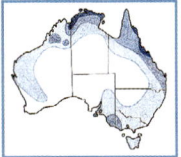

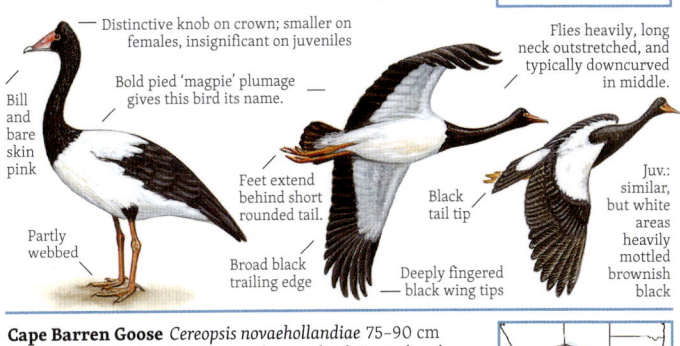

**Cape Barren Goose** *Cereopsis novaehollandiae* 75–90 cm
Offshore islands, Bass Str., SA, nearby mainland. Ocean beaches, headlands, margins of wetlands, pastures. Occasionally on water. On S coast WA most keep to islands (Recherche). Flight strong; rapid wing beats interspersed with glides. Usually in flocks, except when breeding, then pairs. Often call in flight; male has loud, harsh, rapidly trumpeted 'ark-ark, ark-ark'; both give low grunts, hisses. Common in some parts of range.

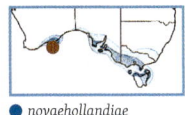

● *novaehollandiae*
● *grisea*

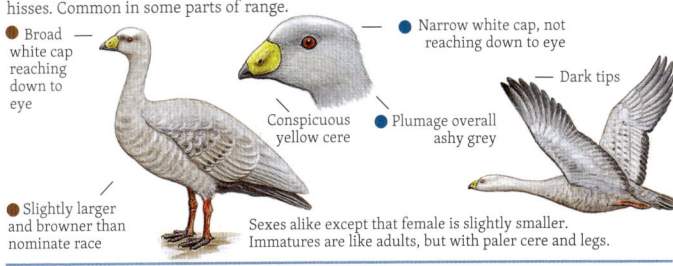

**Mute Swan** *Cygnus olor* 1.3–1.6 m
Feral swan, first introduced in Aust. in 1886 and subsequently released at many locations until about 1920; some colonies still survive, usually where release occurred.

**Black Swan** *Cygnus atratus* 1.1–1.4 m
Utilises diverse habitats, lakes, estuaries, rivers, including temporary wetlands of arid interior. Feeds on aquatic vegetation of shallow waters. The only swan so extensively black. A long, powerfully flapping take-off run is needed to get the heavy body airborne. Has a loud, clear bugling call, given in flight and on water; carries far. Loud hissing defending nest. Common; nomadic.

# WHISTLING DUCKS, AUSTRALIAN WOOD DUCK

**Plumed Whistling-Duck** *Dendrocygna eytoni* 40–60 cm
Grazes on grasslands and margins of wetlands at night; rests by day in flocks near water. Shrill whistlings, high wheezy 'tzwit-tzwit-tzwit-tzwit...'; in roosting flocks, an almost continuous 'jizzing-chittering'. Also, whistling calls from flocks in flight. Common; nomadic.

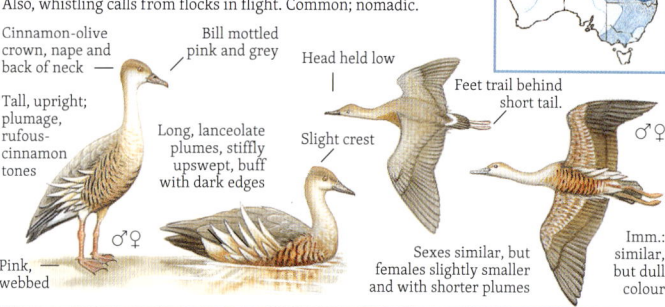

**Wandering Whistling-Duck** *Dendrocygna arcuata* 55–61 cm
Makes greater use of water than does the Plumed Whistling-Duck, swimming and diving for aquatic plants on swampy wetlands, billabongs, floodplain pools, tidal creeks, wet season floodwaters. Twittering, whistling voice; dense flocks; lowered head in flight. Calls mostly from flocks, a confused, high, whistled, rather tremulous 'wit-wit-wit...', slightly slurred. From flocks, much whistling and twittering. Common; migratory or nomadic.

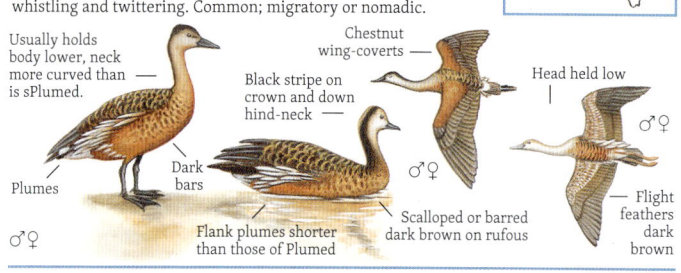

**Spotted Whistling-Duck** *Dendrocygna guttata* 43 cm
A NG species; recorded N coast Aust. May be a breeding resident of E Cape York coastal swamps. Usual habitat the well-vegetated margins of swamps, rivers. Calls differ from above two species; does not have their high twitterings, but rather a nasal 'zziow' and 'whu-wheow-whee', which may continue as a repeated sequence.

**Australian Wood Duck** *Chenonetta jubata* 45–60 cm
Must be seen at close range for the delicately patterned beauty of its plumage to be appreciated. Vicinity of dams, lakes, estuaries; grazes on damp pasture, dabbles in shallows. Flight swift, low, small parties swerving through trees. Female a querulous, nasal 'grouwwk', beginning low, rising; male similar, but higher, abrupt 'nowk!' From flock, rapid clucking, louder when agitated. Abundant.

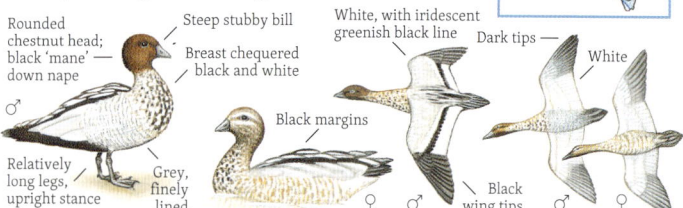

## BLUE-BILLED, MUSK & FRECKLED DUCKS

**Blue-billed Duck** *Oxyura australis* 36–44 cm
Small, compact duck with concave or scooped bill. Deep, densely vegetated freshwater lakes, swamps when breeding; winters on more open waters. Usually shy, secretive. If disturbed, is likely to disappear under water to reappear much further away; rarely on dry land. Usually silent, but, in display, male gives soft, throbbing, 'dunk, dunk-dunkdunk…', and a soft 'chi-chi-dunk-dunk'. Female, soft 'tet-tet-tet…'. Uncommon; sedentary.

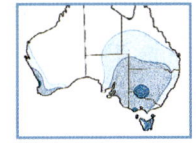

**Musk Duck** *Biziura lobata* male 65 cm; female 55 cm
Lakes and deep swamps with both reeds and open water. Can submerge with scarcely a ripple; occasionally emerges onto land, where it waddles clumsily. In flight, has rapid, shallow wing beats; can travel swiftly and far, although take-offs appear laborious and landings are clumsy. Male attracts attention in breeding season with long periods of whistle-splash display, giving grunt, whistle, splash of water, deep 'plonk' sound. Sedentary; common.

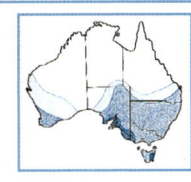

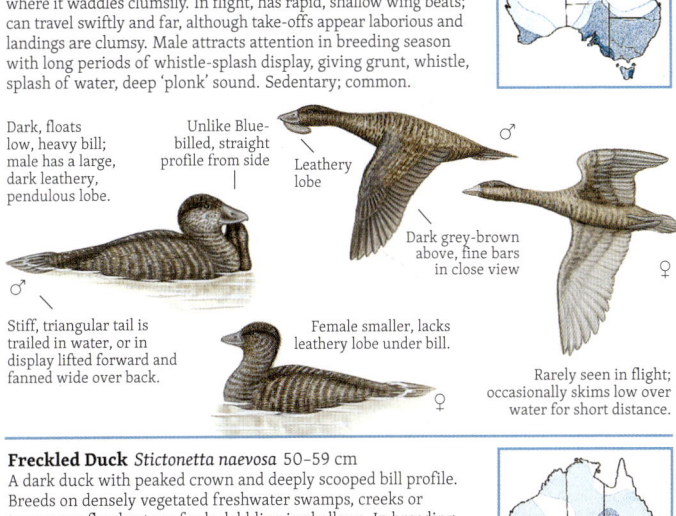

**Freckled Duck** *Stictonetta naevosa* 50–59 cm
A dark duck with peaked crown and deeply scooped bill profile. Breeds on densely vegetated freshwater swamps, creeks or temporary floodwaters; feeds dabbling in shallows. In breeding season, moves to densely vegetated lake or swamp; after breeding, may be seen on more open waters. Both sexes give a querulous groan, often peevish. Male gives an 'axe-grind' low buzz and squeak; female has a throaty chuckle. Rare, endangered.

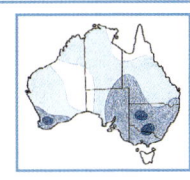

# SHELDUCKS, HARDHEAD

**Australian Shelduck** *Tadorna tadornoides* 56–73 cm
Fresh or brackish lakes, estuaries, dams, flooded paddocks. Largely terrestrial in habits, feeding on pasture and aquatic vegetation obtained in shallow wetlands. Comes frequently to water to loaf, preen, mate. On some large expanses of water, flocks of thousands gather. Later, in the winter breeding season, these flocks disperse, pairs establishing territories where there are tree hollows near water. Male gives a deep, harsh, nasal buzzing grunt or honk; female has a higher call, though with similar nasal tone, 'ank-aank' or 'an-ganker'. Locally common in SE and SW Aust., elsewhere uncommon.

**Radjah Shelduck** *Radjah radjah* 49–61 cm
Large, dumpy shelduck. In tropical dry season, flocks congregate on mangrove-lined river channels, tidal mudflats and beaches, or remain inland on permanent lagoons. In wet season moves from littoral habitat to the shallow margins of the expanding wetlands. Flight swift and powerful; dashes fast and low between trees. Very vocal, calling in flight, on water or land. Male has a rattling whistle; female a deeper, slower, more harsh rattling. Common in optimum habitat areas of NT, scarce to rare elsewhere.

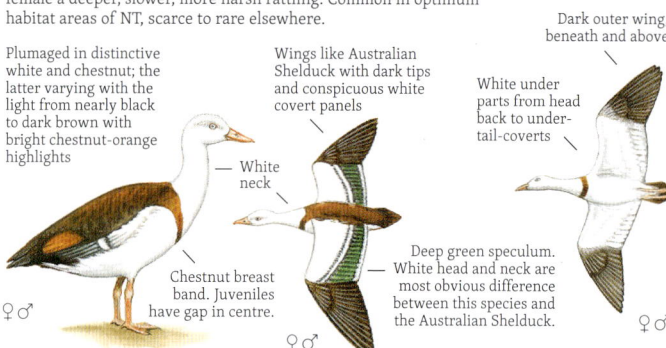

**Hardhead** *Aythya australis* 45–60 cm
Australia's only representative of true diving ducks able to feed in deep water; favours parts of lakes, swamps with abundant aquatic vegetation, but also on creeks, floodplain pools; rarely on coastal lagoons, mangroves. Rises steeply, taking off from water with audible whirr from fast-beating wings. Male gives a soft wheezy whistle and 'whirrr'; female a loud, harsh, rattled 'gaak,gak,gak-gakgak.' Common in S; after rain disperses across usually arid parts of interior and far N.

## PINK-EARED DUCK, SHOVELERS, MALLARD

**Pink-eared Duck** *Malacorhynchus membranaceus* 38–45 cm
Uses shallow, open, muddy wetlands and temporary floodwaters. Small parties to large flocks. On water or loafs on logs and limbs rising from water; seldom on land. Groups work through shallows. More easily approached than most ducks. If disturbed, usually do not fly far: the flock circles to land again nearby. Flocks noisy in flight and on the water, with rapid, whistled twittering; also a sharp 'ti-wit, ti-wit, ti-wit' alarm call and a drawn out 'wheeii-ooo' in display. Widespread, nomadic; common.

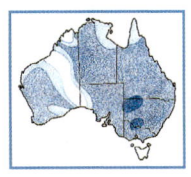

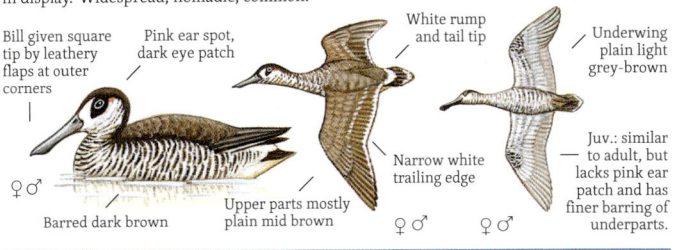

Bill given square tip by leathery flaps at outer corners

Pink ear spot, dark eye patch

White rump and tail tip

Underwing plain light grey-brown

Narrow white trailing edge

Juv.: similar to adult, but lacks pink ear patch and has finer barring of underparts.

♀♂

Barred dark brown

Upper parts mostly plain mid brown

♀♂   ♀♂

---

**Australasian Shoveler** *Spatula rhynchotis* 45–54 cm
Uses wide variety of wetlands; prefers large permanent lakes or swamps that have abundant cover. The massive bill is used to feed on small creatures filtered from the water; a semi-nocturnal feeder; during the day floats with other ducks far out on open water. Wary, takes off steeply; flight swift and direct. Courting male gives a soft 'took-ook..., took-ook...'. Female has a rapid sequence, becoming faster, lower and softer, 'quaaak..., quaak, quak-quak-'. Generally rare, except SW coast of WA; nomadic and dispersive.

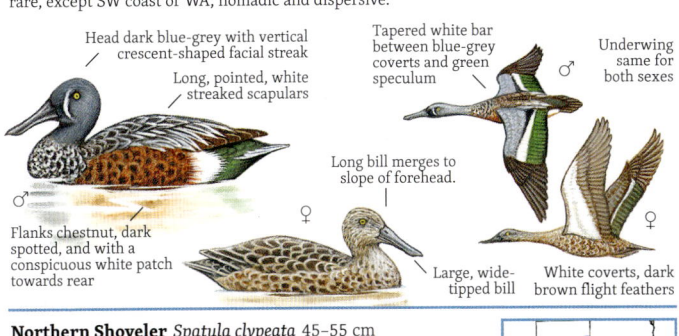

Head dark blue-grey with vertical crescent-shaped facial streak

Long, pointed, white streaked scapulars

Tapered white bar between blue-grey coverts and green speculum

♂ Underwing same for both sexes

Long bill merges to slope of forehead

♂

♀

Flanks chestnut, dark spotted, and with a conspicuous white patch towards rear

Large, wide-tipped bill

White coverts, dark brown flight feathers

♀

---

**Northern Shoveler** *Spatula clypeata* 45–55 cm
A filter-feeder: dabbles, rarely dives, but up-ends to reach bottom mud of freshwater billabongs, lagoons, swamps. Usually silent, calls weak; male gives a soft hoarse 'took', or 'took-took, took-took', distinguishable from calls of Australasian Shoveler. A rare vagrant to Aust., very few records.

Head deep green, usually appears black with green highlights.

White

Wings (dorsal and ventral) like those of Australasian Shoveler.

Bill long, with wide tip, looks heavy.

♀♂

Female, nonbr. male like those of Australasian, but orange lower edge of bill and white edges to tail

White  Cinnamon-chestnut flank patch

---

**Northern Mallard** *Anas platyrhynchos* 50–70 cm
Feral; widespread on lakes, dams, mostly near towns; hybridises with Black Duck. Voice is very close to that of Black Duck.

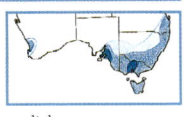

Yellow tint

Glossy dark green

Pale grey-brown, finely vermiculated

Face pattern like Black Duck, but legs orange, not grey

Coverts light grey-brown

Brown with buff feather edges

♂

Blue violet between thin white bars

Chestnut breast, white neck collar. Legs orange

Nonbr. male: bill pale yellow to blue-grey

♂ Female: similar (bill buff edged, greyer breast)

## PYGMY-GEESE, CHESTNUT TEAL

**Green Pygmy-Goose** *Nettapus pulchellus* 30–36 cm
On deep permanent freshwater lakes, dams and lagoons with abundant aquatic vegetation, especially waterlilies; moves to floodplain swamps as they fill in the wet season. This tiny goose-like duck swims and dives strongly, swift in flight; rarely seen perched on land or tree. Male has a sharp, high, whistled contact call, given almost continuously, in flight or on the water. Fluctuates in pitch, strength and rapidity, 'chi-wip, chi-wip, chiwip…'; from the female, a whistle of declining pitch and intensity, 'phee-oo'. On coming in to land, a deeper, softer 'kuk-ka-kadu'. Alarm call from both sexes is a sharp, whistled 'whit!' or 'whit-whit'. Locally common; sedentary and locally nomadic.

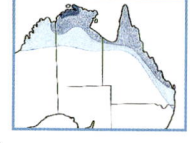

A tiny duck superficially like a miniature goose, largely due to the stubby goose-like bill

Oval, stark white cheek between dark brown of cap and glossy deep green of neck and back

Iridescent green highlights across greenish black coverts

White panels are conspicuous beside dark greens and browns.

Short stubby bill

White brow

Bright white panels in dark underwing

Flanks of both sexes are barred with dark greenish-brown lines following the shapes of feathers.

**Cotton Pygmy-Goose** *Nettapus coromandelianus* 34–38 cm
Coastal wetlands, preferring deep permanent pools and swamps with abundant aquatic grasses; moves out to floodplain swamps and pools as these fill in the wet season. Almost exclusively aquatic, rarely on land. Usually out on deeper water; perches on logs; flies readily. In flight, male makes a continuous, rapid, nasal quacking, 'car-car-carwak'; a cacophony of sound when coming from many birds. The female has a much softer version. Uncommon vagrant over most of range; sedentary or locally nomadic.

Small goose-like duck with white neck and dark breast band; dark green back

White neck, dark band

Dark greenish brown cap

White face and neck

Dark iridescent green, looks almost black in dull light

Olive-brown

White bar almost full length of wing

Thin white rear margin

White neck with dark band

Flanks of both sexes are white, very finely barred with grey-buff

Juv.: like female, has dull brown upper parts.

Underparts mostly shades and tints of grey-brown

**Chestnut Teal** *Anas castanea* 38–48 cm
Wetlands, with preference for salt and brackish coastal estuaries, lakes, salt marshes, tidal mudflats and coastal islands. Feeds mostly at dusk and dawn, dabbling and up-ending in the shallows; small flocks loaf for much of the day beside the water. Voice is, except for minor differences of pitch, almost indistinguishable from Grey Teal; calls of male slightly deeper. Abundant in Tas., common elsewhere in SE Aust. and extreme SW of WA; usually sedentary, but some wander far N and well inland.

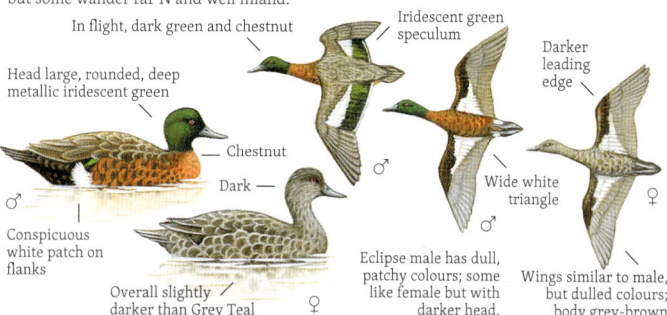

In flight, dark green and chestnut

Iridescent green speculum

Head large, rounded, deep metallic iridescent green

Darker leading edge

Chestnut

Dark

Wide white triangle

Conspicuous white patch on flanks

Overall slightly darker than Grey Teal

Eclipse male has dull, patchy colours; some like female but with darker head.

Wings similar to male, but dulled colours; body grey-brown

## BLACK DUCK, GREY TEAL, GARGANEY, PINTAIL

**Pacific Black Duck** *Anas superciliosa* 48–60 cm
Utilises almost any and every wetland habitat throughout Aust., from small densely vegetated freshwater swamps to sheltered marine waters. Well known; common both in public parks and in the wild. Male gives a quick 'rhaab-rhaab' of varying strength; in warning, a loud, extended 'rhaaaeeb'; in display, a high pitched whistle immediately followed by a deep grunt. Female, typical loud duck quackings, single quacks of varying strength, and a long sequence, descending and fading, 'quaak, quaak, quak, quak-quakquak'. Widespread, nomadic; common.

Large duck with unique and obvious striped head pattern and a conspicuous iridescent green-blue speculum panel in the wing

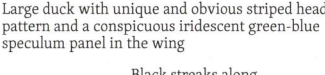

**Grey Teal** *Anas gracilis* 42–45 cm
Varied habitat, uses almost any wetlands whether permanent or temporary. In pairs or small to huge flocks; swims buoyantly, head often high, emphasising slender-necked appearance. Takes flight with explosive upward leap and splash; rises swiftly. Flight fast; shows white centrally on underwings. Mainly aquatic; feeds among floating aquatic vegetation, usually in shallows. Highly mobile and opportunistic; quickly reaches temporary floodwaters on distant inland plains. Male gives a sharp whistle with soft, low grunt and loud, whistled 'gedg-ee-oo'. Female has a loud, chuckled, descending series of quacks; also a slow, harsh, drawn out 'que-aark'. Abundant; nomadic, dispersive.

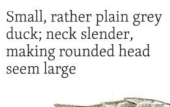

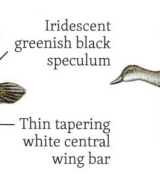

**Garganey** *Spatula querquedula* 37–41 cm
Freshwater wetlands, swamps, shallow lakes, flooded grasslands, floodplains, sewage farms. Uncommon vagrant to N coastal Aust.

**Northern Pintail** *Anas acuta* 50–65 cm (including tail)
Wetlands, estuaries, pastures. Usually silent. Floats high, tall-necked. Flight swift on fast-beating wings. Very rare vagrant: recorded once, SW of WA.

The nonbreeding male, without long streamers and scapulars, is like female.

## WING PATTERN: DUCKS

**Flying ducks:** Waterfowl travel swiftly on fast-beating wings. When only a brief glimpse is possible, details of plumage pattern may not be recognised. But there can be an impression of one or two more conspicuous features, whether the white seen on the flickering wings was on the upper, the lower, or both surfaces of the wing, or perhaps the impression that there was no white at all. A clearer sighting might, for example, give a glimpse of the shape of a white or pale patch, whether wide, thin, short or full length of wing, or reveal that there was green along the wing ahead of the white. This brief impression, though incomplete and obtained in just a few seconds, could be a helpful clue to identity, and probably the only markings to be seen with the duck vanishing into the distance.

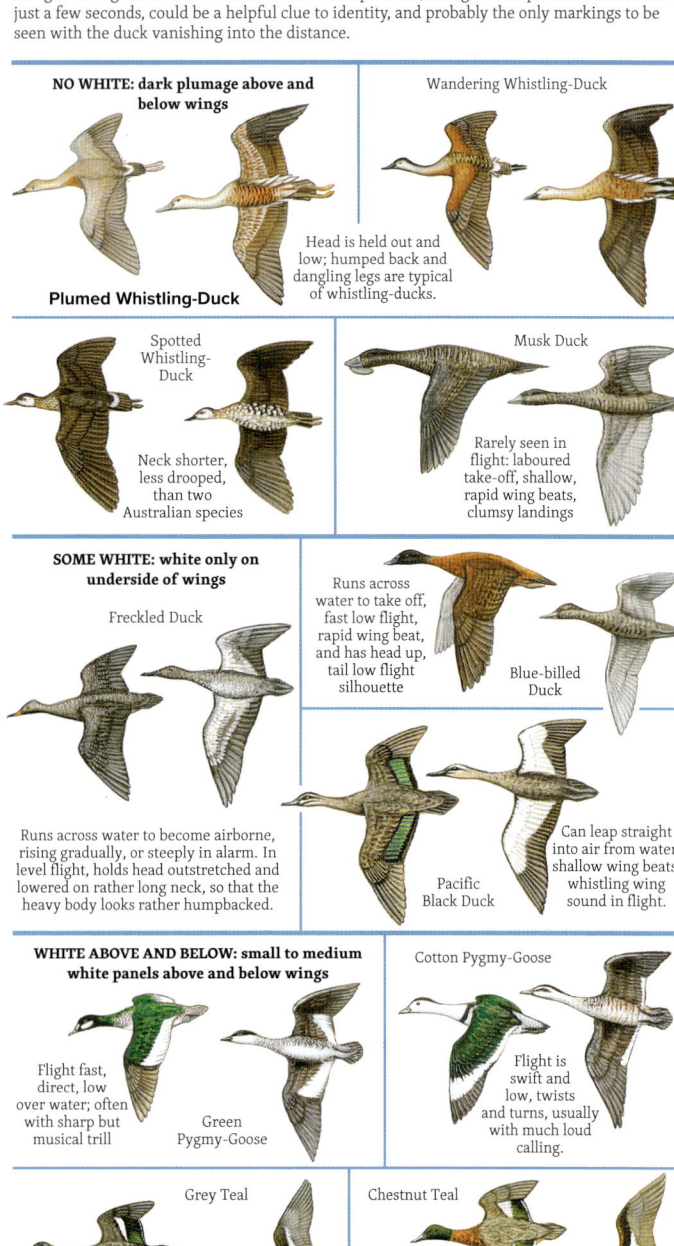

**NO WHITE: dark plumage above and below wings**

Plumed Whistling-Duck

Wandering Whistling-Duck

Head is held out and low; humped back and dangling legs are typical of whistling-ducks.

Spotted Whistling-Duck

Neck shorter, less drooped, than two Australian species

Musk Duck

Rarely seen in flight: laboured take-off, shallow, rapid wing beats, clumsy landings

**SOME WHITE: white only on underside of wings**

Freckled Duck

Runs across water to become airborne, rising gradually, or steeply in alarm. In level flight, holds head outstretched and lowered on rather long neck, so that the heavy body looks rather humpbacked.

Runs across water to take off, fast low flight, rapid wing beat, and has head up, tail low flight silhouette

Blue-billed Duck

Pacific Black Duck

Can leap straight into air from water; shallow wing beats; whistling wing sound in flight.

**WHITE ABOVE AND BELOW: small to medium white panels above and below wings**

Flight fast, direct, low over water; often with sharp but musical trill

Green Pygmy-Goose

Cotton Pygmy-Goose

Flight is swift and low, twists and turns, usually with much loud calling.

Grey Teal

Swift direct flight, often in small to quite large flocks

Chestnut Teal

Often in large flocks that rise suddenly into air when disturbed, their flight swift and agile

## WING PATTERN: DUCKS, GREBES

**A LITTLE ABOVE, MORE BELOW:** small pale panels or bars on upper surface of wing, extensive areas on under surface

Pink-eared Duck

Tail distinctive, dark, and set between white tail tip and a wider white rump

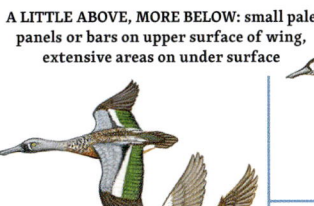

Australasian Shoveler

At take-off, rises abruptly, vertically, flight then swift and direct. The silhouette makes wings appear set too far back along the body, but this, in part, is due to the long bill stretched out in front.

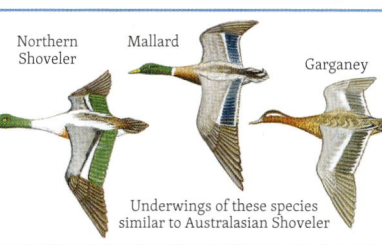

Northern Shoveler   Mallard   Garganey

Underwings of these species similar to Australasian Shoveler

**MUCH WHITE ABOVE AND BELOW:** extensive bands or panels of pale or white plumage on both wing surfaces

Northern Pintail

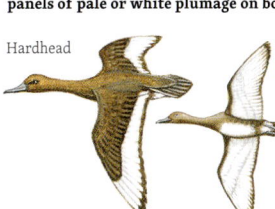

Hardhead

Flight swift and agile. Rapid wing beats, greenish speculum band, narrow white trailing edge, pale underbody ♀

Rises steeply from water; rapid wing beat produces a characteristic whirring. Flight direct and swift; looks long-necked; wings seem set far back.

Australian Wood Duck

Flight swift and agile. Slower wing-beat than other Aust. ducks

Australian Shelduck

Pairs, or small to huge flocks that take flight suddenly and noisily; flight fast, direct. Travelling flock may form long V across sky.

Radjah Shelduck

Strong swift flight, tending to keep quite low over water and weaving between trees. Flocks noisy in flight and on land.

**GREBES:** Seem to fly mostly at night, but occasionally skitter along water surface or fly short distances close above water, when the flickering white of the wing panels is obvious. A third small species is the Little Grebe, a rare visitor to northern Australia. Much larger, and relatively uncommon, is the Great Crested Grebe, again more likely only to be seen undertaking brief daylight flights low across water.

Australasian Grebe

Both species have fast-beating wings with long white patch; fly with neck outstretched, feet protruding, humpbacked shape.

Hoary-headed Grebe

Little Grebe: Rare vagrant to N Aust. Wing bar confined to secondaries, dull white. In breeding plumage has more red on head.

Great Crested Grebe

Rapid wing beats, twin white wing panels, whitish underwing; holds neck extended, head low; has humpbacked silhouette.

# GREBES

**Australasian Grebe** *Tachybaptus novaehollandiae* 23–25 cm
Found on diverse lakes, swamps and dams, usually wetlands with abundant aquatic vegetation. After breeding, often gathers in large numbers on open waters. Floats high, dives with a smooth forward plunge, leaving only a slight ripple. Both sexes give a rapid, sharp chittering; harsher versions in threat or aggression; strongly territorial. Widespread and common.

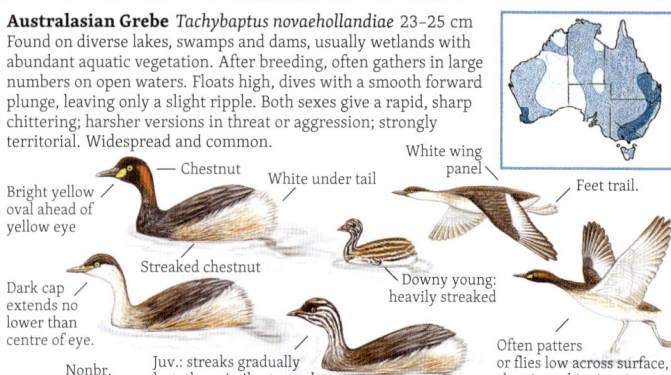

**Little Grebe** *Tachybaptus ruficollis* 24–25 cm
Easily identified in breeding plumage. Throat, cheeks and front of neck rufous-red rather than black; has a much smaller white patch under tail, and eye red rather than yellow. In other plumages – nonbreeding, juvenile and immature – much more like the Australasian Grebe. However, in all plumages the wing bar is dull white and confined to secondaries. On the Australasian Grebe, the bar is brighter white and extends almost the full length of the wing, across both secondaries and primaries. This may be seen if the wings are stretched upwards, or in brief low flights along the water. Reported from near Darwin, Sep.–Nov. 1999. Rare vagrant; a species that has a wide range through Eurasia and Africa.

**Hoary-headed Grebe** *Poliocephalus poliocephalus* 30 cm
Inhabits wetlands including fresh water and estuaries. Usually on large open areas of permanent or temporary water. Prey includes small yabbies and jilgies, bugs and beetles, dragonflies, spiders, small fish and some water plants. Usually silent, gives very muted churring and low guttural sounds in courtship and at the nest. Common to locally abundant, scarce in far N; nomadic.

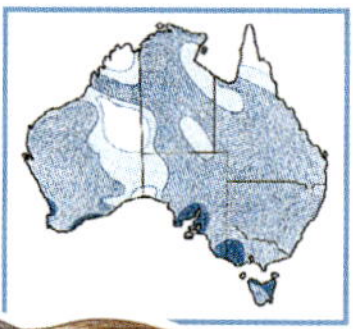

**Great Crested Grebe** *Podiceps cristatus* 47–61 cm
A specialised aquatic species, never on land. Prefers large deep lakes and swamps, usually those with both open water and dense reedbeds or other concealing vegetation. Rarely seen flying; when disturbed, submerges to surface out of sight. Like other grebes, seems to travel between wetlands only at night. Usually silent, but some gurgling and barking sounds in breeding season. Nomadic; uncommon to rare.

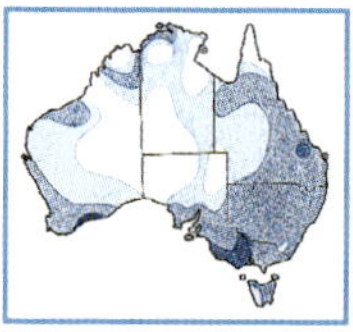

## ADELIE TO LITTLE PENGUINS

**Adelie Penguin** *Pygoscelis adeliae* 70 cm
A slight crest at times gives a peaked contour to the crown; usually in threat and courtship displays. The contact call is a sharp barking; other calls in display include thumping, rasping sounds. Occurs Antarctic coast and seas in vicinity of pack-ice. When dispersing from huge breeding colonies reaches Heard and Macquarie Is.; rare vagrant to Tas. coast and SE Australian seas.

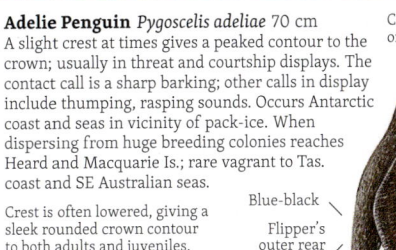

Crown pointed or rounded
Bill looks stubby, with base partly covered by feathers; black and orange-red.
Tapering dark edge
Small dark tip
Pink to pinkish white
Blue-black
Flipper's outer rear edge is white.

Crest is often lowered, giving a sleek rounded crown contour to both adults and juveniles.
Body floats low, head and tail held higher.
Imm.

**Chinstrap Penguin** *Pygoscelis antarcticus* 70–76 cm
Antarctic seas, often around icefloes and light pack-ice; when breeding, forages in shallower waters around its island colonies. Known calls are from nest site and include a loud, cackling 'arh-kauk-kauk', and soft humming and hissing sounds. Rare vagrant to Aust. coasts and seas, mostly Tas. and S Vic.

Bill black
Dorsal surfaces dark grey
Fine black line along leading edge of flipper
Fine white line along trailing edge of outer part of flipper
Appears to have a black cap held on by a slender strap passing conspicuously across the white face and around the throat.
Pale grey tip
Imm.
Smaller, and has dark or greyish mottling on face, chin and throat.
Fleshy pink to yellowish pink

**Magellanic Penguin** *Spheniscus magellanicus* 70 cm
Abundant and apparently increasing within usual range; breeds coasts and islands around southern S America. Normally sedentary or making short nomadic movements. Accidental to Aust.: Phillip Is., Vic.

Bare pink around eye
White band encircles the face, forehead to throat.
A banded penguin of distinctive appearance with eye-ring of bare pink skin and deep, heavy, dark bill
Black bands on throat and upper breast
Brownish black; browner in worn plumage
Flipper has thin white leading edge and broad white trailing edge.
Imm.: greyish cheeks and patchy breast bands

Swims low in water with only part of back showing, occasionally lifts tail; dives through waves.

**Little Penguin** *Eudyptula minor* 35–45 cm
The smallest of all penguins; also known as the Fairy Penguin. Inhabits inshore waters of mainland coast and islands. Sole Aust. resident species. Noisy ashore at night, with sharp yapping, high trilling, a resonant 'quarrr', vibrating, deep grunts and braying. At sea, short sharp yaps. Common around S coast of Aust.

Breeds on islands around southern coast.

Lacks crest, bright colours or bold pattern, but this plain look is in itself unique, and, together with the very small size, makes the Little Penguin easy to identify.
White with silvery sheen
Dark silvery blue-grey; in worn plumage, slightly brownish
Light flesh pink

Well known tourist attraction at Phillip Is., Vic. The sole Aust. breeding species; colonies on many offshore islands from SW of WA to mid-NSW coast.

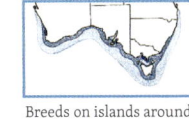

Juv.: like adult, but smaller bill

## KING, GENTOO & MACARONI PENGUINS

### King Penguin *Aptenodytes patagonicus* 85–95 cm
Very tall, colourful; second largest of all penguins; vagrant to Aust. seas, Tas. to WA. Juveniles especially wander far. Breeds on islands of S Atlantic and Indian Oceans. In colonies and at sea, both sexes give far-carrying cooing contact calls. Other calls form part of courtship and nesting behaviour.

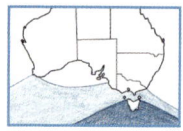

- Distinctive, inverted, tear-drop shaped orange ear patch
- Long, slender, downcurved bill with pinkish or orange streak
- Bill streak varies pink to ivory.
- Back blue-grey, silvery towards nape and shoulders
- Golden orange, fading downwards to faint yellow tint
- Pale lemon
- Imm.
- Wide dark edge tapers towards tip.
- Swims with head held high, orange ear patch well above the water.
- Large dark tip, but some are entirely white.

Highly gregarious at island breeding colonies. At sea, solitary or in small groups. When 'porpoising', launches from waves and re-enters with a smooth, graceful action.

### Gentoo Penguin *Pygoscelis papua* 70–80 cm
Abundant in Antarctic seas, but a rare vagrant to Aust. coasts. Few records, all from Tas. Breeds in colonies, Antarctic Peninsula and N to sub-Antarctic islands including Macquarie Is. In colonies has various loud trumpetings, harsh croakings, vibrant, harsh, throaty raspings, 'aargh', 'aa-aa-argh', hissing sounds. Calls at sea unknown.

Two subspecies: *papua* in Australasian waters; *ellsworthii* in seas between Antarctica and S America

- Has unique white triangle over each eye; these join narrowly across crown to form an hourglass shape when seen from above.
- Bill orange with black top ridge
- Thin black leading edge that often fades to a break midway down the flipper
- Flipper has broad white trailing margin, very thin white leading edge.
- Black tip
- White triangle over eye is smaller than on adult, and usually does not quite reach down to the eye.
- Conspicuous bright orange
- Imm.

### Macaroni Penguin *Eudyptes chrysolophus* 65–75 cm
Includes both dark and light-faced races. After breeding, disperses widely, on rare occasions being reported from Aust. S coasts. In colonies, noisy, with loud trumpetings of greeting and recognition at nest: raucous, rattling, braying 'kaaa-kaa-aargh' calls come from a huge number of birds. Black-faced form visits S coast waters; white-faced (Royal) not yet recorded in Aust. seas.

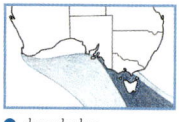

● *chrysolophus*
● *schlegeli*: not near Aust. coast

- Long golden plumes on each side, all arising from the one central area around forehead
- Bill massive, orange or brownish pink
- Imm., near age 2 yrs, after moult: short crest
- Face grey, streaked white
- Flipper blue-black with thin white trailing edge
- Black edge extends half to full length.
- Long golden plumes join on central forehead
- Black near tip
- Black elbow patch

● **Macaroni Penguin**: most are the black-faced morph, a minority white-faced; varies regionally.

● **'Royal Penguin'**: most are the white-faced morph, but ratio varies at different colonies.

# CRESTED PENGUINS

## Southern Rockhopper Penguin
*Eudyptes chrysocome* 45–60 cm

Marine, Antarctic and sub-Antarctic waters; breeding colonies on rugged islands. Gives abrupt barks or croaks as contact calls; noisy in colonies with loud raucous braying.

● *moseleyi*
● *filholi*
● *chrysocome*

Yellow crest starts from bill, spreads wide behind eye; length varies with race

Strongly diverging plumes begin behind bill but do not join on forehead.

Bill deep, bulbous, brownish pink

Yellow plumes few and short; bill dull

Flipper dorsally blue-black with thin white trailing edge

Medium width black edge may fade out before reaching streaky dark tip.

Imm.: after moult

Has longest crest plumes, darkest flipper; lacks pale pink at gape.

● Is frequent straggler, more likely in SW Aust. ● Pale pink fleshy gape, shorter plumes, paler under flipper, is most numerous in SE. ● Very rare

## Fiordland Penguin *Eudyptes pachyrhynchus* 55–65 cm

Breeds around S of NZ, then wanders W to Aust. coasts where most records have been of young birds coming ashore to moult. At sea, feeds in small groups or alone; porpoises from waves when travelling fast. Voice, in breeding colonies, is loud, harsh, deep; there is little calling at sea.

Bill large, deep, bulbous

Crest broad, downcurved, not branched

Imm. or yearling after moult. Before moult, short crest is whitish, bill grey.

White cheek streaks

Grey

Flipper has blue-black dorsal surface with thin white trailing edge.

Tail is often lifted.

At sea, adult shows broad yellow crest streak over eye, flattened when wet; white cheek streaks; no bare pink skin at base of pinkish orange bill.

Broad dark top, broken bottom edges linking large black patches at elbow and tip

## Erect-crested Penguin *Eudyptes sclateri* 65–70 cm

Makes rattling and shrill sounds, deep, throbbing grunts; calling is loud, persistent in breeding season. Rare vagrant, southern coasts.

Crest usually upswept, brush-like

Golden crest is usually upswept towards rear; can be raised in display or flattened in submission.

White at gape

Crest shorter than on adult

Flipper has thin white trailing edge.

Grey

Imm.: yearling, approaching 2 yrs

Black elbow

Black leading edge blends unbroken into black tip.

May porpoise above surface when swimming fast. On land, walks or uses two-legged hop to move more rapidly.

## Snares Penguin *Eudyptes robustus* 50–60 cm

Breeds on Snares Islands S of NZ; immatures, more than adults, then appear to disperse widely. In colony makes loud, harsh, ratchetty squawks, croaks, grunts. Rare near Aust. coasts; most sightings around Tas.

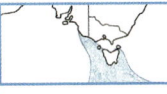

The long, silky golden plumes radiate outwards, most standing far out to each side of the head, slightly drooping; a few, shorter, point upwards.

Broad black leading edge joining, or fading into, a large dark patch across the tip

Imm. at about 2 yrs: shorter crest, bill dull, chin changing grey to black

Flipper: thin white trailing edge

Black patch inside elbow

# 3. ALBATROSSES, PETRELS & SHEARWATERS

Five families make up the Order Procellariiformes. All are seabirds, mostly oceanic, but some keep nearer coasts and others occasionally come in sight of headlands and islands. They are very difficult to identify: many are very alike, and birds are often seen only at a great distance. The best way to start is through joining offshore excursions with a group such as BirdLife Australia. Birds are often closer, and the group leader helps, as do others who are familiar with the species.

Silver Gull = 42 cm

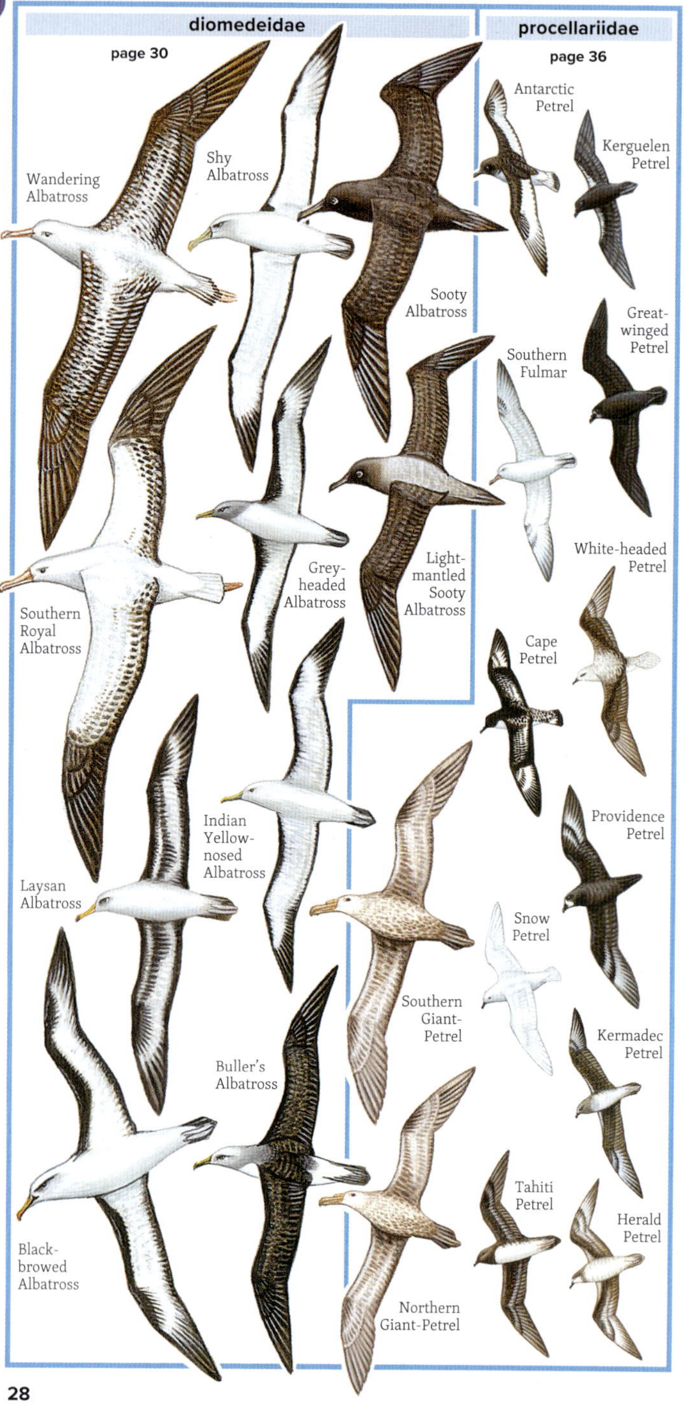

## diomedeidae
page 30

- Wandering Albatross
- Shy Albatross
- Sooty Albatross
- Southern Royal Albatross
- Grey-headed Albatross
- Light-mantled Sooty Albatross
- Laysan Albatross
- Indian Yellow-nosed Albatross
- Buller's Albatross
- Black-browed Albatross

## procellariidae
page 36

- Antarctic Petrel
- Kerguelen Petrel
- Great-winged Petrel
- Southern Fulmar
- White-headed Petrel
- Cape Petrel
- Providence Petrel
- Snow Petrel
- Kermadec Petrel
- Southern Giant-Petrel
- Tahiti Petrel
- Herald Petrel
- Northern Giant-Petrel

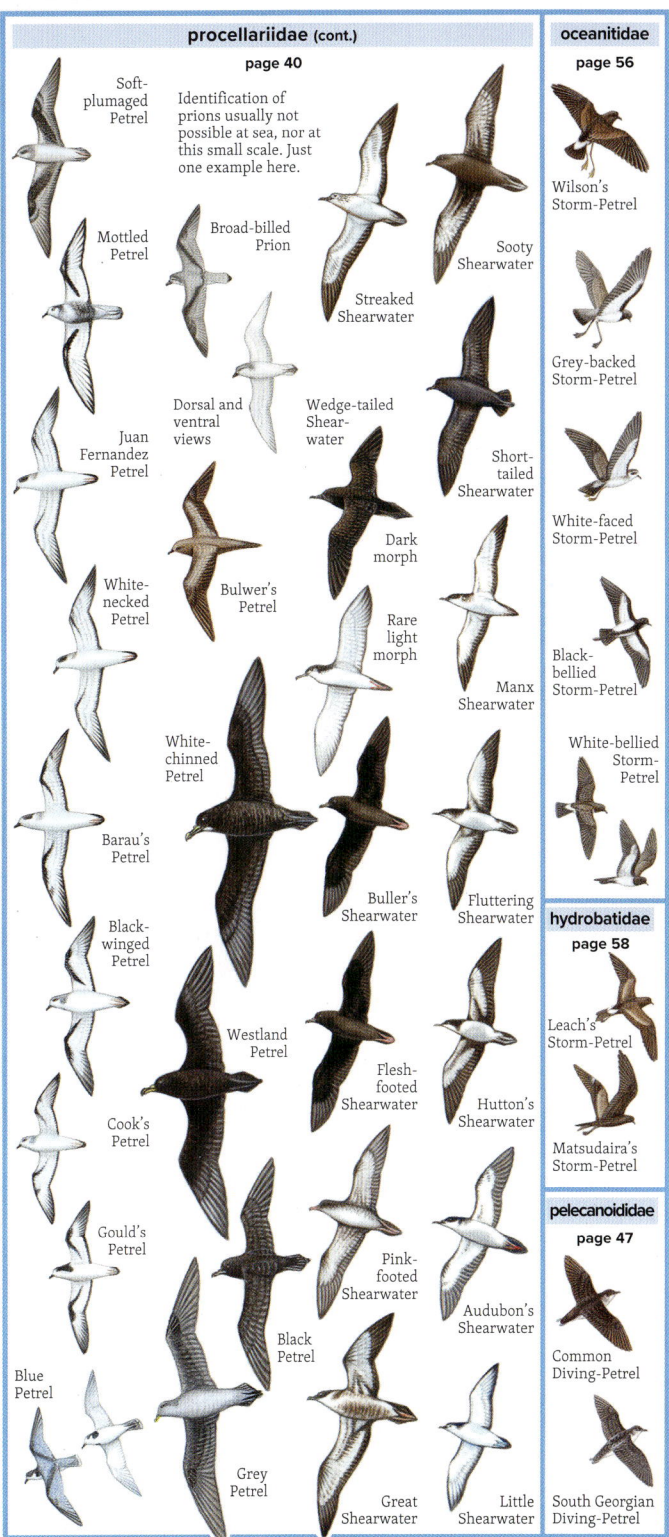

# WANDERING ALBATROSS

**Wandering Albatross** *Diomedea exulans* 1.1–1.2 m
Very large, extremely long wings, span of 2.5–3.5 m. Wings usually stiffly outstretched in graceful, dynamic gliding and soaring, using the updraft from waves to sustain flight for long periods with little need for flapping. In strong winds, wings held bowed and flexed. Habitat is marine: open oceans, edge of pack-ice; feeds over both deep pelagic and shallower continental shelf waters. Usually keeps within 15 m of sea's surface, often rests on sea in calm conditions – requires long take-off run. Follows ships. At sea usually silent, but occasionally makes gurgling, croaking sounds. In colonies, groaning, yapping, croaking interspersed with harsh braying and shrill shrieking. A common visitor to offshore and continental waters of SE Aust., and regular around S coast to SW WA; more often sighted winter to early spring; most belong to race *exulans*. Breeds on Antarctic and sub-Antarctic islands.

● *exulans*
● *chionoptera*
● *gibsoni*

Races share southern oceans; probably overlap across map area.

● Most common race in Aust. seas, snowy white at maturity; ● is slightly smaller, slightly darker underwing; ● slightly browner even at maturity.

These races are not separated in illustrations.

A population of race *gibsoni* on Amsterdam Is. in S Indian Ocean is now described as a separate species, the Amsterdam Albatross, *Diomedea amsterdamensis*. It may occur in Aust. waters.

Old males are entirely white except for wing tips and trailing edge. Females are like males, but average slightly smaller; appear never to become quite as white as old males.

Neat black wing tips

Black edge

Mature male, 8-10 yrs: body and tail entirely white, extent of black on wings reduced to outer wing and trailing edge

♂

Body all white

Aged 5 yrs (sub-adult): white spreading out along centre of wings

The body is all white at about 4 years, but wings do not reach full extent of white until 8-10 years and tail has black around the tip.

Variable intensity of bill colour

All white

♂

Dark wing tip and trailing edge

Underwing remains almost unchanged from juvenile to adult.

White face

Dark patch

Expansion of white begins on back, continues many years, outwards along centre line of each wing.

Imm: underbody, in second year is white except under tail and breast band; after about third year, underbody is entirely white.

White face

Juv. (first plumage)

Uniform dark brown above

Bill shows no fine black cutting-edge line – unlike Royal

All white

Pale

Overhead sunlight makes darker line of shadow under protruding brow.

Close-up, finely barred

Juv.

## SOUTHERN ROYAL ALBATROSS

**Southern Royal Albatross** *Diomedea epomophora* 1.1–1.2 m
Huge, long wings, span 3–3.5 m. Looks heavy bodied, has large, bulbous-tipped bill. In flight, graceful and skilled in using the uplifts of wind across waves. The soaring, banking turns and long, sweeping glides are only infrequently interrupted by any flapping. But flight is more laboured in calm conditions when albatrosses often resort to resting or 'loafing' on the surface. Habitat is marine, subtropical to sub-Antarctic oceans; occasionally further S into the Antarctic. In Aust., occurs over both open ocean and shallower inshore waters. Solitary or gregarious, it follows fishing boats, squabbling over scraps. At sea, gives occasional harsh squawks. Noisy in breeding colonies. Calls are similar to, perhaps not as harsh as, those of the Wandering Albatross, and include loud ratchetting sounds, slow, deep, resonant, gurgling clucks and low vibrant groans. Breeds NZ mainland and nearby islands.

● Map: two races share southern seas, but *epomophora* more common in SE, while *sanfordi* recorded more often SA to WA.

Races' plumage:
● *epomophora*
● *sanfordi*: adults of this smaller, endangered race have unique juvenile-like plumage, a combination of all-white body and all-dark wings on adults.

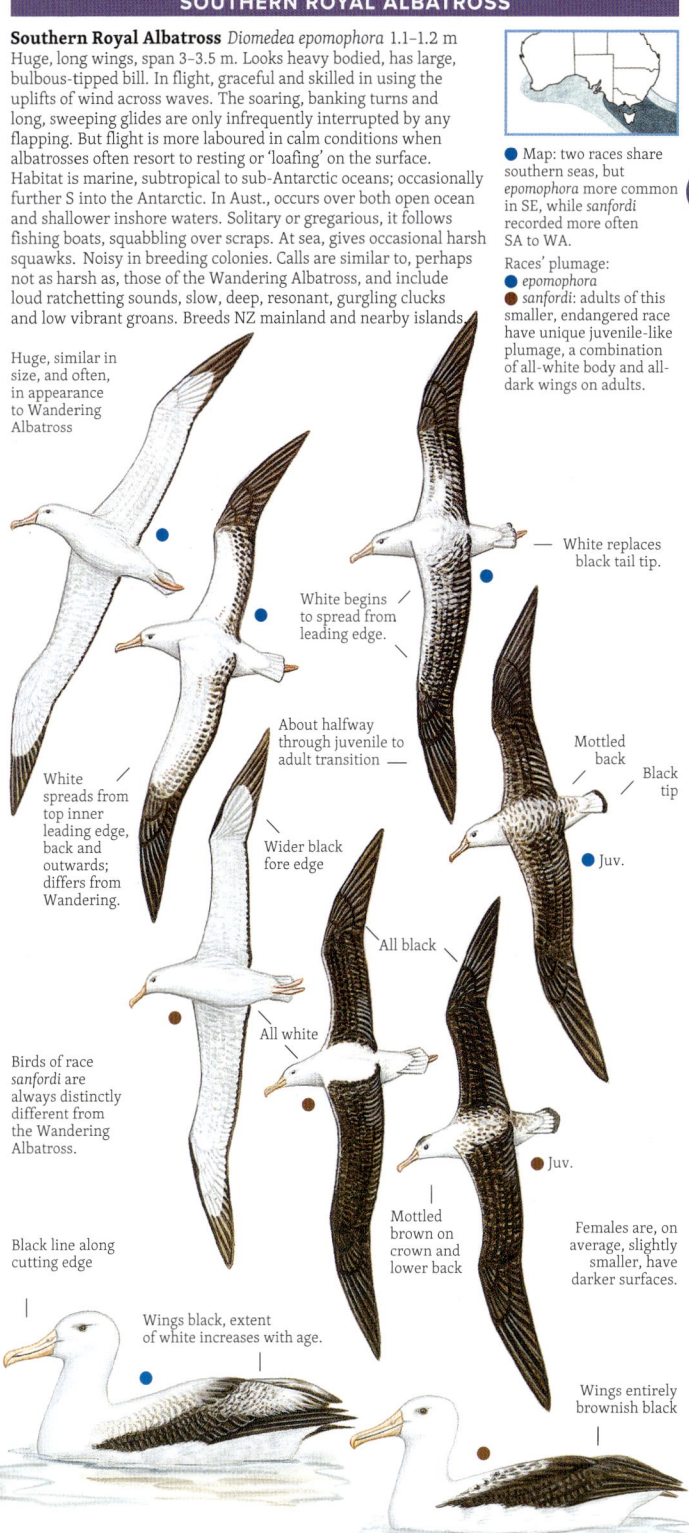

Huge, similar in size, and often, in appearance to Wandering Albatross

White spreads from top inner leading edge, back and outwards; differs from Wandering.

White begins to spread from leading edge.

About halfway through juvenile to adult transition

Wider black fore edge

White replaces black tail tip.

Mottled back

Black tip

Juv.

Birds of race *sanfordi* are always distinctly different from the Wandering Albatross.

All black

All white

Mottled brown on crown and lower back

Juv.

Females are, on average, slightly smaller, have darker surfaces.

Black line along cutting edge

Wings black, extent of white increases with age.

Wings entirely brownish black

## BLACK-BROWED & YELLOW-NOSED ALBATROSSES

**Black-browed Albatross** *Thalassarche melanophrys* 80–90 cm
A medium-sized albatross with unique combination of features. Short dark brow, conspicuous against white head, overhangs and shadows the eye, giving a frowning, penetrating expression. In flight, this albatross is light and graceful. In strong winds, it soars seemingly effortlessly, wheeling high on rigid wings. Is forced to flap in light winds; often waits out calm periods on the water. Usually silent at sea, except for harsh croaks when squabbling over food; noisy in colonies. Uses a range of marine habitats: inshore shallows, bays and channels to the edge of the continental shelf and beyond to pelagic ocean environs. Common May to Aug. around southern coasts.

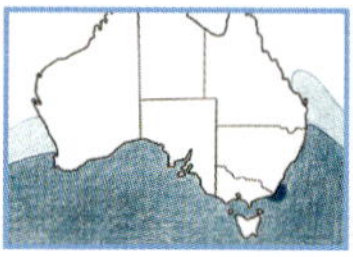

Races overlap widely.
● *melanophris*
● *impavida*

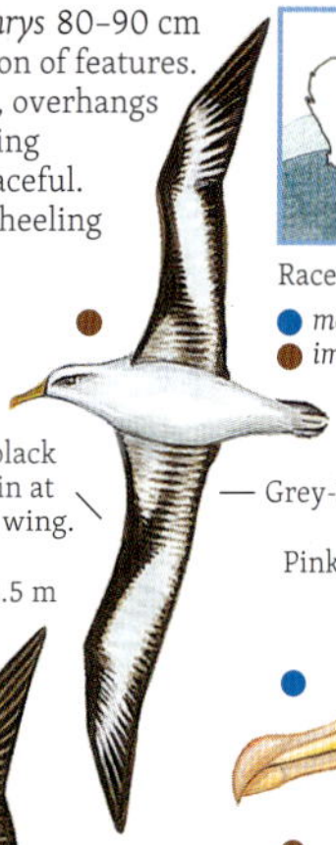

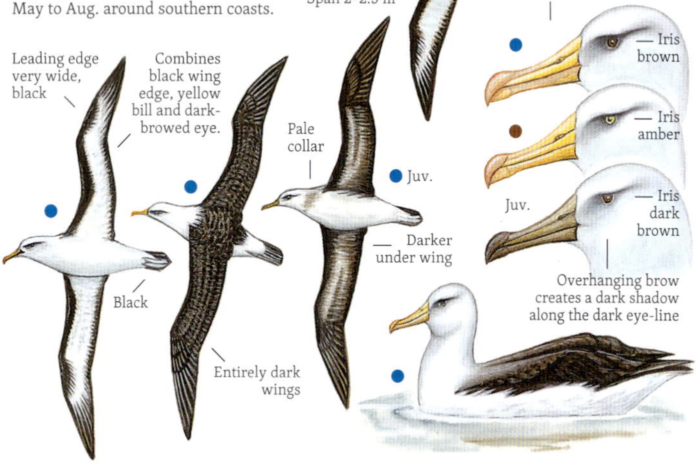

Three black lines join at central wing.
Span 2–2.5 m
Grey-brown tint
Pink tipped, yellowish bill
Iris brown
Iris amber
Juv.
Iris dark brown
Overhanging brow creates a dark shadow along the dark eye-line

Leading edge very wide, black
Combines black wing edge, yellow bill and dark-browed eye.
Pale collar
Juv.
Darker under wing
Black
Entirely dark wings

---

**Indian Yellow-nosed Albatross**
*Thalassarche carteri* 70–80 cm
A small, slender 'mollymawk' with yellow-orange top edge to the long, slender, glossy black bill. Flight is like that of other small albatrosses – elegantly aerobatic in medium to strong winds, but often floats on the surface when it is too calm to get wind uplift from the waves. Solitary, with same species, or mixed with other seabirds around trawlers or other food sources. Habitat covers seas subtropical to sub-Antarctic, but most abundant over warmer waters. Prefers shallower waters of continental shelves and the vicinity of upwellings and confluence of oceanic currents. Usually silent at sea, except when fighting over food. Noisy in colonies with a rapid, high, vibrating, nasal yapping, 'hek--ek--ek-ek-ek-ekek-eg-eg-eeg-eeeg'. Regular and common visitor to S coasts of Aust. and N as far as SE Qld.

● *bassi*: common
● *chlororhynchos*: may be a rare vagrant to SW Aust. from Atlantic.

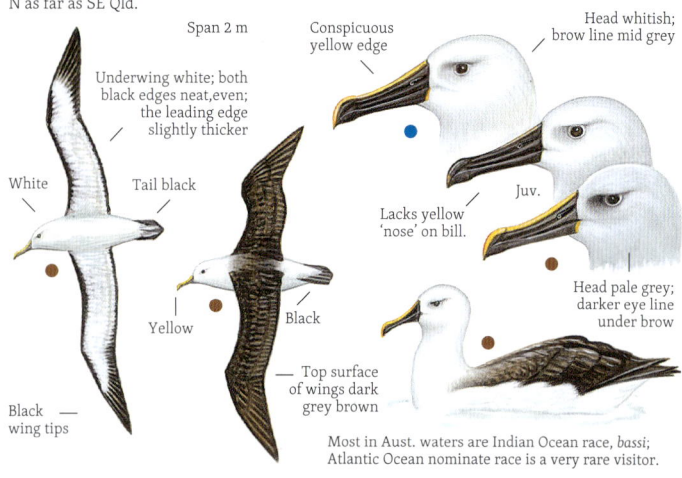

Most in Aust. waters are Indian Ocean race, *bassi*; Atlantic Ocean nominate race is a very rare visitor.

## SHY ALBATROSS

**Shy Albatross** *Thalassarche cauta* 1 m; span 2–2.5 m
Identified by size and extent of white under wings, and, at closer range, by a small black 'notch' on the leading margin of each wing close to the body. The only albatross to breed in Australian waters and breed only within the Australasian region. Wanders from subtropical to sub-Antarctic oceans, often visiting shallower waters on the shelf and around islands. Comes close inshore, entering bays and harbours extending offshore beyond the shelf edge; is scarce further out over pelagic depths. Proportionately very long winged and rather stout bodied. The huge bill adds to an overall heavier, less agile jiz than the other smaller, black-backed mollymawks. In flight, its casual effortless grace resembles the larger Wandering Albatross. It soars and wheels high above the horizon on rigid but slightly drooped wings. In light airs, without the lift and energy of wind across the tall waves of the open ocean, the Shy Albatross must use deep, slow, heavy beats of its long wings, a laboured flight that usually keeps close to the surface. Usually silent at sea, but croaks harshly when squabbling with other birds for food, especially around fishing boats. Noisy in colony, usually a deep, vibrating, nasal croaking, 'argk-argk-argk-argk'; also a harsh, strident wailing. Nominate race common; other races are uncommon or rare vagrants to Aust. seas, endangered.

- *cauta*
- *eremita*
- *salvini*

Races share southern oceans and probably intermingle across Aust. seas. However, only *cauta* is common, present throughout, and breeding in this region. Race *salvini* occurs mostly to the SE of region (uncommon); *eremita* is a rare vagrant.

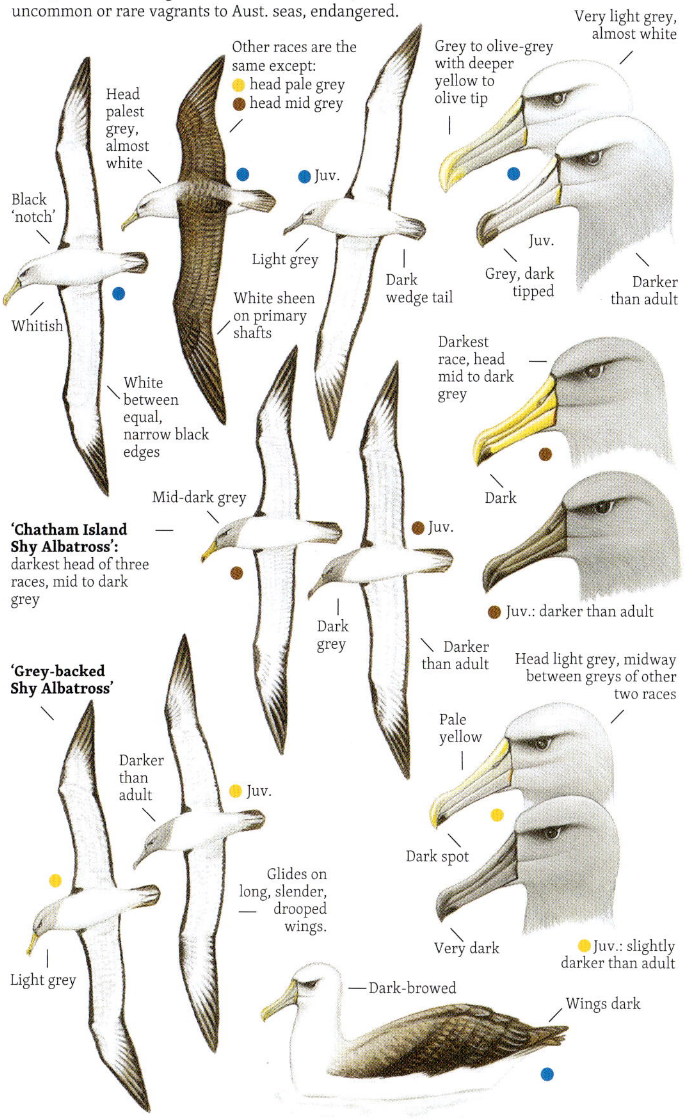

## GREY-HEADED & BULLER'S ALBATROSSES

**Grey-headed Albatross** *Thalassarche chrysostoma* 70–85 cm
Medium-sized albatross with obvious characteristics – entirely grey head and neck; black bill with a yellow line along top and bottom edges; broad black leading edges on underwing. Habitat marine: Antarctic to sub-Antarctic in summer; subtropical in winter, but seeking colder waters of southerly currents. Feeds over deep seas beyond the shelf-edge shallows; only seeks refuge inshore in rough conditions. Soars, glides and wheels effortlessly in broad arcs, only slightly adjusting its stiffly outstretched wings to use the uplift of wind across waves. Often rests on the surface in light or calm conditions. Silent in flight, but, in breeding grounds, gives vibrating, nasal croaking, a drawn out 'aarrrgh, aarrrgh' like a heavy, creaking creaking door, and an abrupt, sharper 'a-a-a-a-a'. Regular winter visitor to Aust.; offshore and coastal waters from SE and Tas. to SW WA. Colonies on islands of sub-Antarctic oceans.

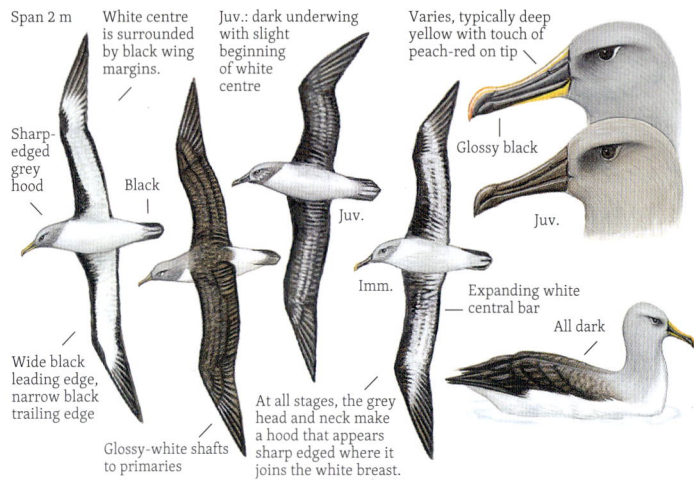

**Buller's Albatross** *Thalassarche bulleri* 75–80 cm
A small, slender albatross with graceful and apparently effortless flight; soars high on rigid wings, wheeling in wide arcs, gliding low into wave troughs. In light airs, often rests on calm water, floating high, upright posture emphasising slender-necked, small-headed appearance. Inhabits tropical to sub-Antarctic waters, shelf edge and pelagic, preferring warmer waters, or S seas where currents from N are warmer. Uses inshore waters; visible from land when sheltering from rough conditions on open seas. Solitary, in small groups or mixed gatherings. Silent at sea; in colony gives low, nasal, grating croaks like a crow, 'argh-argh-argh', abrupt or drawn out. Males give longer croaks, also wail and groan. Regular but uncommon visitor to SE Aust.; usually offshore Apr.-Jul.

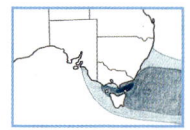

● *bulleri*: common
● *platei*: 'Pacific Albatross', has larger bill, darker hood but paler cap.

## SOOTY, LIGHT-MANTLED SOOTY & LAYSAN ALBATROSSES

**Sooty Albatross** *Phoebetria fusca* 85–90 cm
This small, sooty-brown albatross gives an all-dark impression at sea. Its tapering, streamlined shape enhances its sleek contours. Flight swift, graceful and effortless, especially in strong winds, with aerobatic gliding, soaring and stooping on wings that are always slightly flexed. Rarely resorts to flapping; uses the updrafts over waves and the winds above, rising to 10 or 15 m; often follows ships. Uses pelagic environs, the oceans beyond the shallower continental shelf. Silent at sea except for threat call when feeding – a harsh, throaty 'ghaaaow'. Regular visitor in small numbers to S Aust., Jan–Nov.

**Light-mantled Sooty Albatross** *Phoebetria palpebrata* 85 cm
Flight graceful, elegant; demonstrates complete mastery of wind and waves; soars and glides with little flapping. Range extends S to Antarctic pack-ice in summer; N to subtropical waters in winter. On average, prefers colder seas than the Sooty Albatross, but overlapping. Usually over deep pelagic and shelf-edge seas; occasionally inshore waters. Looks similar to Sooty Albatross, but upper body, from nape to upper-tail-coverts, is pale ash-grey in contrast to dark, sooty brown head. Usually silent at sea except for a harsh 'ghaaaa!' threat call, usually squabbling for food. Rare winter–spring visitor to SE Aust.

**Laysan Albatross** *Phoebastria immutabilis* 80–82 cm
White head with extensive dark lores extending back through eyes to give a very dark browed look. Tail dark, feet project beyond tail tip. Habitat marine, pelagic. Abundant in N Pacific; has expanded its breeding range to the NW Pacific, increasing the chance of sighting vagrant individuals in E Australian seas. One sighting near Wollongong, NSW.

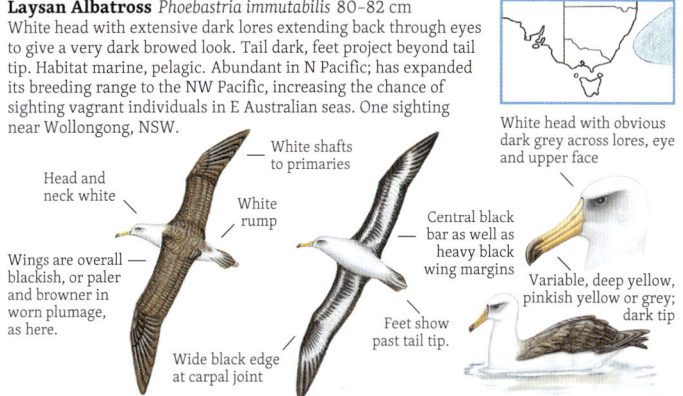

## GIANT-PETRELS

**Southern Giant-Petrel** *Macronectes giganteus* 85–100 cm
In active flight, heavy, humpbacked, laboured. The stiff-winged glides lack the graceful skill of an albatross. In strong winds the gliding improves, with some wheeling and weaving. Habitat marine, over open seas and inshore waters; favours edges of continental shelf and pack-ice. Gathers at carrion, offal, sewage outlets. Usually silent at sea except when squabbling over food. Noisy whenever at the breeding grounds; various sounds, mostly unpleasant. A stuttered sequence of groaning sounds, 'ur-ur-ur-ur', beginning very deep and resonant, becoming high and rapid. Common around S coast of Aust., these predominantly immatures. Breeding colonies located on Antarctic and sub-Antarctic islands and Antarctic mainland.

Only Aust. race, but two colour morphs: either dark grey-brown with whitish head and neck, or overall white with scattered dark feathers

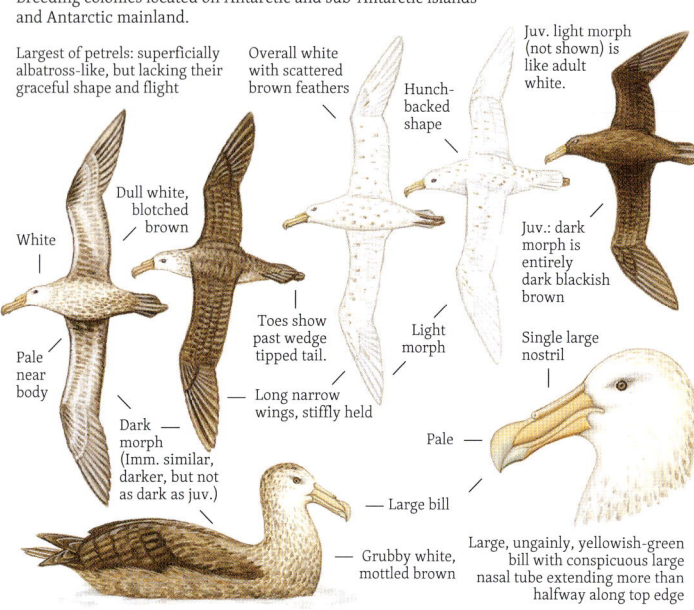

**Northern Giant-Petrel** *Macronectes halli* 81–94 cm
Flight and behaviour similar to the Southern Giant-Petrel. Habitat marine, using temperate and sub-Antarctic seas, frequenting inshore and pelagic seas out from edges of continental shelves. On average tends to prefer slightly warmer seas. Usually silent at sea except in a squabbling mob near carrion or refuse. Generally calls are like those of Southern Giant-Petrel; display calls possibly lower and harsher. Some deep 'argh-argh-argh-' or rising to a whistling, screaming crescendo. Most likely to be confused with dark phase juveniles of Southern Giant-Petrel, but bill details differ. Common along coasts and offshore waters of southern Aust., May–Oct.

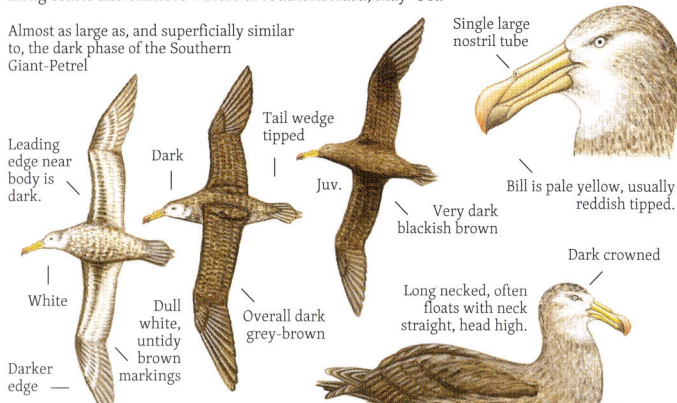

# FULMAR; ANTARCTIC, CAPE & SNOW PETRELS

**Southern Fulmar** *Fulmarus glacialoides* 45–50 cm
Distinctive in southern oceans, where most petrels are dark. At sea a gregarious, quarrelsome scavenger; hangs about fishing boats, trawlers; otherwise rarely follows ships. Periods of graceful wheeling and gliding, but settles on sea to feed or rest. Glides low to snatch or make shallow plunges. Often active at night when small squid are near the surface. Noisy at sea when quarrelling over food, and in nest colonies. Various fowl-like cluckings, croakings. Rare visitor to S coasts.

Dark grey
Pale pearl grey
Dark
Head white, pale grey nape and hind neck
Long, dark, blue-grey
Underparts white, slight grey tint on breast
Light pink; grey on some very dark birds
Dark trailing edge
Dark grey wing tip and trailing edge, on some birds, darker

**Antarctic Petrel** *Thalassoica antarctica* 40–45 cm
Usually a bird of the pack-ice seas around Antarctica. In flight, keeps well above waves. Stiff-winged glides are interspersed with bursts of rapid, shallow wing beats; sometimes hovers before plunging into sea to depths of 1.5 m; occasionally scavenges around ships. Swims well; often floats with wings outstretched. In small to large flocks at sea. Usually silent at sea, but gives rapid, harsh, abrupt 'argh-argh-argh' in colonies – some notes like creaky door. Abundant in Antarctic, very rare vagrant to seas of SE Aust.

Brown
White wing bar
Dark band
Hooded
White
Narrow edge
Broad dark edge
Dark margins of wings completely encircle white of both upper and lower surfaces.
Wings longer than tail

**Cape Petrel** *Daption capense* 35–42 cm; span 80–90 cm
In flight, displays typical fulmar pattern of brief glides interspersed with bursts of shallow fluttering flight. Tends to fly much higher than most other petrels and shearwaters, especially in strong winds, when it swoops and soars buoyantly. Dives for food from on wing or while floating. Noisy when breeding and in squabbling flocks at sea; has a rapid, hoarse 'arghargh-arg', a sharper 'cac-cak-cak', and an excited but softer, almost musical 'cook-cook-cook'. Habitat sub-Antarctic seas; in Aust., inshore and shelf-edge waters; common.

● *capense*: Antarctic
● *australe*: local race has more extensive black across upper parts.

White of back and wings is heavily chequered black.
Very dark, hooded
Black at front edge is slightly wider.
More black on head and upper back
Swims well, floating high. Often gathers in large, noisy, quarrelsome flocks around fishing boats.

**Snow Petrel** *Pagodroma nivea* 30–40 cm; span 75–85 cm
The only small all-white petrel. Flight erratic and fluttering; rapid, shallow wing beats; occasional brief glides, coming close to water, almost hovering, but rarely alighting. Does not usually follow or forage around ships. Usually silent at sea; can give a 'teck-teck' and other calls, guttural to quite high-pitched. Usually confined to the Antarctic pack-ice and nearby seas. A possible or very rare vagrant; several unconfirmed sight records.

Flight fluttering: erratic, shallow wing beats

Two races occur within usual Antarctic range: *nivea* and *confusa*.

## PETRELS: OVERALL VERY DARK

**Great-winged Petrel** *Pterodroma macroptera* 38–43 cm
A dark, thickset, long winged petrel. Flight in light winds is usually a lethargic, buoyant, meandering glide; rises and falls in wide arcs with occasional flaps. In direct travel has strong, steady wing beats, each with high, gull-like upstroke, brief pause, deep, strong downbeat. In high winds, flight is powerful: soars high, plunges down to sweep across the waves. Silent at sea; near nest in display, 'kee-kee-', gruff 'carrrk' or 'quawer'. Habitat marine, oceans and coastal waters. Common; usually *macroptera* WA to SA, both *macroptera* and *gouldi* further E.

● *macroptera*: SW to SE
● *gouldi*: mostly E of Tas.

Pale bases to primaries may gleam in direct sunlight

Almost uniformly dark, above and beneath

Thick, stubby bill

Wings long, narrow,

Heavy bodied, thick necked appearance

Tail wedge-shaped; often folded, then sharply pointed

White facial area larger: includes forehead, lores, cheeks, chin and throat; all juveniles like this.

Rear body tapers towards tail.

Head large, rounded

Wings extend past tail.

Holds carpal joints ('wrists') well forward, and the finely pointed outer wing swept back; a zigzag silhouette.

Small area of pale grey around bill

1.5x

**Providence Petrel** *Pterodroma solandri* 40 cm; span ≈ 1 m
A heavily built gadfly petrel. Flight relaxed with slow, easy wing beats, but in strong winds is fast. Dashes and weaves close to waves; lifts to soar high in great arcs. Habitat is marine, pelagic; E edge of continental shelf is possibly a favoured feeding ground for Lord Howe birds during winter. Silent at sea. In colonies, a loud, rapid, harsh 'kak-kak-kaaak-kak-kak' or 'kir-rer-rer'; on ground a deeper, trilled 'kerr-rer, kuk-kuk-kuk, ker-er-er'. Migratory; dispersive into N Pacific after breeding.

Forehead and face may be slightly scaly or scalloped

Darker underwings

A large, heavily built petrel identified by white beneath outer wing.

Mantle, back and upper wings mottled paler grey

Stubby bill

Wedge tip

Dull white around face

Slaty grey-brown

From beneath, throat and upper breast are darker than pale belly; more so in worn plumage.

A heavily built, long-winged petrel, distinguished in flight by whitish bases to underwing primaries and their coverts that make twin white crescents

**White-chinned Petrel** *Procellaria aequinoctialis* 50–58 cm
Heavily built dark petrel; usually has a tiny patch of white under chin. Shares with Westland and Black Petrels a distinctive 'albatross' pattern of flight: close to the waves, wing beats slow, deliberate and graceful, interspersed with long glides, but also soars high. Only nominate race occurs around Aust. Usually silent at sea, but, around trawlers, high trilling and aggressive, harsh clucking: 'chek-chek'. Habitat is marine, pelagic, but often over shallower coastal water.

Tubed nostrils on upper third of pale bill

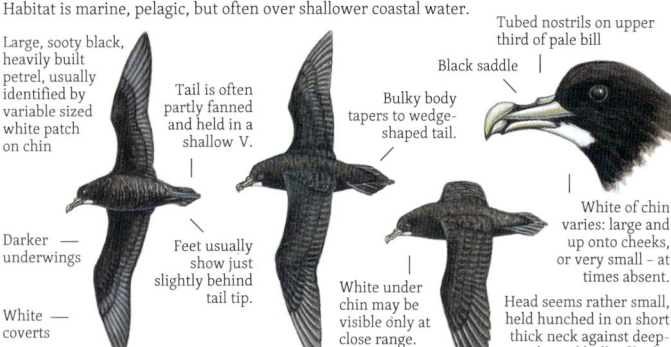

Large, sooty black, heavily built petrel, usually identified by variable sized white patch on chin

Tail is often partly fanned and held in a shallow V.

Black saddle

Bulky body tapers to wedge-shaped tail.

Darker underwings

Feet usually show just slightly behind tail tip.

White of chin varies: large and up onto cheeks, or very small – at times absent.

White coverts

White under chin may be visible only at close range.

Head seems rather small, held hunched in on short thick neck against deep-chested bulk of body.

## PETRELS: OVERALL VERY DARK

### Westland Petrel *Procellaria westlandica* 50–55 cm
Flight is more laboured against light winds, but then glides with wind close to waves with occasional slow wing beats. Action looks more vigorous in the lift of stronger winds: weaves and wheels on strongly flexed wings, swooping low between crests, then rising albatross-like to soar and bank. Usually solitary, but congregates around fishing trawlers. Feeds by snatching from surface or by shallow surface diving. Silent at sea; noisy visiting nest colonies at night; makes various harsh cackles. Inhabits offshore waters when breeding, then disperses over deeper seas. Rare visitor.

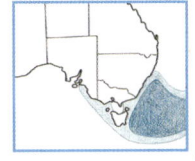

One of the two largest petrel species; heavily built, sooty black

Span 1.3–1.4 m

Side view shows the depth and bulk of body of this species.

Has a high, rounded-head silhouette.

Prominent nostrils over basal third of pale bill

Tips of black feet show behind tail.

Bill large, black with hooked 'nail' at tip

Carpal joint

Inner wing curved strongly forward to carpal joint; outer wing backswept

Flight feathers and greater coverts glossy, which may make these areas gleam paler at times, or reflect silvery highlights

Bill colour varies; usually pale cream, may appear ivory to buff.

### Black Petrel *Procellaria parkinsoni* 46 cm; span 1.1 m
Usually indistinguishable at sea from the larger Westland, but has lighter build, more slender wings and a graceful, buoyant flight. Glides with carpals forward, outer wings backswept, pointed. Prefers shelf-edge pelagic seas; usually avoids inshore waters. Mostly silent, but gives harsh, sharp, clacking, 'ak-ak-ak-ak-' and a throaty squawked, 'argk, argk, argk' in colonies. A rare, but perhaps regular, summer visitor to SE Aust.

Similar to Westland Petrel but smaller, of lighter build

Wings slender, finely pointed

Wedge tipped

Bill mostly pale ochre or greenish horn; off-white at distance.

Bill smaller than Westland or White-chinned

Tips of feet trail behind tail.

Wings gently curve forward to carpal joints, then sweep back to wing tips.

Sooty black in fresh plumage, but with age wears to brownish black.

Floats buoyantly, typically with head high

### Kerguelen Petrel *Lugensa brevirostris* 33–36 cm
Blunt bodied yet slender winged silhouette. Usually far out to sea, rarely near land unless driven inshore or wrecked by gales. In light winds it has a gently weaving flight; glides interspersed with brief bursts of bat-like fluttering. In strong winds soars and circles, tern-like, very high. Calls at sea unknown, possibly silent. In colonies it calls mainly at dusk, a wheezy 'ch-chee-chee-chay'. Uncommon winter visitor to S coasts of Aust.

Large head with high forehead gives a distinctive appearance.

Medium sized 'gadfly' petrel, dark plumaged with large, high head

Holds wings forward at carpal joints, then backswept.

Nostril tube on basal quarter

Bill rather short, stubby, looks small on large head.

Unlike reflective flashes, white of underwing leading edge is a constant field mark.

Glossy areas beneath the outer wing can reflect light, a soft gleam or silvery flash.

Tail quite long, wedge-shaped, usually almost closed, quite pointed

## PETRELS: UNDERWING DUSKY WITH DARKER DIAGONAL

**Herald Petrel** *Pterodroma heraldica* 35–39 cm; span 1 m 
Stockily built, typical gadfly petrel in overall jiz. Distinctive underwing pattern with white patch at base of primaries. Usually glides low with leisurely wing flaps; in strong winds, higher banking and wheeling. Most appear to be the light phase. Apparently silent at sea. At island colonies, pairs and individuals call in display flights, night or day, rarely from the ground. Usual call is a high chattering variously described as 'hi-hi-hi-' or 'chi-chi-chi-'. Pelagic habitat, i.e. uses the ocean surface and the airspace close above. Is found on tropical and subtropical deep water beyond the continental shelf. Rarely seen away from breeding islands; disperses from colonies away from coastal waters; a very few vagrant and beach-washed records for E coast Aust.

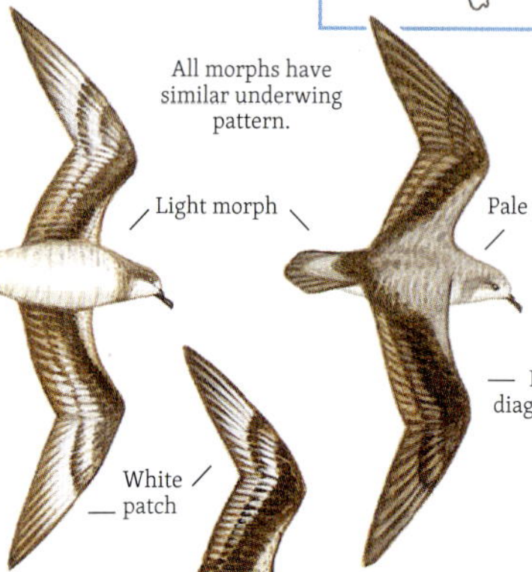

All morphs have similar underwing pattern.

White — Light morph — Pale grey

White underside to primaries

Light grey-brown

Dark diagonal bar

White patch

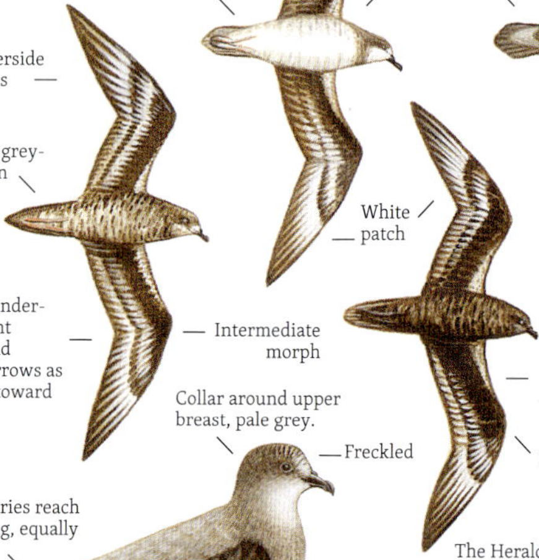

White on underside of flight feathers and coverts narrows as it extends toward the body.

Intermediate morph

White leading edge

Collar around upper breast, pale grey.

Freckled

Dark morph

Dark primaries reach to tip of long, equally dark tail.

The Herald Petrel occurs in a wide range of plumage shades, from light to dark. All morphs have the distinctive underwing pattern, although it is less obvious on the darkest birds.

---

**Soft-plumaged Petrel** *Pterodroma mollis* 32–36 cm
Flight fast with rapid wing beats alternating with glides on angled wings, travelling in wide arcs, zigzag progression. Rarely soars very high. Habitat marine, pelagic, Antarctic to subtropical waters. Gregarious, often found in small parties travelling low and fast, at times in loose flocks of up to a thousand birds. Occasionally follows ships. Usually silent at sea; at colony calls in flight – a low, wavering, musical wail, repeated several times, often ending with a sharp upwards 'whik'. Possibly common in seas S of Aust.; a quite regular visitor to SW WA, less so in SE.

Race reaching Aust. seas is nominate *mollis*, of which a few are the rare dark morph. This form varies from an overall very dark grey-brown on which the breast band and underwing band are barely discernible to birds with a wide dark breast band and heavily streaked underparts.

Light morph — Dark morph — Light morph

Grey-black diagonal almost lost on dark underwing

Dark tips

Dark eye

Indistinct M band across upper wings

Dark grey

Eye is set in a dark patch that merges with grey nape.

Speckled dark grey

Has grey upper parts, grey underwing with darker diagonal band and grey breast band.

Soft grey

Span 84–95 cm

40

## PETRELS: UNDERWING DUSKY WITH DARKER DIAGONAL

**Kermadec Petrel** *Pterodroma neglecta* 38 cm; span 93 cm
Stocky body, long-pointed wings held forward to the carpal joint, outer wing backswept, short tail – all part of the overall jiz of large gadfly petrels of the genus *Pterodroma*. Identification of the species is difficult, with sightings at sea often brief and distant. Adding to the complexity, the Kermadec has three recognised colour phases that grade continuously from light to dark. Fortunately, all have the pale or white area beneath the outer wing at the base of the primaries that show them to be Kermadec. On the upper wing, the same patch shows a white 'flash' – white streaks formed by the reflective, glossy white shafts of the primaries. The pattern of flight also helps identification. Most time in the air is spent gliding, effortlessly riding on the wind, wheeling and banking, occasionally aided by a few deep, leisurely, gull-like wing beats. The call is varied, but usually a loud 'yuk-ker-a-wooo-WUK' – the first part a hoot, the final part loud and abrupt. An extremely rare vagrant, accidental to E coast NSW.

● Aust. race *neglecta*, with light, dark and intermediate morphs

White primaries under wing
White streaks on primaries
White strip along leading edge
Light morph
Intermediate morph
Light morph
White with buff-brown tint
Light morph
Pale
Between light and dark morphs lies a continuous range of colour forms.
Intermediate morph
Mid grey-brown
Grey-brown
Bill black, stubby
Dark brown
White
Dark morph
Dark morph
The dark morph is entirely brownish grey-black. All morphs have the same identifying markings – the pale patch on underwing primaries, and fine white lines of primary shafts on the upper wing.
Primaries black with glossy white shafts
White

**Tahiti Petrel** *Pseudobulweria rostrata* 38-40 cm
Typical gadfly petrel jiz; stocky, bull necked, tapering silhouette; dark hood; large bill; long wings often held relatively straight; pale reflective bar under wing. In light weather, leisurely gliding with occasional languid flaps of wings; keeps close to the surface. In strong winds, a more purposeful, albatross-like soaring and swooping. Usually solitary, at times in loose flocks. Silent at sea; noisy in colony at night. Habitat pelagic; with Australian sightings usually beyond continental shelf. Perhaps a regular visitor; recorded E, N and NW Aust. seas.

Pale narrow bar along wing
Flight feathers have silvery sheen.
Brownish black band
Dark hood
Deep, bulbous tip, black
Hood deep sooty brown
Feet pink and black
White
Span 85–95 cm
Overall dark brown
Neat, sharp edge
No seasonal variation; both sexes and juvenile similar
Wing tips curve back.

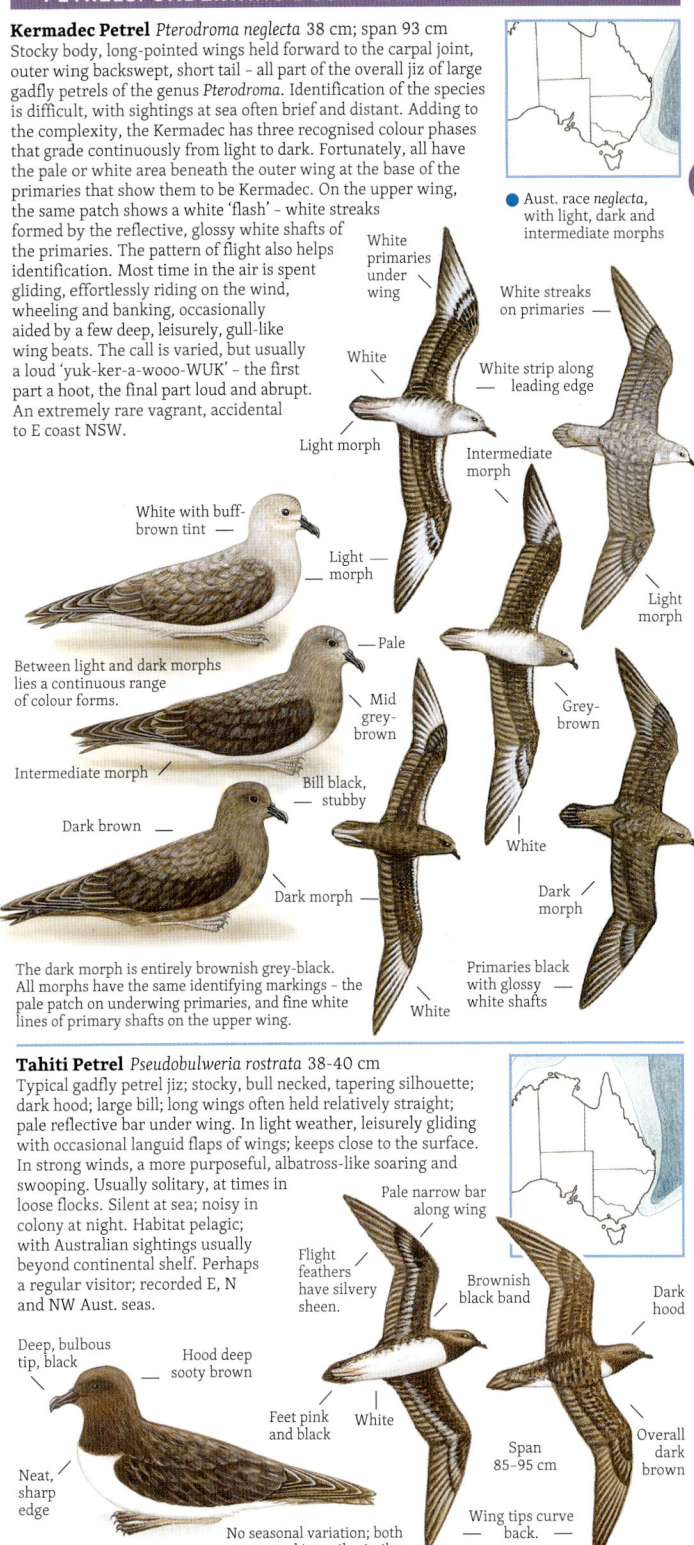

## PETRELS: UNDERWING PREDOMINANTLY DARK

**White-headed Petrel** *Pterodroma lessonii* 40–46 cm
Head, neck and underbody are white, against which background the dark grey area around the eye is most conspicuous and distinctive; at a distance gives impression of a very large eye. In light winds keeps close to the waves with short glides and periods of steady, shallow, stiff-winged beats. In strong winds, little flapping, wheels in high arcs, catching the lift of wind over waves. Usually silent at sea, but colonies on islands of sub-Antarctic seas are noisy. The activity and noise peak just after sunset and again before sunrise, with much calling from birds overhead. Flight calls are 'wit-wit-wit' and 'tiew-ee, tiew-ee-' with an occasional gruff 'ooo-err'. Habitat includes Antarctic and sub-Antarctic oceans as far S as the edge of the pack-ice. Except near breeding grounds, avoids inshore waters. Uncommon winter visitor to seas along S coast of Aust. – edge of continental shelf unless blown in by storms. Breeds on N islands of sub-Antarctic seas.

Span 1.1 m

While the underwing is overall very dark, an indistinct wide band may be discernible depending on underwing lighting, whether shadow, sunlight or reflected infill light from sea surface.

Slightly darker band along overall dark underwing

Faint collar

Distinctive small white triangle

Pointed when closed

Wedge-tipped when spread

Primaries and coverts are, on basal half, highly glossy, reflective, at times giving a flash or gleam of light.

Dark eye obvious on white head

Long narrow wings, backswept to sharp wing tips

Crown and nape are white finely vermiculated pale grey.

Deep, black

Dark primaries reach to tip of long, equally dark tail.

The other species on these two pages have similarly plain dark underwings, but are otherwise obviously different. Also similar is the light phase Soft-plumaged Petrel.

**Grey Petrel** *Procellaria cinerea* 50 cm
A distinctive large petrel plumaged ashy grey on the upper parts, softly blending to white on underbody. In flight, wings are often held rather straight and stiff, which, together with a rather tubby body and slender wings, imparts a rather shearwater-like jiz. Appears swift and strong in flight; long graceful glides intersperse with periods of powered flight with distinctive wing action, quick and powerful with a characteristic jerkiness. Often plunges into the sea from considerable heights. In strong winds the energy in the flight reflects the power of the wind with albatross-like soaring, wheeling and gliding; wings are held more flexed than in the stiff, straight-out position used for lighter airs. Accompanies and scavenges around ships and fishing boats, following for long periods. At sea may be solitary or in small groups, occasionally gathering in large rafts. Apparently silent at sea. Noisy in colonies, especially early in breeding cycle: cackles, moans, resonant alarm calls. Habitat marine, mainly pelagic; tends to keep well offshore, so not often sighted from land. Rare vagrant to S Aust.

Underbody white, very light grey towards tail-coverts, in contrast to dark wings

Short, wedge tipped, dark

Toes just visible

Primary undersides are reflective, flash in sunlight.

Silvery grey in fresh plumage wearing to a slightly brownish grey

Slender, pale greenish yellow

White with touches of faint blue-grey tint, bright white in full sunlight

Span 1.1–1.3 m

Large-bodied with rather small-headed appearance

## PETRELS: UNDERWING PLAIN & DARK

**Bulwer's Petrel** *Bulweria bulwerii* 26–27 cm; span 68–72 cm
Small, dark slender body tapering to fine tail; long narrow wings with carpal joints carried well forward, outer wing backswept. In flight, zigzags low, usually within several metres of the waves. Prion-like twisting and turning, each brief burst of quick wing beats followed by a short glide on bowed wings; tail fans briefly to manoeuvre. When feeding, circles low, dipping and snatching at surface prey. Solitary or in small groups, occasionally resting on calm seas. Silent at sea. Habitat marine, pelagic; typically over warmer waters. May be common from Sep. to Apr. in NW Aust. waters. Elsewhere, a rare vagrant; records for NE Qld, Vic.

Small, dark, short-necked, long-tailed appearance and with long, slender, pointed wings

Carpal joints are held well forward.

Brownish black

Pale grey to grey-brown wing bar

Long tail, usually folded almost to a point, but wedge-tipped when briefly spread

Slender wings, slightly rounded outer wing, or backswept with wing tips sharply pointed

Head and bill appear rather small for the body size.

Dark, sooty brown above and beneath appears black where shadowed.

**Jouanin's Petrel** *Bulweria fallax* 30–32 cm; span 70–75 cm
Very closely resembles Bulwer's Petrel but larger. Like that species, looks overall dark at sea, but may show a paler brown band curving along the upper-wing-coverts as the feathers wear. Keeps bill on a downward angle more noticeably than some other species. Tail is slightly broader than Bulwer's, which has a long steadily tapering rear body that merges almost imperceptibly into a finely pointed tail. Wings are long, slender, backswept, and held with the carpal joints well forward. In flight zigzags low, usually within several metres of the waves.

Disperses southward from breeding grounds in Arabian Sea to northern Indian Ocean, SE to Sumatra and, probably, on occasions well offshore the NW Aust. coast. One record, near Ashmore Reef off the NW Kimberley coast, WA.

**Atlantic Petrel** *Pterodroma incerta* 43 cm; span 1–1.1 m
A large petrel with stocky jiz; dark brown head and upper breast. Wings long, uniform dark brown beneath, held forward to the carpal joint then backswept. This species is similar to the Tahiti Petrel, which, however, differs in having a pale line down the central underside of each wing. Flight described as swift and 'careening'. Inhabits pelagic waters of S Atlantic and rarely the W Indian Ocean. Common in central S Atlantic and is one of the most abundant petrels; possibility of being a very rare vagrant to Aust. waters.

Unlikely or extremely rare accidental occurrence in Aust. seas. A hooded petrel much like the Tahiti with the same sharp contrast between stark white lower breast and belly and the very dark brown wings and hooded head. Under-tail-coverts brown.

Wings long, backswept

Plain underwing without obvious bar

Long, wedge-tipped, dark

Dark under-tail-coverts

Dark hood

Underwings entirely dark brown

The similar Tahiti Petrel has white under-tail-coverts and a narrow, pale, reflective central wing bar.

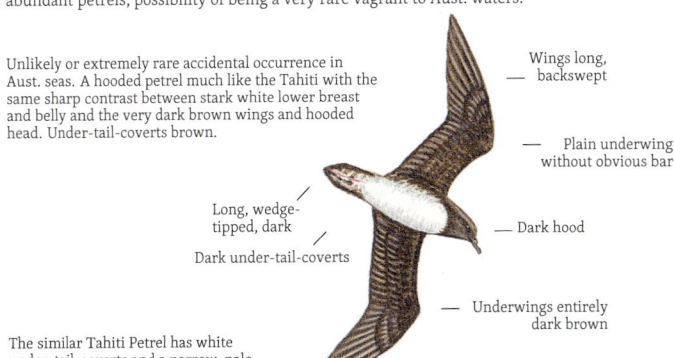

## PETRELS: WHITE UNDERWING, LONG THICK DIAGONAL

**Mottled Petrel** *Pterodroma inexpectata* 33–35 cm; span 85 cm
Medium-sized gadfly petrel. Long, bold black underwing diagonal crosses white secondary coverts. Black M across upper surface of outstretched wings. Flight is swift with bursts of quick, shallow wing beats and long, sweeping glides. Sweeps high in the wind; turns in great arcs; again swoops towards the waves. In colonies, makes rapid, vibrating, metallic 'ki-ki-ki-' and various growling and crooning sounds. Most alike is Soft-plumaged Petrel, but underwing is darker. Uses open seas and Antarctic iceberg belt. Accidental or rare vagrant to coast of SE Aust. Breeds on islands near NZ, including Snares Is.

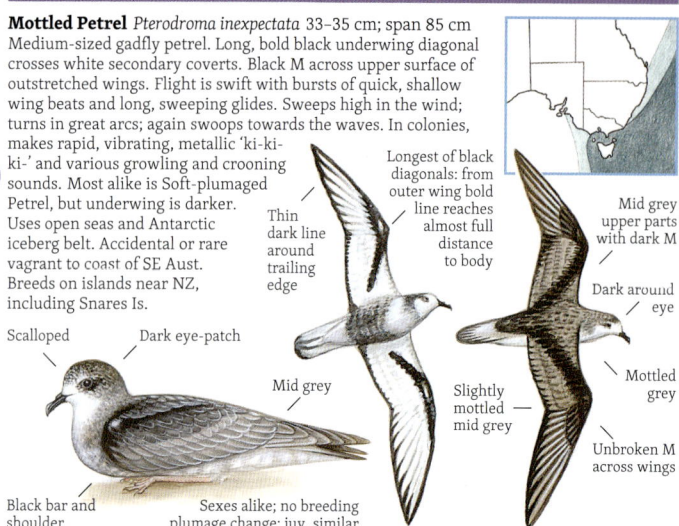

**Black-winged Petrel** *Pterodroma nigripennis* 28–30 cm
A small species with distinctive markings. Flight is fast, powerful; at times dashes and weaves low across the sea. More often, in light winds, follows a leisurely, undulating path; brief bursts of rapid wing beats alternate with short glides. In a strong wind, flight gains vigour; wheels, soars high, swoops low. Habitat is marine, pelagic; usually no closer to land than the edge of the continental shelf, but sometimes driven into coastal waters by storms. In flight over colony, gives rapid, high 'peet-peet-peet'. Records along E coast increasing. There are colonies on Lord Howe and Norfolk Is.

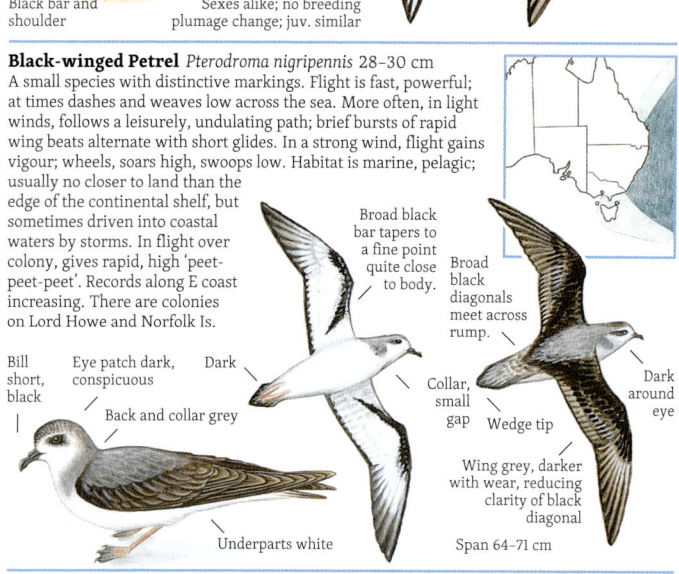

**Gould's Petrel** *Pterodroma leucoptera* 30 cm; span 68–71 cm
Flight is slower than similar petrels, less high towering; banks, weaves and dips, glides on stiff wings. Keeps to offshore waters – the deep open oceans. Usually silent at sea; at breeding grounds, calls in flight and on the ground. After dark, circles above colony giving a high, metallic ticking, 'kik-kik-kik-kik' or 'zit-zit-', more rapidly and excitedly in aerial pursuits. Usual call from birds on ground is a shrill 'pee-pee-peeoo' or a tremulous growling. Population of Aust. race small, breeding pairs on Cabbage Tree Is., NSW.

## PETRELS: WHITE UNDERWING, LONG FINE DIAGONAL

### Barau's Petrel *Pterodroma baraui* 38 cm
A large, dark-capped petrel. Flight is slower, not as agile or aerobatic as most other petrels of same range. Habitat is marine; seems attracted to areas of warmer surface water across the tropical and subtropical parts of the Indian Ocean. Noisy when flying over breeding colonies; calls are described as a high 'ti-ti-ti-ti-' and lower 'kek-kek-kek-kek-'. Probably silent at sea. Species is unique within its usual range; elsewhere, most alike are the Juan Fernandez, Atlantic, Herald and Gould's Petrels. The normal range covers the western half of the Indian Ocean. Further E it becomes rarer and rarer. Only an extremely rare chance of accidental occurrence in coastal waters of WA. Breeds on islands near Madagascar.

Wide M pattern wing tip to wing tip, but in worn plumage, this is less obvious.

Dark crown, nape

Slender body and narrow wedge tail similar to Juan Fernandez and White-necked Petrels

Thinner, shorter, finely pointed black bar extends just over halfway from carpal to body.

Faint bar

Slender, wedge-tipped

Black behind eye

Grey back, darker in worn plumage

White

Dark edge

Span 94–98 cm

Faint grey at sides of breast

Tail is dark; white on outer pair of feathers.

Jiz is generally slender with long, straight wings.

---

### White-necked Petrel *Pterodroma cervicalis* 40–43 cm
Previously a race of the species *Pterodroma externa*; common name is unchanged. White collar around back of neck is unique. Flight is leisurely yet strong, a gentle 'roller-coaster' travel; in strong winds wheels and arcs high. Feeds by dipping and snatching from the surface; solitary, occasionally in loose flocks. Silent at sea, noisy in colonies during the breeding season. In flight, gives a loud, harsh 'ka-ka-ka-ka' and a softer 'tse-tse-tse-tse'. From ground and in flight, it gives a long 'kukoowik-ka-ka'. Habitat marine, oceanic especially along edge of continental shelf. A rare or uncommon vagrant; sightings usually well offshore, occasionally from land, NSW and Qld, summer. Breeds Nov.–May, Kermadec Islands, NZ.

Span about 1 m

Glides with long wings slightly flexed back. The broad white hind-neck collar is an identifying feature.

Dark leading edge to outer wing

Black M across full length of wingspan; unbroken across rump

Broad white hind-neck collar

Narrow dark trailing edge

Broad white hind-neck collar

Black cap extends below eyes.

Thinner, shorter, finely pointed black bar extends just over halfway to body.

Mantle colour varies with feather wear. Silvery when new; as the pale tips wear, the mantle darkens until blackish brown like wings.

Very pale grey

Dark edge continues around wing tip where it is wider, but just under 10% of birds have much more, almost half the area of the outer wing being quite dark.

Fledglings have silvery mantles when adults are dark, late autumn.

## PETRELS: UNDERWING DIAGONAL SMALL OR ABSENT

**Cook's Petrel** *Pterodroma cookii* 25–30 cm; span 66 cm
A gadfly petrel with black underwing diagonals that stop well short of the body. Flight leisurely and rolling in light winds, or vigorous with rapid, jerky wing beats and erratic banking and weaving that can reveal diagnostic markings of both under and upper plumage. Usually silent at sea and on ground. Calls in flight over colonies include high, nasal, rather duck-quacked 'kwek-kwek-kwek', soft or loud and harsh. Habitat pelagic subtropical and, rarely, sub-Antarctic waters. Rare vagrant off the E coast of Aust. Breeds on islands off N NZ.

**Juan Fernandez Petrel** *Pterodroma externa* 40–43 cm
Larger than most 'Cookilarias' – the petrels that are similar to Cook's Petrel. Flight similar to but less energetic than that of White-necked and Barau's, appearing strong, graceful and effortless; wheels in high arcs. Previously a race of the White-necked Petrel. Call is probably similar to that of White-necked Petrel. Similar species include Barau's, Gould's and White-necked Petrels. Habitat pelagic over tropical and sub-tropical parts of central and E Pacific. Extremely rare vagrant or accidental visitor to offshore waters of Australian E coast.

**Blue Petrel** *Halobaena caerulea* 26–32 cm; span 66–71 cm
Small blue-grey petrel; unique square-cut tail with white tip is visible at a distance and in poor light. Flight fast and buoyant; frequently glides low on stiff, bowed wings, dipping between waves, uplifting over crests. High and wheeling in strong winds, shows alternately the white underparts then the steely grey upper pattern. Gregarious, small scattered flocks, often with other species. Silent at sea; in courtship, gives a soft 'ku-ku-ku-COO-COO'. Most alike is Gould's Petrel, but it has no white tail tip. Habitat mostly pelagic, but sometimes recorded over shallow coastal waters. Uncommon but regular winter visitor to SE Aust. waters.

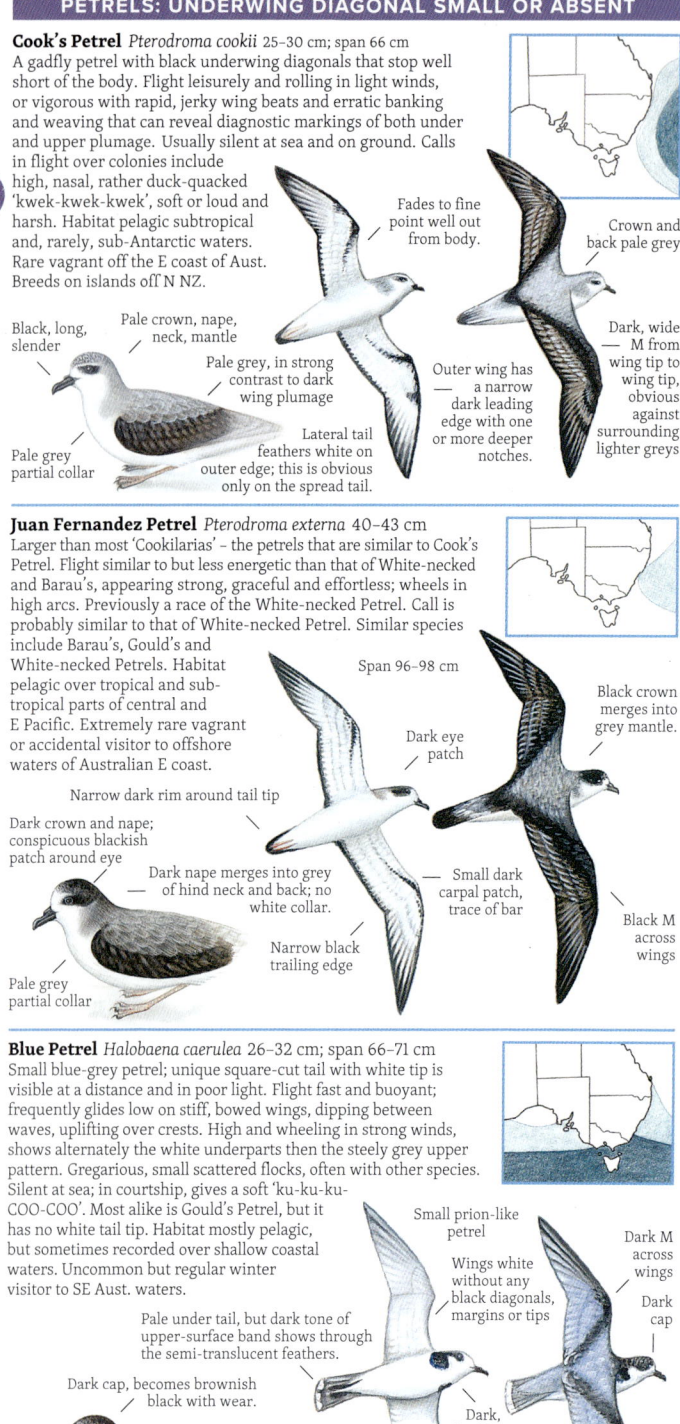

# DIVING-PETRELS

**Common Diving-Petrel** *Pelecanoides urinatrix* 20–25 cm
In common with other diving-petrels, this is small and stocky with a short-winged, short-tailed, tubby, neckless jiz. When floating on the surface, it looks like a miniature penguin. Flight is distinctive, unlike any other type of seabird: always low, fast and direct with quail-like whirring of wings, clipping wave tops. May fly directly into waves to burst out and into flight at the other side; dives deeply, using wings in flight-like action in pursuit of small fish. Uses varied habitats, all marine, including open oceans, shallow continental shelf, inshore waters, bays. Usually in small flocks. Apparently silent at sea, but very noisy at night in breeding colonies. Male gives a slow 'koo-ah', rising in pitch, and 'koo-ah-ka, koo-ah-ka-', also described as 'ku-ku-miaw'. Female returns variations like 'koo-aka-did-a-did'. Common in and near Bass Str.; rare in coastal waters as far N as SE Qld; rare vagrant westwards to SW WA.

Several races, but distribution overlaps; at sea uncertain. Race *urinatrix* seems the usual SE Aust. race; *exsul* a very rare vagrant to SW WA.

Brown edges to primaries

Tubby, short necked; flight fast and low with rapid, whirring wing beats; flies into waves; patters along surface.

Span 33–38 cm

Pale to mid grey

Race *exsul* has slightly broader bill; may also have darker breast band; indistinguishable from *urinatrix* at sea.

Feet, toes and webs are quite bright cobalt blue.

Scapular feathers lightly tipped white in fresh plumage

Glossy black

Nostrils are divided in two by ridge or septum; second transverse divider toward rear of nostrils.

Dives while swimming as well as plunging from air.

Both Common and South Georgian become slightly browner as plumage wears.

**South Georgian Diving-Petrel** *Pelecanoides georgicus* 20 cm
At sea, indistinguishable from Common Diving-Petrel, but so rare that any diving-petrel seen in Aust. waters is almost certainly the Common. Positive identification requires that the bird be in hand to examine details. Previously it was thought that the distribution of the two species did not overlap, but it has been found that both share many island breeding grounds – the calls and colour of downy chicks differ. South Georgian chicks are initially grey; those of the Common are white. Silent at sea. From burrows, a series of 5–10 varied squeaks; in flight, a squeaked 'ku-eeek' at intervals of several seconds. Uses both coastal and offshore waters around islands of Antarctic and sub-Antarctic seas; probably also over open ocean, but identification for confirmation of range is difficult at sea. Extremely rare vagrant to SE Aust.

Scapular tips white in fresh plumage

Slightly smaller and paler than Common Diving-Petrel

Brightness of plumage varies with wear, so of little value in separating the two species.

Feet, toes and webs are quite bright cobalt blue.

White to pale grey

In hand, small details of bill give safer identification. Differ from Common, above, with transverse subdivider midway along nostrils.

Pale grey inner edges to primaries

Ear-coverts, sides of face are very slightly paler than on the Common Diving-Petrel.

47

## PRIONS

**Broad-billed Prion** *Pachyptila vittata* 28 cm; span 60–62 cm
Largest of prions, extremely wide bill. Flight slower; skims low across sea; slower wing beats and more gliding, less erratic, steep banking than smaller narrow billed prions. Wings held well forward; appears short-necked, long-tailed; massive bill and high forehead, looking front-heavy in flight. Habitat deep pelagic seas, occasionally inshore waters in storms; subtropical, tropical and sub-Antarctic waters. Silent at sea, but noisy at night in colonies; gives a rough, rasping 'ku, ku-aah, kuk'. Rare vagrant; some beach-cast on SE and SW coasts.

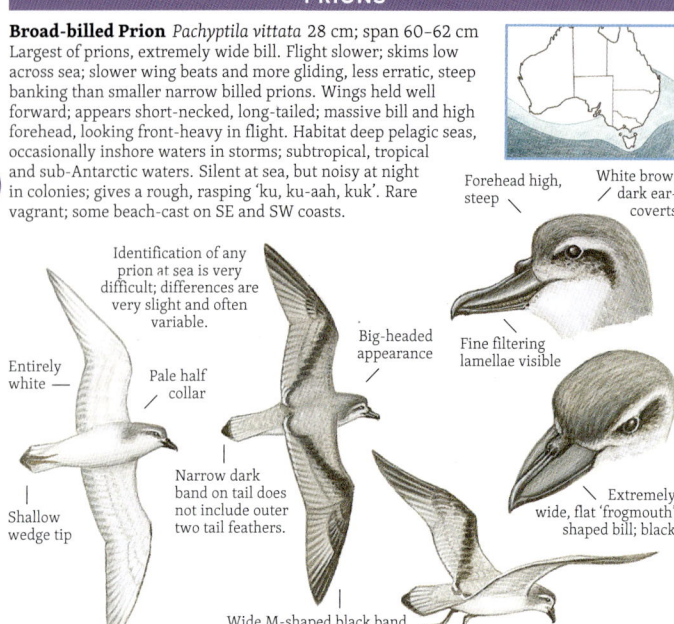

**Slender-billed Prion** *Pachyptila belcheri* 25–26 cm; span 56
Flight fast, erratic, buoyant, keeping close to the waves; lifts over crests with bursts of rapid wing beats and through troughs with long, weaving, twisting glides. Feeds largely on small crustaceans. When feeding, glides and flutters just above the waves, pattering the water occasionally for added lift, often snapping with slender bill at the surface. Silent at sea; in colonies usually a harsh cooing, occasional squawks and trills. Habitat includes sub-Antarctic and Antarctic seas, S to edge of pack-ice; also shallower shelf waters. Common SW coast; uncommon in SE.

**Fulmar Prion** *Pachyptila crassirostris* 28 cm; span 60 cm
A rare species closely resembling the much more abundant Fairy Prion; the two species are usually indistinguishable at sea. Flight similar; Fulmar more erratic, often performs an aerobatic, high, up-and-over loop, returning to its original course. Marine, pelagic; strongly gregarious, feeds in flocks. Females slightly smaller. Noisy at night in colonies; gives stuttering, low, harsh croakings. Vagrant; a beachwashed specimen, Tas.; unconfirmed records Vic. and S coast NSW.

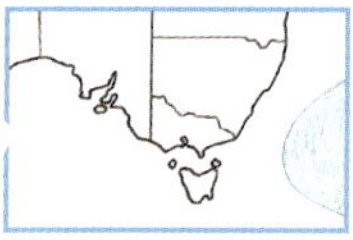

## PRIONS

**Salvin's Prion** *Pachyptila salvini* 25–30 cm; span 58 cm
Is sometimes included as race *salvini* of the Broad-billed Prion, and is probably indistinguishable at sea from that species and the Antarctic Prion. However, recent studies show the Broad-billed is a distinct species more closely related to the Antarctic; flight of Salvin's is similar to both related species. Gregarious, often in dense flocks. Voice not well known; probably calls only in colonies, loud 'KA-kaka-DU'. Sub-Antarctic habitat, extending into adjacent Antarctic and temperate waters. Common visitor to Aust. offshore waters Jun.–Oct.; presence revealed by beached specimens.

Salvin's Prion usually indistinguishable from Broad-billed Prion at sea.

Very faint short collar

Very short dark collar

Lamellae

Narrow dark tail band does not include outer two tail feathers.

Wide M-shaped black band across blue-grey wings

In close view, bill narrower, straighter, but, like the Broad-billed, has fine filtering lamellae comb visible along the side of the bill.

Bill wide, flattened; dark blue-grey

**Antarctic Prion** *Pachyptila desolata* 25–28 cm; span 64 cm
At sea, similar in appearance and flight to above species. Has same feeding tactics: dense flocks work together across the water into the wind, wings outstretched, feet paddling – shallow plunges just below the surface to snatch small marine creatures. Antarctic to sub-Antarctic, favouring cold waters just N of the zone of pack-ice. Silent at sea; noisy in colonies with its throaty, dove-like cooing: 'uk-coo-uk-cooo-uk-uk-u-cooo'. Moderately common in S Aust. seas; often beach-cast winter-spring; considerable numbers are wrecked after storms.

Wide M-shaped black band spread across pale blue-grey upper parts, wing tip to wing tip

At sea, looks identical to Salvin's Prion.

Larger but incomplete dark collar

Lamellae hidden, or very nearly so, with bill closed

Darker partial collar extends lower onto neck than on either Broad-billed or Salvin's. Bill longer, slightly less broad than Salvin's.

Medium width, straight sided, blue-grey

Narrow dark tail band does not include outer two tail feathers.

**Fairy Prion** *Pachyptila turtur* 24–28 cm; span 56–60 cm
Smallest prion; keeps close to the waves, often huge flocks weaving and banking in unison, displaying blue-grey then white surfaces. Flight is buoyant and erratic; feeds by dancing lightly across the surface, wings fluttering and feet pattering, dipping head to surface, then plunging beneath it. Usually silent at sea, but noisy at night in breeding colonies. Habitat is sub-Antarctic seas, islands, while breeding, then wanders to subtropical regions, then keeping to open oceans, rarely close inshore except when sheltering from storms. Commonly seen, but usually well offshore. In the SE this is the most abundant of the prions; breeding colonies can have many thousands of breeding pairs.

Wide M-shaped black band spread across pale blue-grey upper parts, wing tip to wing tip.

Dark diagonal wing bar is slightly softer or less distinct.

Larger, whiter brow line

Wider dark tail-tip band

Wedge tipped

Insignificant breast band

More extensive, brighter white in face; very faint grey wash for breast band

## LARGE SHEARWATERS

**Buller's Shearwater** *Ardenna bulleri* 46–47 cm
Flight leisurely, buoyant, banking, gliding; slow, deliberate wing beat; arcs higher in strong winds. Tropical seas; preference for areas of strong upwelling along continental shelves and the food-rich waters where warm and cooler currents meet. Forages low over surface; hunts by low shallow plunges. Silent at sea; noisy in colonies after dark and pre-dawn; varied wailing, screaming, howling sounds, like other shearwaters. Similar species are pale morph of Wedge-tailed Shearwater, and Juan Fernandez, White-necked and Barau's Petrels. Regular but scarce summer visitor to SE Aust.

Long-necked, long-billed, slender-bodied jiz.

**Wedge-tailed Shearwater** *Ardenna pacifica* 38–45 cm
The common 'Muttonbird'. Flight is slow and leisurely; seems to drift buoyantly close to the water, carpal joints well forward, tail usually tapering to a long point. Inhabits tropical and subtropical seas; pelagic, frequenting and feeding across the ocean surface, often at junction of warm and nutrient-rich cool waters, and also where inshore water meets deep oceanic water. Silent at sea, but noisy in colonies at night; a wailing 'ka-wooo-ah', repeatedly, becoming faster, louder, rising to an almost hysterical climax. Common in coastal and oceanic waters of E and W Aust.

Most are dark morph, the light being rare off E Aust. coast, this rising to about 25% of population off the W coast.

## LARGE SHEARWATERS

**Flesh-footed Shearwater** *Ardenna carneipes* 40–45 cm
A large, burly, brownish-black shearwater with characteristic lethargic flight; long slow glides, banking with one wing tip then the other towards the waves, interspersed with deep, slow wing beats. In strong winds the glides become more swift and purposeful, banking, lifting and swooping between the swells, but without the erratic energy of the smaller and lighter shearwaters. Solitary or in flocks; follows boats. Usually silent at sea; noisy in colonies with its repeated, hoarse 'ku-KOOO-uhg', rising to a strident scream. Habitat marine, usually beyond edge of continental shelf. Common in southern Australian seas.

A large, heavily built, broad-winged shearwater

Body looks bulky or heavy, head flat crowned; seems slender on rather long neck.

Wing often held gently downcurved

Head large, black

Toes do not extend beyond tip of tail.

Wings are straighter with carpal joint not so far forward as some other shearwaters.

Fanned tail makes broad wedge.

Legs and feet pale pink or pinkish brown

Bill long, pale pinkish or greenish horn, and darker brown tip

Underside of primaries reflective silvery towards their bases

Flies low over water, plunging from heights of 1 to 5 m.

Span 1–1.5 m

Combines completely dark plumage with deep, bulky, pale bill.

**Sooty Shearwater** *Ardenna grisea* 40–46 cm; span 1 m
Flight fast and direct: mixes bursts of rapid, stiff-winged beats with long, banking glides when one wing tip brushes the water's surface. In strong winds, arcs higher on flexed wings. Silent at sea, but noisy on arrival and prior to the pre-dawn departure. The usual call is a loud, regularly repeated, rather dove-like wailing coo; begins slowly, softly, then rises through the sequence, building to a loud frenzy of excitement before dying away again: 'awook, awook, awow, awow-AWOW-awow-awook, awook'. Most like Short-tailed and Flesh-footed Shearwaters. Marine habitat, Antarctic to subtropical seas; occasionally coastal waters. Common to abundant around S coasts.

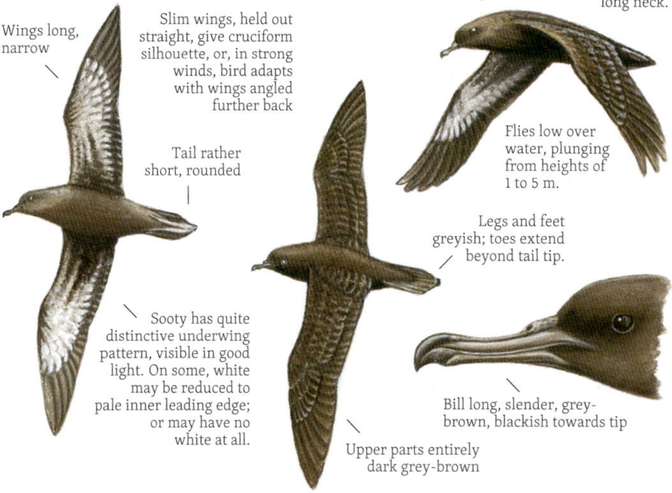

Body looks bulky or heavy. Head flat-crowned; seems slender on rather long neck.

Wings long, narrow

Slim wings, held out straight, give cruciform silhouette, or, in strong winds, bird adapts with wings angled further back

Flies low over water, plunging from heights of 1 to 5 m.

Tail rather short, rounded

Legs and feet greyish; toes extend beyond tail tip.

Sooty has quite distinctive underwing pattern, visible in good light. On some, white may be reduced to pale inner leading edge; or may have no white at all.

Bill long, slender, grey-brown, blackish towards tip

Upper parts entirely dark grey-brown

# LARGE SHEARWATERS

**Short-tailed Shearwater** *Ardenna tenuirostris* 40–45 cm
The 'muttonbird' harvested on islands of Bass Str. May be seen in huge flocks at sea around colonies, some estimated to contain millions of birds. These may form large rafts on the sea surface near breeding islands in late afternoon; birds await the cover of darkness before approaching nests. Flight is slow, appears leisurely; floats over ocean on stiffly held wings. In stronger winds the gliding birds skim the waves, dipping into the hollows and lifting over the crests. Habitat marine, often over continental shelf waters. Common, the Tas. population has been estimated at about 16 million birds.

Three forms: dark morph; light morph with pale underwing; rare light morph with white central underwing

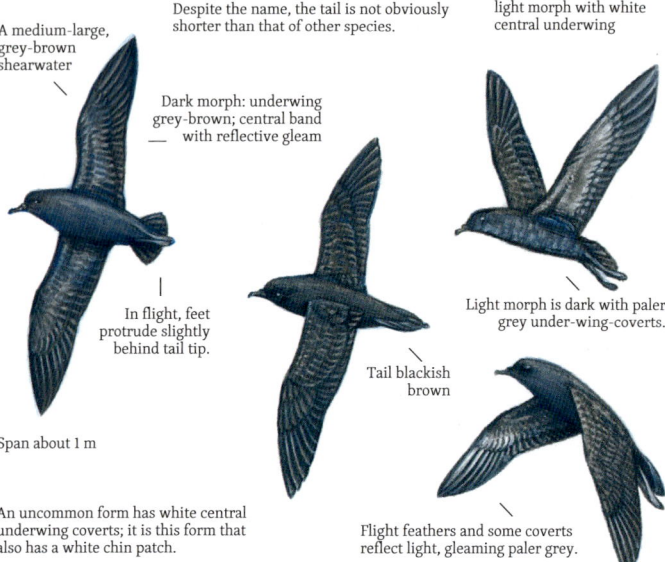

A medium-large, grey-brown shearwater

Despite the name, the tail is not obviously shorter than that of other species.

Dark morph: underwing grey-brown; central band with reflective gleam

In flight, feet protrude slightly behind tail tip.

Light morph is dark with paler grey under-wing-coverts.

Tail blackish brown

Span about 1 m

An uncommon form has white central underwing coverts; it is this form that also has a white chin patch.

Flight feathers and some coverts reflect light, gleaming paler grey.

**Pink-footed Shearwater** *Puffinus creatopus* 48 cm
A large shearwater with grey-brown plumage; variable extent of mottling on flanks and underwing. Buoyant, languid flight with rather heavy, laboured flapping action alternating with long glides on downcurved wings. In strong winds, the flight becomes invigorated with higher wheeling and banking. Most alike is the light morph of Wedge-tailed Shearwater, but it has longer, more pointed tail and other minor differences. Most often seen over shallower continental shelf waters. Rare vagrant to Aust.; several possible sightings.

A large, heavily built shearwater: white underbody, variable underwing, pink bill and feet

Upper parts dark grey-brown; light morphs may show a faint dark M across wings.

Has light, intermediate and dark morphs; intermediate is the most common.

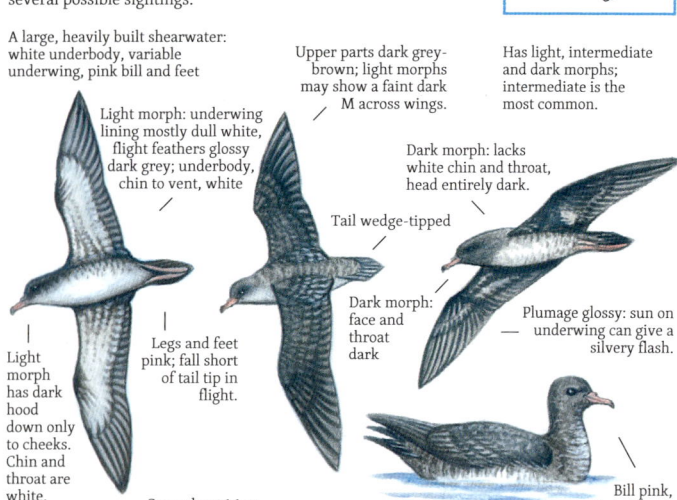

Light morph: underwing lining mostly dull white, flight feathers glossy dark grey; underbody, chin to vent, white

Dark morph: lacks white chin and throat, head entirely dark.

Tail wedge-tipped

Dark morph: face and throat dark

Plumage glossy: sun on underwing can give a silvery flash.

Light morph has dark hood down only to cheeks. Chin and throat are white.

Legs and feet pink; fall short of tail tip in flight.

Span about 1.1 m

Bill pink, dark-tipped, heavy

# LARGE SHEARWATERS

**Great Shearwater** *Puffinus gravis* 43–51 cm; span 1–1.2 m
Large with clear-cut dark cap. Flight swift and deliberate with powerful, rapid beating of stiff, straight wings; glides and banks close to the waves, plunging into the water or rising 4 to 10 m to dive at a shoal of fish or other marine life. Usually solitary, often follows ships. Makes a noise like fighting cats when squabbling around trawlers. Inhabits open seas; rarely inshore. Rare vagrant, several sightings of a single bird at sea, Robe, SA, Jan.–Feb. 1989.

**Streaked Shearwater** *Calonectris leucomelas* 47–49 cm
Large shearwater; face and crown white with black streaks. Flight is slow and graceful in leisurely arcs; glides on bowed wings, occasional deep, rather heavy flaps. In strong winds, flaps less; lifts and banks higher, faster. Follows fishing boats, solitary or in small groups, or mixed with other seabirds. Habitat is the pelagic oceans, shelf waters and edges; rarely close inshore. Usually silent at sea. Common summer–autumn visitor to N, W and E coasts of Aust.

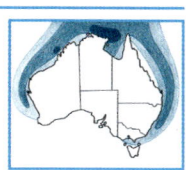

## SMALL SHEARWATERS

**Manx Shearwater** *Puffinus puffinus* 30–38 cm
Medium-sized; black upper parts in sharp contrast with white underparts. Glides and banks low above the waves with brief flurries of rapid, shallow, stiff wing beats. Against stronger winds, lifts, banks and glides with little flapping. Favours edge of continental shelf: infrequent further out; rarely over inshore waters. Silent at sea, but calls at colonies. Similar species are Fluttering and Hutton's Shearwaters. Extremely rare vagrant to Aust. seas; one dead on beach, SA, several possible sightings SE Aust.

**Fluttering Shearwater** *Puffinus gavia* 32–37 cm
Dark brown upper parts, white beneath. Flight is swift and low, short glides on stiff wings. Slightly slower, deeper wing action than other small shearwaters, which gives a fluttering effect. Its flight follows a slightly less direct course with more banking, lifting and falling. Hutton's Shearwater is almost identical; Little Shearwater smaller. Silent at sea, noisy over island colonies. More frequently seen inshore in estuaries and harbours than are other shearwaters. Common over Aust. inshore waters Apr.–Oct.

## SMALL SHEARWATERS

**Hutton's Shearwater** *Puffinus huttoni* 36–38 cm
Flight like Fluttering Shearwater; slightly longer necked, shorter tailed appearance. Often in flocks, small to huge. Prefers waters of continental shelf; at times comes inshore to estuaries, bays and channels. Floats high when swimming. Usually silent at sea, but occasionally some cackling sounds. Very noisy in breeding colonies; a juddering, squawking 'kouw-kouwkee-kee-aah' with variations. Migratory around most of Aust. coast, mainly birds under 2 years. Adults probably sedentary near NZ breeding grounds.

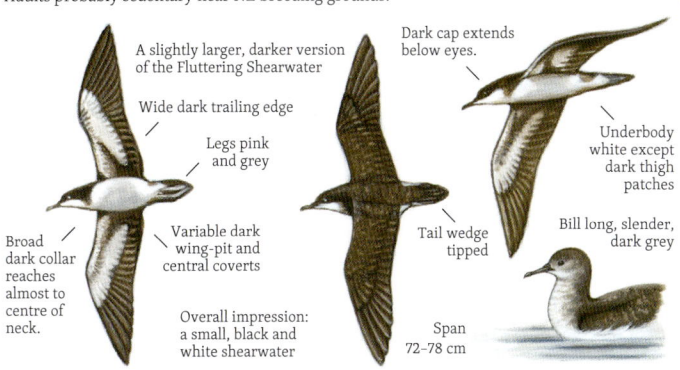

**Audubon's Shearwater** *Puffinus lherminieri* 27–33 cm
Small, stocky; long tail with black under-tail-coverts and underside of tail, which is often spread. In light winds it glides more than other fluttering shearwaters; flight looks smoother. In strong winds has high, wheeling glides. Inhabitant of tropical and subtropical oceans; occasionally inshore bays, estuaries. Noisy in breeding colonies; usual call 'shooo-kree', either rasping or sharply screeched. Similar are Fluttering and Little Shearwaters. Several possible sightings off E coast.

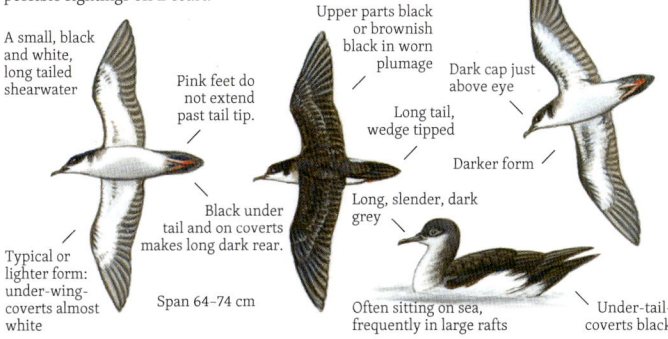

**Little Shearwater** *Puffinus assimilis* 25–30 cm; span 58–67
Smaller, more slender than Audubon's Shearwater. Flight is 'flutter and glide' – shallow, whirring wing beats followed by brief low glides; flat, not banking or twisting. In strong winds, long banking glides over waves and troughs, fluttering to catch the wind on crests. Alights on water; swims under water. Habitat is oceanic and continental shelf waters. Silent at sea; in breeding colonies, a growling, sobbing 'wah-i-wah-i-wah-ooo'. Common offshore S WA; uncommon in SE Aust.

● *assimilis*
● *tunneyi*: smaller, shorter wings; breeds on islands of SW Aust. coast. Both races could be found in map area. Possible vagrants might include other races of broader Australasian region, *elegans* and *haurakiensis*.

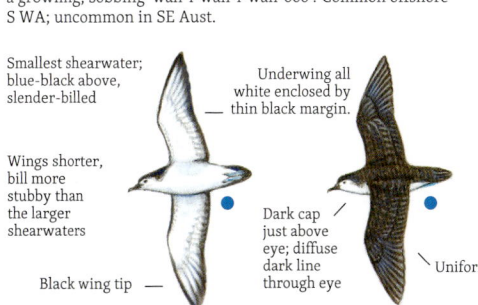

## WILSON'S & GREY-BACKED STORM-PETRELS

**Wilson's Storm-Petrel** *Oceanites oceanicus* 15–19 cm
When travelling, flight is swift and swallow-like with bursts of rapid wing beats interspersed with short glides on upswept wings close to the waves. When feeding, switches to a slow, fluttering flight; works across the surface against the wind, skimming along the windward slopes of wave troughs. Often hovers with rapid, shallow, fluttering beats of uplifted wings, feet pattering the waves, head bobbing down to snatch small fish, crustaceans or carrion on or close beneath the surface. Varied habitat: marine, deep pelagic seas, shelf slopes and shallower shelf and inshore waters; ranges from Antarctic pack-ice to subtropics. Aust. records usually from edge of continental shelf. At sea, silent except for occasional querulous, sparrow-like chattering while a flock feeds; chirping calls at night. In colonies, churring and peeping sounds. Widespread, abundant. Migrates from Antarctic breeding grounds. Winters across tropical seas, passing northward along W and E coasts of Aust. in autumn; remains around N coast, returns S in spring.

● *oceanicus*
● *exasperatus*: has slightly longer wings and tail, plumage identical. Races have separate Antarctic or sub-Antarctic breeding grounds, at other times intermingle.

**Grey-backed Storm-Petrel** *Garrodia nereis* 16–19 cm
Travelling flight is swift and direct with continuous rapid fluttering – rather bat-like wing beats uninterrupted by glides. Feeding flight is much slower; patters across the surface using feet to kick and bounce across the water. Observations around SE Aust. show species is most common along the shelf edge and further out; rarely closer inshore. Silent at sea; in colonies, a wheezy, scratchy chirping or twittering. Birds from the NZ population are regular visitors to the Aust. coast in winter.

## BLACK- & WHITE-BELLIED STORM-PETRELS

**Black-bellied Storm-Petrel** *Fregetta tropica* 20 cm; span 46
In flight distinguished from beneath by rough edged black band in centre of white belly; this is harder to see from side view. Keeps low with erratic changes of direction, skimming and hugging contours of waves, legs dangling and feet dipping, occasionally plunging breast against water then pushing clear with feet and wings. Silent at sea; in colonies makes a distinctive, whistled 'hieuuuw' and softer piping. Antarctic, sub-Antarctic seas in summer; tropical and subtropical in winter, usually beyond the shelf-edge; rarely enters inshore waters. Uncommon winter visitor to W, S and E coasts of Aust.

Race *melanoleuca* is doubtful: white bellied

Neat, contrasted and unique underbody plumage

Head to upper breast always black, but belly varied; most have central black band.

Blackish brown

Unlike White-bellied Storm-Petrel, feet project beyond tip of tail

White crescent

Wide, black leading edge

White

Male, female, juvenile and immature alike

White spot on chin and throat

Faint brownish diagonal is most evident in fresh plumage.

Black band

Outer wing curves back.

The black belly strip usually separates this species from the White-bellied Storm-Petrel; otherwise difficult to separate.

Underside of body extremely varied. The central band, breast to under tail, may be narrow, wide, or spread out to cover most of the underbody black. This black band may be broken, consisting of large or small flecks of black, or may be absent altogether, breast and belly then entirely white.

**White-bellied Storm-Petrel** *Fregetta grallaria* 18–22 cm
Occurs in two colour morphs, light and dark, and a range of intermediate plumages. Distinctive flight using the wind – glides into the wind with uplifted wings, touching the surface with dangling feet, hugging the wave contours – brief bursts of fluttering, shallow wing beats. In Aust. is found along edge of continental shelf and further out; only occasionally over inshore waters. Silent at sea and in flight over colonies. Calls are usually from within the nest burrow, a long series of high 'pew-pew-pew-pew' notes. After breeding, disperses across Pacific and over Tasman Sea. Possible Aust. range is from S Tas. N to Coral Sea. Most claimed sightings are from S coast NSW to S coast Qld.

Four subspecies; probably only the Australasian-breeding nominate, *grallaria*, occurs in Aust. seas. Plumage varies from light through to dark phases.

Light morph plumage pattern is similar to the Black-bellied Storm-Petrel, but without the distinctive longitudinal black band.

Span 46–48 cm

Marks that distinguish the White-bellied from any form of the Black-bellied:
1. feet do not project beyond tail tip;
2. chin and throat are always black.

Conspicuous white crescent

Black chin

Feet not visible

White between wide dark margins

In fresh plumage, inner coverts and scapulars have white edging.

Light morph: there is also an intermediate morph.

Black-hooded

Black chin

Underwing pattern of broad black leading edge and wide dark trailing edge is conspicuous on light phase (above) but much less obvious on intermediate and (at left) darkest forms.

Dark morph

## LEACH'S, SWINHOE'S & TRISTRAM'S STORM-PETRELS

**Leach's Storm-Petrel** *Hydrobates leucorhoa* 19–22 cm
A rare vagrant similar to Wilson's Storm-Petrel; overall jiz slender, graceful. Flight distinctive: when travelling is fast and direct with sudden changes of direction. Bursts of powered flight with rapid, shallow but strong wing beats alternate with zigzag glides on bowed wings. When foraging, the flight is slow and buoyant; flutters with wings slightly raised, almost hovering against the wind, moving gradually forward, feet paddling the surface, snatching with the bill at small surface creatures. Does not usually follow ships, but is occasionally attracted to boats during fishing. Calls at sea, in flight and over the colony. Most common call is a staccato ticking that ends with a slurred trill; also makes chuckling sounds. Most alike is the White-rumped Storm-Petrel. Inhabits offshore waters. Rare vagrant to Aust.: several records from SW WA and an old record from W Vic. Breeds N Pacific and Atlantic – colonies of millions on islands off Nova Scotia, Maine, Labrador and Newfoundland, with smaller colonies in N Scotland.

Nominate race in NE Pacific shows less white on rump; birds of southernmost colonies have entirely dark rumps.

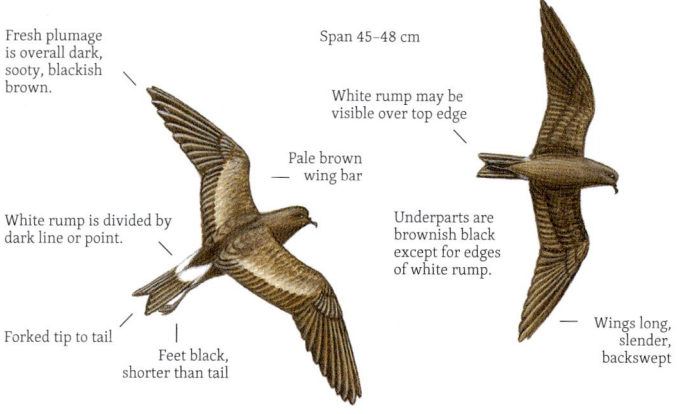

Fresh plumage is overall dark, sooty, blackish brown.

Span 45–48 cm

White rump may be visible over top edge

Pale brown wing bar

White rump is divided by dark line or point.

Underparts are brownish black except for edges of white rump.

Forked tip to tail

Feet black, shorter than tail

Wings long, slender, backswept

Worn plumage is paler with more obvious wing bar.

Dances across the waves, webbed feet pattering at surface, wings out or up to lift or balance as wave crests surge underneath. Snatches small creatures from the water's surface with fine, hooked bill.

High steep forehead, slender, hooked bill

---

**Swinhoe's Storm-Petrel** *Hydrobates monorhis* 19–22 cm
A rare vagrant. Overall jiz slender, graceful; flight distinctive. When travelling, is fast and direct, but with sudden changes of direction. Medium sized, overall dark grey-brown above and below. No white on rump, but has slightly paler bar along upper-wing-coverts, more prominent as feather tips wear, as on Leach's Storm-Petrel. Also similar is the slight fork at the tail tip. Habitat is oceanic; breeds in the N Pacific, then disperses S and to N of Indian Ocean. Recorded off WA's NW coast near Broome on several occasions, 1999 to 2001.

---

**Tristram's Storm-Petrel** *Hydrobates tristrami* 19–22 cm
A rare vagrant. Similar to Bulwer's Petrel and other completely dark storm-petrels. Head is blackish brown, body is slightly lighter brown with pale brown bars along the upper-wing-coverts and a pale brown rump. Overall jiz is slender, graceful; tail is quite deeply forked. Oceanic habitat, N and central Pacific, breeding on Midway and other islands. One Aust. record only, off coast near Sydney, Nov. 2000.

## MATSUDAIRA'S & WHITE-FACED STORM-PETRELS

### Matsudaira's Storm-Petrel
*Hydrobates matsudairae* 25 cm; span 54–58 cm

In flight, appears slender with long, pointed wings. Compared with smaller species, flight is relaxed: slower wing beats, more time spent in flapping flight, fewer erratic changes of direction, but is capable of bursts of speed. Glides low, sometimes tipping the water. To feed, alights on the surface holding wings up with an occasional flap for stability as waves pass, or in calm conditions feeds with wings folded. Follows ships. Inhabits tropical and subtropical pelagic waters, possibly at upwellings of cooler water. Migratory; travels from NW Pacific to Indian Ocean via NW Aust. Records of sightings from Kimberley coast to Indonesia and Christmas Is., Jul.–Nov.

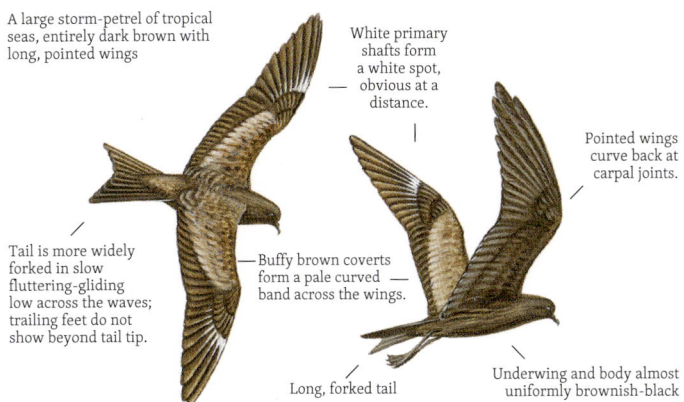

A large storm-petrel of tropical seas, entirely dark brown with long, pointed wings

White primary shafts form a white spot, obvious at a distance.

Pointed wings curve back at carpal joints.

Tail is more widely forked in slow fluttering-gliding low across the waves; trailing feet do not show beyond tail tip.

Buffy brown coverts form a pale curved band across the wings.

Long, forked tail

Underwing and body almost uniformly brownish-black

### White-faced Storm-Petrel
*Pelagodroma marina* 18–21 cm; span 42–43 cm

The only storm-petrel breeding in Aust. waters. Distinctive facial pattern; common. Uses continental shelf waters when breeding, but is usually well out of sight of land; otherwise, much further out over deep oceanic waters. Flight when travelling is erratic and weaving – low, level glides on rigid bowed down wings, steep banking turns, brief bursts of jerky wing beats. In low foraging flight, glides slowly into the wind, wings outstretched parallel to waves, long legs dangling. Unlike some other storm-petrels, hits surface with feet together, bounding rather than striding. Call is a soft, repetitive 'peeoo-peeoo' or 'woooo, wooo'. Numerous subspecies around world; 3 in Australasian region.

Race *dulciae* common; breeds around S and SE coasts of Aust.

Curved back

Feet visible beyond tail tip

Rump and upper-tail-coverts pale grey

Brow and face, white; crown and line through eye, dark grey

Very thin black edge

White

Tail square-tipped or slightly forked

Feet extend well to rear of tail in travelling, or down to touch water when feeding.

Grey, slightly browner in worn plumage

Wings fold to similar length as tail.

Grey, webs pale yellow

# 4 PELICAN & TROPICBIRDS TO FRIGATEBIRDS

This group encompasses the 'Pelican' Order, Pelecaniformes, six families of medium-sized to very large birds. Most live in marine habitats, but a few also or mostly use inland waters. They eat aquatic life, from quite large fish to small molluscs and arthropods. All six families occur in Australia or its nearby seas. All share certain features; among them are webbed feet with clawed toes that help to climb about the cliffs and trees often used as nest sites. Eggs of most are plain, chalky; chicks of most species hatch naked.

Silver Gull = 42 cm

## pelecanidae
page 61

Australian Pelican

## anhingidae
page 67

Australasian Darter

## fregatidae
page 66

Great Frigatebird

Christmas Island Frigatebird (not illustrated)

Lesser Frigatebird

## sulidae
page 62

Australasian Gannet

(Cape Gannet is similar.)

Masked Booby

Red-footed Booby (white morph)

Red-footed Booby (white-headed brown morph)

Brown Booby

Abbott's Booby

## phalacrocoracidae
page 68

Great Cormorant

Little Black Cormorant

Black-faced Shag

Australian Pied Cormorant

Little Pied Cormorant

## phaethontidae
page 63

Red-tailed Tropicbird

White-tailed Tropicbird

# FLIGHT: TROPICBIRDS TO FRIGATEBIRDS

From almost entirely white wings through black and white to almost entirely black. (Not to scale)

Red-tailed Tropicbird · White-tailed Tropicbird · For Pelican flight, see foot of page · Masked Booby · Australasian Gannet

Red-footed Booby · Brown Booby · Brown Booby (juv.) · Masked Booby (juv.)

Red-footed Booby (White-tailed morph, Brown morph) · Australasian Darter ♀ ♂ · Australian Pied Cormorant · Little Pied Cormorant

Great Cormorant · Little Black Cormorant · Lesser Frigatebird ♂ · Great Frigatebird ♂ ♀ · Christmas Frigatebird

# AUSTRALIAN PELICAN

**Australian Pelican** *Pelecanus conspicillatus* 1.6–1.8 m
Familiar huge bird with massive bill; will use almost any large or small area of water, from sheltered coastal bays and estuaries to temporary pools in the desert. Head tucks back in flight giving tubby shape. Soars on flat wings, circles in thermals to great heights, then travels far in long glides, helped by occasional slow flapping. Flocks travel in formation, often V shaped. Gregarious: pairs to large flocks in flight, loafing or fishing. At times, birds work cooperatively to round up fish. Voice, in display or when squabbling, is deep with resonant croaks and guttural grunting. Common; nomadic, dispersive.

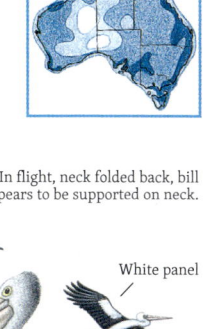

Bill deep pink at courtship · Yellow tint · Black and white underwing pattern · Br. · Span 2.3–2.5 m · Sexes similar, but females have a shorter bill. · In flight, neck folded back, bill appears to be supported on neck. · Nonbr.: bill pale · White panel · White V

# GANNETS

**Australasian Gannet** *Morus serrator* 85–90 cm; span 1.8 m
Sleek, with pointed bill, wing tips and tail. Hunts by plunge-diving; groups or large flocks wheel 20–30 m above schools of fish, spectacularly plunging bill first on half-closed wings. Gregarious: lines of birds travel close to waves. Casual flap-and-glide flight; gather in 'rafts' on calm seas. Marine habitat, usually continental shelf waters; prefers seas near coastal islets, capes, channels. Enters estuaries, bays and harbours to shelter from rough seas. Noisy in colonies; raucous, sharp squawks intermingled with deep, barked 'urragh-urragh'. Common mid WA coast to mid Qld coast; has a far larger NZ population.

Sleek from sharp, straight bill that merges into smoothly contoured body, back to tapering tail and slender wings

Bill is large, straight, grey with black edges; tapers evenly to sharp point.

White

Central black

Pale yellow tint on crown fades away to white down over nape, face and neck.

Iris grey, eye-ring blue

Bill blends smoothly to streamlined body.

Black wing tips and wing bar

Black

Juv.

Tail variable: just central pair black; more often, central two pairs; some have tail entirely black.

Primaries, primary coverts and secondaries are black, in bold contrast to the white that dominates in the plumage. Underwing similar except that primary coverts are white.

Transition to adult plumage occurs over several years.

Span 1.7–2 m

**Cape Gannet** *Morus capensis* 85–92 cm; span 1.7–1.8 m
Usual range confined to continental shelf waters of W and S Africa. First recorded in Aust. 1980; a single bird found nesting, mated to an Australasian Gannet in a small colony on Wedge Light, Port Phillip Bay, Vic. Another, banded in S Africa, was recaptured off Cape Leeuwin, WA. Accidental to Aust. but would be difficult to separate from the Australasian Gannet at sea.

Like Australasian Gannet except for throat and tail

Tail is entirely black, unlike that of the Australasian Gannet, which usually has only central one or two pairs black.

Iris silvery cream, paler than the grey iris of the Australasian Gannet

Long thin black line

Thin black streak from underside of base of bill extends down throat much further than the similar short chin streak of the Australasian Gannet.

Juvenile and immature are like those of Australasian Gannet except for the much longer throat stripe.

## TROPICBIRDS

**Red-tailed Tropicbird** *Phaethon rubricauda* 90–100 cm
Long, fine tail streamers account for 30 to 40 cm of total length. Tropical and subtropical seas, often far from land. Open ocean for plunge-dive fishing; cliffs and forest canopy of islands for rest and shelter, breeding display, nesting. White plumage conspicuous against blues of sea and sky; less obvious at a distance are the long red tail streamers. In flight, usually keeps quite high; 'butterfly' flight of fluttering wing beats, alternating glides. Dives vertically; disappears entirely beneath surface; takes flying fish and small squid. Has loud, sharp, harsh, cackling calls during aerial displays; squawks and purring at nest. Breeds on islets and occasionally the mainland coast of SW WA; otherwise regular to rare accidental visitor to mainland coast within its range.

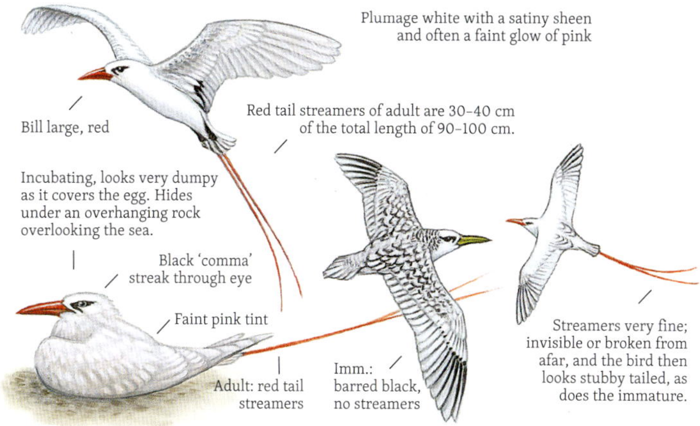

Plumage white with a satiny sheen and often a faint glow of pink

Bill large, red

Red tail streamers of adult are 30–40 cm of the total length of 90–100 cm.

Incubating, looks very dumpy as it covers the egg. Hides under an overhanging rock overlooking the sea.

Black 'comma' streak through eye

Streamers very fine; invisible or broken from afar, and the bird then looks stubby tailed, as does the immature.

Faint pink tint

Adult: red tail streamers

Imm.: barred black, no streamers

**White-tailed Tropicbird** *Phaethon lepturus* 65–80 cm
Small, slender and graceful; has white tail streamers. A largely oceanic species, probably pelagic; rarely seen over inshore waters, except at nesting colony and nearby waters. At sea, is usually solitary; flies high and strongly with rapid tern-like wing beats. Takes fish and other marine life in vertical dives from 10 to 20 m height; plunges deep below the surface. Often hovers before descending. May swoop low across surface to take flying fish. In courtship, in aerial displays, gives a 'kek-kek-kek' call while flying in high circles. White morph is a regular but rare visitor, mostly to N coastal seas.

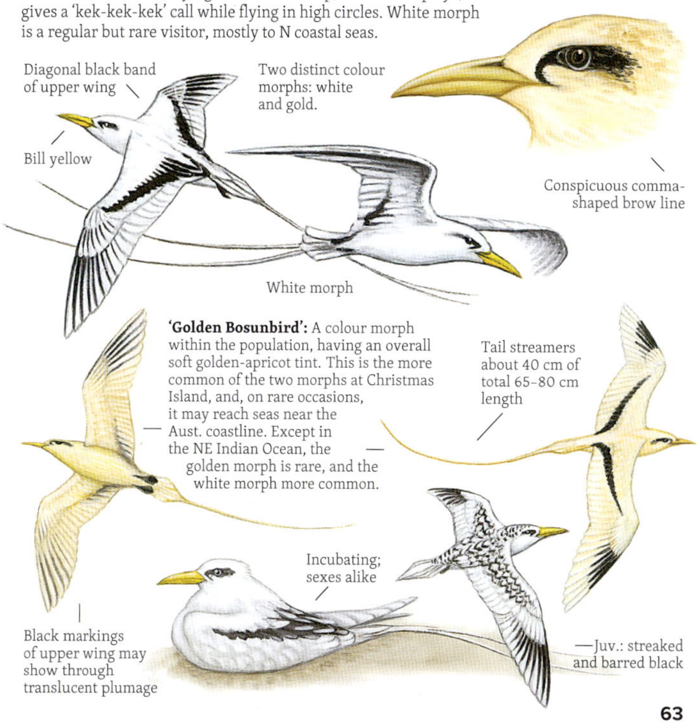

Diagonal black band of upper wing

Two distinct colour morphs: white and gold.

Bill yellow

Conspicuous comma-shaped brow line

White morph

**'Golden Bosunbird':** A colour morph within the population, having an overall soft golden-apricot tint. This is the more common of the two morphs at Christmas Island, and, on rare occasions, it may reach seas near the Aust. coastline. Except in the NE Indian Ocean, the golden morph is rare, and the white morph more common.

Tail streamers about 40 cm of total 65–80 cm length

Incubating; sexes alike

Black markings of upper wing may show through translucent plumage

Juv.: streaked and barred black

## RED-FOOTED & BROWN BOOBIES

**Red-footed Booby** *Sula sula* 70–80 cm; span 1.4 m
Smallest booby; flies swiftly in long glides that skim the waves; travels far out from land. Habitat tropical, pelagic; often travels far from breeding island to deep-water feeding grounds, but also feeds in shallows of lagoons around islands and reefs. Has the agility to catch flying fish in aerial pursuit, but generally feeds by plunge-diving vertically from 8 m and higher. Feeds in flocks; often follows boats. Silent at sea except for alarm call, a loud, grating 'karrak, karrak'. Noisy at breeding colony, a grating 'kurr-uk, kurr-uk'; in flight over colony, a loud 'rarh, rarh'. Common, widespread; breeds only on Qld offshore islands – not often recorded around other parts of Aust. coast.

Plumage extremely varied: most near Aust. are white, a few are entirely brown, and many intermediate between these morphs.

Dark morph: dark grey-brown to brownish buff with darker flight feathers

White-tailed brown morph

White morph: crisp contrast of dark flight feathers against white

All white

Bill light blue-grey; eye-ring pale violet; lores blue-grey, skin at base of bill pink, brighter in breeding season

Feet large, deep or light carmine

White morph

Because the plumage is so varied, the large red feet, long fine tail and small, slender build are better identifiers.

White-headed brown morph

Juv.: overall very dark brown, feet grey

**Brown Booby** *Sula leucogaster* 65–75 cm; span 1.3–1.5 m
Flies in bursts of smooth, easy wing beats alternating with low glides; alone or in flocks. Often travels low over waves. May plunge into water to take fish while travelling low, as well as diving from greater heights. Habitat marine, largely tropical; deep waters and inshore shallows. Calls at sea when squabbling and fishing; noisy in colonies. Male has soft hiss; female a harsh, deep honk or quack. Common N coast of Aust. from North West Cape in WA to SE Qld. Breeds on islands off NW of WA and on Barrier Reef.

Most alike is the juvenile Masked Booby, which has a white collar and breast, and back and wings mottled brown; juvenile of Red-footed Booby is somewhat similar, but entirely brown.

Sleek, slender booby with clear-cut brown and white plumage; sharp demarcation between brown neck and upper breast and white belly

Male has a creamy grey bill grading to blue-grey near base and on adjoining bare skin; female has yellowish bill and facial skin.

Sharply defined colour edges, clean whites ♂

Tail and rump brown

Very dark

Sharp edge to brown ♂

Juv.

Mottled grey-brown

♀ White neatly enclosed by brown all sides

## MASKED & ABBOTT'S BOOBIES

**Masked Booby** *Sula dactylatra* 75–85 cm; span 1.6–1.7 m
Feeding habitat is marine, pelagic; often far from land beyond the shelf-edge over deep water. Breeding birds travel far out from island colonies to deep seas. Flies fast and high; strong, steady wing beats alternate with glides; solitary or in loose groups when returning to colony. Dives vertically from heights of 15–100 m to hunt, plunging several metres beneath surface. Boobies are built for such plunges; the tapered bill breaks the surface, and their velocity and sleek lines carry them deep under water. Usually silent at sea; noisy in breeding colonies. Male gives a high, descending, whistled 'whieeoooo', female a loud, harsh, honking 'aarh-aarh-a-yah'. Abundant on islands; common breeding species of Aust. coast and islands.

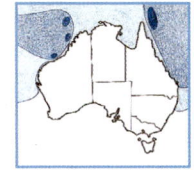

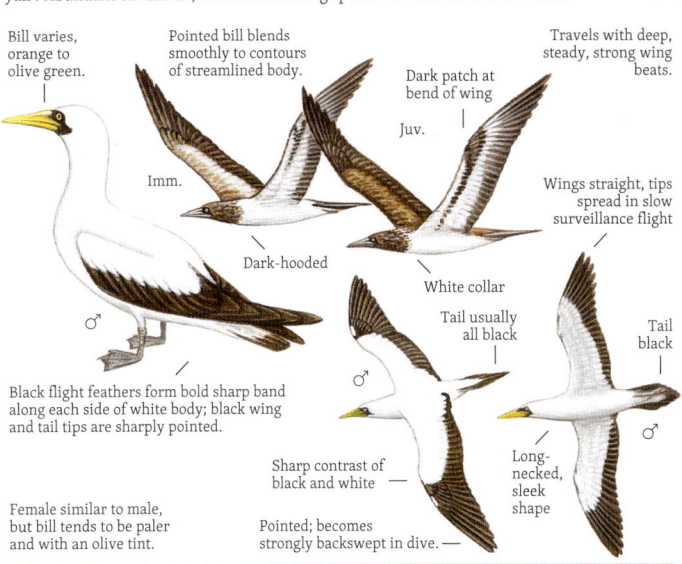

Bill varies, orange to olive green.

Pointed bill blends smoothly to contours of streamlined body.

Dark patch at bend of wing

Juv.

Travels with deep, steady, strong wing beats.

Imm.

Dark-hooded

Wings straight, tips spread in slow surveillance flight

White collar

Tail usually all black

Tail black

Black flight feathers form bold sharp band along each side of white body; black wing and tail tips are sharply pointed.

Sharp contrast of black and white

Long-necked, sleek shape

Female similar to male, but bill tends to be paler and with an olive tint.

Pointed; becomes strongly backswept in dive.

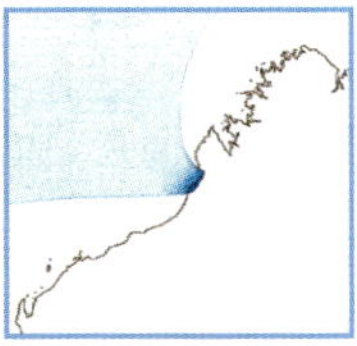

**Abbott's Booby** *Papasula abbotti* 78–80 cm
Unique to Christmas Is. and its surrounding seas; little is known of its movements away from the island, but may reach seas off NW Aust. coast. A large, slender, very long winged booby; dives for fish, plummeting bill-first into the ocean; prey includes squid and flying fish. This booby's flight is distinctive: slow, relaxed wing-flaps and glides, a more leisurely action than other smaller, shorter winged boobies. Its long necked body and the great reach of its narrow wings give a distinctive, 'flying cross' shape. Voice is a bellowing sound, with brief male-female duets. Young have monotonous, quavering begging calls. Usual habitat is the Christmas Is. plateau and high limbs of rainforest canopy for resting and nesting; open ocean for fishing. Endemic species, threatened, endangered; only 2500–3500 pairs. Breeding habitat restricted to Christmas Island's high plateau between 160 and 260 m above sea level. Parts of this habitat were destroyed by phosphate mining, but are now being replanted and included in national park areas.

Sparse white feathers down centre-line of wing

Tail black, pointed

Head and neck white; feathers lift in display.

Pale grey-buff or blue-grey

Very large booby; unique plumage pattern and flight action

Radiating white lines formed by white inner webs of outer primaries

Rump may be flecked black.

Long slender wings – a distinctive 'flying cross' shape

Dark flank patch behind legs

Female is like male except bill is deep to dull pink.

Large, black with hooked tip and serrated cutting edge

## LESSER & CHRISTMAS ISLAND FRIGATEBIRDS

**Lesser Frigatebird** *Fregata ariel* 70–80 cm
Flight graceful: soars, glides, dives; deep, easy wing beats; can stay airborne all day. Habitat mostly marine: the airspace over tropical seas, usually pelagic. Often far from land, but also over shelf waters, in places close inshore, and inland over continental coasts. Feeding and harassing behaviour similar to other frigatebirds. Perches on trees, rarely on ground. Both sexes give landing calls: males a whistled 'wheees-wheees-', females a high 'chip-ah, chip-ah' followed by sharp shrieks. Common N Aust. seas; breeding species. Sites include Barrier Reef islands and Ashmore Reef, WA.

Extent and shape of white on breast and nearby part of underwings separates the three species.

Zigzag wing shape typical of all frigatebirds ♀

Wingspan 1.8–1.9 m

Juv.: head rufous-orange

♂

With throat pouch not inflated

Black centre to belly; white breast, collar, 'armpits' of innermost part of wing

Juv.

Red throat pouch develops in courtship.

White collar

White extends into 'armpits', both sexes.

♂

Rusty orange-brown

♂

♀

Throat pouch inflated in display

Juv.

Forked tail is much longer than tips of folded wings.

**Christmas Island Frigatebird** *Fregata andrewsi* 90–100 cm
Breeds only on Christmas Is. as a small population with restricted range. Ranges over pelagic tropical seas, favouring warm waters of S equatorial current. Both sexes have larger areas of white underneath than either of the other species. Like others, avoids settling in low sheltered situations or on water – it has difficulty taking off except from trees, cliffs or into the wind. Calls are restricted to the vicinity of nests in breeding season. Male has a landing call, a descending 'i-eer, i-eer'; female a squeaky chatter. Usually confined to NE Indian Ocean around breeding sites on Christmas Is. Vagrant to N Aust., Darwin, 1974.

Juv.: head and neck rufous-orange

The only male with a large area of white on lower breast and belly and without white extending out to 'armpit 'area of inner wing

Female has a large area of white on breast and belly, extending to 'armpits'; unlike Lesser Frigatebird, no black patch up through central belly.

Juv.

♂

Partial collar

♀

Juvenile difficult to separate from those of other species; has some white in armpit area.

Male is shown here with throat not inflated; in display expands balloon-like as with other frigatebird species.

All frigatebirds have highly recognisable silhouettes: wings are narrow at body, widen at carpal joint and are sharply angular.

## GREAT FRIGATEBIRD & AUSTRALIAN DARTER

**Great Frigatebird** *Fregata minor* 85–105 cm
Large, predominantly black with long pointed wings, deeply forked tail. Over tropical seas, pelagic; occasionally inshore shelf waters. Wings typically held far forward, strongly bowed. Soars high on thermal uplifts or on updrafts above cliffs of oceanic islands; uses only occasional deep wing beats. Feeds by snatching flying fish and squid from surface of sea, and, to some extent, by harassing other seabirds arriving from fishing grounds until they drop food for the frigatebird to catch. Usually silent in flight. On landing at nest, male gives a repeated, yelping 'tjiew-tjiew-'; from nest, call is a braying 'wah-hoo-hoo-hoo'. Occurs often around N coasts, Pt Cloates, WA, to N Stradbroke Is., Qld.

Of five subspecies: *F. m. minor* breeds Christmas Is., *F. m. palmerstoni* in W Pacific and Coral Sea.

**Australasian Darter** *Anhinga novaehollandiae* 85–90 cm; span 1.2 m
Occurs in estuaries, sheltered bays, wetlands (fresh or brackish if half a metre or more in depth), usually with trees, logs and well-vegetated banks. Floats very low, body often submerged, only head and neck visible, snake-like. Slips beneath surface with scarcely a ripple. Hunts under water with a spearing action as the kinked neck suddenly straightens. Voice is a harsh 'kar, kar, ka, ka-ka-kakaka', loud and slow, gradually more rapid, then fading. At nest, loud, brassy cacklings and clicking sounds. Generally common; vagrant to Tas.

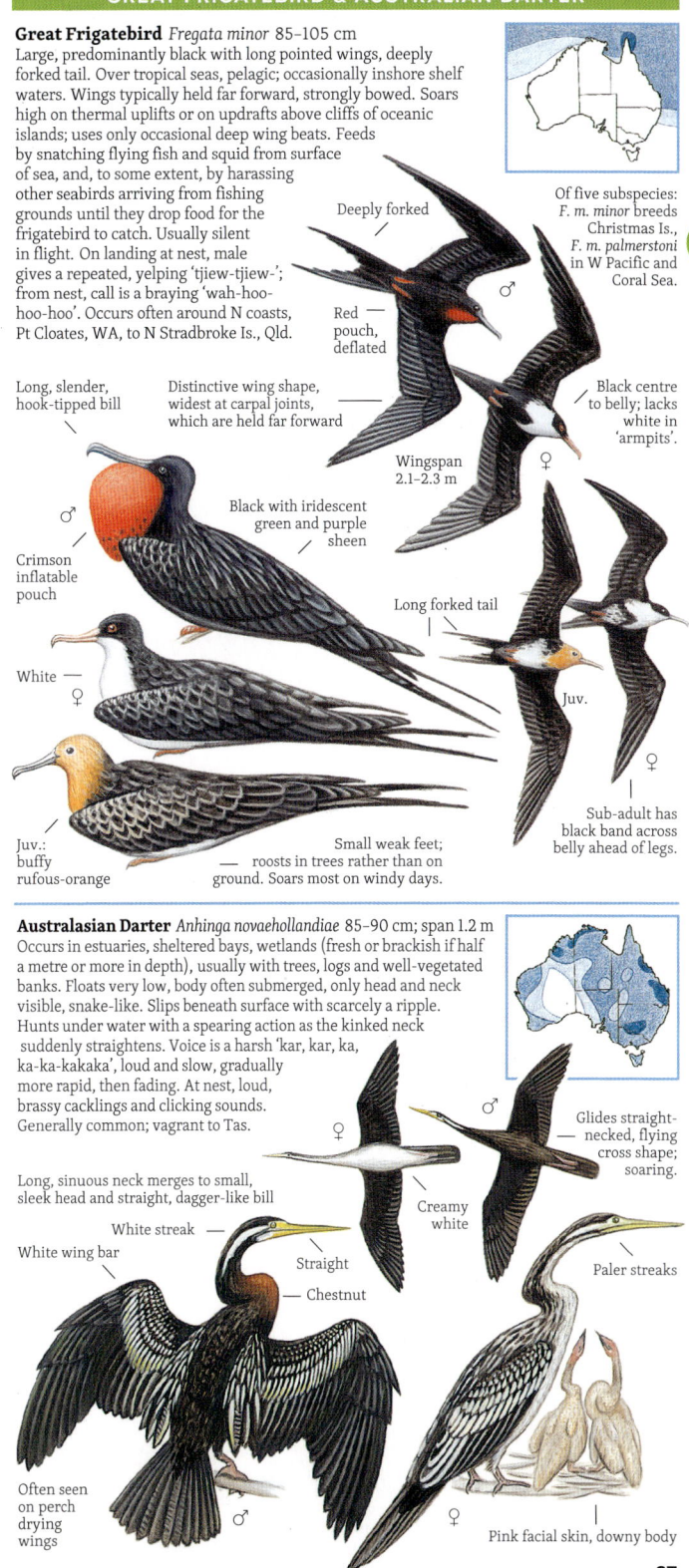

## LARGE CORMORANTS

**Great Cormorant** *Phalacrocorax carbo* 80–85 cm
On large expanses of fresh or salt water: most abundant on estuaries, bays, lagoons, inland on deep rivers, lakes, swamps and floodwaters. Uncommon on small or shallow waters. Has white on face and flanks when breeding. In flight, rapid wing beats alternating with glides. Solitary, or in small to large flocks. Usually silent. In breeding season: male gives croaking, stuttering groans like a heavy, creaking door, and raucous, barking threat calls; hoarse hissing from female. Common almost throughout Aust.

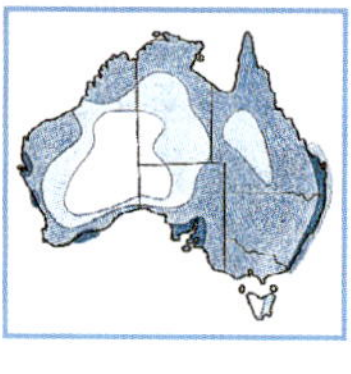

In flight, head high, neck kinked

Span 1.3–1.5 m

In flight, wings are held straight out, cross shaped.

No crest, dull black

Largest Australian cormorant. When breeding, glossy black with blue-green sheen

Slight crest

Bright white edge to yellow facial skin

Small, dull white edge

Br.

White bar across thighs

Nonbr.

Imm.

**Australian Pied Cormorant** *Phalacrocorax varius* 70–80 cm
Uses both marine and inland waters – coastal lagoons, estuaries, lakes, rivers, billabongs, and the more open and deep areas of swamps. May be solitary, in small groups or in thousands. Flight strong, direct; slower wing beats than Little Pied; head high, neck often kinked. When travelling in numbers, forms V pattern. Rests, roosts and nests in trees; often perches with wings extended. Usually silent, but various cacklings, loud ticking, deep guttural grunting in breeding colonies. Moderately common to abundant in all states.

Yellow, blue and pink skin on bare face

Wide white sides to neck. The black strip on back of neck to nape and crown can appear quite narrow.

In flight, keeps head slightly above body. Neck usually not as straight as the Little Pied Cormorant

Black with bronze-green sheen

Br.

Black bar down thigh

Span 1–1.3 m

Nonbreeding: facial colours are less intense than when breeding.

Imm.

Back mottled brown

Breast white, smudged brown

## SMALLER CORMORANTS

**Little Black Cormorant** *Phalacrocorax sulcirostris* 55–65 cm
Uses smaller bodies of water such as farm dams. In flight, bursts of quick wing beats alternate with glides; flocks take up V formation. Gregarious: large flocks fish cooperatively. Usually silent, but some ticking and croaking among birds in fishing flocks; guttural croaks and tickings at nest. Most common on inland waters – lakes, river pools, deep open swamps, estuaries and lagoons. More coastal in Tas. Common.

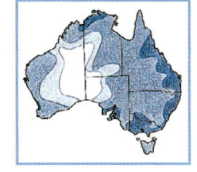

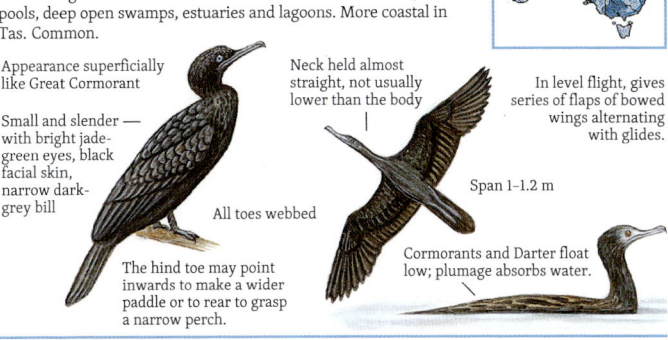

**Little Pied Cormorant** *Microcarbo melanoleucos* 57 cm
Uses sheltered coastal lagoons, harbours, bays. Inland, uses small lakes, dams, billabongs, swamps, floodwaters. Follows creeks well inland. Uses dams in forest and farm country and feeds on introduced fish. Flight strong; can take off and rise more steeply than other cormorants. Near nest has a cooing 'keh-keh-keh' and a harsh, deep 'uk-uk-urk' by male arriving with nest materials or food for young. A sharp croak is used as alarm call. Common almost throughout Aust., including the interior after heavy rain.

**Black-faced Cormorant** *Phalacrocorax fuscescens* 60–70 cm
Habitat marine: feeds predominantly over inshore waters and reefs, around offshore islands; occasionally enters estuaries, tidal reaches of large rivers. Does not perch on trees. Usually silent, but within the breeding colony, makes loud, guttural croakings from the male and soft, hoarse hissing from the female. Common to abundant.

# 5 HERONS & EGRETS TO CRANES

A group of four families with several very large species. Most are tall, long-necked, long-billed, long-legged: herons, egrets, bitterns, one stork, ibises, spoonbills and cranes. Several, Striated and Nankeen Night Heron, hold head hunched down, looking short-necked. Birds in this group have similar habits, stabbing small prey in shallows or damp surrounds of swamps and lakes. The spoonbills differ, having wide bills adapted to filter small aquatic prey.

Magpie = 40 cm

## ardeidae
page 73

## ciconiidae
page 81

## threskiornithidae
page 78

## gruidae
page 80

Eastern Reef Egret (dark)

Great Egret (light)

Pied Heron

Little Egret

Intermediate Egret

White-faced Heron

Grey Heron

Striated Heron

Nankeen Night Heron

White-Necked Heron

Australian Little Bittern

Eastern Cattle Egret

Yellow Bittern

Black Bittern

Great-Billed Heron

Australasian Bittern

Black-necked Stork

Glossy Ibis

Australian White Ibis

Straw-necked Ibis

Royal Spoonbill

Yellow-billed Spoonbill

Sarus Crane

Brolga

## WHITE-NECKED & PIED HERONS

**White-necked Heron** *Ardea pacifica* 75–105 cm
Large; has white neck, dark wings, bold white spot to edge of upper wing. Folds neck back in flight; slow, deep wing beats. Solitary or in small to large flocks; soars high on thermals. Usual habitat includes shallow wetland, swamp, flood water, wet grassland, shallows of lakes: mainly fresh water, but occasionally coastal mudflats. Gives a loud single or double croak in alarm or flight: 'argh, aarrgh'. At nest gives a deep, loud 'oomph!' and a raucous cackle at changeover. Most likely to be confused with Pied Heron, White-faced Heron or Grey Heron. Common, widespread, migratory; dispersive or irruptive after widespread rain.

In breeding season has longer plumes; white neck has two lines of spots that usually reduce or disappear with wear of dark tips to feathers.

Conspicuous white spots at carpal joints

Breeding plumage: long plumes on back; neck all white or may have some spots.

Nonbr.: few short back plumes; dark spots down centre of neck

Juv.: more rows of dark neck spots; no plumes

Imm.: like nonbreeding adult, but slight grey cap, no plumes.

Long legs trail far behind the short tail.

In flight, neck is tightly folded, head pulled back.

Long lanceolate plumes have an iridescent maroon and green sheen.

White neck

Turquoise tinted

Blue-grey

Nonbreeding has heavier spotting.

Imm.: has wider, more conspicuous bands of dark spots.

May have a few spots.

Sparse grey plumes

Br.  Nonbr.  Juv.

Long white down

**Pied Heron** *Egretta picata* 43–52 cm
Small, dainty, dark capped heron with neat, pied plumage. In breeding season, has long, dark plumes at nape, mantle; white plumes around base of neck. Forages actively, dashing after prey; hunts and travels in flocks. Habitat is coastal wetland of N Aust.: floodplain pools, billabong shallows, swamps, wet pasture, tidal and mangrove mudflats. Voice, in flight, an abrupt, deep, rough croak, 'orrrk!', given as alarm or when feeding in groups; at nest, soft cooing. The White-necked Heron is superficially like immature and nonbreeding Pied Heron, but much larger with white wing spots, grey or black bill, and, usually, dark neck spots. Across N, abundant only in areas of extensive wetland; dispersive.

Feet trail far behind tail.

In flight, keeps neck folded back, typical of herons and egrets; spoonbills fly with neck outstretched.

Long black nuchal crest

Dark blue-grey plumes cascade down the back in breeding season.

During courtship, bill bright yellow and iris red

Imm.: crown and nape entirely white

Plumage preceding juvenile stage may be mottled black or brown on crown and nape.

Br.  Long white plumes in the breeding season  Nonbr.  Downy young  Imm.

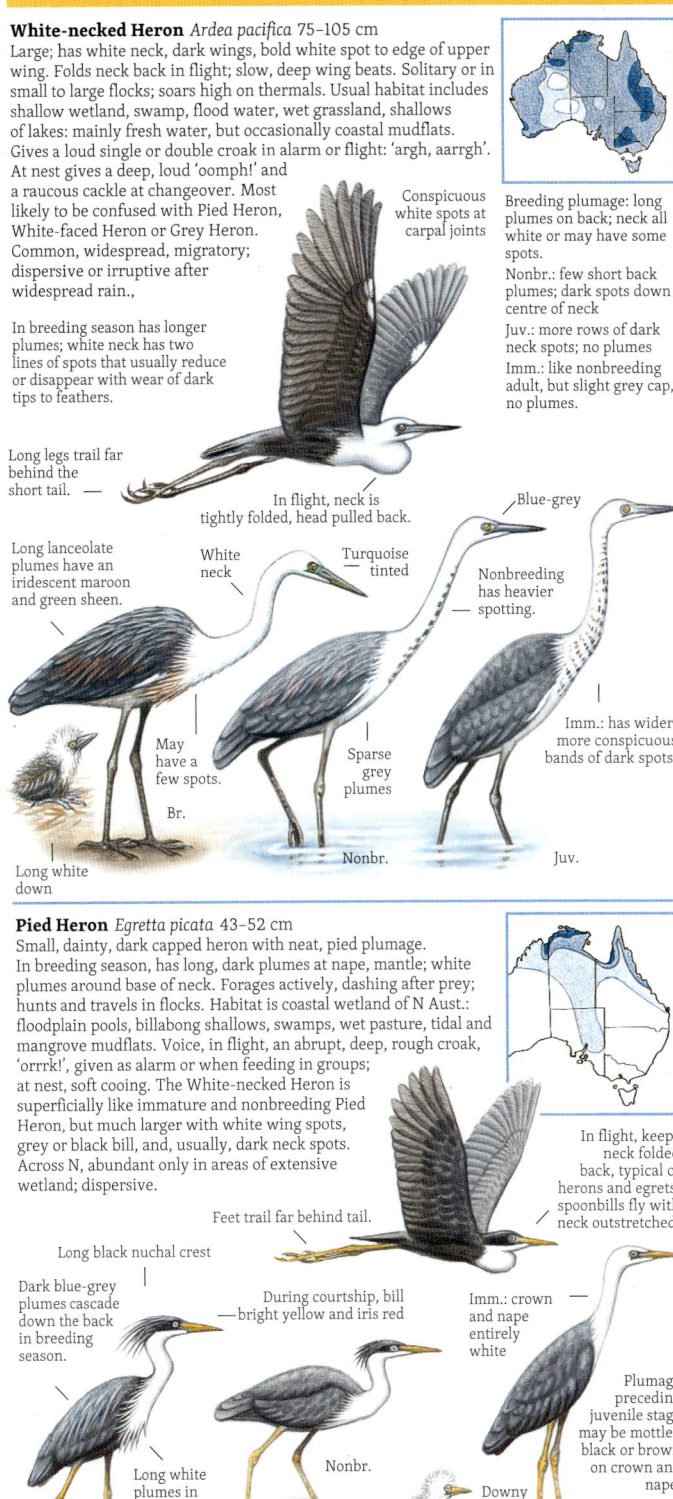

71

## GREY & GREAT-BILLED HERONS

**Grey Heron** *Ardea cinerea* 75–100 cm
White head with black crest, white neck with black stripes. In flight, shows contrasting black flight feathers. Nearest in appearance in Aust. are the smaller White-faced and White-necked Herons. Inhabits coastal and freshwater wetland. Voice is a deep croaking and a harsh, barked 'kraak'. Looks bulky, bittern-like. Common from Europe S to Indonesia; rare visitor to Aust. In May 2002, a single Grey Heron was identified in SW WA, on the New River wetland, Busselton.

Heavy black line through eye

Black plumes in breeding plumage

White plumes in breeding plumage

When on a perch, the Grey Heron adopts a more hunched posture, giving a bulkier effect, described as being more bittern-like.

Black streaks down front of neck

In flight, holds neck folded back.

Imm.: dull greyish head and neck

The sole Aust. individual was identified as an adult, most likely to have come from Indonesia. The extensive migrations of the species makes it possible that this individual may have come from further afield. Many subspecies occur across the Grey Heron's wide range from Europe, Africa and Asia S to Sumatra and Java. Differences between subspecies are slight.

**Great-billed Heron** *Ardea sumatrana* 1–1.1 m
Very large, heavily built, wary, solitary; stalks with slow, deliberate movements. Elusive, rarely in open; retreats into cover of mangrove or paperbark swamp while any observer is still distant. Low, heavy flight; slow, deep beats of dark wings; keeps below the tall canopy of mangroves and follows the water's edge and airspace above channels in swamps; rarely seen overhead. Inhabitant of dense, gloomy swamp-edge forest of tropical coastal estuaries; mudflats, river-edge paperbark thickets, billabongs; well inland on major rivers. Call is a slow, hollow, drawn out, croaking 'a-arr-argk'. Also attributed with a deep, rumbling 'bull roar' or 'crocodile roar', usually at night. Most alike is juvenile Black-necked Stork, but it flies with neck extended and has longer legs, white trailing edge to wings. Uncommon; sedentary.

Upper parts entirely brownish-grey with exception of silvery grey plumes in breeding season

In flight, long legs and feet trail far behind tail; head is hunched back.

Tall, heavily built, brownish grey heron with rather large head and heavy bill

Face blue-grey

Huge dark bill, some with a trace of yellow along the lower base

When breeding, has silvery plumes on nape, neck and back.

In nonbreeding plumage, some neck plumes are retained.

Overall more brown than adult with rufous streaks or edging to many feathers; no plumes

Br.    Juv.

Juv. and nonbr.: face yellow

Legs grey-brown to olive-grey

Grey, pale on head

## WHITE-FACED HERON; REEF & CATTLE EGRETS

**White-faced Heron** *Egretta novaehollandiae* 66–69 cm
A very common small heron with white face and throat, and yellow legs. May be seen in any of many habitats; usually in or near shallow wetland: margins of swamps, dams and lakes, damp or flooded pasture. Also salt and brackish shallows – estuarine, mudflat, mangrove, saltpan, reef and beach. In breeding season, has long, lanceolate plumes on nape, mantle and back; shorter plumes on lower fore-neck and breast. Calls at, and vicinity of, nest, while in flight, landing at roost, in courtship flights, and in contact or alarm. Various croaking, grunting sounds: 'urrk-urrk-urrk'; 'arrrgh, arrrgh, arrrgh'; 'graaow'; grunted 'urgk-urg-urgh'. Abundant almost throughout Aust. in temporary habitats. Favoured by man-made wetlands.

Overall grey, flight feathers slightly darker

Plumes

Nonbr.

White face

Plumes pale brown

Juv.

Feet extend beyond tail.

White is obvious.

Br.: plumes on back, nape, neck

Underwing: strong contrast between dark flight feathers and pale underwing-coverts

**Eastern Reef Egret** *Egretta sacra* 60–65 cm
Medium-sized heron of the shoreline; habitat also includes estuarine mudflats and inshore reefs. Shape distinctive: long-necked but short-legged; lacks visually balanced neck-to-leg ratio typical of other egrets. This is emphasised by the thick legs, and is most obvious when it stands alert, neck upstretched. Two colour morphs: white and very dark grey. Gives a harsh, abrupt 'yowk, yowk' in alarm. At nest during courtship, a deep, abrupt, guttural, frog-like 'yrok, yroak' or 'yok-yok'. Dark morph is unique – no significant white in plumage. White morph is superficially like other white egrets, which all have longer legs and different bill colours. Common on N and NE coasts; uncommon to rare in SE and SW; now absent from or vagrant to most of Vic. and Tas.

Wispy plumes

Varies, grey to yellow.

Juv. and imm.: dark morph

Neck and bill long in contrast to short legs

Dark morph

Very dark sooty or slaty grey

Legs short compared with other egrets

Juv. and imm.: light morph

Light morph

**Eastern Cattle Egret** *Bubulcus coromandus* 48–53 cm
Small, squat egret, often in flocks with livestock. Habitats include moist pasture with tall grass, shallow open wetland and margins, mudflats. Posture is usually hunched; walks with forward-back swing of head; darts forward, neck outstretched, to take prey. Flight is swift with rapid wing beats. Highly sociable, small groups to huge flocks. Arrived in N Aust. about 1950; now widespread. In colonies, gives a very deep croak, 'krok, krok', and aggressive, harsh 'krow'. Intermediate and Little Egrets are like nonbreeding Cattle Egret but stand tall. Common across N Aust., uncommon in S.

Bill red at courtship; iris briefly red

Spiky orange-buff plumes

Some still grey at fledging

Br.

Nonbr.

Juv.

Typically squat, hunched, but neck is actually quite long, reaching out to jab at prey.

In flight, legs trail with tarsal joint just short of tail tip.

## GREAT & LITTLE EGRETS

**Great Egret** *Ardea alba* 0.85–1.05 m 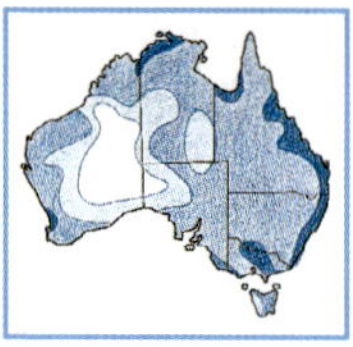
Tall, graceful, long legs, long kinked neck, snowy white plumage. In flight, slow, deliberate wing beats, neck folded. Hunts with slow, stealthy movements; often poises motionless, neck extended, but kinked ready for a thrust of the straight, spearing bill. Solitary or small groups; rarely large flocks. When alarmed, on taking flight, gives a low, hollow, croaking 'argh-argh-arrgh-aargh'. If disturbed at nest, gives a series of abrupt, harsh guttural croaks, 'grok-grok-grok-grok-grok'. Greeting call on landing at nest, a slow succession of deep, rasping croaks, 'gor-rork, grok-gro-grkgrok', fading. Habitat includes wetland, flooded crops, pasture, dams, roadside ditches, estuarine mudflats, mangroves and reefs. Common, very widespread in suitable permanent or temporary habitat.

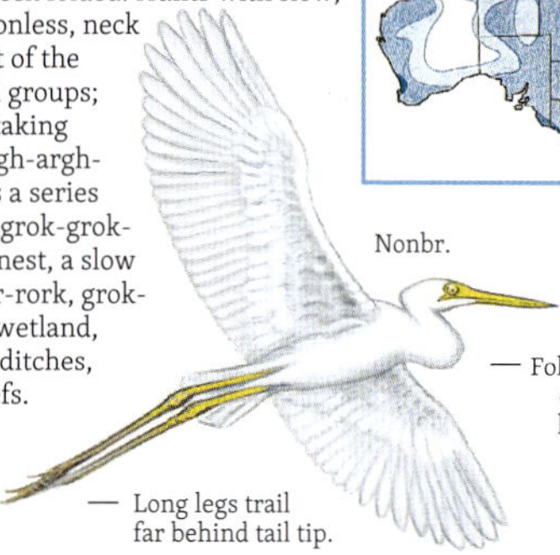

Nonbr.

Folds with obvious kinks at joints

Long legs trail far behind tail tip.

Tallest egret with very long, kinked neck. Length of head and neck together, outstretched, is greater than body length.

Plumes are only on back.

Lores turquoise, bill black

Bill and lores yellow

Kinks

Nonbr.

Plumage entirely white

Br.

Nonbr.

Great Egret: gape extends well behind eye.

Breeding plumage: long aigrettes hang in a profusion of fine white filaments.

Intermediate Egret: gape is no further back than rear of eye.

**Little Egret** *Egretta garzetta* 55–65 cm 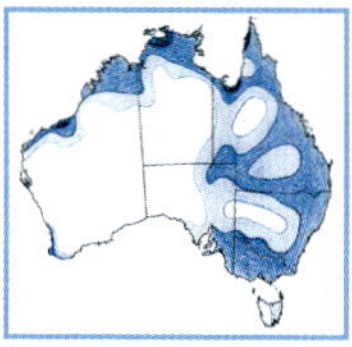
Light, delicate build evident in this egret's sleek, slender body and emphasised by a long neck. Flight is light and buoyant, uninterrupted by glides. Dashes about in erratic pursuits; lifts wings to startle water creatures. When breeding, long, slender plumes curve from rear of crown; fine filamentous plumes grow from breast and over the back to tail tip. Silent away from colonies except for a croak of alarm. Noisy when breeding; harsh croaking 'argk-argk-argk', squawked 'kiaw', 'kurik-kurik', gurgling sounds. Is similar to other all-white egrets and white morph of Eastern Reef Egret. Inhabits wetland, both fresh and marine. Usually forages in shallows of open waters – swamps, billabongs, floodplain pools, mudflats and mangrove channels. Common around N coasts, uncommon to rare in S of range; nomadic or migratory.

Always looks sleek and slender; may retain a few plumes in nonbreeding plumage.

Two long, slender white plumes from nape

Fine, lacy filamentous plumes spread in display.

Fine lacy plumes

Br.

Black

Nonbr.

Early in courtship-nesting period, the bare skin between base of bill and eye changes briefly from the usual yellow to deep pink or red. The eye also changes from buff or yellow to red.

In flight, neck is always folded back.

Legs trail behind tail tip.

## INTERMEDIATE EGRET & NANKEEN NIGHT-HERON

**Intermediate Egret** *Ardea intermedia* 55–70 cm
Long, fine filamentous plumes hang in a lacy veil from breast and back, cascading across folded wings and over the tail, enhanced in courtship and greeting displays when lifted and spread. Colours of bare parts intensify during courtship, fading by time eggs are laid. In flight, head is drawn back, feet extend beyond tail tip. Least vocal of the white egrets; silent away from colony. If startled at nest, alarm call is a hollow croak, 'glok-glok-glok'; if threatened, a loud, throaty 'kroooo-krooo'. Greeting at nest is a soft, rasping croak, 'grrrawk-grrrawk'. Found on freshwater wetland, especially lake margins, billabongs and swamps with abundant emergent vegetation; also occasionally mangrove swamps, tidal mudflats. Common in N; uncommon in SE; vagrant to Tas.

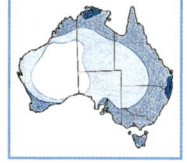

The long train of lacy aigrettes, raised and displayed in courtship, extends well beyond the tail.

Courtship colours

Yellow

Fine, lacy plumes

Neck curves smoothly, unlike the series of kinked segments obvious on the neck of the Great Egret.

A sleek, slender egret

Br.

Nonbr.

Head is rounded compared with the rather flattened crown of the Great Egret.

---

**Nankeen Night Heron** *Nycticorax caledonicus* 55–65 cm
Sometimes called 'Rufous Night Heron'. Mainly nocturnal, it is most likely to be seen if flushed from dense foliage of its daytime roost. Takes off with a croak of alarm; circles overhead with rapid wing beats, neck hunched back, flight feathers translucent red under sunlight. Solitary or in groups; often large numbers roost together. Gives various harsh and mostly deep croaks if disturbed at night when feeding or when flushed from nest. Other croaks on arrival and departure from roosts or wetland feeding sites. The alarm is a hoarse but not deep croak, 'ow-uk' or 'qwu-ok'; at roost and nest a more abrupt, nasal 'auk-auk-ak'. No similar birds; adults are unique. Juveniles show some likeness to Striated Heron, which is smaller, and Australasian Bittern, which is much larger, browner. Habitat includes most secluded wetlands: billabongs, flooded grassland, damp fields, estuarine environs such as mangroves, tidal channels. Prefers sites with cover of tall vegetation and dense trees nearby for roosting. Common in SE and SW.

Adult has several long, slender, white plumes from nape, not only when breeding, but retained through the year.

Eye yellow, or orange-red in display

Bare face is pale green, becomes blue in courtship.

Underwing: pinkish brown, coverts nearly white

White, heavily streaked brown

Legs are thick, yellow; brighter pink early in breeding season.

Feet trail behind short tail.

Cap black, no plumes from nape

Dull yellow

Cinnamon-rufous, bright in sunlight

Partly yellow; black on adults

Entirely heavily spotted and streaked with brown, black, buff and white

Post-juvenile moult

Juv.

## STRIATED HERON; AUSTRALIAN LITTLE & YELLOW BITTERNS

**Striated Heron** *Butorides striatus* 45–50 cm
Small stocky bittern-like heron. Secretive – skulks in shadowy mangroves, emerging at low tide to forage along pools and channels of mudflats. In flight, keeps low with fast-beating, rounded wings; folds neck, legs trailing. Habitat is tidal strip around coast and adjoining wetland, typically in mangroves, estuaries, tidal rivers. In courtship display, a scratchy, sneeze-like 'tsch-aar' and abrupt, loud 'hooh!' If flushed or disturbed at nest, squawks loudly; stays in vicinity making 'chuk-chuk-' cluckings. Rest of year, silent unless alarmed or flushed; harsh, scratchy screech, 'tchew-chit-chit'. Common in N; uncommon in SE.

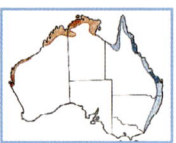
● *macrorhynchus*
● *stagnatilis*
Both races have grey and rufous morphs.

**Australian Little Bittern** *Ixobrychus dubius* 25–35 cm
A small secretive bittern, may be glimpsed, but only rarely, among reeds. Skulks through dense vegetation at water's edge; rarely emerges into the open. When disturbed, stands with neck vertical, bill skywards, mimicking the surrounding reeds. Male has a repetitive, resonant 'ook-ook-ook-ook-ook-' or 'corr-orr-orr-orr-'. At the nest, female gives soft crooning sounds when flushed from nest, a sharp 'cra-aa-ak, khok-kuk-kuk-kuk-'. Inhabits freshwater swamps, lakes and rivers with dense reedbeds, tall sedges and well-vegetated margins, also in brackish-saline mangroves, salt marsh and coastal lagoons. Perhaps more common than sightings suggest.

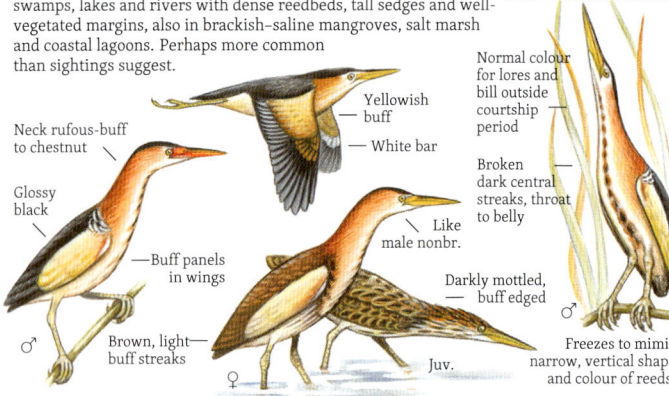

**Yellow Bittern** *Ixobrychus sinensis* 30–40 cm
Occurs S Asia, Indonesia, NG. One confirmed Aust. record – accidental, windblown immature after a cyclone, 1967. Several very old unconfirmed records. Habitat in usual range is reedbeds, densely vegetated margins of lakes, mangroves and ricefields. Voice a deep croak; possibly a softer 'kak-ak, kak-ak' in flight.

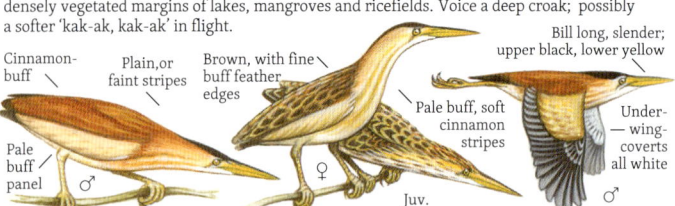

## BLACK & AUSTRALASIAN BITTERNS

**Black Bittern** *Ixobrychus flavicollis* 55–65 cm
Usually seen when accidentally put to flight; may perch before dropping down to cover. Occurs singly or in pairs, hunting in dense water-edge vegetation. Habitat is wetland, estuarine and littoral – needs dense water-edge vegetation, even if only a narrow fringe – also dense surrounds of freshwater springs and billabongs, and tidal reaches of creeks and rivers. Presence may be revealed by loud, drawn-out booming 'whOOOm, whOOm' at intervals of 10–20 seconds; often answered by similar calls from afar. Also moaning sounds, soft, low hissing, and a repetitive 'e-eh, e-eh' at the nest. Moderately common N coast; uncommon in SE and SW.

Appears mostly black.

- Sooty black
- Dusky brownish-black
- Lower bill and bare lores are reddish brown.
- ♂
- Buff to deep yellow streak, narrow at throat, wider down each side of neck; black and white streaks down centre of neck to breast
- ♀
- Bill, both sexes, yellow to grey-brown; dark top edge
- Legs and feet extend past tail tip.
- Legs olive-brown
- If disturbed, may freeze in upright pose with streaked neck matching the surrounding reeds.
- Brown, edged rufous and buff
- In flight, looks compact, with head pulled back, long black wings and stumpy tail.
- Juv. and imm.
- Browner, paler yellow, dull streaking

**Australasian Bittern** *Botaurus poiciloptilus* 65–75 cm
Large, powerfully built bittern. If startled into flight, rises heavily on broad wings, neck outstretched, legs dangling. Uses slow, shallow wing beats on longer flights; holds neck hunched back and appears short; feet trail behind tail tip; usually keeps low, but may circle high. Forages in shallows or hunts in deeper water from platforms of bent-over reeds. Habitat includes heavily vegetated freshwater wetland, sometimes estuarine – flooded shrubbery, reedbeds, sedges. Calls in spring and summer. Males give a deep, repeated booming 'oo-OOM, oo-OOM'. If flushed, an abrupt, harsh 'craaak!'. Uncommon to rare.

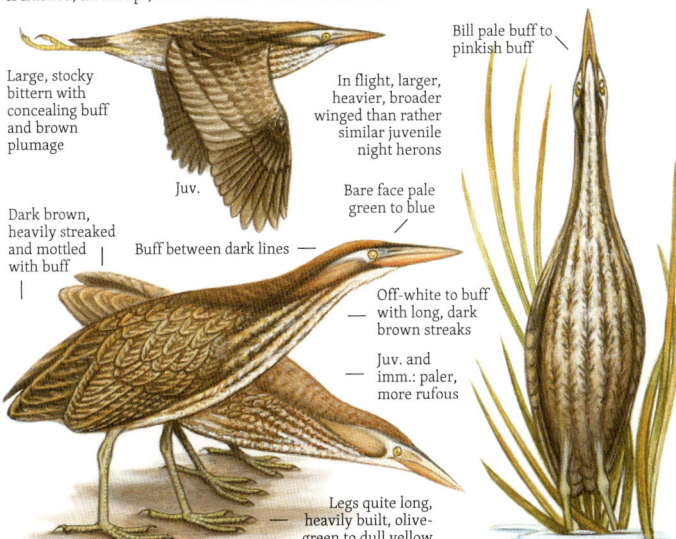

- Large, stocky bittern with concealing buff and brown plumage
- Juv.
- In flight, larger, heavier, broader winged than rather similar juvenile night herons
- Bill pale buff to pinkish buff
- Dark brown, heavily streaked and mottled with buff
- Buff between dark lines
- Bare face pale green to blue
- Off-white to buff with long, dark brown streaks
- Juv. and imm.: paler, more rufous
- Legs quite long, heavily built, olive-green to dull yellow

## WHITE, STRAW-NECKED & GLOSSY IBIS

**Australian White Ibis** *Threskiornis molucca* 65–75 cm
Bare black skin of head, neck, legs and lacy black of tertials is in clear-cut contrast to white plumage; long downcurved bill is distinctive. Flocks circle, soar, travel in undulating lines or rough V formation; appear clean white against deep blue sky. Inhabits shallow fresh and tidal wetland and pasture. Only usual call a deep, grunted 'urrrk'; flocks are noisy when settling to roost – and noisier in colonies, with deep croaked and grunted honkings. Common to abundant N, SE and SW; sedentary, dispersive.

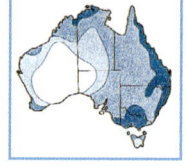

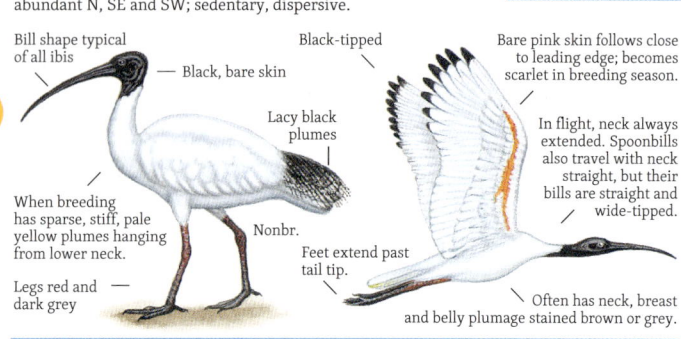

**Straw-necked Ibis** *Threskiornis spinicollis* 60–70 cm
Neck extended in flight, typical of ibis. Flocks often perch conspicuously on dead trees. Habitat is grassland, wet or dry, often on cultivated and irrigated pasture. Occasionally uses shallows of wetland; rarely arid or marine. Large flocks wherever food such as grasshoppers abound. Loud croaks when intending to fly or if alarmed, and repeatedly on take off; during flight, occasional long, harsh croak. When joining flock, announces arrival with rapid series of croaks. Common to locally abundant; nomadic.

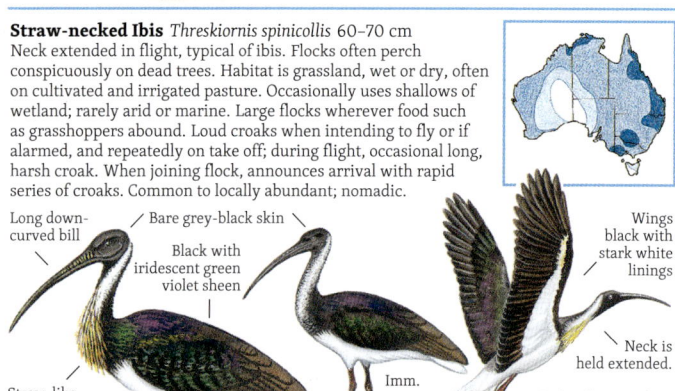

**Glossy Ibis** *Plegadis falcinellus* 50–54 cm
Small and dark with a glossy iridescence that is most obvious in the breeding season. Travels in clusters, V formation or spread out in wavering lines. Flocks congregate and roost on dead trees near water. When feeding, bird walks or wades with slow, deliberate movements, probing mud with long bill. Uses shallows of swamps, floodwaters, sewage ponds, flooded, moist or irrigated pasture; occasionally feeds in sheltered marine habitats. Call, in flight, is an occasional deep, grunted croak; if startled, gives loud, hoarse croaks. Common across coastal N, less common elsewhere; nomadic.

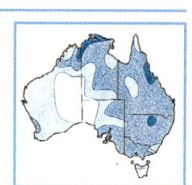

## ROYAL & YELLOW-BILLED SPOONBILLS

**Royal Spoonbill** *Platalea regia* 75–80 cm
Tall, white-plumaged. The long bill tapers to distinctive wide, flat tip that gives the two species of spoonbill unique use of wetland food resources. These birds stride through the shallows sweeping the slightly opened bill in broad arcs side to side. Any small creature – tiny fish, crustacean or insect – that touches the inside of the broad tip triggers it to shut instantly. Birds usually form small parties flying in lines or V formation. In breeding season, white plumes sprout from nape. Silent except at nest site, when it gives soft, low grunts, groans and hisses. Also makes non-vocal sounds in displays: a loud 'whoof' made by the wings and soft bill-snapping. The Yellow-billed Spoonbill is similar, but has dull yellowish bill and legs. Usual habitat is shallow wetland and margins of deeper water – fresh or saline coastal lagoons, mangroves. Common except in arid parts of range; rare in Tas.

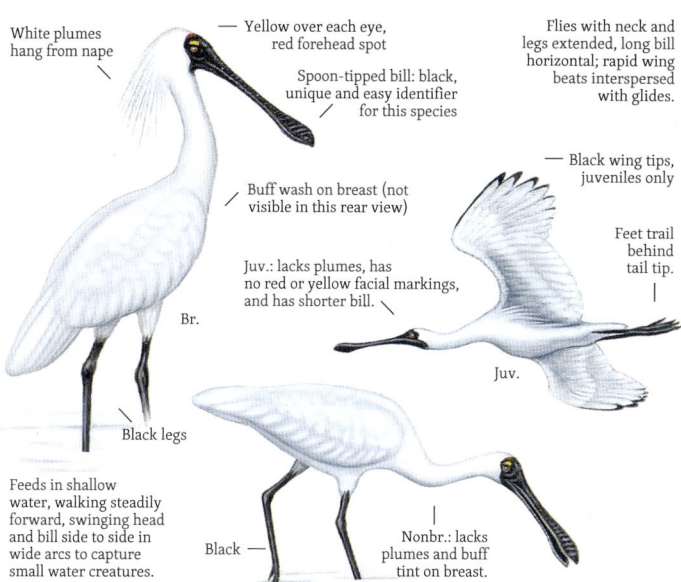

**Yellow-billed Spoonbill** *Platalea flavipes* 75–90 cm
Distinguished from Royal by its dull yellow to off-white bill and legs. Flies with neck outstretched, bill level; steady, shallow wing beats broken by short glides; travels in flocks; soars very high. Usually silent, but in threat displays gives soft, nasal coughs or grunts, makes clattering sounds with the bill. Similar to Royal Spoonbill (black bill and legs). Inhabits shallow swamps of fresh or brackish water, flooded pasture, small pools and dams, rarely tidal. Common on coast and inland after heavy rain.

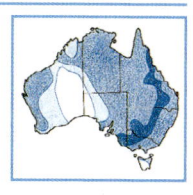

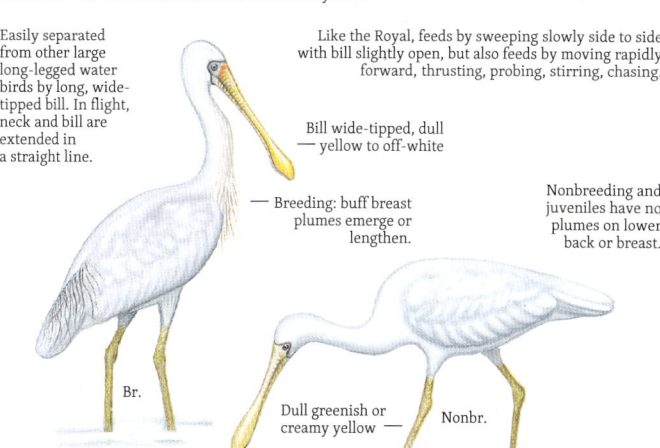

## BROLGA & SARUS CRANE

**Brolga** *Antigone rubicunda* 0.8–1.3 m; span 1.7–2.4 m
Tall, elegant, graceful in every movement; well-known dancing displays, loud bugling call. Usually near water in the dry season, moving out over greater areas of wet, green grassland and shallow flood water in the wet. Usually in flocks, of thousands through the dry season when the Brolgas congregate on the last shrinking wetlands. When the wet season arrives, flocks disperse and pairs move inland, probably to traditional breeding sites. Holds elaborate dancing displays as pairs form; groups and pairs leap gracefully, wings outspread, head thrown back, bill skywards to give the wild, far-carrying, bugling call. This bugling, given as a duet by a pair, is the best known call. It carries as far as 2 km, often the first indication of Brolgas' presence. Stimulates other distant pairs to call. Contact call is a low, purring 'grruw', alarm call, a high 'ga-r-r-oo'. In flight, a hoarse, grating 'graough'. Habitat includes freshwater swamp, flooded grassland, margins of billabongs, lagoons, dry grassland, floodplain, irrigated pasture; occasionally estuaries, mangroves. Widespread, common in N and NE Aust., uncommon in S. Dispersive and seasonally migratory.

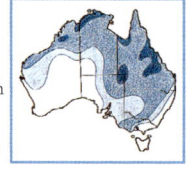

**Sarus Crane** *Antigone Antigone* 1.2–1.5 m; span to 2.4 m 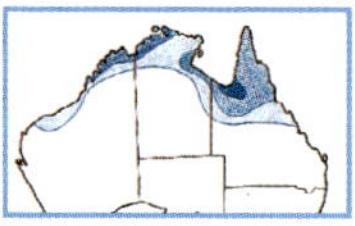
So closely resembles the Brolga that, for many years, it occurred as an established breeding species in NE Qld without being recognised. Further sightings indicated that the species had an expanding range across most of N Aust. Although occupying the same territory and habitats, Sarus and Brolga do not interbreed. Behaviour appears little different, but the Sarus tends to be wary, flying quickly from disturbance. Sarus occur in pairs, family parties or larger groups, roosting in large numbers in swamps. Occupies drier habitat than the Brolga provided there is water in the area for drinking, bathing, and secure roosting. Dry season habitats are tidal flats, edges of billabongs and dams. In the wet, is common in woodland around flooded areas. Expanding, occasionally recorded in the Kimberley. Most abundant NE Qld.

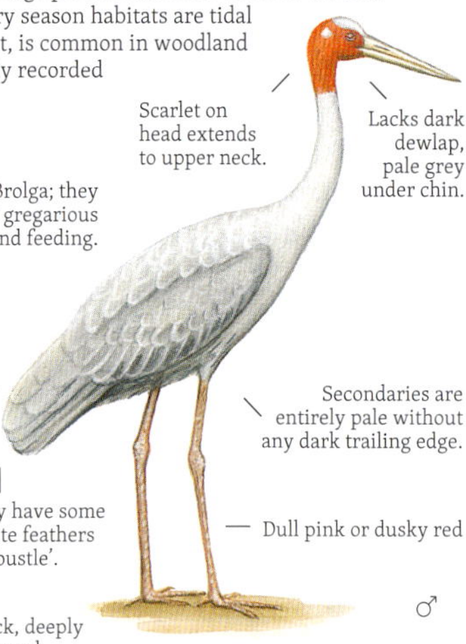

## BLACK-NECKED STORK

**Black-necked Stork** *Ephippiorhynchus asiaticus* 1.1–1.3 m
This tall, stately stork with massive black bill is still commonly known as the 'Jabiru'. Head and neck are highlighted by an iridescent shimmering green and purplish sheen. Movements are slow and deliberate as it stalks the shallows of tropical swamps; suddenly it unleashes a spearing jab of the powerful bill or dashes after prey with long strides and flapping wings. In flight, soars on long, broad wings. Habitats diverse, but often wetlands and their vicinity; prefers open freshwater environs, including the margins of billabongs, swamps, shallow flood waters over grassland, wet heath, watercourse pools, sewage farms, dams, adjacent grassland and savannah woodland. Less often on inter-tidal shoreline, margins of mangroves, mudflats and estuaries. Clacks the bill; sometimes gives guttural grunts, usually in threat or dancing displays. Moderately common across coastal N Aust.; becomes increasingly scarce southward on both E and NW coasts until it becomes a rare vagrant at southern and inland extremities of its range.

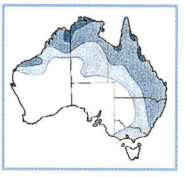

One of Australia's larger birds, tall and stately, usually rather slow and deliberate in its movements

Square tipped, deeply fingered

Female has yellow eye.

Black panel obvious along both upper and under side of white wings

Legs trail far behind tail tip.

♀

Male has brown eye.

Distinctive silhouette is apparent at a distance: extended neck, domed crown and long, heavy bill with straight upper edge

Massive straight bill spears or stabs at prey, such as frogs, water snakes, fish.

Neck is not black except under dull lighting. In direct sunlight usually shows an overlying iridescent sheen with highlights of blue, purple and green.

♂

Brown without iridescence; fades gradually towards pale belly.

Deep pink

Juv.

Juv.: legs brown

In transition to adult pattern and colour

Imm.

Immature birds here leap in exuberant play or perhaps greeting display. Juveniles change plumage gradually over about four years, acquiring glossy adult colours first on head and neck, last on primaries.

Imm.

Adult leg colour developing

This action is from a photo sequence; scale about 60% of other illustrations of species.

# RAPTORS

Raptors are often seen at a considerable distance, and then, most likely, in flight. Then plumage details are not visible, and the overall silhouette, position and shape of wings, pattern of flight and use of habitat are more helpful. These pages shows relative size, while later pages show jiz identifiers (pp. 98–101) as well as the detail that is only seen when a raptor is unusually close. Australia has two of the three families that make up the order Falconiformes: Accipitridae and Falconidae.

Magpie = 40 cm

## accipitridae
### page 83

- Pacific Baza
- Spotted Harrier
- Wedge-tailed Eagle
- Black-shouldered kite
- Swamp Harrier
- Eastern Osprey
- Letter-winged Kite
- Papuan Harrier
- Little Eagle
- Black Kite
- Brahminy Kite
- Grey Goshawk
- Black-breasted Buzzard
- Whistling Kite
- Brown Goshawk
- Square-tailed Kite
- Collared Sparrow-hawk
- White-bellied Sea-Eagle
- Gurney's Eagle
- Red Goshawk

## falconidae
### page 94

- Brown Falcon
- Nankeen Kestrel
- Australian Hobby
- Grey Falcon
- Black Falcon
- Peregrine Falcon

# PACIFIC BAZA & ORIENTAL HONEY-BUZZARD

**Pacific Baza** *Aviceda subcristata* 35–45 cm

Habitats include margins and spaces of gallery forest, monsoon forest, swamp forest, rainforest and woodland, and tropical and subtropical open forests, preferably adjoining closed forest areas and often near water; occasionally uses suburban gardens in leafy tropical suburbs. Weaves and circles around the treetops, often in family parties, occasionally in flocks of 30 or more; plunges into canopy foliage, snatching at stick insects, mantids, frogs or small reptiles on branches or foliage or, at times, the ground. Gliding and soaring flight is graceful; effortlessly floats on long, wide wings. Has spectacular aerial display flights: plunges down with wings held in steep dihedral showing underwing colour; then swoops up in a climb. Usually repeats the sequence several times, all the while calling loudly. Has a distinctive, rising and falling 'whiech-yoo, whiech-yoo, whiech-yoo'; the first part scratchy yet musical, rising strongly; the final 'yoo' low and mellow; repeated many times in succession, the calls a few seconds apart. Also has a rather mellow and musical 'kaka-kaka-kak-a kak-ak-ak', at times by both birds together, very rapid. Uncommon across N and NE Aust., rare in NSW.

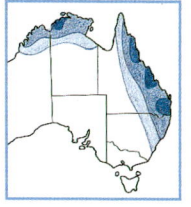

Small pointed crest
Eyes deep yellow
Pale slaty grey

A slender, colourful crested hawk with bold dark bars across white belly and flanks. When perched, wing tips almost reach the end of the long, broad tail.

Heavily barred

When gliding, primaries are backswept, closed, giving narrower wing tip than when soaring.

Rounded tail when soaring

White, barred rufous or brown

Tail square-tipped when folded

Broad dark terminal band

Dark S-curved trailing edge

Brown barring across belly continues around flanks to axillaries.

Brown

Soars and glides on bowed wings

Span 0.8–1.1 m

Outer wing broad, inner wing narrow near body. Wing tips rounded, deeply fingered; primaries heavily barred.

Female slightly larger; weighs 340–350 grams compared with male at 305–310 g.

Juv.

Orange-buff underwing and under-tail-coverts

Takes frogs, large insects, lizards from foliage of trees.

**Oriental Honey-Buzzard** *Pernis ptilorhynchus* 45–55 cm; span 90–100 cm

The first Australian mainland record of this species was at Leinster, in NW WA. It was found on 22 Jan. 2003 in a very weak condition, apparently swept south and inland by a cyclonic disturbance a few days before. This bird was subsequently identified at the Western Australian Museum as an immature Oriental Honey-Buzzard, race *orientalis*. In Dec. 2001, an Oriental Honey-Buzzard was recorded at Christmas Island, 1400 kilometres W of the Australian mainland, but just 350 kilometres S of Java. That bird was identified as a juvenile of the same race. Several species of honey-buzzard and other similar raptors live in Indonesia. The Oriental Honey-Buzzard is moderately large, slightly bigger than the Little Eagle. It soars with wings held at a shallow dihedral and with upswept wing tips. The three races vary from light to quite dark. All have a whitish throat patch and a dark crest toward the rear of the crown. Distinctive scale-like feathers in front of the eyes are visible up close. In flight, the undersurface pattern is distinctive, the long whitish wings being encircled by black along leading and trailing edges and across the primary tips. Most distinctive are the three fine black lines from a black patch near the carpal joint, down and back along each wing to the rufous-barred body. The large tail has a thick black subterminal band and two thinner bands beside the rufous-barred under-tail-coverts. In flight, the head looks small and the neck quite long. The upper parts are heavily mottled chestnut and barred dark brown.

## BLACK-SHOULDERED & LETTER-WINGED KITES

**Black-shouldered Kite** *Elanus axillaris* 35–38 cm
Elegant, small, pale grey hawk with white head and black shoulder patch. Typically hunts over natural grassland or low farmland stubble that has just enough height to harbour mice or other prey; also over heaths or saltbush with only sparsely scattered trees. In flight shows all-white underparts except for a small black patch at each 'wrist' joint and dark wing tips; superficially gull-like. Often hovers skilfully while hunting, then the wings move through a wide arc, and it drops to take mouse, small lizard or ground bird. Soars on strongly upcurved wings; primaries slightly spread, blunt wing tips, tail widely fanned, tip rounded. Often hunts at daybreak and dusk when mice most active. Call is a short, plaintive piping 'siep', repeated regularly at intervals of about 5 sec; a drawn out, wheezy, husky or scraping 'scrair' at intervals of 5-10 sec. Also a 'chek-chek-chek' contact call and a sharp 'kik-kik-kik' distress call given aggressively when defending nest. Common coastal Aust.; scarce semi-arid and arid regions, vagrant to Tas. Irruptive after rodent plagues.

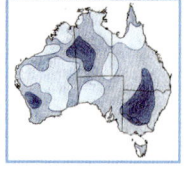

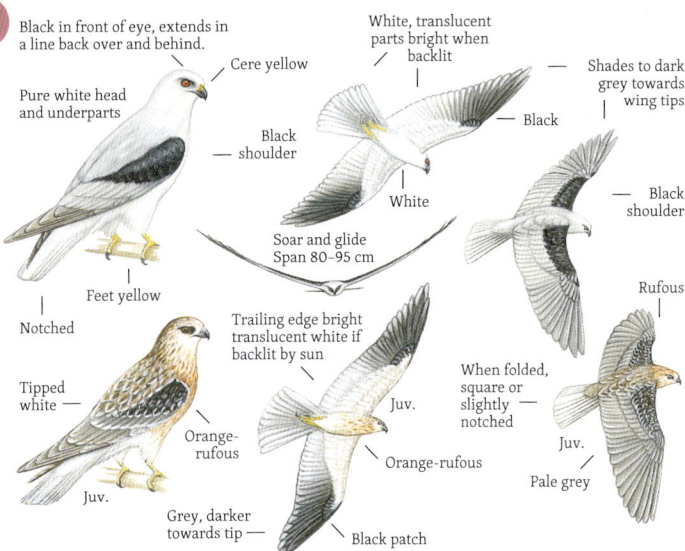

**Letter-winged Kite** *Elanus scriptus* 30–36 cm
Usually in semi-desert or desert along tree-lined creeks; hunts over grassland and other low vegetation. Largely nocturnal; roosts by day, often in flocks; most hunting at night, seeking rodents and small marsupials. Flight similar to Black-shouldered Kite, but more erratic, indirect, with sudden changes of direction; brief glides and downward swoops, and long, slow sweeps back and forth while hovering. This slower, more agile flight may be advantageous in nocturnal hunting. Gregarious, often in flocks; nests in colonies. Calls sharp, clear, penetrating, whistled 'pseep', like that of Black-shouldered, but louder. Also harsh rasping, louder, more penetrating than from Black-shouldered Kite, rather like a distant cockatoo screech. Usually rare, at times locally abundant, then soon disperse widely.

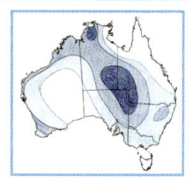

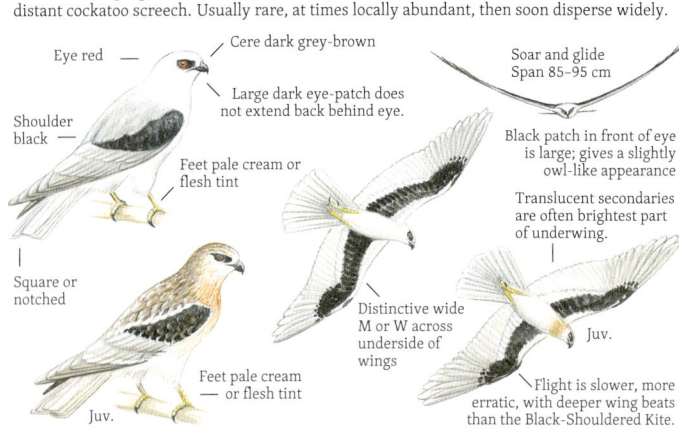

## WHISTLING & BLACK KITES

**Whistling Kite** *Haliastur sphenurus* 50–60 cm
Flight buoyant, with easy languid soaring and gliding, wings slightly drooped, held far forward. Soars in haphazard, untidy circles with widely fingered wing tips, fanned tail. In fast glide, wing tips back, tail narrowed, often glides low around treetops. Often over wetlands, but also arid regions, open woodland, scrub. In flight and at nest gives a loud 'whistling' call, first a drawn out, descending 'peee-arrgh' followed by a burst of short, sharp, harsh, upward notes, 'ka-ke-ki-ki'. The full call: 'peee-aa-rgh, ka ke-ki-kiki... peee-arrgh, ka ke-ki-ki-ki'; sequence repeated several times. Harsh 'eeargh' in defending nest; a loud 'kaairr' as warning near nest. Solitary or gregarious; often at carrion, roadside kills; takes some live prey. Common in N, rare in S of range; sedentary and migratory, part of N population moves S in summer.

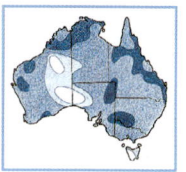

- Distinctive pale sandy buff M-shaped pattern under wings
- Scruffy looking, gingery brown kite
- Soars and glides on arched, slightly drooped wings. Span 1.2–1.5 m
- Sandy fawn, pale streaks
- Throws head back to give whistling call.
- Darker than adult; buff or white spots
- Juv.
- Heavier streaks
- Tail long, round-tipped, plain, pale sandy
- Black, deeply fingered wing tips
- Long tail extends past tips of folded wings.

**Black Kite** *Milvus migrans* M, 50 cm; F, 55 cm; span 1.2–1.5 m
Overall blackish-brown; slow floating glide, occasional lazy wing flaps; widely fingered wing tips, forked tail. Leisurely gliding and soaring with tail often fanned and twisted to gain lift in the breezes and updrafts. Wings are usually held slightly bowed, highest at the carpals, outer wings slightly drooped. But often, possibly more in calm airs, holds wings almost flat. When soaring and gaining height, may lift the profile to a very shallow V dihedral. Often glides just above or between treetops; skilfully maintains a glide in still airs with only an occasional flap of the wings. Woodland, scrub, tree-lined watercourses, mangroves, mudflats, swamps; often around homesteads, grass fires. Upright posture on perch; looks very long-winged, the tips of the folded wings reaching almost to the tip of the long tail. Rarely solitary, often in gatherings of hundreds. Principally scavenges carrion and rubbish around coastal towns and stations. Call is a plaintive, peevish, descending, quavering 'kwe-ee-ier'; also a sharp, staccato 'kee-ee-ki-ki-ki'. Common across coastal N Aust.; elsewhere nomadic or vagrant and uncommon.

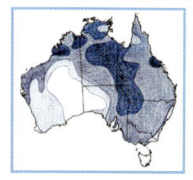

- Dark behind eye
- Soars and glides on slightly drooped, arched wings.
- Dusky blackish-brown
- Tail often twisted, especially when gliding low among trees or station buildings
- Pale, lightly barred
- S curved trailing edge
- Long tail, forked
- Fork of tail is not always evident when folded, worn or in moult.
- Pale streaks
- Juv.
- When spread, makes square-tipped triangle; more distinctly forked when partly closed.

## SQUARE-TAILED KITE & SWAMP HARRIER

**Square-tailed Kite** *Lophoictinia isura* 50–55 cm
In flight, broad wings with long, deeply fingered primaries derive lift from slightest updraft or thermal; needs only an occasional slow, casual wing beat. Often glides at treetop level, or low over heath or grassland – harrier-like with similar buoyant, hesitant, almost hovering quartering of low vegetation in search of small birds or other prey. In slow glide, has a slight sideways rocking motion. Has a yelped call, clear, plaintive and rather musical, 'yip-yip-yip-', rising in pitch to an excited 'yeep-yeep-yeep' or a hoarse, yelped 'airk-ek-k'. Food begging call by female is a high, whistled 'wheee'. In S Aust. uses eucalypt woodland, open forest and heath-woodland; in N Aust., diverse habitats dominated by eucalypts, pandanus, gallery forest, heath. Sedentary or partly migratory; generally rare; uncommon SW of WA.

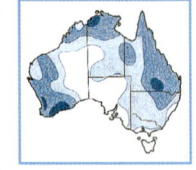

**Swamp Harrier** *Circus approximans* 50–60 cm
Glides low over reed beds and open water of swamps on long, broad, upswept wings. Flight action is buoyant, but not as light and floating as the Spotted Harrier. Habitats include wetlands, swamps and lakes, either vegetated or with open waters, salt marshes, mangroves, temporary floodwaters. In breeding season, calls are given near the nest or while soaring high over the nest territory, a loud, brief, whistled 'ki-yoo'. At other times, may give a chattered 'kik-kik-kik' in competition for food. Common; sedentary and nomadic or migratory.

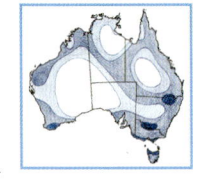

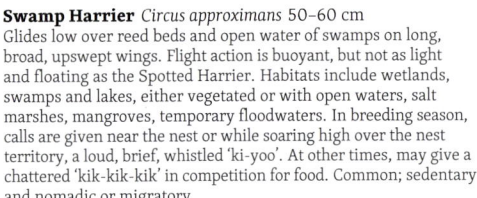

# SPOTTED & PAPUAN HARRIERS

**Spotted Harrier** *Circus assimilis* 50–60 cm
A large, slender raptor with long, widely fingered, broad wings; colourful chestnut and white underparts and chestnut face with rather owl-like facial disc. Often seen gliding low over grassland, spinifex or farmland; wings upswept, long tail fanned widely, and with characteristic slow, side to side rocking motion. In faster gliding, wings are held lower, slightly bowed. Most commonly seen over open country with dense groundcover, often low, harrying with trailing talons almost brushing the tops of grass or shrubs. Sometimes will soar higher, wings held up in a wide V; then, when high enough, travels fast on long, downward glide with wings held lower, tips backswept and pointed. Habitat includes grassland, spinifex, open shrubland, saltbush, very open woodland, crops and similar low, mostly inland vegetation. Much of the range is across semi-arid country, including vast areas of spinifex, relatively undeveloped. Also in regions of broad-acre wheat farming. Nomadic, part migratory or dispersive; movements linked to abundance of prey species.

Grey, barred darker — Soar and slow glide — Widely fingered
Chestnut face edged with grey, forming a facial disc
Legs long, yellow
Chestnut, spotted white
Long, barred, rounded
Tail long, barred
Cinnamon-buff streaked brown
Widely fingered black wing tips are conspicuous.
Soars and sails on steeply upswept wings. Span 1.2–1.45 m
Backswept in fast glide
Juv.
Juv.
Tail long, rounded tip
Chestnut shoulder
Outer wing lowered in fast glide
'Harrying', almost hovering over low vegetation; ready to snatch at prey
Patrols low over grassland in a slow floating glide, wings held upswept except to give an occasional deep, slow flap.

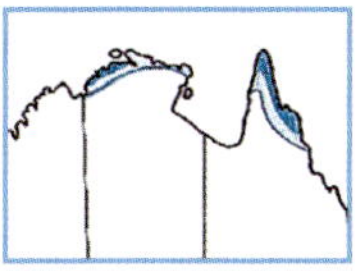

**Papuan Harrier** *Circus spilonotus* 50–55 cm; span 1.3–1.4 m
In flight the pattern of bold black and white is distinctive. Glides and soars rather slowly, wings held at steep angle. Often alone, quartering ground for small prey; often stands on ground or low perch. Usually silent, but a shrill 'whieeeuw' has been recorded. Habitat in NG includes grassland, floodplains and swamp margins. Probably quite common in NG; Torres Str. islands; unconfirmed records for NT and NE Qld.

Male distinctive: black hood, back, shoulders in bold contrast against white underparts
Held steeply up in slow soar and glide
Bold pattern, black wing tips and hood
Black streaks
Pale bars
♂
♂ Tail pale, unbarred
Rounded, rather owl-like facial disk
Narrow black trailing edge
Legs long, slender, orange-yellow
Long narrow tail
Wing tips backswept in glide; more widely spread and forward when soaring and very slow gliding
Pale rump
♀ Extensive streaking of rufous and white
Tail more heavily barred than that of Swamp Harrier
Juv. (not shown): dark brown like juv. Swamp Harrier with larger white patch on nape

# BRAHMINY KITE & EASTERN OSPREY

**Brahminy Kite** *Haliastur indus* M, 45 cm; F, 50 cm
Glides and soars low along shoreline, shallows, mangroves and mudflats. Usually the more sheltered waters of tropics and subtropics; prefers coasts with islands, mangroves, estuaries, mudflats, harbours, coastal towns; penetrates far inland along rivers. Scavenges for carrion along shoreline and shallows, but also an opportunistic predator of small fish and other marine creatures of reefs and beaches – insects and small reptiles from adjoining land can be taken in a shallow swoop. Will perch on dead trees by the shoreline, from which vantage point it scans the water's edge. Call is a high, harsh, wheezy, drawn out, descending 'pei-ir-ah'; mainly heard in spring breeding season. At the nest, harsh, wheezy whistling sounds. Although adults are distinctive, other species have similar juveniles – the Whistling Kite, Little Eagle (light morph) and White-bellied Sea-Eagle. Common around Australia's N coastline; uncommon in S of range; widespread through SE Asia.

**Eastern Osprey** *Pandion cristatus* M, 50 cm; F, 65 cm
Coastal waters and estuaries – but usually not far out to sea except on islets or exposed reefs. Follows major rivers far inland from the coast, even to arid regions where large pools lie in gorges. Distinctive in flight with slow, leisurely, shallow wing beats, but powerful enough to cause noticeable undulating of the body. When hunting, often plunges deep into the sea, at times completely submerging. Usually silent, but noisy near the nest in the breeding season. Usual call is a drawn out, plaintive or peevish whistled 'pee-ieer', but also screams and an anxious, sharp 'tchip-tchip'. Alarm calls are harsher, especially near the nest. Common around N coast, especially on rocky shorelines, islands and reefs; uncommon to rare or absent from closely settled parts of SE Aust.

# LITTLE EAGLE

**Little Eagle** *Hieraaetus morphnoides* 45–55 cm
Widespread over diverse habitats, including coastal forest, woodland, open scrub, tree-lined watercourses of interior. Most abundant where open country intermixes with wooded or forested hills, as in farmland, irrigated land. In hilly country often soars on the updrafts generated by wind deflected up the slope. Avoids dense forest, but will use clearings and margins of dense eucalypt and rainforest. Although not large, has the appearance of a true eagle: stocky powerful build; long legs fully and sleekly feathered in buff or tawny brown; feet and talons large and powerful. An eagle also in behaviour, posture and presence, and in ability to hunt quite large prey. In flight, the Little Eagle is distinctive, usually seen soaring in tight circles that reveal the flat wing position. From directly beneath, overall shape is compact and neat, wings long and broad, of even width, rounded wing tips deeply fingered, and with distinctive underwing pattern. To hunt, soars, circling; dives steeply to strike prey on ground or in trees. Prey is mainly rabbits, but also birds as large as ducks, and possums, cats and foxes. Although itself only 0.65–1 kg, the Little Eagle kills prey to 1.5 kg and can lift and fly with 500 g. The call is a far-carrying, musical, yelping whistle in distinctive double- or triple-note sequence: a very rapid 'chik-a-chuk' or 'chik-a-chuk, kuk'; the first 'chik' strong, sharp and high, the '-a-chuk' softer, lower, mellow; sometimes a soft low fourth note, 'kuk', at the end. The call is most often given in spring, usually as part of display, when the eagle, very high, dives steeply, wings closed, then swoops upwards, repeatedly. The Little Eagle, depending on its colour phase, may be confused with the Whistling Kite, Black-breasted Buzzard, or perhaps juvenile Brahminy Kite. Widespread, uncommon; vagrant to Tas. Adults sedentary; young dispersive.

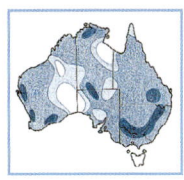

**Light morph:** Head, neck cinnamon-buff, underbody buffy white finely streaked cinnamon; underwings have a pale buff diagonal band to a large pale window on the outer wing.

Soars and glides on flat wings.

May display small dark crest.

Buffy white, finely streaked cinnamon

Pale band

Pale bar across wings

Span 1.1–1.35 m

Juv.

Legs fully feathered

Barred

Square-tipped when folded

Rounded when spread while soaring

Rufous wing linings

Juv.

Wide diagonal pale bars across underwings link large white panels at base of primaries.

Rounded wing tip when primaries are spread; soaring, slow glide

Glides with wing tips back, tail folded.

**Dark morph:** White and buff are replaced by light reddish brown streaked dark brown; underwing pattern remains, but darker. Northern birds average smaller than southern.

Mid-brown streaked darker

Deep rufous

Juv.

Fully feathered

Juv.

**Collared Sparrowhawk** *Accipiter cirrocephalus* 29–39 cm

Small, lightly built hawk, typically solitary and secretive; perches among foliage and darts out in fast, short pursuits of small birds. At other times, a low glide can take prey by surprise. Usually a rapid tail-waggle on landing. Mostly hunts small passerines, occasionally birds to rosella size, all taken in flight. Patrols territory in fast, low dashes, flushing prey that is taken after twisting, turning pursuit. In low flight, it has agile bursts of rapidly flickering wing beats, and sometimes soars and glides higher, showing a rounded wing silhouette. Calls near nest, and in display, a shrill, chattered 'ki-ki-ki-ki' and a slower, mellow 'kwiek-kwiek-kwieek-'. Calls are like those of the Brown Goshawk, but higher, weaker, and more rapid. Inhabits forest and woodland almost Aust. wide, including tree lines fringing watercourses of arid interior and coastal forest. Widespread; quite common.

Male is much smaller than female, length about 29 cm compared with her 39 cm. Wingspan: male 55; female 78. Weight: male 126 g; female 220 g. The female is about the same size as a male goshawk, which is also smaller than the female.

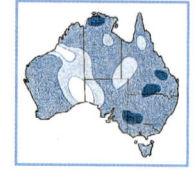

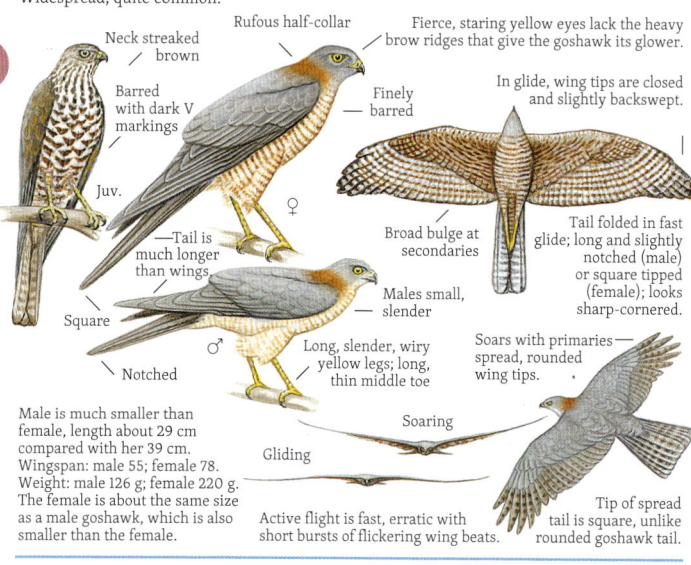

**Grey Goshawk** *Accipiter novaehollandiae* 40–55

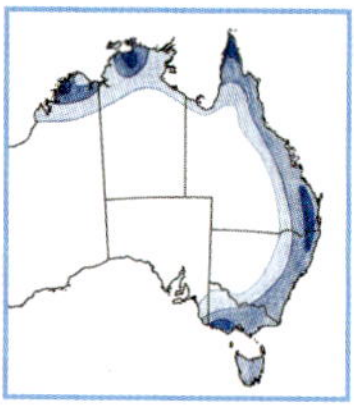

Sleek, heavily built goshawk plumaged white or delicate pale greys; cere and legs deep orange, eyes crimson. Skulks in foliage, bursts out with speed and aerial agility to take prey, whether bird in flight or rabbit. Powered flight fast, direct with strong, deep wing beats alternating with glides. Makes short flights among trees or soars above canopy. Male has slow, high, piercing 'kieek-kieeek-kieeek-', 10–20 times; female's call is similar but slower, a drawn out, mellow 'kuuwieek-kuuweik-kuuweik-'. Inhabits rainforest, gallery forest, mangroves, eucalypt forest, woodland, river edge forest. Prefers mature forest with open understorey that suits hunting technique. Uncommon, patchy; in many places rare due to forest clearing.

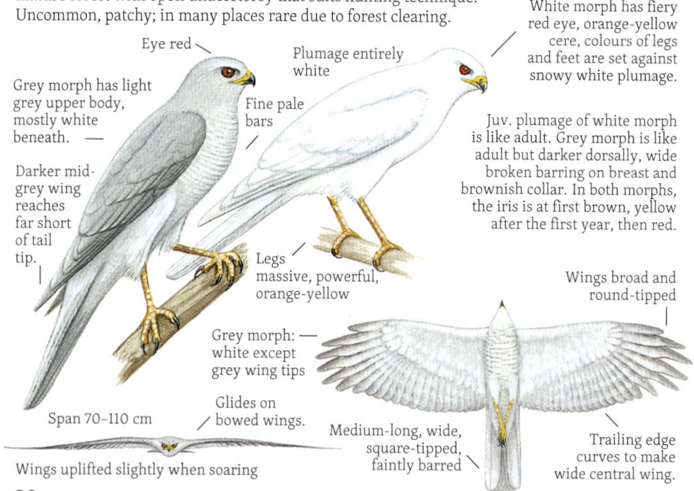

## BROWN & RED GOSHAWKS

**Brown Goshawk** *Accipiter fasciatus* 40–50 cm
Fast, agile, powerful hunter in forest and woodland, dry scrub and farms, temperate and tropical Aust. Often hunts from perch, waiting half concealed to dash or dive suddenly at still or moving prey; less frequently pursues very small birds in flight. Takes much more of its prey from the ground than does the Sparrowhawk, especially rabbits where these are numerous. Soars high, rounded tail broadly fanned, wings gently upswept, primaries spread. On landing, slow, wide waggle of tail. Calls loudly in vicinity of nest, a high 'keek-keek-keek', rising in pitch. Also a rapid, excited, descending 'kik-kik-ki-ki-kikik', possibly in defence of nest site. At times uses a slow, drawn out 'youwick, youwick'; female deeper than male. Common; sedentary and part migratory.

● *fasciatus*
● *didimus*

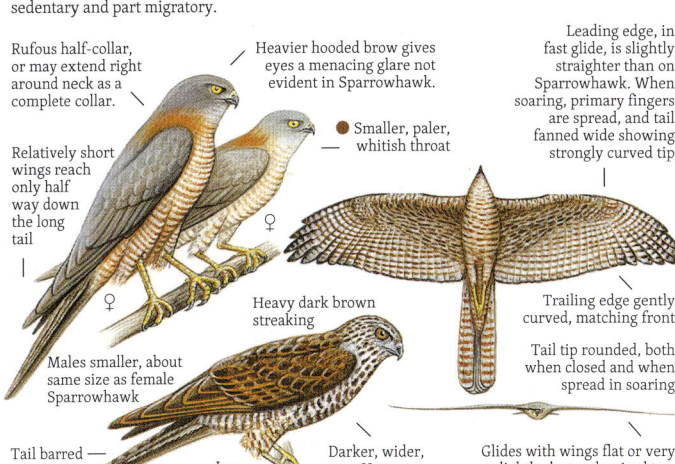

**Red Goshawk** *Erythrotriorchis radiatus* 45–60 cm
Largest Australian goshawk, and unique in many respects. Appears to have features similar to both goshawks and harriers, with its face somewhat harrier-like. Probably always rare or uncommon, it was not studied in detail until recent years. Habitat is undisturbed forest or woodland with mosaic of mixed vegetation, especially patches including river, billabong or swamp wetland with large bird populations. Usually hunts from foliage-screened perches – a typical goshawk attack, fast and from ambush – or will fly low and fast with deep, powerful beats and brief flat-winged glides to take by surprise. When soaring, shows long, heavily barred wings and tail; rounded wing tips; holds wings slightly uplifted. Calls unlike those of other goshawks. Female has harsh, strident, crowing call, 'arhk, arhk, awk', repeated for up to 30 min. Male's call similar, higher, yelped. Uncommon to rare, even in far N; rare and declining in S part of range.

# WHITE-BELLIED SEA-EAGLE & GURNEY'S EAGLE

**White-bellied Sea-Eagle** *Haliaeetus leucogaster* 75–85 cm
Often conspicuously perched on a high limb or cliff crag overlooking coastal or inland waters. Habitats include vicinity of islands, reefs, bays, headlands, beaches, estuaries, mangroves, swamps, lagoons and floodplains; often far inland along major rivers. Soars skilfully; circles on rising air columns above sun-heated islands or headlands, or above coastal ranges or cliffs where wind is deflected upwards. Sometimes soars in company with the similarly sized Wedge-tailed Eagle. Call is a harsh, nasal, carrying, goose-honked 'ank,ak-ak,ank,ak-ak-', or two birds in chorus – a confused, rapid 'ank-ank-arkakak-ank-akakak-ak'. Established pairs usually sedentary, immatures dispersive. Common around most of the coastline, scarce near major coastal cities.

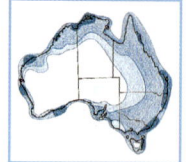

**Gurney's Eagle** *Aquila gurneyi* 74–87 cm
Large, solidly built, blackish brown eagle, similar in size to the Wedge-tailed Eagle and White-bellied Sea-Eagle; fully feathered legs. In flight, primaries are deeply fingered. Soaring and gliding it looks slow and lethargic, but has powerful, deep wing beats in active flight. When gliding, wings are almost level, but are lifted slightly higher to a shallow V dihedral when soaring or in fast glide. Soars high, hunts low over forest canopy and open ground. Call is a nasal, downward, piping whistle. Inhabits lowland and coastal rainforest, offshore islands: Boigu Is., Torres Str.; also NG and Moluccas.

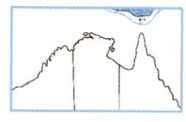

# BLACK-BREASTED BUZZARD & WEDGE-TAILED EAGLE

**Black-breasted Buzzard** *Hamirostra melanosternon* 55 cm
Soars high; also glides low, rather like a harrier, across plains seeking small reptiles, ground birds, carrion. When sailing low, often sways or tilts side to side; in flapping flight uses deep, deliberate, powerful wing beats. Usually solitary or in pairs, unlikely to be in a group larger than family party. Usually silent but quite vocal near nest. On return to nest, excited yelping, 'kyik-kyik-kyik'; as alarm, high, long 'screee'. Also a variety of harsh, scratchy, grating sounds. Mostly inhabits semi-arid and arid regions; nests in large trees along inland watercourses. Widespread but scattered; all mainland states, most common through semi-arid inland and in N. Sedentary.

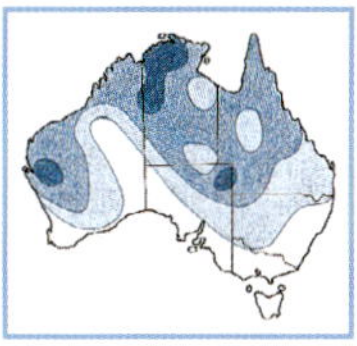

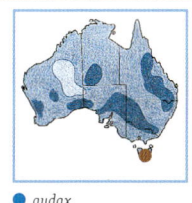

Span 1.4–1.55 m

**Wedge-tailed Eagle** *Aquila audax* 0.9–1.1 m
Takes flight heavily with slow, deep, powerful wing beats, but once aloft, skilfully uses updrafts or thermals to rise effortlessly. Soars very high on upswept wings in great circles, showing widely spread flight feathers of deeply fingered wing tips. Active, flapping flight used only close to ground; deep, slow powerful beats interspersed with flat glides. Usually silent, but calls in breeding season, perhaps only in display or near nest: 'tsIET-you, tsIET-you'. Also a hoarse, drawn out yelp. Occupies diverse climatic and vegetation types: forest, woodland, scrub, alpine, mallee, coastline, wetland, farmland. Quite common except in closely settled regions.

● *audax*
● *fleayi*

# AUSTRALIAN HOBBY & NANKEEN KESTREL

**Australian Hobby** *Falco longipennis* 30–35 cm
Flight swift, direct with rapid, shallow wing beats; occasional brief glides on flat or slightly lowered wings. In pursuit of birds the wing action becomes vigorous, deeper; the wing silhouette curves back, becomes scythe shaped, almost swift-like. Soaring, wings are held fully outstretched so that the trailing edge is almost straight. Uses woodland and open forest, surrounds of lakes and swamps, watercourse trees of interior, scrub, heath, farmland, gardens. Call is a sharp, harsh, metallic 'kiek-kiek-kiek-', accelerating to 'kir-kie-kie', or 'kikikikiki' with greater anger or anguish. Calls are similar to those of Kestrel, but more harsh and metallic; also similar to calls of Peregrine Falcon, but not as deep, harsh or powerful. Calls in territorial defence, defending nest, or by male approaching nest with food. Uncommon; scarce in Tas.; widely distributed.

● *longipennis*
● *murchisonianus*

These races overlap at times, perhaps due to dispersive movements; some records of ● E of Great Dividing Range.

**Nankeen Kestrel** *Falco cenchroides* 30–35 cm
Compared with other falcons, flight is lighter, wandering, with erratic changes in direction, often stopping to hover, glide; wing beats are rapid and less powerful. Hovers skilfully into wind with quick, shallow wing beats, or hangs with wings flexed, uplifted; drops suddenly to take prey in groundcover. Usually silent, but quite noisy in early breeding season. The sharp, high, almost metallic 'ki-ki-ki' has many variations, territorial defence, displays, fighting, approaching nest with food: ranges from fast, shrill, chattered 'kikikik-' to slower 'kee-kee-kee,' and very slow, metallic, tapping 'kik, kik, kik'. Also has a drawn out, screaming, rising 'keeeiir, keeiir' at food exchange and copulation.
Open habitats: woods, grassland, sparse scrub, heath, farms, roadsides and coastal dunes. Common, mainland Aust.; nonbreeding visitor to Tas.

## PEREGRINE FALCON & EURASIAN HOBBY

**Peregrine Falcon** *Falco peregrinus* 35–50 cm
In flight, powerful, fast. In pursuit, the wing beats become deeper, powerful, lashing strokes, soon overtaking most prey. Often hunts from above, stooping with tremendous velocity on half-closed wings. When soaring, circling high, the silhouette changes – wings are held straight out with the trailing edge almost a straight line, wing to body; tail fanned, rounded. Habitat is extremely diverse, from rainforest to arid scrub, from coastal heath to alpine. Calls higher from male; deeper and very harsh from the larger female. Noisy in breeding season and when nest is approached, with variations of the 'hek-hek' or 'chek-chek' call, depending on the excitement or agitation of the birds. In alarm or aggression, becomes a harsh but drawn out, loud 'kiek-kiek-kiek-kiek-' or 'hiek-hiek-hiek'. Peregrines may swoop aggressively at intruders, and calls will then become extremely loud, grating, harsh, rises to a machine-gun-fire crescendo in diving attack, 'ekekekekekek-'; slows and fades as attacking birds circle and climb for another attack. Also quiet sounds between pair at nest, or female to young – frog-like croaks, plaintive 'chip-chip-' and 'chirrup' sounds. Requires abundant prey, secure nest sites, lack of human interference. Hunts over rainforest, estuaries, offshore island seabird colonies. Widespread; generally uncommon to rare.

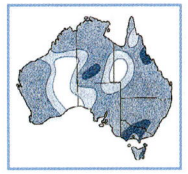

**Eurasian Hobby** *Falco subbuteo* 30–35 cm
A falcon of this species, an immature, was recorded on Ashmore Reef on 2 November 2003, the first confirmed record from Aust. Previously there was a sighting of a falcon, tentatively identified as this species, from between Scott Reef and Rowley Shoals, on 25 November 1997. All three locations are in the Timor Sea, around 350 to 400 km W to NW of the Kimberley coast, NW WA. In its usual range, the Eurasian Hobby uses woodland and open forest, surrounds of lakes and swamps, watercourse trees of interior, scrub, heath, farms and gardens. The places where it was observed in Aust. were low sand-capped reefs with sparse vegetation including some very small trees. The call is a sharp, harsh, metallic 'kiek-kiek-kiek-', accelerating to 'kir-kie-kie', or 'kikikikiki' with greater anger or anguish. Flight is swift, direct with rapid, shallow wing beats; occasional brief glides on flat or slightly lowered wings. In pursuit of birds, the wing action becomes vigorous and deeper; the wing silhouette curves back, becomes scythe-shaped, almost swift-like. When soaring, wings are fully outstretched so that the trailing edge is almost straight. The thighs and under-tail-coverts are distinctive: bright rufous on adults, or streaked over pale rufous-buff on immatures.

## BLACK & GREY FALCONS

**Black Falcon** *Falco subniger* 45–55 cm
Core habitat, semi-arid and arid interior; uses tree-lined watercourses, isolated stands of trees; hunts over low vegetation of surrounding plains, grassland, saltbush and blue-bush. Also often hunts over wetlands, temporary waters or bore drains in arid regions, taking advantage of birdlife attracted to water. Disperses to coastal regions, but keeps to open country. Flight variable, leisurely with crow-like wing beats, occasional glides on slightly drooped wings. Accelerates with deep, powerful, flickering wing beats to pursue prey. Call is like Peregrine's, but deeper, slower, 'gaak-gaar-gaak-', becoming an excited 'gak-gak-gak-' if an intruder nears the nest tree. In sudden alarm, a single 'gaaark!'. Also calls quite unlike other falcons in courtship and display flights – a loud, high, sharp, scratchy 'eeik…eeik…' every 3 to 5 sec. Uncommon; migratory; main stronghold and breeding region is inland Qld and NW Vic. After breeding, disperses across most of E Aust. and across NT into WA.

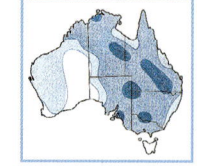

Largest Australian falcon: power of wide shoulders obvious from front; when perched, bulky shoulders make head look quite small.

Uniformly sooty brown to sooty black, rarely sooty grey

Span 95–110 cm

With age, streaky white throat

Cere, eye-ring, pale blue-grey

Soar and glide, slightly drooped

When spread, tail square-tipped but with notched corners

Strongly curved trailing edge, straight when soaring

Juv.

Juv. darker; pale edges to dorsal feathers

Legs are partly feathered; feet large, powerful, pale grey

Long, round-tipped tail, notched at corners; longer than folded wings

Soars and glides with carpal joints well forward so that the trailing edge of the sharply pointed wing is almost straight.

Often uses steep stoop from height to gain speed for fast low dash to pluck a bird from the air or small mammal from the ground.

Similar: dark morph of Brown Falcon

**Grey Falcon** *Falco hypoleucos* 30–45 cm
A sleek, grey-plumaged falcon of the interior plains; flight swift and direct; patrols low over groundcover below treetop level, propelled by easy, shallow wing beats and brief glides, when dark wing tips are most obvious. Soars with wings held close to level, dark-tipped primaries slightly spread to give a blunt wing tip. When gliding, outer wing is backswept. Small birds flitting between trees or flushed from the ground are caught by surprise, taken in a brief burst of speed generated by deep, strong wing beats. Inhabitant of lightly timbered country, especially stony plains and acacia scrub. Usually silent, but gives a loud, slow 'kek-kek-kek' or 'kak-ak-ak-ak' in breeding season, similar to the call of the Peregrine Falcon but slower, deeper, harsher. Rare; resident or nomadic to most of semi-arid interior.

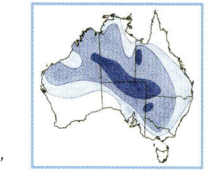

Grey bill, cere, eye-ring

Juv.

Grey with fine brown streaks

Streaked darker

Yellow bill, cere and eye-ring

In flight, the dark wing tips are conspicuous against pale grey or white of upper or lower surfaces, reliably identifying this species.

Soaring, wings are held straight, well forward, trailing edge straight.

Fast glide

Medium length, slightly rounded, closed or spread

Fine grey streaks

In soaring, primaries are slightly parted.

Large, powerful, deep yellow feet

Wings reach tail tip.

Span 85–95 cm

Soaring and gliding

Dark grey

# BROWN FALCON

**Brown Falcon** *Falco berigora* 40–50 cm

One of Australia's most common birds of prey, seen across the continent and conspicuous through most of the interior. It is widespread in most open habitats – woodland, lightly treed farmland, mulga scrub, watercourse tree lines, alpine areas, heath, coastal dunes, crops. Once its unique, loud, carrying, raucous, crowing and cackling calls are known, any traveller or camper will know of its presence where it would otherwise be unnoticed. Unlike Australia's other large falcons, the Brown Falcon feeds mostly on the ground; it hunts reptiles, grasshoppers, beetles and mice, and also feeds on carrion. This versatile raptor is well equipped for terrestrial hunting: its long, heavily scaled legs and short talons give it the ability to run, leap and snatch at such agile small prey as mice, dragon lizards, snakes up to a metre in length and small ground birds. The Brown Falcon's flight, speed and agility, while not comparable to a Peregrine's, are adequate to catch small birds that it may take by surprise as it glides low across scrub or grassland. While it usually hunts from perches, the Brown also quarters open country at 5–20 m height, hovering low in search of prey. In active flight the Brown Falcon is unmistakeable once its peculiar wing action is recognised: a deep, laboured, overarm rowing action, interrupted by brief, wobbling, side-slipping glides, meanwhile attracting attention with loud, raucous cacklings, the whole performance unique among Australian raptors. Often calls kilometres from the nest and continues until in the nest tree; together with equally noisy behaviour at the nest, this can make the presence of a nest, even if well concealed, impossible to overlook. Plumage ranges from almost uniform brownish black to white fronted with sandy brown back, and some that are much more rufous. The call is a loud raucous cackling, somewhat like a laying hen but louder and harsher, 'karairk-kuk-kukkuk', the first part raucous, rising, the following 'kuk-kuk-' as a low clucking. Also 'karark','kar-r-rak', 'kairrrk' as single calls a few seconds apart. Common; sedentary, at times irruptive.

There are three colour morphs, with plumage varying greatly, from light rufous to nearly black. All colour morphs have double dark vertical streaks enclosing paler cheeks, and a dark brown thigh feathering.

Flight action varies from rapid shallow beats in pursuit to its more usual deep, erratic, rather wobbly and very distinctive rowing action.

Glides rather erratically on uplifted wings, rather like a harrier; occasionally hovers, but without the skill of a kestrel, as it searches the ground for its prey.

Light morph — Darker thigh — Buff to creamy white

Brown morph, fast glide

Pale between dark ear patch and dark vertical streak down from eye.

Brown morph

Juveniles are darker versions of each morph; have same distinctive markings.

Juv., brown

Adult, brown morph; the intermediate between light and dark morphs

Barred

Brown

Wide rounded fan when soaring

When soaring, wings are held straight out, trailing edge is a long gentle curve; primaries spread enough to give a rounded wing tip.

Fast glide

Rufous morph

Wings slightly short of tail-tip

Fast glide    Span: 90–120 cm

Dark 'trousers', on all morphs, but harder to see on darkest forms.

Dark morph

Brown Falcons attract attention in noisy display flights: hold wings steeply upwards; sway and tumble towards nest tree; all the while cackling loudly.

## OVERHEAD: BAZA, LARGE KITES

**Baza:** Floating, slow-flapping, gliding and soaring flight over and around tree canopy; takes small prey from foliage of trees.
**Large kites:** Slow floating or sailing flight, often low over foliage or ground; snatch food from vegetation, ground or water.
**Small white kites:** Swift flight, sailing glides, skilful hovering, dropping to ground on uplifted wings to take small prey.
**Goshawks:** Hunt by ambush and pursuit of birds in flight, small mammals on ground; usually seen gliding or soaring.
**Falcons:** Most are very fast hunters of birds taken in steep stoops, fast pursuit on pointed backswept wings, except Brown Falcon, slower, usually hunts taking terrestrial prey from perch.

Magpie
Span = 60 cm

**Pacific Baza** preys on frogs and large insects such as mantids; circles close around treetop foliage, snatching prey from leaves.

Orange-buff wing linings, barred body

Slow floating glide close to treetop canopy with only occasional leisurely beats of wings; soars higher, circling with broad, slightly bowed wings held almost flat.

Strong central bulge makes outer secondaries the broadest part of wing; most obvious when wing tip flight feathers are held tightly closed.

Tail tip rounded, dark-banded

Wing tips white, heavily and conspicuously barred

Tail long, dark tipped, obviously square-tipped with sharp corners

**Square-tailed Kite** has long wings, broadest at outer wing, where widely spread primaries usually greatly increase the width, giving paddle-shaped appearance.

**Whistling Kite:** Scruffy gingery plumage; buoyant, languid soaring and gliding on slightly drooped wings. Soars in loose circles, wing tips and rounded tail widely spread, or wing tips backswept and tail folded in glide.

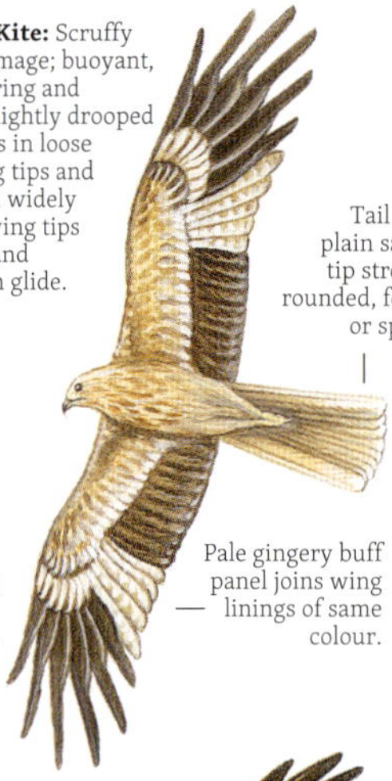

Primaries long, dark, usually held widely splayed

Tail long, plain sandy, tip strongly rounded, folded or spread

Pale gingery buff panel joins wing linings of same colour.

**Black Kite:** Slow floating glide, occasional slow wing flaps, primaries and tail usually spread wide

Dull blackish brown coverts

Dull blackish-brown secondaries

Tail plain. Dark, shallow fork when fully spread, deeper when partly closed

**Brahminy Kite:** Soars and glides over beaches, mangrove swamps, estuaries, mudflats, at times high, often low glides to snatch food from beach or water surface.

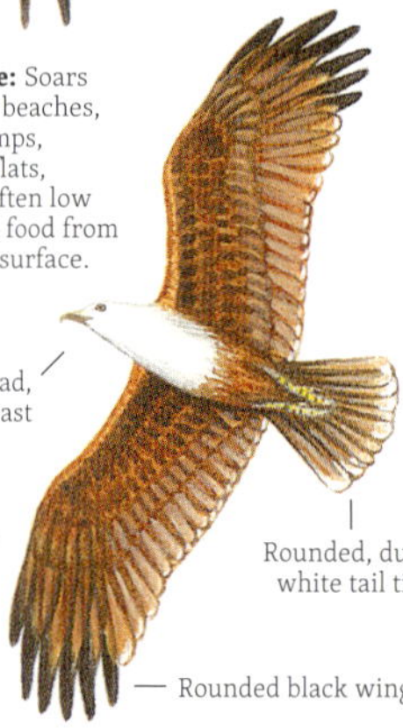

White head, neck, breast

Deep chestnut, brighter in sun, dull dark brown when shadowed

Rounded, dull white tail tip

Rounded black wing tips

**Brahminy Kite (juv.):** Similar to some pale juvenile Little Eagles and immatures of Black-breasted Buzzard

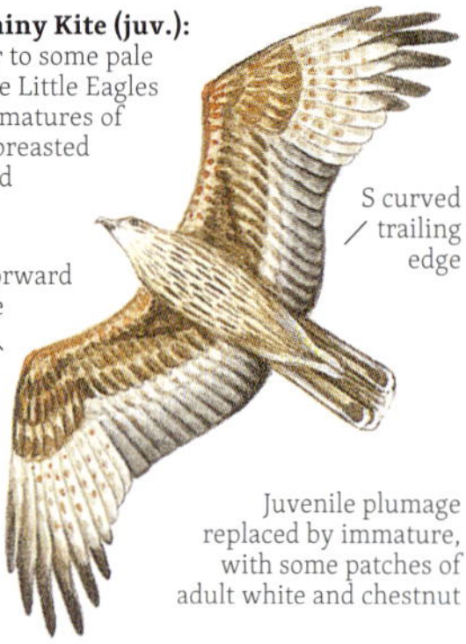

Held forward in glide

S curved trailing edge

Juvenile plumage replaced by immature, with some patches of adult white and chestnut

# OVERHEAD: SMALL KITES, GOSHAWKS, FALCONS

**Black-shouldered Kite:** Glides on upcurved wings. Hovers skilfully holding a perfectly stationary position.

- Tail tip rounded
- Grey darker towards wing tip
- White
- Black near 'wrist'
- Juv.: Orange-rufous
- Translucent secondaries, white, bright under sun

**Letter-winged Kite:** Flight slow, rather harrier-like, glides on steeply uplifted wings, hangs motionless against the wind.

- Adult: White
- Juv: Orange-rufous
- Wide black M or W across wings

**Brown Goshawk:** Flight powered by bursts of deep, quick wing beats

- Broad slight bulge, a fairly straight trailing edge
- Long rounded tail, spread when soaring

**Collared Sparrowhawk:** Active fast flight on flickering wings

- Rounded outer wing profile
- Gently curved trailing edge
- Slightly notched, or more square than tail of Brown Goshawk

**Grey Goshawk:** White or grey beneath

- Bulging trailing edge; central wing broad
- Rounded wing tips
- Tail medium length, wide, square-cut tip

**Red Goshawk:** Soars on rounded wings, long tail spread; rufous wing linings and body.

- Long tail is gently rounded when spread, almost square when closed.
- Bulging curve to secondaries

**Australian Hobby**

- Wings long, scythe-shaped, backswept, pointed
- Tail long, barred, gently rounded tip

**Grey Falcon:** Deep strong wing beats in pursuit

- Dark grey, barred wing tips
- Out-curved secondaries, straight when soaring
- In soaring, wings held forward, wing tips rounded

**Nankeen Kestrel:** Often hovers.

- Tail long, barred, bold black band
- Tail barred, slightly curved

**Peregrine Falcon:** Trailing edge curved in glides; almost straight when soaring

- White bib
- Short, square; or, in soaring, wide, rounded
- Wings set for glide; when soaring, straight out with blunt tips

**Black Falcon:** Very fast in pursuit, deep rapid wing-beats

- Tail long, narrow; or, when spread, gently rounded with notched corners
- Backswept wings in fast flight and glide; in soaring and slower glide, wings straight, trailing edge straight, wing tips rounded

**Brown Falcon:** Brown morph

- Trailing edge curved in glides, almost straight when soaring
- Shown in glide. Soaring, wings are straight, tail spread wide, rounded tip

99

## OVERHEAD: HARRIERS, EASTERN OSPREY

**Harriers:** Typically hunt with slow 'harrying' flight, low over land or water, hovering, long bare legs dangling to snatch prey.
**Eastern Osprey:** Hunts over coastal waters, estuaries; soars, glides, plunge-dives to take fish, sometimes fully submerging.
**Eagles:** Include the relatively small Little Eagle, light and dark, and the very large, blackish Wedge-tailed Eagle.
**Buzzard:** Easily identified, stark white patches in outer wing brighter than on any other local raptor with white wing spots.
**Sea-eagle:** Large, broad-winged hunter along coasts and rivers.

Magpie
Span = 60 cm

**Spotted Harrier:** Usually seen gliding or soaring low over spinifex or crops with occasional slow wingflaps. Wings are held low in glide, in a steeper V shape when soaring.

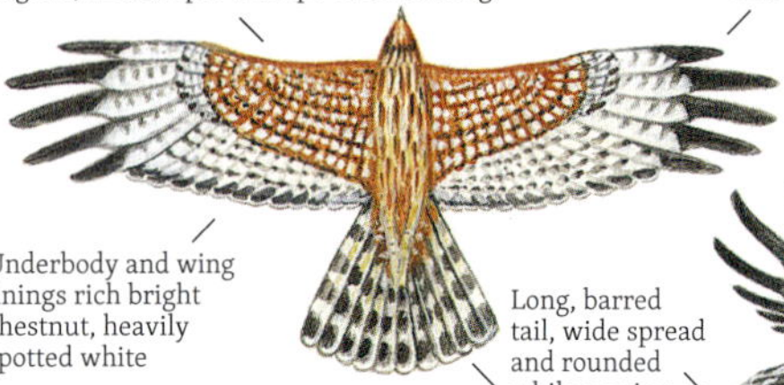

Unbarred black wing tips very widely fingered when soaring and in slow glide

Underbody and wing linings rich bright chestnut, heavily spotted white

Long, barred tail, wide spread and rounded while soaring

**Spotted Harrier (juv.):** Underbody and wing linings sandy buff

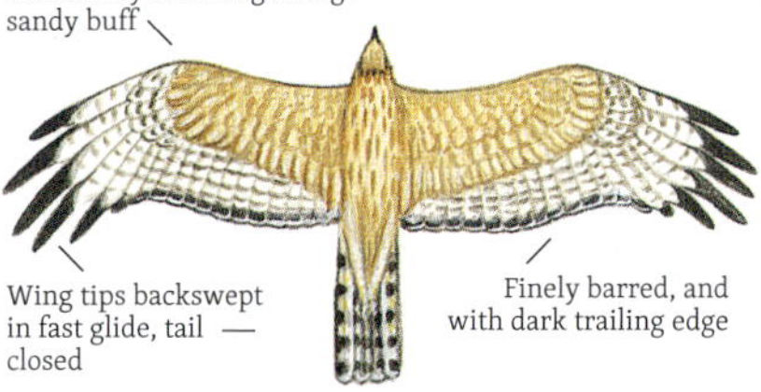

When seen 'harrying' low over vegetation in distance, usually it is a side view, rather than from beneath. This gives a glimpse of blue-grey of upper surfaces, as well as the bright chestnut of underparts.

Wing tips backswept in fast glide, tail closed

Finely barred, and with dark trailing edge

Males paler than females. Breast and wing lining buffy white with dark brown streaks; female browner beneath

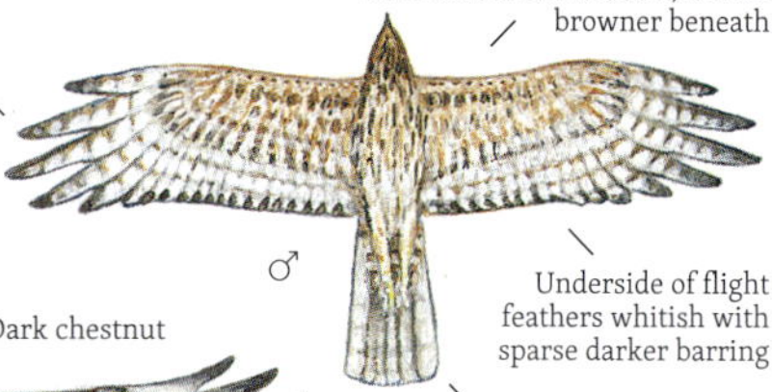

**Swamp Harrier (male):** Glides low over reed beds and open water, sometimes hovers rather clumsily; harries small waterbirds or lands among reeds.

Wing shape as in flapping flight

♂

Dark chestnut

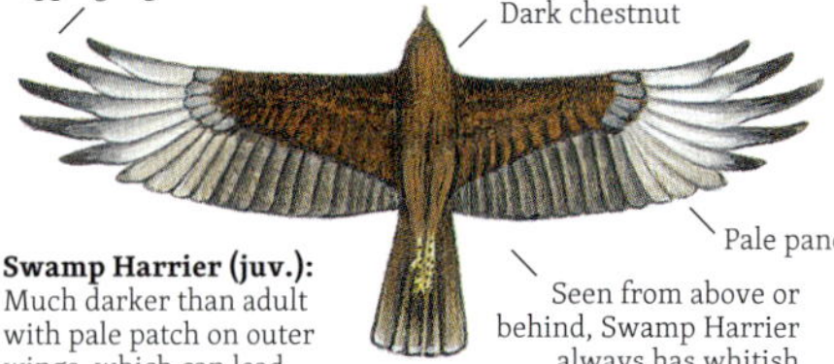

Underside of flight feathers whitish with sparse darker barring

In glide the tail is closed, almost square-tipped; when spread in soaring or hovering, tail tip is rounded.

**Swamp Harrier (juv.):** Much darker than adult with pale patch on outer wings, which can lead to confusion with Black-breasted Buzzard

Pale panel

Seen from above or behind, Swamp Harrier always has whitish rump; buff on juvenile.

**Papuan Harrier (male):** White underparts, black hood and wing tips

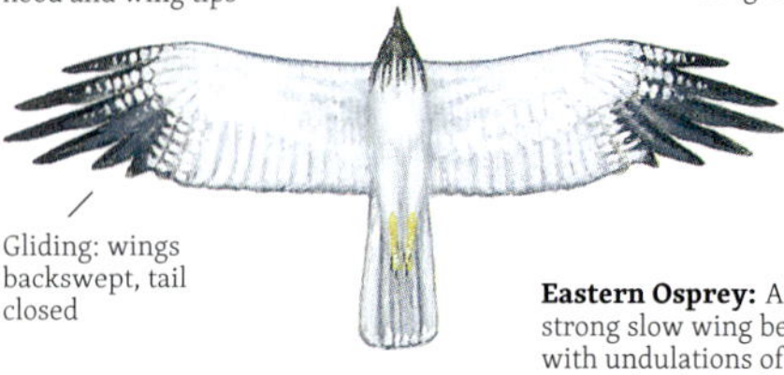

Wings barred except white wing linings

Black patch at 'wrist'

♀ has streaked collar.

Gliding: wings backswept, tail closed

**Eastern Osprey:** Active flight of strong slow wing beats, steady but with undulations of body. Glides on nearly flat but kinked wings.

Long wings follow zigzag shape, with carpal (wrist) joints held forward, wing tips back even in slow glide and soaring.

Juvenile is like that of Swamp Harrier, but of a lighter, brighter chestnut with head and much of underparts streaked white.

## OVERHEAD: EAGLES, BUZZARD

**Little Eagle (light morph):** Occurs in light and dark colour morphs. A mid-sized, stocky eagle. Soars and glides on wings held almost flat. Powered flight by deep, powerful wing beats

In soaring, wings are held straight out, leading and trailing edges roughly parallel and primary fingers spread, making a rather square wing tip.

In gliding, as speed of glide increases, carpal joints are pushed forward, outer wings angled further back

**Little Eagle (light morph, juv.)**

Pale diagonal bars link white primary panels making a broad M shape

Dark tips

Tail fanned when soaring; rounded

Body and wing linings rich rufous

Tail closed in glide

Lacks pale diagonal bar

**Little Eagle (dark morph)**

**Little Eagle (dark morph, juv.)**

Dark rufous to deep brown

Body and wing linings dark, varying from dark sandy to deep rufous or dark chocolate brown

Dull white panel

Dull white panel. Lacks pale diagonal of light morph.

**Black-breasted Buzzard:** White panels in outer wing are far brighter than on any other Aust. raptor.

Underwing will appear darker when shadowed.

Wings long, broad, straight; deeply fingered

Body black, leg feathering rufous

Shown in glide. When soaring, wings held straighter, with fingered tips splayed

Black; usually has patchy rufous and whitish underwing diagonal lines.

Short square-tipped tail

**Gurney's Eagle:** Soaring or slow glide

Tail long, diamond-shaped

**Wedge-tailed Eagle:** Very large, adults mostly black

Tail medium length, rounded

**White-bellied Sea-Eagle (juv.)**

**White-bellied Sea-Eagle:** Unique black and white plumage. Soars on steeply upcurved wings. In glide, wings slightly uplifted

Long-necked; looks small-headed

Edge strongly S curved, wing narrow at body

Tail very short, a rounded wedge tip, quite rounded when fully spread in soaring

**Gurney's Eagle (juv.)**

Inner wing, mid-secondaries, much broader than outer wing

Curved back in glide

Dull white base to primaries

# 7 NATIVE-HENS, CRAKES, RAILS, COOT & BUSTARD

Two families in Australia: Rallidae with 16 species, and Otididae with a single species, the Australian Bustard. The former are small to quite big hen-like birds, inhabitants of waterways, wetlands or lush damp vegetation such as rainforest. Food for some is mostly aquatic vegetation; for others, wetland invertebrates, frogs and other small creatures. The very large Australian Bustard keeps to open, often arid, grassland and scrub, and takes small terrestrial prey.

Willie Wagtail = 20 cm

### rallidae
page 102

- Red-necked Crake
- Baillon's Crake
- Tasmanian Native-hen
- Red-legged Crake
- Australian Spotted Crake
- Australasian Swamphen
- Eurasian Coot
- Buff-banded Rail
- Spotless Crake
- Dusky Moorhen
- White-browed Crake
- Lewin's Rail
- Black-tailed Native-hen
- Corn-crake
- Chestnut Rail
- Pale-vented Bush-hen

### otididae
page 109

- Australian Bustard

---

## RED-LEGGED CRAKE

**Red-legged Crake** *Rallina fasciata* 20–25 cm
Rare vagrant; keeps to cover; partly nocturnal. Usual range, India to Lesser Sundas. Inhabits wettest parts of rainforest, regrowth, swamps, paddyfields. Noisy, often calls at night: screams, grinding sounds, loud, regular 'kek'. In Aust., a single weak male found on a Broome lugger in 1958; storm-blown or off-course migrant?
Possible vagrant to NW coast of Aust.

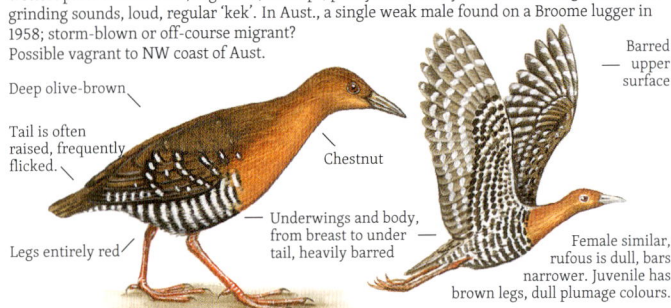

Deep olive-brown
Tail is often raised, frequently flicked.
Chestnut
Legs entirely red
Underwings and body, from breast to under tail, heavily barred
Barred upper surface
Female similar, rufous is dull, bars narrower. Juvenile has brown legs, dull plumage colours.

## LARGE CRAKES

### Red-necked Crake *Rallina tricolor* 24–29 cm
Distinctive: often heard, but secretive, difficult to sight or observe. Forages along edges and shallows of rainforest streams and pools, and in nearby leaf litter. Most active and calling around dawn and dusk. Often flicks tail; stands tall, alert; quick to dash for cover. Voice is a harsh, sharp 'kark' or 'karrark' as a single note; often a sequence, a rasping, aggressive 'karr-ak, karrrk, kak, kak, kak-kakakakak'. Its monotonous 'klok, klok, klok' may continue for hours, often at night. Habitat is vicinity of streams and swamps, in or near dense thickets of rainforest and vine scrub. Mainly sedentary, but probably some migration between Cape York and NG. Has lost much of its lowland rainforest habitat.

Within Aust. very little variation: those at N of range have slightly shorter wings than those furthest S. Two subspecies in NG and nearby islands

(Illustration at 50% scale) Nest may be a shallow depression lined with leaves on the ground beside a tree stump or other shelter, or may be a bowl higher in a stump cavity or in dense foliage.

### Buff-banded Rail *Hypotaenidia philippensis* 28–32 cm
Colourful; wary; easily recognised. Often emerges from dense cover at dawn to feed before full sun reaches exposed mudflats or open marshy ground. Feeds again towards dusk. Calls include a squeaky 'swiit' and a loud, creaky, harsh 'kiek' or 'priep'; often answered by others. Also reported are triple noted 'tchuk-e teika' and braying sounds. Makes soft grunting sounds to chicks and growling hisses when chicks are threatened. Habitat is of dense, damp vegetation around swamps, lakes, creeks, coastal lagoons, tidal mudflats, rainforest margins, moist paddocks and sewage farms. Probably more common on coast than sightings would suggest; sedentary, nomadic, dispersive.

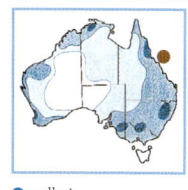

● *mellori*
● *tounelieri*: a form on Great Barrier Reef islands, may not be suffficiently distinctive to justify subspecies status; wing length, on average, is shorter. Many races on islands of SW Pacific

## SMALL UNSPOTTED CRAKES

**White-browed Crake** *Poliolimnas cinereus* 18–20 cm
On wetlands with waterlilies or other floating plants – billabongs, swamps, floodwaters – also surrounding forest or woodland. Unlike other secretive crakes of dense reedbeds, the White-browed forages across lily pads and other floating vegetation, helped by its very long toes. It occasionally flutters low across intervening water. Most active after sunrise, near sunset and beneath overcast skies. Calls include a chattering 'kiak-kiak-kiak', 'chika-chika', a frog-croaked 'kak', and a soft 'charr-ar, charr-ar' in danger. Common in optimum habitat.

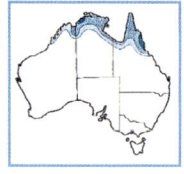

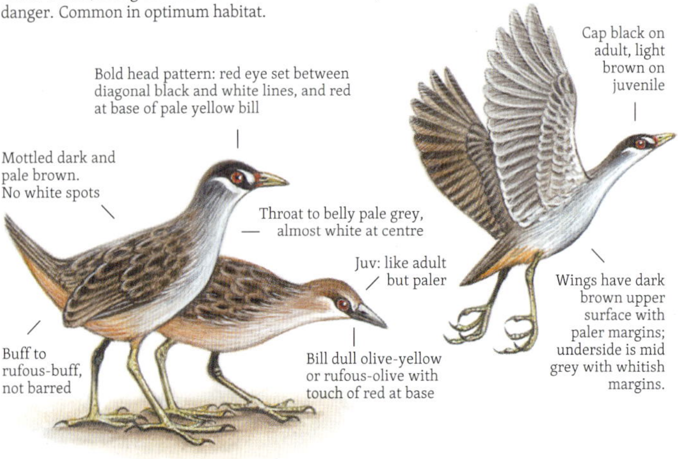

Bold head pattern: red eye set between diagonal black and white lines, and red at base of pale yellow bill

Mottled dark and pale brown. No white spots

Throat to belly pale grey, almost white at centre

Juv: like adult but paler

Buff to rufous-buff, not barred

Bill dull olive-yellow or rufous-olive with touch of red at base

Cap black on adult, light brown on juvenile

Wings have dark brown upper surface with paler margins; underside is mid grey with whitish margins

**Spotless Crake** *Zapornia tabuensis* 17–20 cm
A plain, dark crake with bright red eyes and pink to red legs. Rarely seen out of dense cover; forages in shallows, wading, swimming, climbing over fallen and floating vegetation. Not often seen in flight, but in any brief glimpse would appear dark with pink dangling or trailing legs. Usual habitat the reedbeds and other dense vegetation in and around lakes, swamps, saltmarshes, mangroves. Call is a sharp, harsh 'chaik'; also a fast, rattling, descending and fading 'chak, chak-chakchak-'; bubbling and trilling whistling. Uncommon; nomadic, migratory.

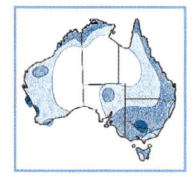

Under-tail-coverts are heavily barred almost to tail tip.

Plain dark rufous-brown with no white spots or tips

Eye and eye-ring crimson

Dusky blue-black

Legs and feet deep pink to red

Long red legs trail behind.

Bill black, pinkish white at base of upper then lower mandible

Pale grey to white

Juv. First brown, then dull pink, some to adult colour

Young are downy black with silvery tips for several weeks.

Nest is a cup or bowl of reeds lined with rootlets, grass; surrounding reeds are bent over as a hood.

# SMALL SPOTTED CRAKES

**Baillon's Crake** *Zapornia pusilla* 15–16 cm
Smallest of Aust. rails; secretive, but like other crakes may forage out of cover early and late in the day; probably less shy than other species. Uses permanent or temporary fresh or saline wetland, swamp, vegetated lake margins, floating vegetation in deep and shallow water. Forages over floating vegetation, tail upright, constantly flicked. Weak, fluttering flight among reeds, legs dangling. Calls loud, sharp, very rapid, ratchetting 'kar-r-r-r-r-r-ak'; also a sharp 'chak' or 'krek' in alarm. Not often seen but probably common; migratory.

**Australian Spotted Crake** *Porzana fluminea* 19–22 cm
Habitat is dense cover, fresh or salt wetland, lakes, swamps, saltmarsh; at times far from water. Usually keeps to dense reedbeds, but early and late in the day will venture out onto nearby shallow open water, mudflats or floating vegetation; constantly flicks tail. Swims across intervening water; rarely flies. The usual call is an abrupt, sharp, metallic 'chaik-chaik, chaik-chaik', more musical than harsh; also a rapid descending sequence, 'chak-ak-ak-ak-akakakak'. Common SE and SW; nomadic.

**Lewin's Rail** *Lewinia pectoralis* 21–25 cm
Secretive and difficult to sight even momentarily in the swamps where it forages; tends to come into the open less than other rails. Voice is a sharp, loud 'krek' given as a burst of 10 to 20 calls; becomes faster, louder, then fades away. Also a series of loud whistles. Usual habitat: swamps, lakes, tidal creeks, salt marsh, lush wet pasture, paperbarks. Uncommon; nomadic.

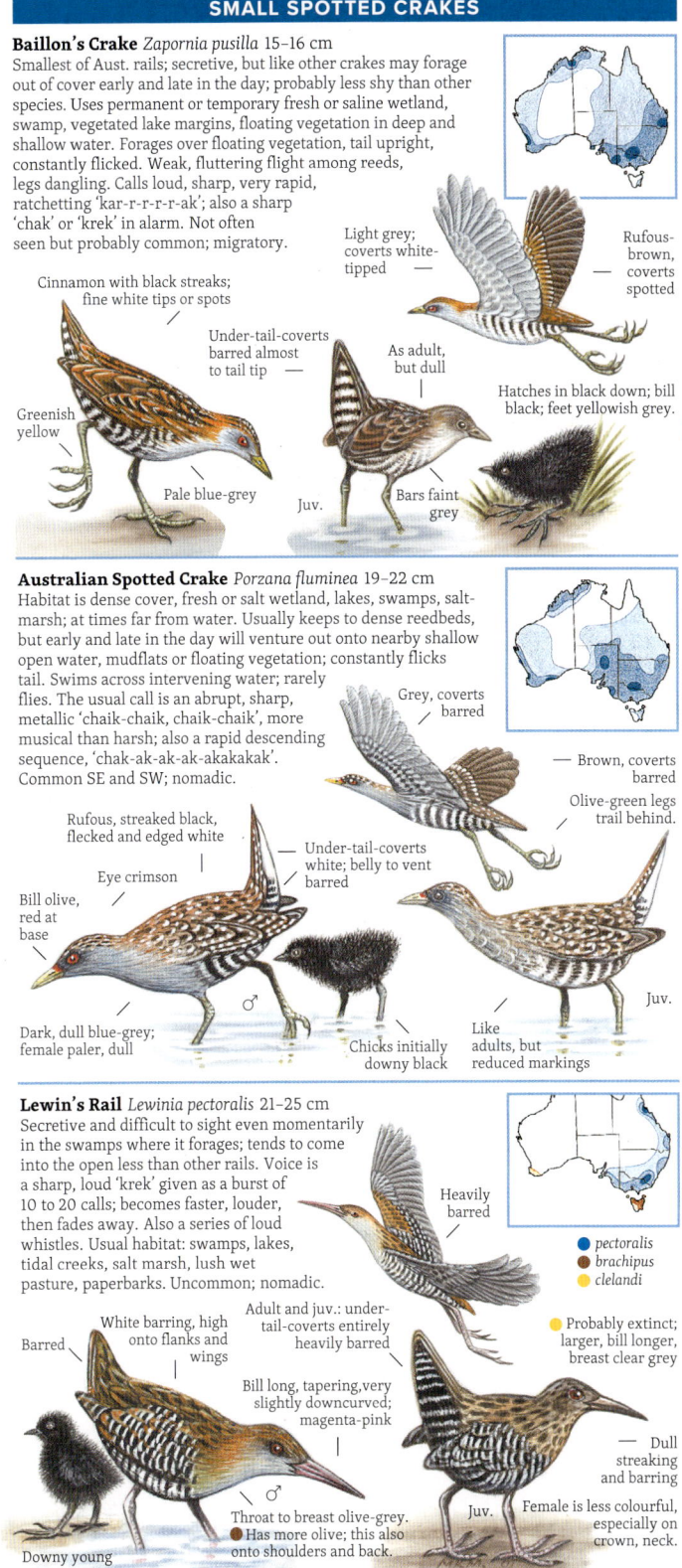

● pectoralis
● brachipus
● clelandi

● Probably extinct; larger, bill longer, breast clear grey

## PALE-VENTED BUSH-HEN, DUSKY MOORHEN & CORNCRAKE

**Pale-vented Bush-hen** *Amaurornis moluccana* 24–26 cm
Noisy; secretive inhabitant of dense margins of freshwater wetland, rainforest and regrowth, dense tall crops, mangrove edges. Tends to emerge from cover early and late or when heavily overcast. Calls are varied: include loud, shrieked, cat-like 'ki-arr-rk, ki-arr-rk', often a long succession, fading away, or with two birds in duet; a clear, piping 'keek, keek, keek', all at same pitch, for long periods; sometimes drawn out, higher 'kee-eek, kee-eek'; soft ticking sounds. Sedentary or partly nomadic.

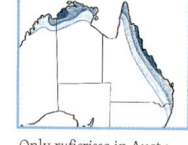

Only *ruficrissa* in Aust.; others in NG, Solomons, Philippines, other islands

Breeding: bill light olive-green or olive-yellow with top edge of shield vermilion

Warm chestnut brown

Overall impression: dark with trailing yellow legs

Rarely seen in flight, but possible if startled far from cover

Dull yellow

Deep buff to rufous

Dark olive-brown

Juv.: paler; whitish throat, bill brown

At times, holds tail erect showing rufous-buff beneath. Often stands tall, nervously alert, flicking tail.

Downy young, first few days

**Dusky Moorhen** *Gallinula tenebrosa* 35–40 cm
Widespread, common, uses diverse wetlands. Prefers open waters with well-vegetated margins; usually fresh water, some brackish including swamps, lakes, estuaries. Often around lakes in suburban parks. Runs and swims well; always wary, quick to dash to cover. Swims jerkily, paddling with jerks of head and flicks of tail. On land forages with slow deliberate step; in alarm runs with flapping wings. Territorial call is an abrupt, raucous 'krurk!' or 'krruk-uk-uk'. Also a repeated, resonant 'krek' or 'krok' and loud, harsh screeches. Common in SE and SW Aust.; uncommon to vagrant in N.

Forehead shield bright red

In flight, plumage is all dark except under tail edges. Legs trail behind tail.

Bill bright red, tipped yellow

Olive, some smudged black, dull red, smaller shield

Nonbreeding in browner, paler, worn plumage

White

Dark, slaty grey fresh plumage wears to slightly paler brownish grey.

White

Br.

Downy young: bare orange skin shows through sparse, wiry black plumage.

Older males appear to retain all or some of bill and shield's breeding colours throughout the year.

Red, tipped yellow

**Corncrake** *Crex crex* 26–30 cm
Extremely shy; keeps out of sight in low, dense vegetation; inhabitant of grassland – usually meadows, hayfields, irrigated lush pastures. Presence most likely to be revealed by its call, or accidental flushing. Call, in breeding season, a persistent, rasping 'crrek, crrek, crrek'. Just two very old records: at Randwick, NSW, 1893, and on a ship near Jurien Bay, WA, 1944.

Bold, dark centred, buff-edged pattern

Bright chestnut

Bill short, stout, pinkish

Legs pinkish buff

Chestnut shoulder, barred flanks

Legs hang low at take off, then trail behind tip of short tail.

# CHESTNUT RAIL & COOT

**Chestnut Rail** *Eulabeornis castaneoventris* 44–52 cm
A very large rail, predominantly chestnut, infrequently sighted; loud calls may reveal its presence. Habitat is dense mangrove vegetation in bays, estuaries and tidal creeks. Occasionally ventures out to search adjacent mudflats or grassy woodland. Feeds principally on crabs and other small creatures probed from the mud. Struts across mangrove mudflats with tail flicking nervously; runs swiftly through tangled vegetation rather than flying. There are many descriptions of calls, but seems to include pig-like grunts or drumming intermixed or alternating with louder, raucous screeches, 'whuh-WHAIKA, whu-WHAIKA', repeated often to a steady rhythm. Population scattered around far NW and N coasts; difficult access protects habitat but makes population assessment difficult.

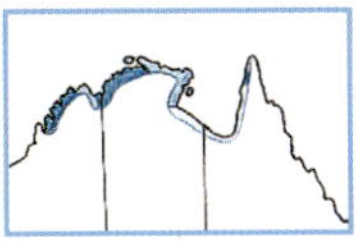

No subspecies in Aust., but three colour morphs:
N Qld, olive-brown back;
NT, chestnut back;
WA, olive-grey back

Large rail. Back colour variable, but head always grey, bill olive-yellow, underparts always chestnut

Underwing all brown

Head partly tucked in

Pale under chin

Feet trail behind tail tip.

Chestnut backed morph, NT

Crown grey, all morphs

Bill pale yellow

Underparts always deep chestnut on all colour morphs

Pointed tail, often raised, often flicked

Olive-grey backed form, NW Kimberley coast, has dusky olive-grey back and wings.

Juvenile has not been described.

**Eurasian Coot** *Fulica atra* 35–38 cm
A common, well-known and easily recognised waterbird with dumpy, dark, slate-grey body, conspicuous white forehead shield and bill. Occurs widely on wetland including rivers, lakes, swamps. Also, rarely, marine wetland such as estuaries. Principally in coastal regions, but has been widely recorded on temporary lakes and floodwater in the semi-arid interior. The Coot forages in shallow to deep water where it up-ends or dives for plant material. At times it emerges onto land; then looks rather clumsy and unbalanced. The legs are set far back to give better thrust in water, but the Coot actually walks quite well; individuals and groups may emerge to feed on grassland near water. Call most often heard is an abrupt 'krek,' and sharper 'krik'; also a grating, sharp 'kiek-kiek-kiek'. Common.

In Aust., just one race, *australis*

Eye brilliant red in black surrounds

Bill and shield white

Wings entirely dark

Rounded, plump body shape tapers to pointed wisp of tail.

Downy young: sparse wispy feathers reveal red skin.

Chases, patters and skitters low across surface, leaving long trail of splashes, often in aggressive pursuits.

Swims high on water; head jerks to match thrusting of feet.

Juv.: pale grey beneath, throat whitish

Flattened, lobed toes give strong propulsion under water and on the surface.

# SWAMPHEN & BLACK-TAILED NATIVE-HEN

**Australasian Swamphen** *Porphyrio melanotus* 45–50 cm
Large, colourful, common waterhen. Uses diverse wetlands, typically swamps, well-vegetated lake and river margins, adjacent grassland, agricultural land, lawns, also estuarine wetland. Deep tones of plumage need full sunlight to bring out intensity of colour. Breast can appear dark slaty blue-grey in dull light, but direct sunlight transforms this to a bright, intense blue with azure sheen, or, on the SW race, reveals an emerald hue down the central breast. Male and female are alike; full colour is kept all year. Aggressive and bullying towards other waterbirds; kills ducklings. Has strength to pull up reeds as food. Clumsy, leg-dangling, crash-landing flight. Great variety of sounds: harsh, abrupt 'kak, kak' rising to sharp, grating 'kiark, ki-aark'; also querulous, grating 'qua-ark' and loud, harsh squawks of warning when with small chicks. At night often gives wild shrieks and boomings, perhaps basis of bunyip stories. Deep thudding sounds from beating wings against body. Widespread, common.

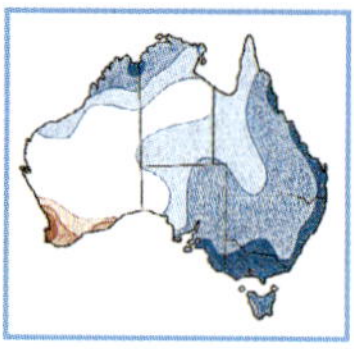

- ● *melanotus*
- ● *bellus*

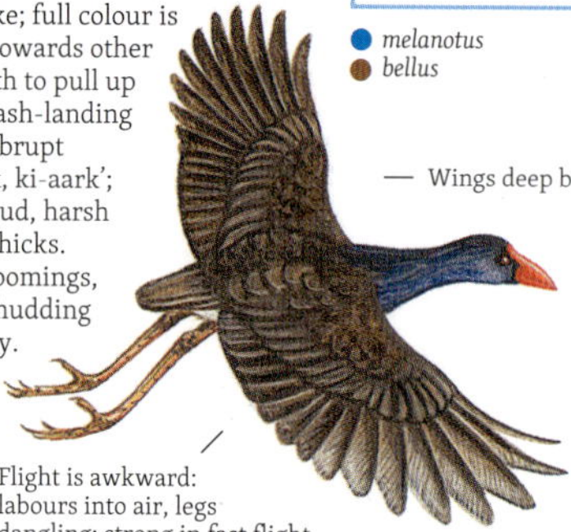

— Wings deep brown

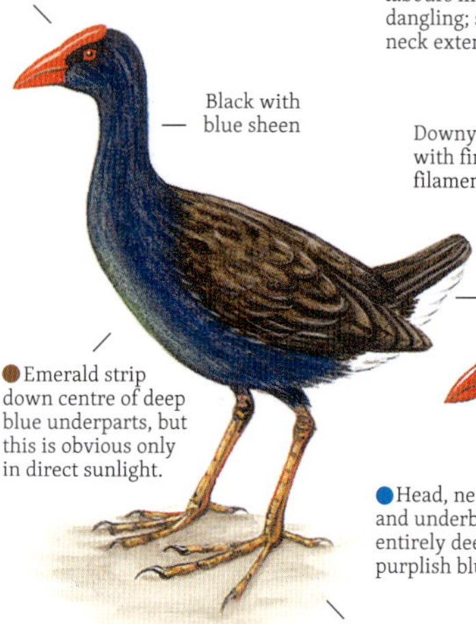

Massive scarlet bill and forehead shield

Flight is awkward: labours into air, legs dangling; strong in fast flight, neck extended, legs trailing.

Black with blue sheen

Rufous skin visible through sparse coarse filaments

Downy young: black with fine white filament tips

— White

● Emerald strip down centre of deep blue underparts, but this is obvious only in direct sunlight.

Dark brown

● Head, neck and underbody entirely deep purplish blue

Powerful legs, long toes, dull to quite bright orange-buff, occasionally greenish ochre

Many displays, including showing of white undertail in a 'wing-up, tail-up' posture, swimming or on land

**Black-tailed Native-hen** *Tribonyx ventralis* 32–38 cm
Highly adapted to an arid continent; responds to rain in semi-desert by breeding rapidly to large populations. On permanent or temporary, fresh or saline wetland of semi-arid regions; often around shallow claypan pools, saltbush surrounds of lakes. Large new flocks suddenly appear in vicinity of any pool or green patch of ground. They vanish as quickly as they arrived when drought returns. Runs fast. Flies with rapid, shallow wing beats, feet trailing. Usually silent, even in large flocks, but at times a sharp 'kak', perhaps in alarm, or a rapid, continuous 'kak-kak-ak-ak-'. Common; nomadic-irruptive.

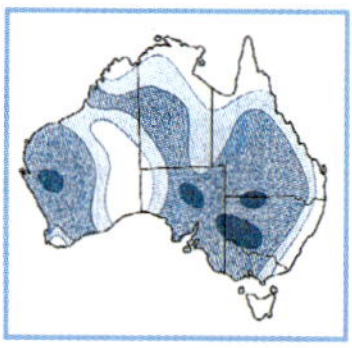

Wings entirely dark brown

Yellow eye has direct, penetrating stare.

Here in alert, tall, upright stance, but often has back horizontal in a more 'hen-like' shape.

Green shield above green and red bill

Deep blue-grey; long black feathers of flanks tipped white

The large, black-tipped tail is usually held steeply upwards.

Carmine-pink

Sexes alike; juvenile like adult; no seasonal plumage change

If disturbed, is quick to dash for cover, half running, half flying.

# TASMANIAN NATIVE-HEN & BUSTARD

**Tasmanian Native-hen** *Tribonyx mortierii* 42–50 cm
Large, heavily built native-hen, almost as large as the well-known Australasian Swamphen. One of only three flightless Australian birds, the others being the Emu and Cassowary. In and near swamps, marshes, lakes; feeds out across surrounding grassland, farmland. Usually in small family parties: breeding groups of 2 or 3 adults together with young in their first year or two. Typical breeding group: two males and one female. Gives drum-like grunts, probably contact calls, higher pitched in aggression or alarm. Males and female combine in see-sawing, harsh, rasping sounds, rising to crescendo in pitch and loudness. Endemic to Tas.; common; sedentary.

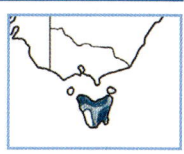

Species found only in Tas.; no races

Large, heavily built, very short wings, flightless

Olive-brown

Large, triangular, greenish yellow bill; bright red eye

Sparse black down is silvery tipped, looks grey.

Tail black; may be steeply upright, but often lower..

Blue-grey

White patch on flanks

Blue-black

Juvenile is like adult but paler brown above, and brownish grey rather than blue-grey underparts; bill grey; eye brown.

**Australian Bustard** *Ardeotis australis* 0.8–1.3 m; span to 2 m
Tall, stately bird of open inland plains. On grassland, spinifex, arid scrub with saltbush and bluebush, open dry woodland of mulga, mallee and heath. Stands neck upstretched, bill uplifted, freezes if an intruder approaches; initially walks away, eventually takes flight with deep, slow, powerful wing beats. Loud calls in breeding season, deep booming, rather like the roar of a distant lion, rising then falling. Closer, sounds include an abrupt, hoarse exhalation 'huhh!', often leading into a hoarse throaty growling – 'huhh!, huhh! -aa-a-r-r-rgh, aa a-r-r-rrrgh'. Female gives similar but higher calls. Nomadic; quite common away from heavily settled regions.

Black

Broad wings, widely fingered

Male has remarkable courtship display: foreneck and gular sac are inflated; long feathers of upper breast spread in fan shape; tail lifts and fans over the back; male begins strutting and booming. This display continues for long periods of time.

Chestnut, finely barred darker

Feet trail behind short tail.

♀

Female lacks black breast band.

Runs into wind for heavy take off; slow, powerful wing beat.

♂ Breast band

Juv.: adult face pattern but with heavier barring around neck

Black and white patterned wing-coverts

Juv.

Nest is a slight hollow on ground, often on a slightly elevated site.

# 8 BUTTON-QUAIL & MIGRATORY WADERS

This group has four families: button-quail in Turnicidae; the Plains-wanderer in Pedionomidae; sandpipers and allies in Scolopacidae; and the Painted Snipe in Rostratulidae. The last three families are part of Order Charadriiformes, members of which are so numerous that they are divided between chapters 8 and 9. This order is dominated by migratory wading birds in very plain nonbreeding plumage. While the main text lists all species, the chart below shows only one or two from each genus for size comparison.

Willie Wagtail = 20 cm

## turnicidae
### page 111

Red-backed Button-quail

Black-breasted Button-quail

## pedionomidae
### page 111

Plains-wanderer

## scolopacidae
### page 114

Lesser Yellowlegs

Asian Dowitcher

Marsh Sandpiper

Red Knot

Sanderling

Latham's Snipe

Wood Sandpiper

Pin-tailed Snipe

Terek Sandpiper

Stilt Sandpiper

Broad-billed Sandpiper

Black-tailed Godwit

Juv.

Common Sandpiper

Buff-breasted Sandpiper

Wandering Tattler

Ruff

Whimbrel

Upland Sandpiper

Ruddy Turnstone

## rostratulidae
### page 115

Australian Painted Snipe

## PLAINS-WANDERER; BLACK-BREASTED BUTTON-QUAIL

**Plains-wanderer** *Pedionomus torquatus* 17–18 cm 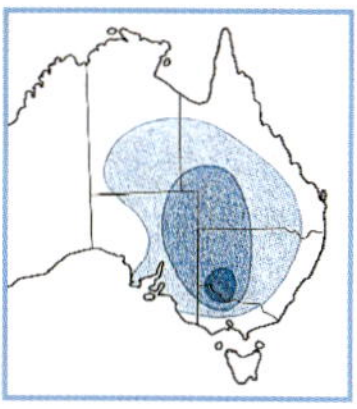
Distinctive quail-like ground bird, the sole member of its family, which is closer to the waders than to quail. Use of similar habitat seems to be reflected not only in appearance, but also in polyandrous breeding. Habitat is predominantly natural open grassland, treeless with patches of open ground; may be lightly grazed. Avoids country where grass is too tall or dense, or too sparse, low or heavily grazed. Infrequent reports from saltbush and similar low sparse shrubland, stubble or other low crops. Feeds on seeds on bare ground at base of grass clumps. A heavy sward of grass would conceal seeds and hinder escape from predators; on the other hand, extremely low, sparse or heavily grazed grass offers little concealment. Call is a repetitive, low, hollow 'coo', given only in spring, perhaps only by female. Described as dove-like, or, more prosaically, like the moo of a distant cow. Maintains contact with chicks using a soft 'tchuk!'. The spotted neck and wing pattern of the Plains-wanderer in flight is unlike any quail. Most common in the Riverina of NSW and in NW Vic. Elsewhere sparse, patchy occurrence; generally rare, vulnerable. Usually sedentary, may move or disperse as habitat changes occur, such as heavy grazing or growth of dense grass.

Outer wing pattern unlike any quail

Legs longer than tail; hang low at take-off and on landing approach.

Often adopts tall, upright stance showing thin neck.

Slender bill

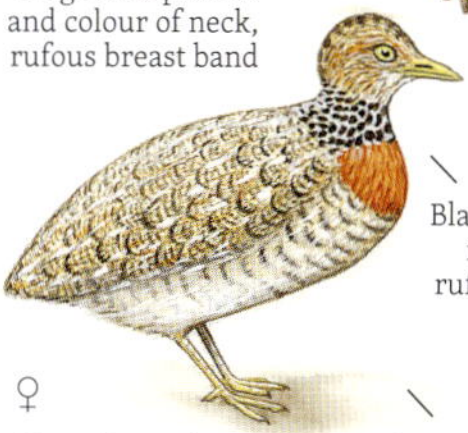

Diagnostic pattern and colour of neck, rufous breast band

Black-spotted neck band, rufous breast band

Legs are longer than those of quail.

Although quail are similar to this species, they lack the tall, narrow necked, angular headed stance and silhouette of the Plains-wanderer.

**Black-breasted Button-quail** *Turnix melanogaster* 17–19 cm 
Distinctive, with female's head black-hooded; breasts of both sexes are black, spotted white. Usually in pairs or small parties of about five. Squats and freezes, or runs when approached. Flies only if forced; takes off in a sudden whirring. Flight is short and fast; glides to landing. Tends to inhabit low closed-canopy rainforest or similar monsoon forest and vine thickets, and drier shrubby scrub such as hoop pine, brigalow, belah and bottletree thickets where there is a deep layer of leaf litter. Also in eucalypt forest such as spotted gum, especially where there is a dense understorey such as lantana with grass groundcover. Feeds on seeds and insects, scratching about in litter, leaving small, bare circular depressions. Call, from female, a low drumming or booming 'oo-ooom, oo-ooom', quickly repeated many times, rising and falling through the series, difficult to establish direction of calls. Also a quick 'ook' when disturbed. Sedentary; formerly common but populations reduced or eliminated by habitat destruction or alteration – now rare, vulnerable.

Black, white spots along brow; chestnut streak on crown and nape

Chestnut, streaked black and white

Heavily spotted white

Faintly barred

Black hooded head, yellow eye

Back, coverts and tail are rich chestnut heavily patterned with black, grey and white streaks; outer wing and secondaries contrasting blackish-brown.

Breast black, heavily spotted white each side; breast pattern of male is similar, but comparatively dull.

# BUTTON-QUAIL

**Red-backed Button-quail** *Turnix maculosa* 12–15 cm
An inhabitant of tussock grassland on black-soil plains and clay flats, grassy tropical woodland, rainforest margins, spinifex, crops, sedges. Usually not far from water, especially in breeding season. Typically in small coveys in dense vegetation; difficult to flush – sitting tight, and then exploding into fast flight, to drop again at a distance. Gives a tremulous 'oo, oo, oo, …' or 'whoo, whoo, whoo,' each note ending with a slight lift. Often a long sequence – begins very softly, resonant and musical, repeats many times, rises, strengthens, until finally quite loud, clear, rather vibrant. Most alike is the Red-chested Button-quail, but has lighter back and rufous collar. Common in parts of tropical N Aust.; scarce or vagrant in SE.

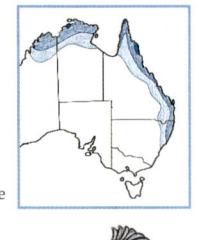

**Chestnut-backed Button-quail** *Turnix castanota* 15–18 cm
Usually on dry, elevated sandstone terrain – ridges, escarpments, plateaus with eucalypt woodland and open forest, groundcover of spinifex, sparse grass. If alarmed, runs with head high. Initially difficult to flush, then whole group bursts up, but drops back to cover very quickly. Has darker rufous undertone to the upper parts than Buff-breasted, and has more black markings, which sometimes extend over mantle. Can have fine black barring on the rump. Voice is a low, moaning 'oom', perhaps a contact call when a group becomes separated. Sedentary; population uncertain.

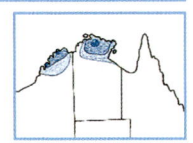

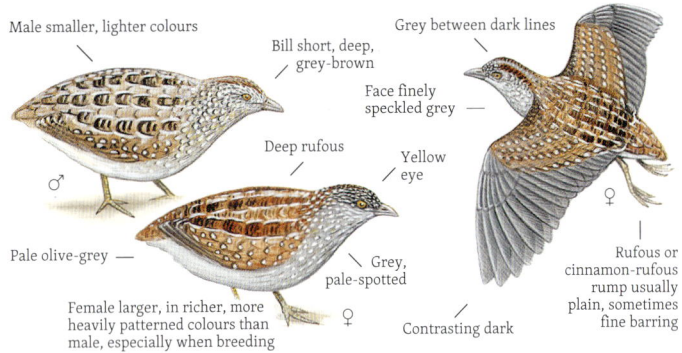

**Buff-breasted Button-quail** *Turnix olivii* 18–22 cm
Open tropical woodland with sparse tussocky grass. Can remain hidden even in typically sparsely covered habitat; difficult to sight. Prefers to walk or run from danger. Finally flushes when hard pressed, rises steadily on whirring wings rather than exploding upwards; usually travels 50–100 m before dropping to cover. Female gives a deep, slow booming, begins softly, a long sequence with increasing intensity and frequency. Male gives a softly whistled 'chiew-chiew'. Rare; locally and seasonally nomadic.

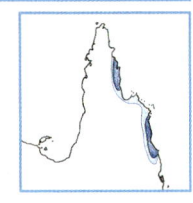

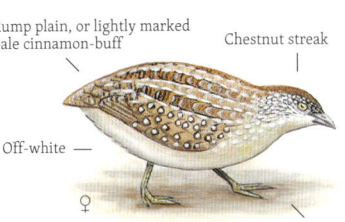

## BUTTON-QUAIL

**Little Button-quail** *Turnix velox* 13–16 cm
Inhabitant of grassland and open woodland, including mallee, mulga, tussock grassland, spinifex, crops. A small, rufous and cinnamon, dark-winged quail with short, deep grey bill. Occurs in small coveys or pairs. When disturbed, squats or scuttles through grass, or rises on whirring wings – keeps low, often turns to show white flanks before it drops into distant cover. Soft, high, resonant, musical 'whoo, whoo, whoo…'. Squeaky chatter when flushed. In flight, differs from other small quail: the cinnamon-rufous panel of secondary coverts contrasts strongly with the dark grey outer wing and secondaries. Common; population density varies locally; irruptive, nomadic.

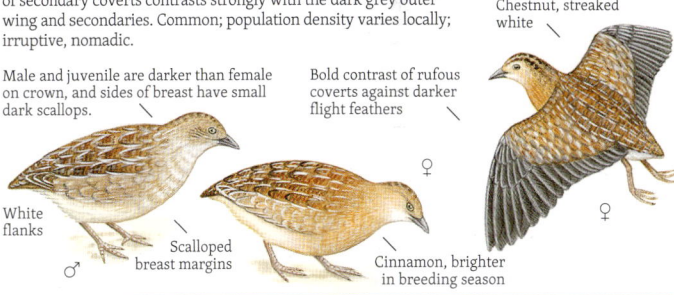

Chestnut, streaked white

Male and juvenile are darker than female on crown, and sides of breast have small dark scallops.

Bold contrast of rufous coverts against darker flight feathers

White flanks

♂

Scalloped breast margins

♀

Cinnamon, brighter in breeding season

♀

**Red-chested Button-quail** *Turnix pyrrhothorax* 12–16 cm
Abundant in good seasons, then, when conditions deteriorate, forced to disperse widely, often to greener coastal regions. Some birds stay all year, others are seasonal or vagrant. Occupies grassland on black-soil plains and other flat, heavy soil country, grassy woodland, rainforest margins, spinifex, crops; in breeding season, grassland and sedgeland near water. Occurs in pairs or small coveys. When approached, tends to squat rather than fly; eventually explodes from cover; flies fast, low and far before dropping to ground and running. Female apparently chooses nest site and initiates construction. Male probably does most of the work. Call is a soft, quite high, booming 'oom, oom' at one second intervals; notes slightly slurred and rising through a sequence of 20-30 calls. Common; populations fluctuate greatly with varying climatic conditions; sedentary or nomadic.

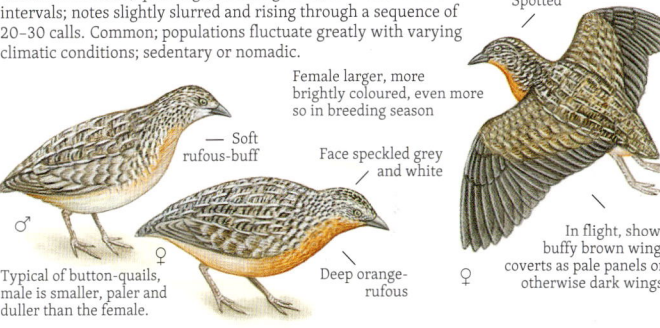

Spotted

Female larger, more brightly coloured, even more so in breeding season

Soft rufous-buff

Face speckled grey and white

♂

Typical of button-quails, male is smaller, paler and duller than the female.

♀

Deep orange-rufous

In flight, shows buffy brown wing-coverts as pale panels on otherwise dark wings.

♀

**Painted Button-quail** *Turnix varius* 17–23 cm
Habitat includes open forest and woodland, banksia woodland, mulga and brigalow, mallee. Prefers stony ridges, abundant leaf litter but sparse grass. Often in small coveys, pairs or alone. Usually walks away when approached; then runs with head high, or squats and freezes. Will finally explode into flight. Like many button-quails, is active day or night; forages among litter for seeds and insects. Call is a long series, up to 30 seconds, of even pitched booms from the female, with notes more rapid towards the end. Widespread and common, but habitat reduced.

● *varius*
● *scintillans*

Black, grey streaked, spotted white

Grey streak

Deep bright chestnut

Eye red

♂

Shoulders bright chestnut

● Confined to islands of the Houtman Abrolhos, mid W coast of WA.; a distinctly smaller race.

In flight, rump and tail appear rufous grey.

● ♀ Paler, cinnamon rather than rufous, finer black markings

When flushed, bursts up with whirr of wings; darts away; keeps low, weaving course; drops to cover again at distance; then runs.

## LATHAM'S & PIN-TAILED SNIPE

**Latham's Snipe** *Gallinago hardwickii* 27–30 cm

As a group, the three species of *Gallinago* snipe are easily recognisable, having extremely long straight bills and a distinctive high head with a steep forehead, flattened crown and eyes set further up and back than usual to give a wide field of vision (probably almost as much to the rear and above as to the front). While these snipe differ distinctly from other birds, differences among the three species are so slight that positive identification is usually based on details like tail structure and requires an in-hand specimen. Typical habitat is low vegetation around wetland and in shallows: sedges, reeds, heath, salt marsh, irrigated crops. Latham's Snipe is the largest; wings and tail appear even longer proportionately, giving a tapering, pointed rear end. Snipe are active around dusk and dawn, remaining concealed through the day, often seen only if flushed. They burst up to fly a fast zigzag course and then drop to cover again. Call is a harsh, rasping 'kzek-kzek-kzek', usually at take-off and in flight. Latham's Snipe is a regular summer migrant; NE to SE Aust.; rare visitor to SW of WA. Locally common in optimum habitat.

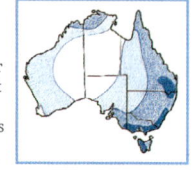

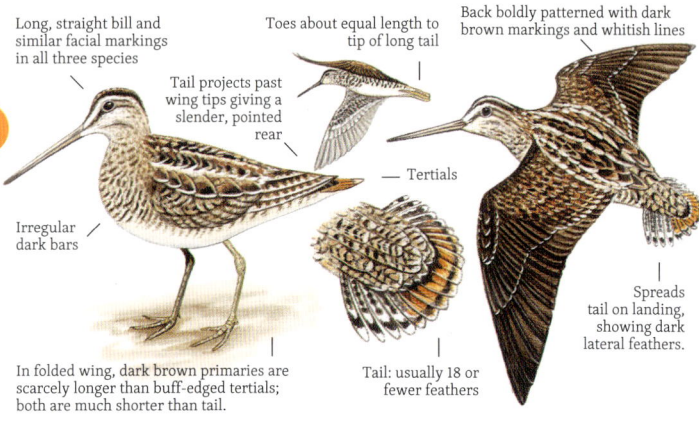

Long, straight bill and similar facial markings in all three species

Toes about equal length to tip of long tail

Back boldly patterned with dark brown markings and whitish lines

Tail projects past wing tips giving a slender, pointed rear

Tertials

Irregular dark bars

Spreads tail on landing, showing dark lateral feathers.

In folded wing, dark brown primaries are scarcely longer than buff-edged tertials; both are much shorter than tail.

Tail: usually 18 or fewer feathers

**Pin-tailed Snipe** *Gallinago stenura* 25–27 cm

The few recorded habitats in Australia have been coastal freshwater wetland – swamps, river pools, sewage ponds, usually with grass. Has a squat rear end in comparison with the other two species, whose longer, projecting tails give a single finely tapered point. Wing tips, tertials, tail tip do not usually fall neatly together as one point, but spread apart to give a truncated look to the rear end. Usually sighted when flushed from cover unexpectedly. Explodes into flight, usually fast, low, direct. May drop quickly to cover, or circle higher overhead before descending steeply to the ground. Calls when put to flight, a rather high, startled sound, unlike usual flight calls, which tend to be more throaty or nasal: 'tchaa' or 'tchet'. Uncommon; probably a regular winter visitor.

Toes extend behind tail.

Smallest of the three species recorded for Aust.: has shorter, rounded wings and stumpy tail.

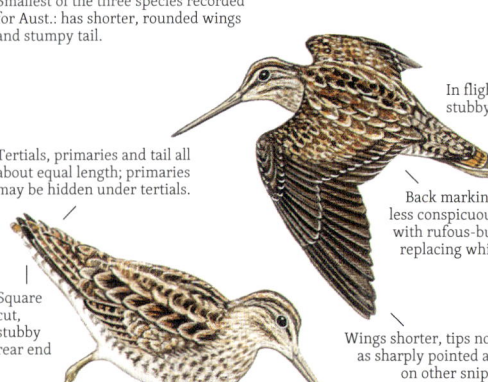

Tertials, primaries and tail all about equal length; primaries may be hidden under tertials.

In flight looks stubby tailed.

Back markings less conspicuous, with rufous-buff replacing white

Square cut, stubby rear end

Tail: usually 26; varies 24 to 28 rectrices with outer 6 to 8 pairs fine, pin-like.

Wings shorter, tips not as sharply pointed as on other snipe

Dark speckled rather than barred

Feet trail well behind short tail; those of Swinhoe's do not extend as far.

## SWINHOE'S & PAINTED SNIPE

### Swinhoe's Snipe *Gallinago megala* 27–29 cm

Extremely difficult to separate from other species, especially the Pin-tailed. Over most of Aust., only one species has so far been found, which makes it easier for observers; however, the overlap will probably increase with further observations; almost any sighting could be of any of the three species. In Aust., recorded near billabongs, swamps, flooded grassland, sewage ponds, claypans. When flushed, Swinhoe's dashes away in characteristic zigzag flight; sometimes keeps low, but may circle high overhead before diving into low, dense vegetation. All these snipe use their unusual tails in display flights over their N hemisphere breeding grounds. The narrow outer feathers are strong and stiff, and, when the tail is fully fanned, stand out at right angles from each side of the body. In display the bird dives, wings half folded, so that air blasts across the series of slots between the stiff feathers to make various unusual sounds – whistling, humming, winnowing – perhaps supplemented by vocal sounds. Such display is most unlikely ever to be seen in Aust. Call, when flushed, is an abrupt, rasping 'skaik!'; has other calls in display. Scattered records in W and NW coastal Aust.; regular visitor to Kimberley region; rare vagrant to SW Aust.

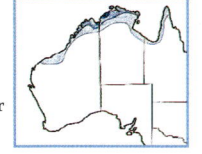

In folded wing, dark brown primaries extend beyond buff-edged tertials; bright rufous tail projects even further.

Buff-edged tertials

Dark stripes

Very long, straight bill

Feet trail behind tail tip.

With tail longer than folded wings, Swinhoe's has a rear end that tapers to a fine point rather than being stubby.

Tail feathers usually total 20; vary 18 to 26 with outer 6 on each side narrow, but not fine pins like those of Pin-tailed Snipe.

### Australian Painted Snipe *Rostratula australis* 23–26 cm

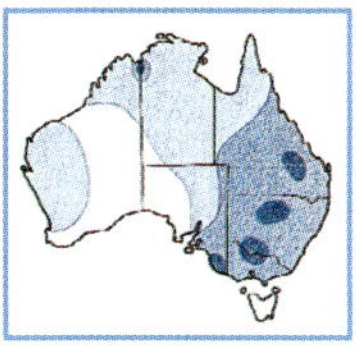

Bold and colourful plumage, but extremely secretive, keeping to dense vegetation of swamps, emerging only in subdued light of dawn and dusk. Prefers the surrounds and shallows of wetlands that are well vegetated with dense low cover. Silent except when breeding. Freezes when approached. Finally, with intruder very close, will flush, but flies only a short distance. Airborne, has broad, rounded wings and slow, erratic wing beats, legs initially dangling. As with both jacanas and snipe, the female is larger and more vivid than the male, and the roles are reversed. Calls in breeding season, the 'advertisement' calls of female, typically around dusk, are a long series of 'kot, kot, kot…' and soft, resonant 'whoo' sounds. When flushed, gives an abrupt, harsh 'krek!' Breeding resident of inland SE Aust.; uncommon

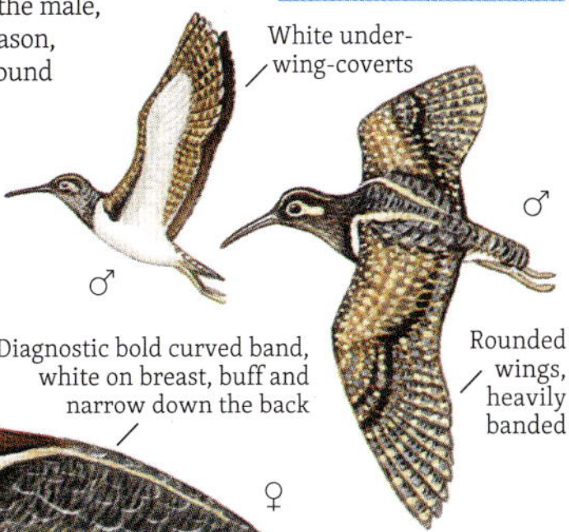

White under-wing-coverts

♂

Distinctive white eye patch

Diagnostic bold curved band, white on breast, buff and narrow down the back

Rounded wings, heavily banded

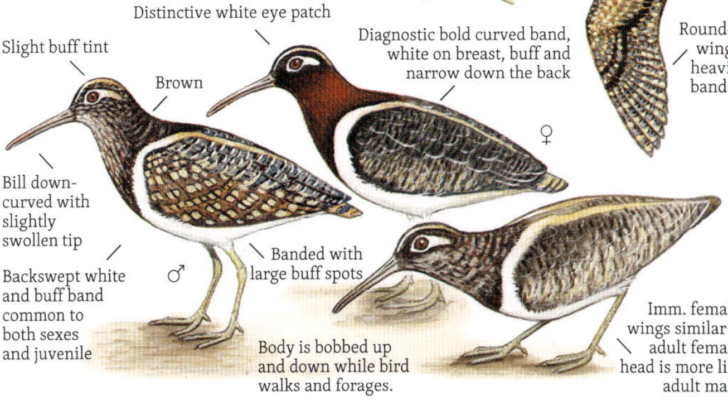

Slight buff tint

Brown

♀

Bill down-curved with slightly swollen tip

Backswept white and buff band common to both sexes and juvenile

♂

Banded with large buff spots

Body is bobbed up and down while bird walks and forages.

Imm. female: wings similar to adult female; head is more like adult male.

# DOWITCHERS & HUDSONIAN GODWIT

**Short-billed Dowitcher** *Limnodromus griseus* 25–30 cm
Dowitchers are migratory shorebirds with long, almost straight, snipe-like bills with a slightly bulbous, sensitive tip. With these they probe deeply into soft mud with rapid vertical vibrations, seeking small invertebrates. Habitat, in Aust., has been lagoons, pools with soft mud, moist grass and semi-aquatic vegetation. Call is usually a mellow 'tsew-tsew-'. Similar species are the Asian Dowitcher, larger and with longer, slightly downcurved bill, and the Long-billed Dowitcher, but this has not yet been recorded in Aust. The Short-billed was recorded, in adult nonbreeding plumage, at the Corner Inlet, Vic., 1995, and was initially identified as being the similar Long-billed Dowitcher.

Darker underwing than Asian Dowitcher, finely barred, looks grey.

Nonbr.
Faintly streaked or mottled
Bill is long, but slightly shorter than other dowitchers; straight.
Like nonbr.: grey with white brow
Greenish yellow
Faintly barred
Juv.
Back slightly rufous, underparts tinted buff
Faintly speckled white triangle
Narrow white margins
Nonbr.

**Asian Dowitcher** *Limnodromus semipalmatus* 33–36 cm
Distinctive very long bill used in rapid-action probing of mud or sand as bird moves forward. Uses beaches, mudflats, sewage ponds, saltfields. In breeding plumage, has extensive chestnut on head, neck, breast; back feathers are darker, edged orange. Nonbreeding has dark-capped crown, white brow line, dark-grey back with white margins, and white breast mottled grey-brown. Gives a yelping 'chiewp' and softer 'kriow'. Most alike is the Bar-tailed Godwit. A regular visitor in small numbers to NW Aust.; elsewhere a rare vagrant.

Bright chestnut
Bill extremely long; has swollen, slightly drooped tip.
Dark grey-brown
White brow
White of barred tail and faintly barred rump extends in a triangle between dark wings.
Deep chestnut and dark brown, cinnamon margins
Br.
Always barred
Heavily streaked
White
Barred flanks
Nonbr.
Spotted or mottled
Juv.
Buff tint
Wing and tail tips equal length
All plumages display a pale wedge between dark wings.

**Hudsonian Godwit** *Limosa haemastica* 38–41 cm
Habitat, in Aust., coastal lagoons, estuaries, freshwater lakes, salt ponds. Usually silent away from the N American breeding grounds, except for the alarm call, a sharp 'weit-weit-weit', higher pitched than call of the Bar-tailed Godwit. Very few accepted records – SA and NSW, usually June–Sept. Accidental or rare vagrant; NSW 1982, SA 1986.

Extremely rare vagrant; shorter legs, overall darker in all plumages than the more common godwits
Grey head, neck; pale face, white brow
Long white central bar
Long bar of black coverts, widest at wing pits
Slightly upturned
Feet trail.
♂ Br.
Barred black and rufous
Chestnut, barred black; female has white as well as black bars.
Nonbr.
Legs shorter than Bar- and Black-tailed
White bar, wider on outer wing
White rump, black tail

# BAR-TAILED & BLACK-TAILED GODWITS

**Bar-tailed Godwit** *Limosa lapponica* 37–39 cm
This godwit is one of the most common Australian migratory waders; occurs in huge numbers at wintering grounds, with peak counts as high as 60 000 at Broome where waders gather to rest and feed. Habitat in Aust. includes coastal mudflats, sandbars, shores of estuaries, salt marsh and sewage ponds. In flocks, have calls of contact and alarm. The former is a sharp 'kak' or 'kerk'; the alarm 'kirrik' or 'kirrark'. Most likely to be confused with Whimbrel and Black-tailed Godwit. Common around coastline; scattered records over much of interior.

Although in drab colours for most of their stay in Aust., the colourful breeding plumage is evident on many birds as they move through northern departure points in March and April. Mottled grey-browns of nonbreeding begin to give way to rich cinnamon and rufous tones, gradually enveloping most of the plumage.

**Black-tailed Godwit** *Limosa limosa* 36–44 cm
Compared with Bar-tailed, smaller and more slender. Longer neck, slimmer body and longer legs that trail substantially behind the tail tip combine to give a more slender, delicate jiz. Bill is straighter, very slightly bulbous tip. Habitat usually coastal: estuaries, sheltered bays, lagoons with extensive tidal mudflats or sand bars, shores and islets of large, ephemeral inland lakes; infrequently on rocky coasts, islets, sewage farms. Quite vocal with a tuneful song at its northern hemisphere breeding grounds, but in Aust. gives only contact and alarm calls. Some soft chattering sounds when feeding and some calling from flocks when flushed. Sounds recorded include sharp 'witta-wit', a rather harsh, strident 'wieka-wiek-wieka', and soft 'kek' or 'kuk'. Similar are Hudsonian Godwit, but with black underwing-coverts; Bar-tailed Godwit (barred underwing, tail). Common; most abundant on N coast E of Darwin.

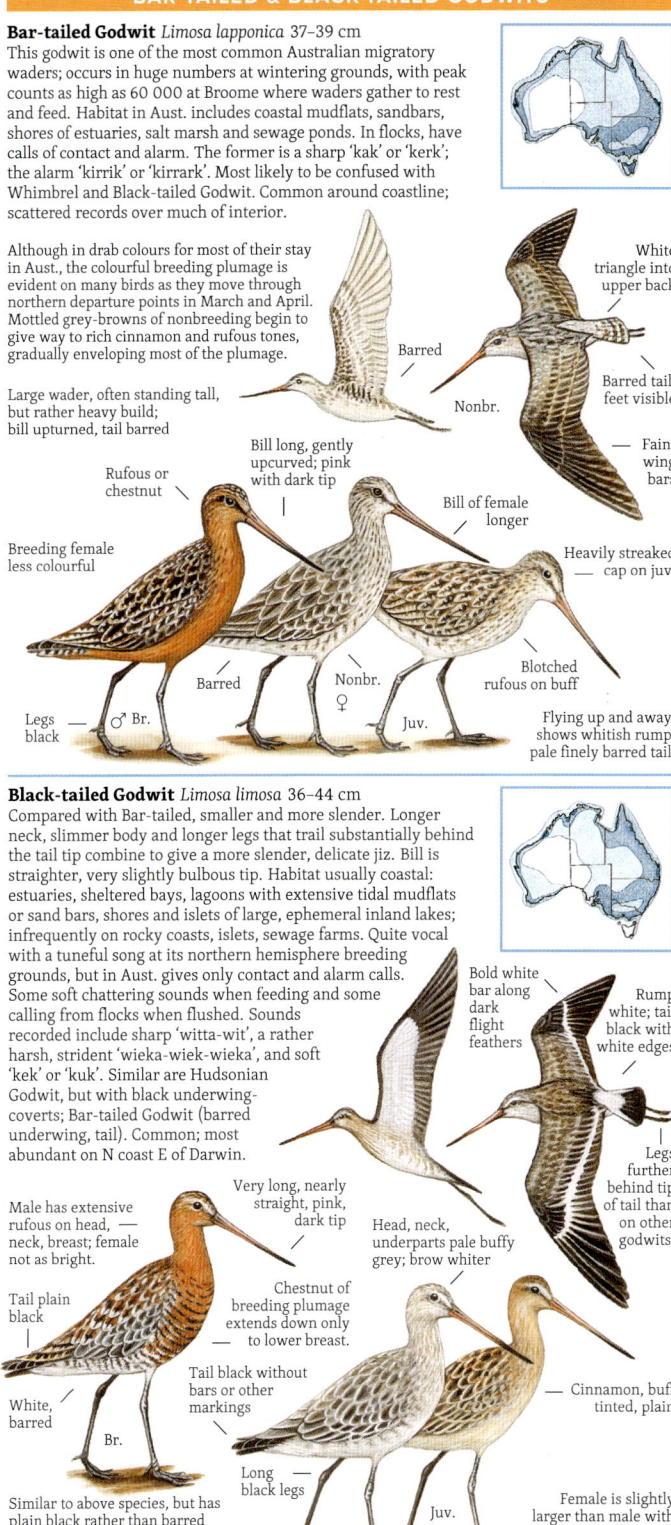

117

## LITTLE CURLEW & WHIMBREL

**Little Curlew** *Numenius minutus* 30–36 cm 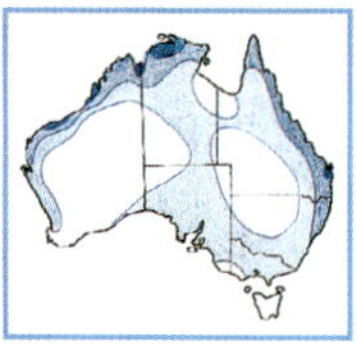
Flocks of hundreds, at times thousands, of these birds may be seen on the extensive swamps and billabongs of the coastal black-soil plains in N Aust. Habitat includes dry grassland of clay and black soil plains, river floodplain, woodland with grassy understorey, margins of billabongs and freshwater swamps, also similar artificial environs – pasture, airfields, sports fields, lawns. Often forages over recently burnt grassland or open woodland. Stands erect, alert; forages busily, actively probing at ground, usually in groups. Flies with easy, relaxed wing action; shows brownish rump, whitish shafts to outer primaries, but no wing bars; holds wings up briefly on landing. Tips of toes just visible beyond tail tip. Call usually three notes in flight – sharp, rising 'kee-kee-kee', like, but without the clarity of the Greenshank's call. In alarm, husky, high 'tchiew-tchiew-tchiew' and rasping 'kwiekek'. Abundant across coastal N; scattered through inland to southern regions; vagrant to Tas.

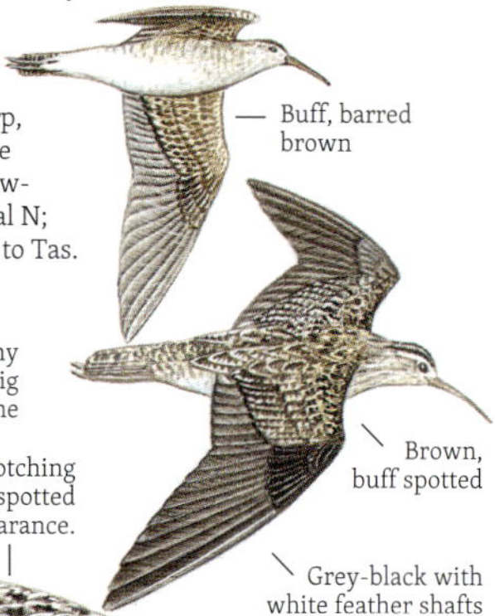

- Buff, barred brown
- Head looks large on skinny neck. Peaked dark crown, big eyes, broad, pale buff brow line
- Brown, buff spotted
- Lower base pink; slight downcurve towards bill tip
- Long thin neck
- Heavier notching gives spotted appearance.
- Grey-black with white feather shafts
- Streaked buff and brown
- Overall warm buff tones
- In flight, toes usually protrude slightly past tip of tail.
- Br.
- Long, slender, tapering shape
- Juv.
- Greyish buff

**Whimbrel** *Numenius phaeopus* 39–44 cm 
Habitat includes mudflats of estuaries, lagoons, preferably with mangroves; less often sandy beaches, reefs, salt lakes. Gregarious, feeds in small flocks or mixed with other waders on coastal mangrove and estuary mudflats, beaches and reefs; takes molluscs and crustaceans. Feeds with energetic dashing about – short runs, rapid jabbing of bill – rather than deep probing. Wing beats seem more rapid than those of curlews. In Aust., call is a rather musical, even pitched 'ti-ti-ti-', but so rapid that the notes run together in almost unbroken tittering, 'ti-titititititi...', usually given in flight. Also a high, clear 'keer-keer-keer' and slow, tremulous 'ke-ee-r'. Common along N Aust. coast; uncommon to rare further S.

- 🟡 *variegatus*: Siberian race with white to light brown barred wedge from rump to back
- 🟢 *hudsonicus*: North American race with dark rump. Very few of this race reach Aust., but either race may occur in any part of the Aust. range of this species.

- A medium-sized curlew with twin dark streaks along the crown and mid length bill.
- Barred
- 🟡 In flight, shows a white wedge from rump to lower back; underwings are entirely barred, brown on buff.
- Underwing is entirely buff, barred brown.
- Dark crown divided by pale central line
- 🟢 In flight, dark brown rump, tail-coverts and lower back are similar to surrounding plumage.
- Bill down-curved; pink lower base
- Wings reach tail tip.
- Juv.: upper parts darker with larger white spots and margins
- Streaked brown
- Plumage darkens as white tips wear away to leave buff portion.
- Barred tail
- Broad wavy barring on flanks
- Buff tinted, lightly streaked
- Rapid jabbing into mud

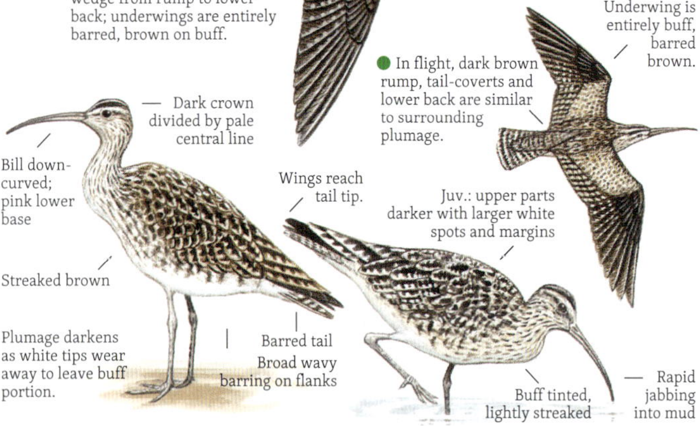

## FAR EASTERN & EURASIAN CURLEWS

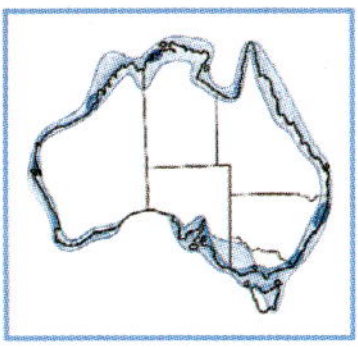

**Far Eastern Curlew** *Numenius madagascariensis* 60–65 cm
Largest wader in Aust. with very long, strongly downcurved bill. In small groups or alone, or sometimes very large flocks; much more wary and quick to take flight than other waders. Most need a short take-off run; wings beat slowly compared with the rapid action typical of most waders. Uses tidal mudflats, sand spits of estuaries, mangroves, lake shores, ocean beaches; feeds by probing deeply into mud or sand. Calls are haunting, melancholic yet melodious and beautiful sounds. Given in flight or from ground, a high, drawn out, two-part call, the second lifting higher, attenuated, 'coor-lee' or 'cur-eek'; also has a rapid, softer, melodious 'curee-cree-cureecuree', and a more strident version as alarm. Most alike are the Eurasian Curlew, which has white wedge from rump to back and white underwings, and Whimbrel, which is much smaller with short bill, usually a white rump. Common migrant to N, NE and SE, Tas.; occasional to SW WA and SA.

**Eurasian Curlew** *Numenius arquata* 50–60 cm
Extremely rare vagrant or accidental visitor; probably birds that travelled too far S or joined with Far Eastern Curlews that travel through to Aust. In flight, wing beats are slow, deliberate, rather gull-like, similar to action of Far Eastern Curlew. Habitat is estuarine mudflats, beaches. Uses the long downcurved bill to probe deep into mud of tidal flats on estuaries; seeks crustaceans, molluscs, worms. Birds from north-central Russian breeding grounds travel to locations as far apart as southern Africa, Japan, Malaysia, Borneo and the Philippines. Call is a loud, carrying, ringing 'cour-liu, cour-liu', and, in alarm, 'tiu-yiu-yiu-yiu-yiu'. Similar are the Far Eastern Curlew, which has longer bill, dark rump and back, barred underwing, and the Whimbrel, which is much smaller, shorter bill. Only two reports for Aust. – these not accepted as certain – near Darwin, 1948, coast near Perth, 1969, race unknown.

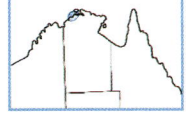

● *arquata*: Europe to Urals
● *orientalis*: from the Urals through central Russia, also intermediate forms. Vagrants, either race possible

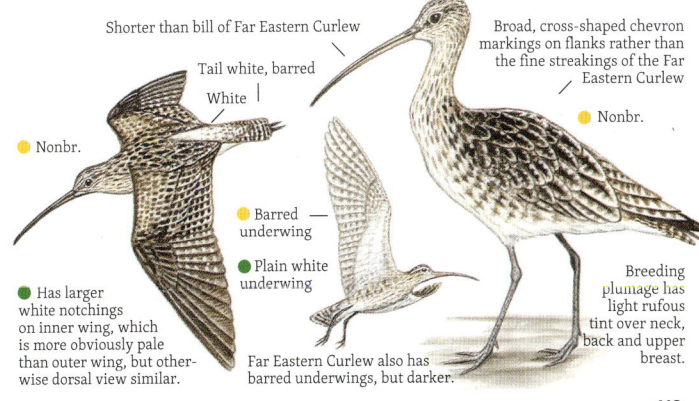

## REDSHANKS

**Common Redshank** *Tringa totanus* 27–29 cm
Often intermixed with other waders; forages briskly; bobs head and body when suspicious; always alert and quick to give alarm calls; if alarmed, departs with strong but erratic flight. Habitat is coastal wetland, including estuaries and lagoons where there are open areas of shallows with mudflats and sandbars. Noisy any time of year. In flight, usually two rapid, high, ringing notes, a third extremely abrupt: 'kier-kier-kp, kier-kier-kp, kier-kier-kp', the three-note sequence repeated several times. Alarm call on the ground is a piercing 'kieer'. Uncommon but probably regular summer visitor to scattered sites around coast.

Many slightly differing races and colour morphs. Those visiting Aust. probably race *ussuriensis*, quite dark upper parts with cinnamon and black markings

**Spotted Redshank** *Tringa erythropus* 29–32 cm
A tall, elegant wader; long-necked with long, slender legs. Preferred habitat of salt marshes, shallow freshwater swamps and lagoons. Shy and wary, it frequently bobs head and body. Forages by delicately probing mud, sometimes working slightly deeper water where it swims and up-ends to reach bottom. Flight is fast and direct; on longer travels may tuck legs up into plumage. The few seen in Aust. have been alone, but in regular wintering regions occurs in small groups to large flocks. Call is a distinctive, flute-like 'tchuet'. Similar is the Common Redshank, which has a wide white panel along the upper wing's trailing edge and a straight-tipped bill. This is an extremely rare vagrant to Aust. with few accepted records of sightings, all from Oct. to Apr.; NW and Kimberley, WA, Top End, NT, Hunter R., NSW, Seaford, Vic.

## YELLOWLEGS

### Lesser Yellowlegs *Tringa flavipes* 23–25 cm

Tall, graceful stance and movement on long legs; frequents margins of muddy wetland, marshy edges of swamps, mudflats. The few records in Aust. are of single birds, but in more popular wintering grounds it gathers in flocks, sometimes of hundreds of birds. Moves about mudflats and shallows with brisk movements, jabbing and probing the surface. Shares the wary, nervous bobbing of head and tail typical of the *Tringa* species; noisy, excitable. If put to flight, has easy, relaxed, rather slow wing action, but can travel fast. Legs trail by almost the full length of the foot. Usual flight call a whistled 'tiew-tiew-', 'ti-' lifting, final 'ew' dropping in pitch; louder in alarm. On ground, loud 'tiew', each note of long series accompanied by bobbing of head and tail. Similar are the Wood Sandpiper, especially juvenile, and Marsh Sandpiper. Rare vagrant, few sightings.

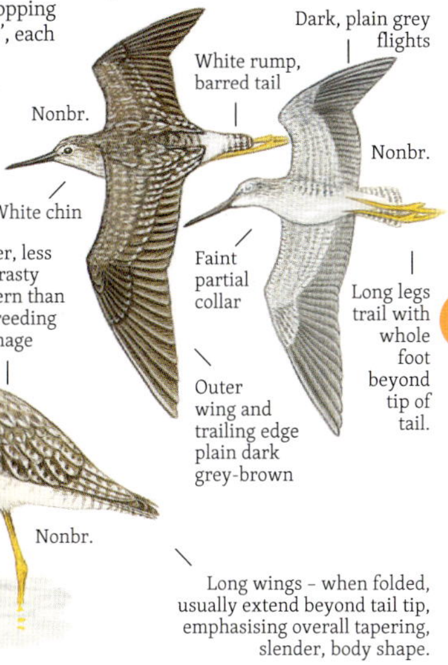

Dark, plain grey flights
White rump, barred tail
Nonbr.
Nonbr.
Heavily streaked black; white eye-ring
Bill almost straight; slender, only slightly longer than head
White chin
Softer, less contrasty pattern than in breeding plumage
Faint partial collar
Long legs trail with whole foot beyond tip of tail.
White
Outer wing and trailing edge plain dark grey-brown
Breast streaked, flanks scalloped
Nonbr.
Long, deep yellow legs; retain colour throughout the year.
Br.
Upper parts dark grey-brown with black and white spots and margins
Long wings – when folded, usually extend beyond tail tip, emphasising overall tapering, slender, body shape.

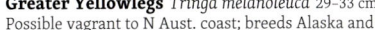

### Greater Yellowlegs *Tringa melanoleuca* 29–33 cm

Possible vagrant to N Aust. coast; breeds Alaska and Canada, winters S of North America and to southernmost coasts of South America. Vagrant to Europe; records from South Africa, Hawaii, possibly Cook Islands. Usual habitat coastal mudflats, salt marsh; possibly inland on fresh or brackish muddy lakes. Tall, elegant wader with bright yellow or orange legs; bill slightly upcurved and more than 1.5 times the length of head from base of bill through to nape. At usual wintering grounds occurs in small flocks or singly, probing mud, and, like Greenshank, wading quite deeply; often the first species in mixed gatherings to give alarm calls and take flight. Calls almost identical to Greenshank, perhaps higher pitched and faster, a rapid 'kier-kier-kier, kier-ker-kier-'. Not yet a substantiated record for Aust., but remains a possibility.

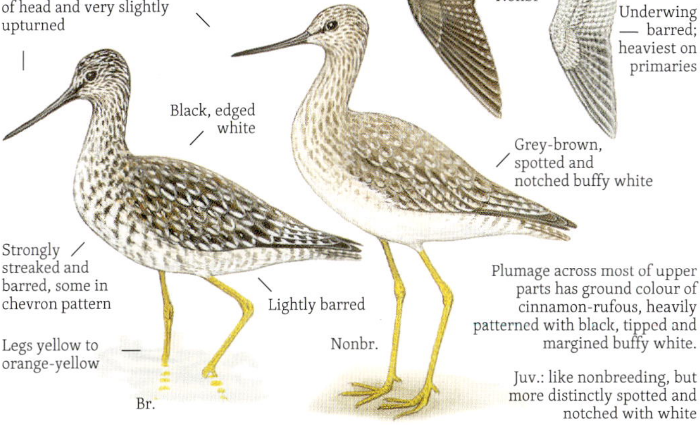

White rump, barred tail
Legs trail by length of foot.
Nonbr.
Nonbr
Underwing barred; heaviest on primaries
Bill more than 1.5 times length of head and very slightly upturned
Black, edged white
Grey-brown, spotted and notched buffy white
Strongly streaked and barred, some in chevron pattern
Lightly barred
Plumage across most of upper parts has ground colour of cinnamon-rufous, heavily patterned with black, tipped and margined buffy white.
Legs yellow to orange-yellow
Nonbr.
Br.
Juv.: like nonbreeding, but more distinctly spotted and notched with white

# GREENSHANKS

**Common Greenshank** *Tringa nebularia* 30–35 cm
Stands tall, erect; bobs head in alarm; leaps into flight with ringing calls; is very timid and wary. Occurs alone, or in small or large flocks, often with other waders. Active and excitable; forages briskly with sudden dashes, erratic changes in direction. Habitat is diverse, both inland and coastal. Found inland on both permanent and temporary wetland – billabongs, swamps, lakes, floodplains, sewage farms and saltworks ponds, flooded irrigated crops. On the coast, uses sheltered estuaries and bays with extensive mudflats, mangrove swamps, muddy shallows of harbours and lagoons, occasionally rocky tidal ledges. Generally prefers wet and flooded mud and clay to sand. The loud, ringing alarm call is one of the most familiar sounds in wader habitat – a clear, rapid 'tiew-tiew-tiew' with a slight drop in pitch on the last note, usually three quick, evenly spaced notes that together last less than a second. Notes in a series may range from one to six with a pause of a second before the next set. A widespread, common migrant Sep.–Apr.; a few remain through winter.

Obvious white rump and back
Feet project slightly past tail tip.
Dark
Barred, pale
Dark grey-brown, plain
Heavily streaked
Long, slightly upturned
Darker than nonbreeding
Chestnut, heavily marked with black central streaks: most feathers are black centred with buffy white margins and notches.
Bold black chevrons
Juv.
Darker streaking than on nonbreeding
Long, greenish yellow to grey-green
Folded wings extend slightly beyond the tail tip.
Grey-brown, edged white, notched black
Dark
Br.
Nonbr.
Pale streaking extends down to sides of breast.

**Nordmann's Greenshank** *Tringa guttifer* 30–33 cm
Not only rare in Aust., where it remains but a possible visitor, but also uncommon: has very restricted breeding grounds on Sakhalin Is., NE Japan. Migration follows a long SW path via Japan, Korea, Hong Kong, Malaysia, then NW to Burma and India's E coast. Any that may stray further S to Aust. would be easily overlooked as its appearance is so like the Common Greenshank. Unique is the slight webbing between the toes, but this identifier requires a bird in hand. Usual habitat is tidal mudflats of estuaries and coastal lagoons. Calls are distinctly different from the Common Greenshank; a less musical, but rather loud, sharp, piercing 'kiyiew'. A rare possible vagrant; uncertain sight record in coastal NT.

Light grey-brown
Nonbr.
Short legs only just reach tip of tail.
Yellow, thick based
Lightly streaked
Greenish or olive-yellow
Slightly smaller than Common Greenshank. Legs quite short
Nonbr.
Juv.
Very slightly upcurved

# SANDPIPERS

**Marsh Sandpiper** *Tringa stagnatilis* 22–26 cm
Tall, elegant, long-necked and very long-legged sandpiper, like a greenshank but much smaller, more slender. Also distinctive in flight; long legs trail so that all toes are completely behind the tail tip, and the deep white wedge that extends from rump far up the back is clearly visible. Habitat is coastal and inland wetlands, salt or fresh; typically estuarine and mangrove mudflats, beaches, shallows of swamps, lakes, billabongs, temporary flood waters, sewage farms and saltworks ponds. Occurs alone, in small groups or with other wader species. Forages in shallows; has rather upright stance, often with quick dashes to take some small creature; wades out to depth limit of long legs, searching and feeling for prey; probes mud and around marshy vegetation. Usual call is a quick, soft, mellow, musical 'kier' or 'teoo'. Alarmed, excited or put to flight, gives short, very high, thin, rather metallic and rapid 'kier, kier-kp, kier-kp, kier-kp, kier-kier', becoming a confused mass of chittering from even small flocks. Most alike is the Greenshank, with larger, heavier, upturned bill and relatively shorter and thicker legs. A regular summer migrant to Aust., Aug.–May. Quite common across far N; more scattered around other coastal parts; sparse through inland regions.

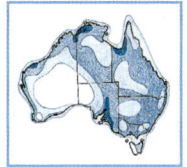

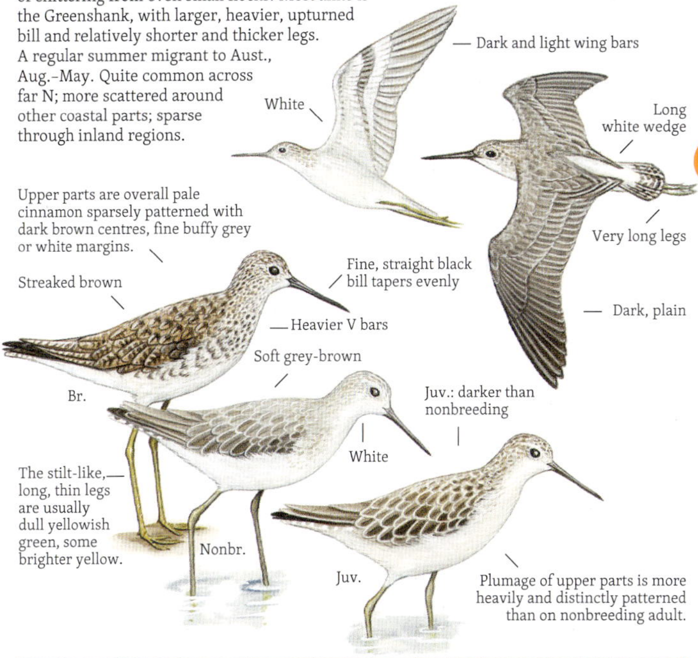

Upper parts are overall pale cinnamon sparsely patterned with dark brown centres, fine buffy grey or white margins.

Dark and light wing bars

White

Long white wedge

Very long legs

Dark, plain

Streaked brown

Fine, straight black bill tapers evenly

Heavier V bars

Soft grey-brown

Juv.: darker than nonbreeding

White

Br.

The stilt-like, long, thin legs are usually dull yellowish green, some brighter yellow.

Nonbr.

Juv.

Plumage of upper parts is more heavily and distinctly patterned than on nonbreeding adult.

**Green Sandpiper** *Tringa ochropus* 21–24 cm
An extremely rare vagrant. Usual habitat small freshwater wetlands, where it forages rather secretively. Tends to have a horizontal, rather hunched posture. If put to flight, rises into air on an erratic flight path with deep, abrupt wing beats. Call is a sharp, ringing 'tlee-it-wit-wit', usually on being put to flight. Similar are the Wood Sandpiper, but that species has some white under the wings, warmer browns with heavier white flecking and long white brow; also the Common Sandpiper, but that has an obvious wing bar, dark centred rump, white peak at sides of breast ahead of folded wings and warmer browns. Several unconfirmed sightings (1979–1988), and one confirmed record (1998); all in Top End, NT.

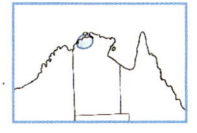

In flight, white rump in strong contrast beside dark wings; several thick black bars across white tail

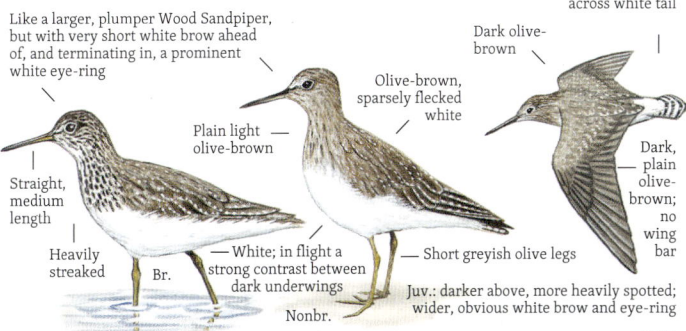

Like a larger, plumper Wood Sandpiper, but with very short white brow ahead of, and terminating in, a prominent white eye-ring

Dark olive-brown

Olive-brown, sparsely flecked white

Plain light olive-brown

Dark, plain olive-brown; no wing bar

Straight, medium length

Heavily streaked

Br.

White; in flight a strong contrast between dark underwings

Short greyish olive legs

Nonbr.

Juv.: darker above, more heavily spotted; wider, obvious white brow and eye-ring

## SANDPIPERS

**Common Sandpiper** *Actitis hypoleucos* 19–21 cm
Almost constant teetering and bobbing of tail. If put to flight, darts away level and close to the water with distinctive spasmodic wing action, glides to land and teeters vigorously. Habitat is varied: coastal and interior wetlands – narrow muddy edges of billabongs, river pools, mangroves, among rocks and snags, reefs or rocky beaches. Avoids open mudflats. Perches on branches, posts, boats. In flight, gives high squeaks, abrupt and penetrating; may be as two rapid squeaks, the second fractionally lower, 'tsie-tsiep', or as a long series, slightly lower pitch in last notes: 'tsie-tsie-tsie-tsie-tsiep-tsiep'. Widespread, scattered; quite common on N and W coasts; uncommon in SE and interior.

**Wood Sandpiper** *Tringa glareola* 20–22 cm
Graceful, active wader, prefers shallows of wooded lakes or swamps with trees, often foraging among fallen trees and vegetation. Habitat includes freshwater swamps, lakes, flooded pasture; less frequently brackish waters, occasionally mangroves. Singly, pairs, small to large flocks; often with other waders. Wary, quickly agitated; then holds head high on long, slender neck and bobs lower body. When feeding, busy but graceful with high-stepping movements. Calls are sharp, high, piercing, rapid: 'chi-chi-chip', 'chi-chi-chi-chip', the last 'chip' slightly lower, abrupt. An uncommon migrant.

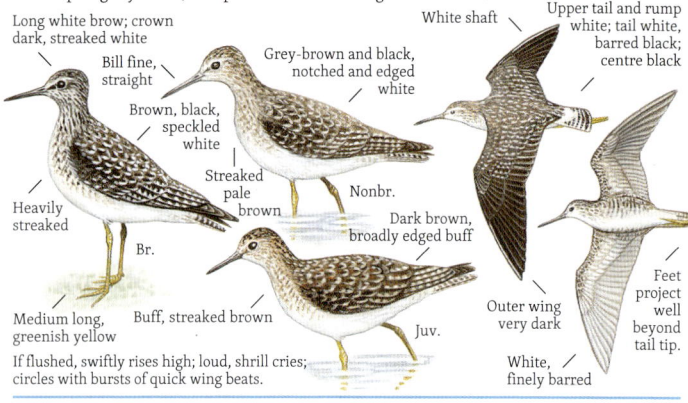

**Terek Sandpiper** *Xenus cinereus* 22–24 cm
Inhabits coastal mudflats in sheltered estuaries and lagoons as well as sandbars, reefs, coastal swamps, saltfields. Teeters, with nervous, exaggerated, bobbing action. A lively, active feeder: dashes erratically about mudflats with head down, pecking and probing in mud and sand, or in shallows with sideways sweeps of the curved bill. Alarm call a musical 'pee-peeweer, peeweer, peewit', and 'teeu-duey, duey, wi-wi-wi-yu'. Common migrant on N coasts, rare in S.

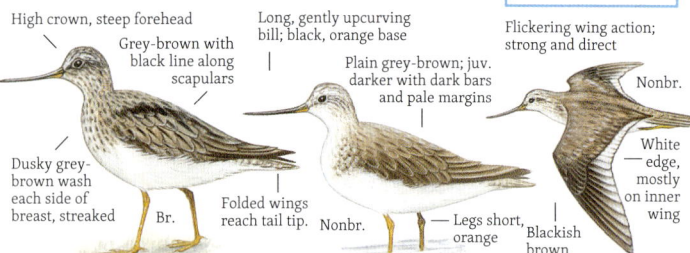

## TATTLERS

### Wandering Tattler *Tringa incana* 26–28 cm

Almost entirely confined to rocky coasts – wave-washed tidal platforms and exposed reefs around headlands or high islands; usually the coral cays and reef-fringed islands of the Great Barrier Reef and similar situations. Likely to use rocky sites, or occasionally jetties, to loaf and roost. While the two plainly plumaged tattler species are quite distinctive, distinguishing between these two species in nonbreeding plumage is more difficult. Usually solitary when feeding or resting. Feeds almost exclusively on rocky shores; sneaks about rocks, probing, teetering; tends to be alert and wary. Flight call is a sharp rippling trill of 5 to 10 piercing notes lasting 0.5-1 sec, evenly pitched, accelerating but fading slightly in strength. Also a flute-like alarm call of just one or two notes. Uncommon but probably regular summer migrant to NE and parts of N coast.

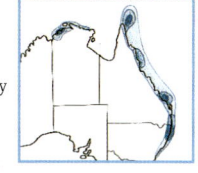

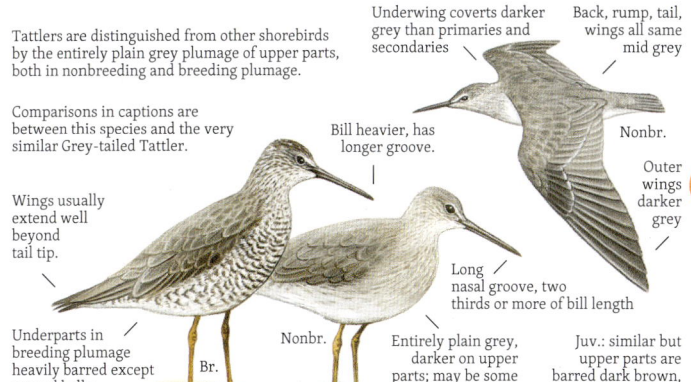

Tattlers are distinguished from other shorebirds by the entirely plain grey plumage of upper parts, both in nonbreeding and breeding plumage.

Comparisons in captions are between this species and the very similar Grey-tailed Tattler.

Underwing coverts darker grey than primaries and secondaries

Back, rump, tail, wings all same mid grey

Nonbr.

Outer wings darker grey

Bill heavier, has longer groove.

Wings usually extend well beyond tail tip.

Long nasal groove, two thirds or more of bill length

Underparts in breeding plumage heavily barred except central belly

Br.

Nonbr.

Entirely plain grey, darker on upper parts; may be some light barring on rear of flanks.

Juv.: similar but upper parts are barred dark brown, edged and spotted buff.

### Grey-tailed Tattler *Tringa brevipes* 24–27 cm

Habitat coastal; forages in inter-tidal pools, shallows, soft surfaces of mudflats and sand beaches as well as rock ledges, reefs. Often perches on branches, posts or jetties; roosts in groups, same or mixed species. Far more common than the Wandering Tattler; in comparisons best not to rely on any one feature, but rather the whole range of observable plumage and behaviour characteristics. Some features can, alone, be unreliable. The name refers to the slightly paler grey rump, where, in fresh plumage, feathers are pale fringed, but not obviously paler, tending to wear to a grey similar to the upper parts. The extent to which folded primary or flight feathers of the wings project beyond the tail tip differs between the two tattlers. Many other features such as the length of the brow (supercilium) and length of nasal groove may be visible using a spotting scope. The Grey-tailed Tattler darts about mudflats, sandbars and beaches, bobbing and teetering between dashes. Flight call is distinctive – fluid, musical, but slightly mournful, drawn out 'too-weet', initially falling in pitch, then rising sharply in the final 'eet'. Also as a rapid sequence, slightly sharper, rising in pitch and accelerating, 'weit-weit-weet-weet-weetweetweet'. Common summer migrant to N Aust.; uncommon in S.

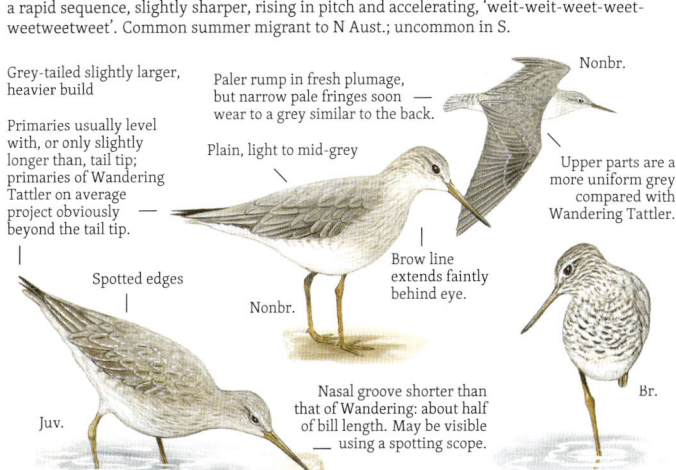

Grey-tailed slightly larger, heavier build

Primaries usually level with, or only slightly longer than, tail tip; primaries of Wandering Tattler on average project obviously beyond the tail tip.

Paler rump in fresh plumage, but narrow pale fringes soon wear to a grey similar to the back.

Nonbr.

Plain, light to mid-grey

Upper parts are a more uniform grey compared with Wandering Tattler.

Spotted edges

Brow line extends faintly behind eye.

Nonbr.

Br.

Juv.

Nasal groove shorter than that of Wandering: about half of bill length. May be visible using a spotting scope.

## KNOTS & DUNLIN

**Red Knot** *Calidris canutus* 23–25 cm
The male's breeding plumage has extensive chestnut-red. In Aust., this will be seen only on some birds near arrival or departure. Habitat includes sheltered coasts on mudflats and sandbars of estuaries, harbours, lagoons; occasionally on beaches, reefs. Soft calls from feeding flocks – low, throaty, harsh 'knut' or 'knot'; from flocks on migration, 'nyup-nyup' sounds. In alarm, whistled 'qwik-ick', 'twit-wit' or 'twit-twit-twit'. Summer migrant to Aust.; abundant in N, less common in S.

**Great Knot** *Calidris tenuirostris* 26–28 cm
Colourful in breeding plumage. Highly gregarious, often large flocks, sometimes of hundreds or thousands of birds. Forages on inter-tidal flats, usually in shallows and at waterline, moving forward slowly, deliberately, thoroughly probing the mud. Flies with slow beats of long wings. Uses sheltered coastal mudflats of estuaries, inlets, harbours, lagoons, mangrove swamps. Occasionally seen in salt lakes, lagoons and saltworks ponds, rarely inland waters. Usually silent, but has a rapid, hollow 'krok-kok' or 'knut-nut'. Taken up by a flock, becomes a continuous, rapid babble of 'krok-knut-kok-nut-nok' sounds. Abundant across N, less common in S.

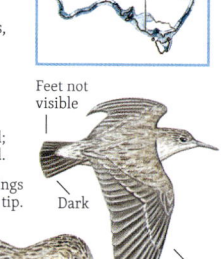

**Dunlin** *Calidris alpina* 16–22 cm
Small, dumpy, hunched shape. Uses sheltered coasts around estuaries, lagoons, mudflats, sandbars; inland on edges of lakes, flood waters. To feed, moves slowly, head down, jabbing rapidly. Upright stance in alarm. Flight call a harsh 'chzee', 'chzeep', 'trzeep' and 'treep', higher than call of Curlew Sandpiper. While feeding or roosting in flocks, birds make faint, soft titterings. Rare vagrant; most sightings in Aust. are uncertain, but two confirmed in Qld: Cairns 1993, and Cape Bowling Green, 1999.

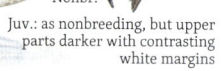

Many races: *sakhalina* and *pacifica* most likely

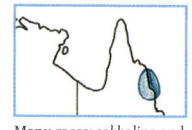

## SMALL SANDPIPERS

### Broad-billed Sandpiper *Calidris falcinellus* 16–18 cm

Shows preference for sheltered coastal estuaries, lagoons with soft inter-tidal mudflats, muddy coastal creeks, swamps, sewage ponds, occasionally reefs. Rather stint-like in behaviour, although more deliberate and persistent as its long bill probes vertically in mud and shallow water. Seems less timid than most waders; tends to crouch when disturbed. Alone or in small loose flocks, often accompanies Red-necked Stints or Curlew Sandpipers. Flight call a high, buzzing 'chzeeip' and an abrupt 'tzit'. A migrant species in nonbreeding plumage in Aust.; generally uncommon, more often seen on N coast, rare inland.

Small sandpiper; superficially like Curlew Sandpiper, but smaller with distinctive wide-tipped, heavy bill. From side view, the bill tapers from thick base to quite fine tip; but from front view, the bill remains rather broad over its full length and becomes quite flattened towards the tip.

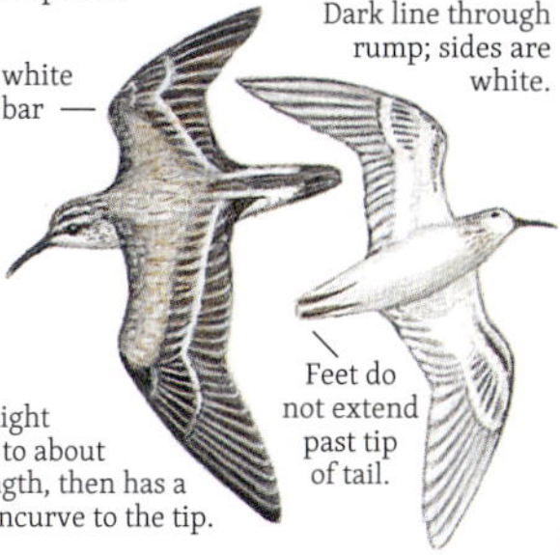

Dark line through rump; sides are white.

Long white wing bar

Feet do not extend past tip of tail.

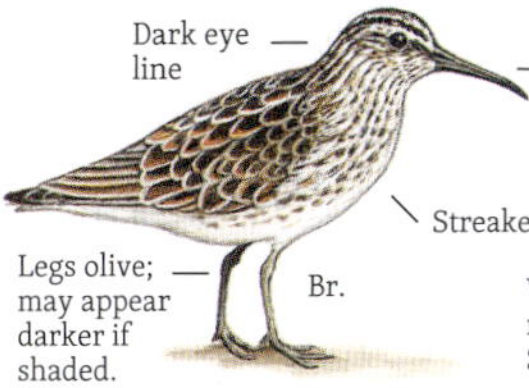

Dark eye line

Bill is almost straight from its base out to about two-thirds its length, then has a pronounced downcurve to the tip.

Streaked, neat edge

Legs olive; may appear darker if shaded.

Br.

White brow line splits into two near forehead, but this may occasionally occur on other waders, including Curlew Sandpiper, at various plumage stages or changes.

Double brow streaks join at forehead.

Bill is almost straight from its base out to about two-thirds its length, then has a pronounced downcurve to its tip; the Curlew Sandpiper's bill has a gradual downcurve over its entire length.

Legs short: a low, dumpy appearance

Nonbr.

Juv.

In distance, a side view of the downcurved bill tip of the Broad-billed, compared with the uniform gradual downcurve of the Curlew's bill, is usually an easier guide to identity than the former's slightly broader bill.

### Curlew Sandpiper *Calidris ferruginea* 18–23 cm

Medium-sized sandpiper, slender if standing tall, plump if hunched down or resting. Usual habitat the inter-tidal mudflats of estuaries, lagoons, mangrove channels, around lakes, dams, flood waters, flooded saltbush surrounds of inland lakes. Call, repeated at intervals of about a second, is a mellow, rippling 'chirrip, chirrip, chirrup' like some Budgerigar calls: given by flock becomes a pleasant confusion of twitterings. Widespread, common summer migrant to Aust. coast; also some scattered across suitable interior sites, depending on occurrence of rain.

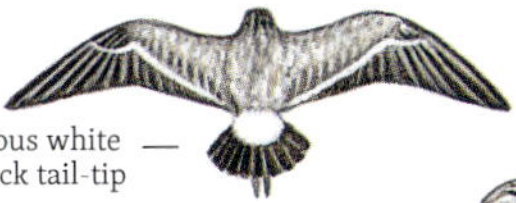

Bold white rump, dark tail tip

Conspicuous white rump, black tail-tip

When breeding, much bright chestnut-red in plumage, patterned black, fringed white

Bill long and gently downcurved along its full length, unlike that of the Broad-billed, which has most of its downcurve towards the tip

White wing bar

Whitish underwing

Wings extend beyond tail tip to give long, slender rear profile.

Pale mottled

Br.

Nonbr.

White

Juv.

Black legs longer than those of Broad-billed Sandpiper

Juv.: upper parts warm brown tones, buff-inted breast

127

## STINTS

### Red-necked Stint *Calidris ruficollis* 13–16 cm
An extremely small wader of diverse habitats, tidal and inland, on mudflats, salt marshes, beaches, saltfields, temporary floodwaters. Highly sociable, in flocks and intermixed with other waders; darts about the mud or sand, stopping to peck and probe. Flocks frequently burst into flight, swift on long wings, white under surfaces flashing against sea or sky. Call is a fast, extremely high, disyllabic 'chirit' or 'chrit', often so abrupt that it sounds more like a single 'chit' or 'prip'. Very common migrant – up to 100 000 birds on the most favoured sites, and scattered widely elsewhere.

### Little Stint *Calidris minuta* 12–14 cm
A tiny wader, rarely sighted, then always as individuals in flocks of other small waders. Typical habitats are mudflats, salt marshes, beaches, saltfields: most places where Red-necked Stints are found. Foraging and general behaviour are similar to the Red-necked Stint: scurries about the sand or mud, pecking and probing; may stand more upright when alarmed. Flight action like Red-necked Stint. Calls are extremely high, a quick 'chit' or 'stit', more abrupt, higher than the 'chrit' of the Red-necked; also a quieter, high 'tsee-tse'. Rare; few sightings scattered across coastal S Aust. and NT.

## STINTS

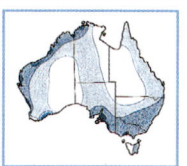

### Long-toed Stint *Calidris subminuta* 13–15 cm
Appears to be less gregarious than other stints; seen in pairs, singly or in flocks at favoured sites. Prefers shallow, fresh water and brackish swamps, lakes with muddy edges; often among low vegetation rather than on open mudflats. Alert, secretive; crouches low, hunched; pecks crake-like around vegetation. At times stands tall, neck extended. Perches on logs, low branches close to mud or water. If flushed, may rise high; or may zigzag low, calling. Gives a sharp trilled 'chirrip', so quick that the sound is more like 'trrp', 'prrp' or 'chrrp'; slightly lower, less metallic than other stints. Regular visitor but scarce; more often seen in WA.

If alarmed stands alert, tall, neck extended.

Mantle rufous and black with fine buff margins

Contrasting white coverts set dark beside flight feathers

Br.

Very dark outer wings both above and beneath; bold contrast against white

Br.

Legs long, dull yellow, but dark if shadowed

Slightly smaller, lighter build than the far more common, rather hump-backed, busily active Red-necked Stint. Looks more slender, creeps and skulks with body held almost horizontal.

Thin white bar on inner wing

Nonbr.

Buff line around mantle, a distinctive V shape from rear

Tips of folded wings about level with tip of tail

Juv.

Nonbr.

Long toes make feet appear large.

Bill fine; tip slightly downcurved

A rather round bellied body shape

### Temminck's Stint *Calidris temminckii* 13–15 cm
One of just two Aust. stints with yellowish legs. Tail is unique among stints – has pure white outer feathers. It is slightly longer than folded wing tips, and longer than trailing feet when in flight. White of tail is conspicuous on take off and landing, and with sudden changes of direction. Call is a distinctive, high-pitched, thin, cricket-like trilling, 'tirr-r' and 'trrr-it'; as flight call, 'tir-ir-ir-irir'. Also has a display song, not likely to be given in Aust. Found on freshwater wetlands – muddy, partly vegetated margins of swamps and lakes, flooded irrigated crops. Winters as far S as Borneo; a possible vagrant to Aust.

Only other stint with yellow legs is the Long-toed; legs may appear dark grey unless in bright light.

Underwing white and pale grey

Br.

Mottled grey-brown

Longest tail of all stints, projects beyond tips of folded wings.

Bold white wing bar, becoming fine on primaries

Black

Plain grey-brown breast band, otherwise underparts white

Nonbr.

Outer feathers of tail white, unique among stints

Nonbr.

Upper parts lightly scaled dark brown and cinnamon

Bill quite short, tapering to a fine, very slightly drooped tip

Br.

Juv.

Plain grey-brown breast band, otherwise underparts white

Rufous-buff streaked darker brown

Legs long, yellowish

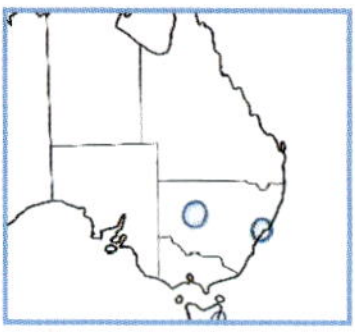

## SMALL SANDPIPERS

**Western Sandpiper** *Calidris mauri* 15–17 cm
Often in company with other waders, but may be solitary. Habitat is on mudflats and shallow wetlands. Not very wary or timid; more approachable than most other waders. Feeds with jabbing and probing actions; may wade quite deep. Females average slightly larger, and have a proportionately longer, more drooped bill tip. Call is a thin, sharp 'jee-it' or 'chee-it'. Similar are Dunlin, but it is larger, and nonbreeding Red-necked Stint, but its bill is straighter and shorter, grey tone each side of breast. Possible vagrant; unconfirmed sight records on NSW coast. If this species occurs in Aust., it is very rare.

Looks rather 'front heavy' or 'chesty'; long tapered body; appearance similar to Dunlin; stands taller than stints; larger head and bill, longer legs.

Wing bar narrow; extends only to inner primaries.

Upper parts almost uniform pale grey with fine dark streaks

Underwing mostly white

Br.

Humpbacked foraging posture, similar to that of Red-necked Stint

Nonbr.

Nonbr. White sides to dark rump

Finely streaked breast band

Male: bill shorter

Bill tapers from broad base to fine, slightly drooped tip; length rather variable, female's typically longer.

Crown, mantle, rufous with black markings

Scapulars: rufous upper rows, grey on lower

Short wings, about level with tail tip

Dark streaks on breast and arrowheads along flanks

Br.

Unique among Australasian stints: toes are partly webbed.

Juv.

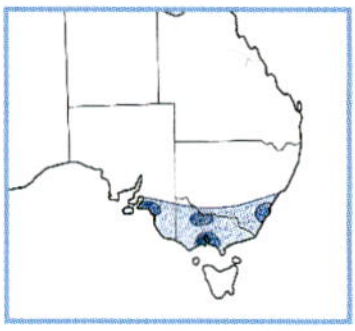

---

**White-rumped Sandpiper** *Calidris fuscicollis* 15–17 cm
This small wader has to travel further than most others to reach Aust. – it breeds only in N Canada and Alaska. The species is known for its wanderings, which extend to Europe, South Africa and South America. Uses inland rather than marine wetlands – margins of swamps, lagoons, lakes, muddy pools and, only occasionally, tidal mudflats. The flight call is a high, thin squeaked 'tzreit', more like that of a bat, but with a slightly vibrant quality. Also described as 'jeeit' or 'eeit'. Most alike is Baird's Sandpiper, distinguished in flight by its dark centred grey tail and thin white wing bar. A rare vagrant to Aust.: only a few accepted records among sightings, mostly from SW and SE Aust.

Crown chestnut, streaked darker

Grey-brown

Underparts white

White rump is visible only in flight.

Streaked brown

Large white rump

Very long wings, primaries well beyond tail tip

Br.

Short, thin wing bar

Nonbr.

Dark centred grey-brown edged light grey: wears to be more uniform.

Yellowish base, slight downcurve, blunt tip

Long wings give attenuated, slender, graceful jiz.

Nonbr.

Juv.

Fine streaks

## BAIRD'S SANDPIPER & SANDERLING

**Baird's Sandpiper** *Calidris bairdii* 14–16 cm
Similar to White-rumped, but even more slender with long, tapering rear profile; usually a horizontal stance. Tends to be alone or in small groups. When alarmed, crouches or stands tall; feeds briskly, darting about, probing mud, wading and sometimes dunking head to reach deeper. In flight, the long wings give slower wing action than stints. Usual habitat is coastal and some inland sites, the margins of freshwater and brackish wetlands. Often feeds on outer edge in drier vegetation as well as damp areas and mudflats; rarely in dense swamp vegetation. The flight call is low-pitched with touch of harshness, a rolling trill, 'preeet' or 'kyrrrp'; also a sharp 'tsiek'. A rare vagrant to Aust.; a few sightings scattered around coast in most states and NT; lone birds in with other small waders.

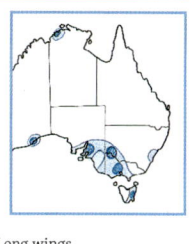

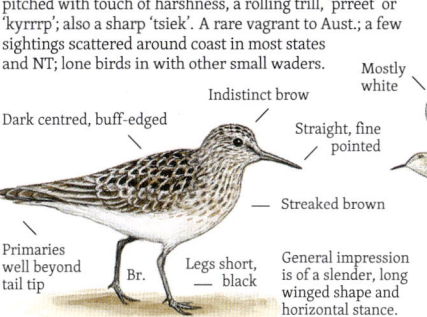

**Sanderling** *Calidris alba* 19–20 cm
This common, small, pale wader of open, sandy beaches washed by ocean swells, runs behind receding waves, pecking and chasing, darting up the beach as each wave breaks. In flight shows a flickering wing action, emphasised by wide, bold, white bar across dark wings. Congregates in small or large flocks, sometimes hundreds on favoured beaches, often with other waders. In flight, a sharp, quick, liquid 'tlick-tlick'; also a very soft twittering within flocks. Most like are Red-necked and Little Stints, but both smaller. Regular migrant; abundant on some beaches.

## SANDPIPERS

**Sharp-tailed Sandpiper** *Calidris acuminata* 17–22 cm
Plump, medium-sized sandpiper. In flight, white sides to dark centre of rump; thin wing stripe. Habitat fresh or salt wetlands – the muddy edges of lagoons, swamps, lakes, dams, soaks, sewage farms, temporary floodwaters. Calls, when flushed, a quick 'pliep', and rapid, high, scratchy, squeaky trills like the highest of fairy-wren trills. Also has chatterings with intermixed soft, low and high squeaky sounds like the chatter of Welcome Swallows. Most alike is the Pectoral Sandpiper, but that species has slimmer look, longer neck, shorter legs, more upright stance and longer, more slender, downcurved bill. Abundant in SE, common elsewhere.

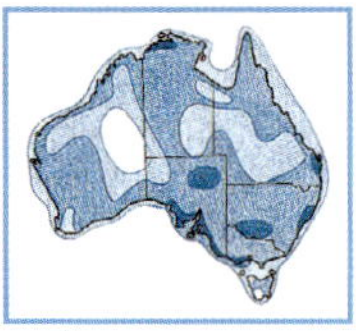

White eye-ring and long white brow that is widest behind the eye

Rump: white each side of dark centre line

Feathers pointed, chestnut with centres blackish, edges pale orange

Downcurved

Streaked, brown over rufous-buff tint

Long, fine wing bar

Dark

Entirely white and pale grey except rufous tint on breast

Br.

Primaries level with tail tip

Olive

Rufous, white margins

Rufous

Orange-buff tint

Rather plump, round bellied; body tapers to quite a fine, pointed tail.

Nonbr.

Breast tinted grey with fine streaks

Juv.: bright rufous and buff

**Cox's Sandpiper** *Calidris paramelanotos* 18–20 cm
Described as a full species in 1982 on the basis of a collection of specimens from Aust., but not known from any likely breeding areas of N hemisphere. Intermingles with other waders. In flight shows thin white wing bar; tail and central rump darker with white edges. DNA testing now shows it to be a natural hybrid, probably with Curlew Sandpiper and Pectoral Sandpiper as parent species. Breeding plumage not known except from birds partially moulted into a pre-breeding plumage. Calls are like those of the Pectoral Sandpiper, but higher, shrill, as 'trilt!'. Most alike are the Curlew, Pectoral, and Sharp-tailed Sandpipers. Habitat includes fresh and marine wetlands, including tidal mudflats, salt farms, sewage ponds. A rare summer migrant.

Thin white wing bar

Rufous tone

Moderately long downcurved bill

White edge to rump

Chestnut-buff fringes

Buff, streaked brown

Dark centred grey tail

Olive-green or brownish green

White

Light grey-brown, streaked darker

Pre-breeding plumage

Pre-breeding plumage

Streaked brown

Folded wings longer than tip of tail

White without any streaking on flanks

Under surfaces white with streaked breast band

Nonbr.

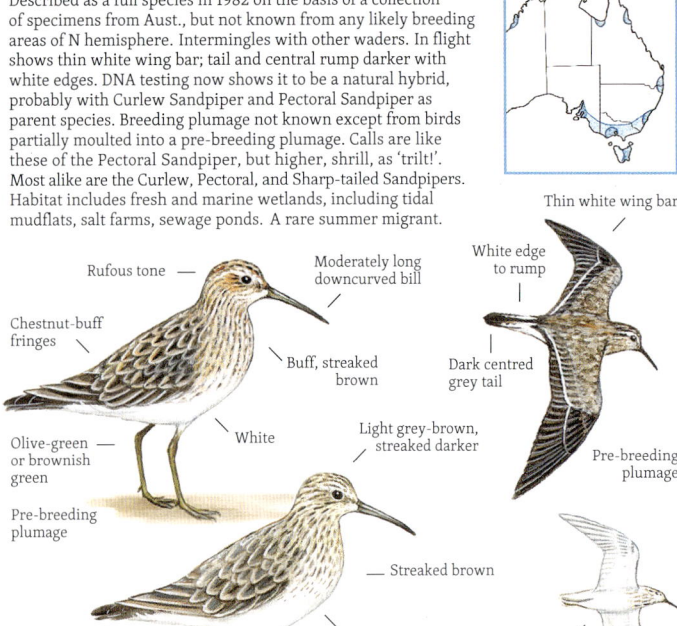

## SANDPIPERS

**Pectoral Sandpiper** *Calidris melanotos* 18–24 cm
Medium-sized sandpiper. All plumages have heavily streaked breast sharply demarcated from clean, white belly. Often adopts upright stance, but also hunches low, compact. When flushed, flies a fast, low, twisting course. Forages in shallows and soft mud; solitary or small groups, or mixed with other waders. Call, when flushed, a repeated loud, harsh 'tirrit' or 'prrip'. Uses coastal wetland, both fresh and saline, but also inland on permanent and temporary wetland. Prefers sites with mudflats, fringing vegetation, swamps with heavy overgrowth of vegetation. A regular but uncommon visitor to Aust., mostly to SE, but scattered through most coastal and inland areas.

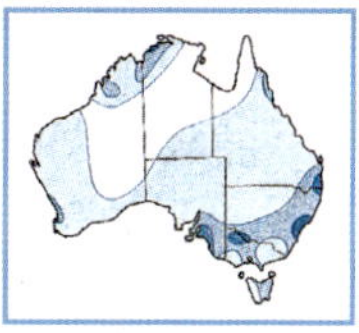

Long neck, rather short legs; often upright stance; breast always heavily streaked

Edged rufous-buff

Primaries level with tail tip

Legs dull olive to yellow

White

Faint, fine white wing bar

Dark outer wing

Neat, sharp edge between streaked breast and clean white of belly

Brown, streaked

Rufous margins

Bill longer, more downcurved than Sharp-tailed

Buff streaked with brown

Grey-buff breast heavily streaked; clear-cut lower junction with white belly

Olive-yellow legs, rather short

Br.  Nonbr.  Juv.

---

**Stilt Sandpiper** *Calidris himantopus* 19–23 cm
Habitat usually shallow waters of estuaries, tidal mudflats, swamps, ponds of saltworks and sewage farms; at times among flooded or swampy vegetation. The long neck may be extended when bird is alert, or pulled compactly onto shoulders in a stubby, hunched body shape – an example of effect of posture on shape. Usually feeds in shallows to depth of legs; uses a rapid jabbing action of bill, sometimes immersing head in deeper water. Flies powerfully; long pointed wings, long bill and trailing legs give a distinctive silhouette. Usually occurs as a solitary vagrant in company with Sharp-tailed or Curlew Sandpipers, or other waders. Flight call a soft, rather rattling, trilled 'kirrt' and clearer 'whiuw'. Rare vagrant; very few sightings in NT and Vic., these on either sewage or saltworks ponds.

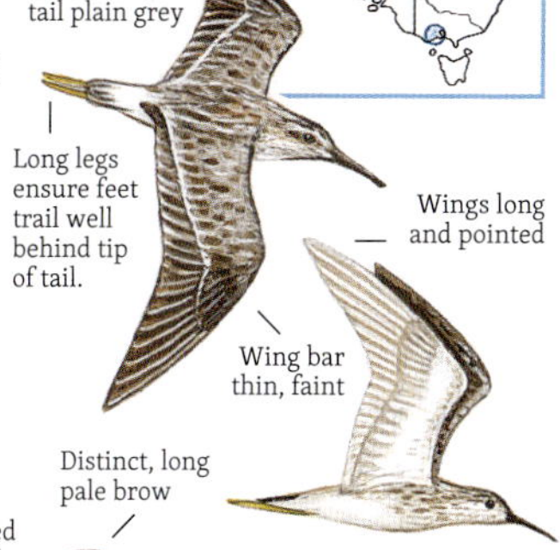

Rump white, tail plain grey

Long legs ensure feet trail well behind tip of tail.

Wings long and pointed

Wing bar thin, faint

Medium-sized sandpiper with long, thin stilt-like legs that seem to hold the rather slender body strangely high above the ground

Tall, with distinctive, long, tapering, slightly drooped bill

Dark brown, edged rufous and white

Downturned with slightly swollen tip

Distinct, long pale brow

Thick base

Tip slightly drooped, bulbous

Plain grey-brown

Fringed rufous

Buff tint, faint streaks

Barred

Long, spindly, stilt-like, greenish yellow legs

Br.  Wing tips extend beyond tip of tail.  Grey streaked neck, breast, flanks  Juv.

Nonbr.

## RUDDY TURNSTONE & UPLAND SANDPIPER

**Ruddy Turnstone** *Arenaria interpres* 22–24 cm
Named for its foraging technique of turning over beach stones, seaweed and other objects to feed on tiny crustaceans, sand-hoppers and other small creatures hiding beneath. Gregarious, usually in small groups that rush about prodding, probing and turning. Short legs give the impression this wader is tubby, dumpy or stocky. The body is actually quite slender, tapering towards the tail. The breeding plumage is easily recognised. Nonbreeding plumage is only slightly less distinctive; the juvenile's indistinct and rather variable pattern is most likely to cause identification difficulty. Voice is a rapid, irregular, weak chattering and fast twittering in short bursts; a high 'trit-tit-tit-tit-tit...' or 'tritititi...', fast enough to be a rippling trill. Also gives a clear, whistled 'kiew'. Usually on ocean coasts with exposed rock, stony or shell beaches, mudflats, exposed reefs and wave platforms; occasionally inland on shallow pools. Common; entire Aust. coast.

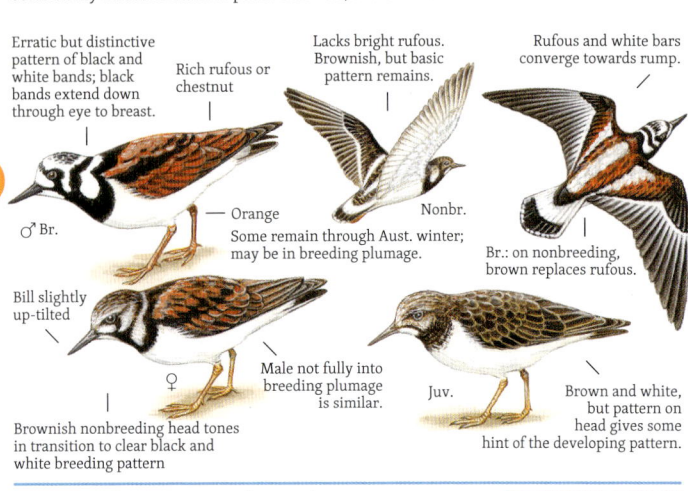

Erratic but distinctive pattern of black and white bands; black bands extend down through eye to breast.

Rich rufous or chestnut

Lacks bright rufous. Brownish, but basic pattern remains.

Rufous and white bars converge towards rump.

♂ Br.

Orange
Some remain through Aust. winter; may be in breeding plumage.

Nonbr.

Br.: on nonbreeding, brown replaces rufous.

Bill slightly up-tilted

♀

Male not fully into breeding plumage is similar.

Juv.

Brown and white, but pattern on head gives some hint of the developing pattern.

Brownish nonbreeding head tones in transition to clear black and white breeding pattern

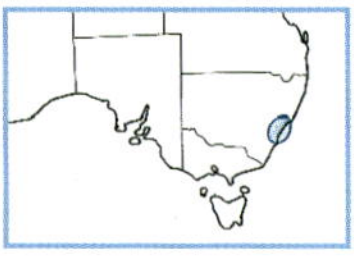

---

**Upland Sandpiper** *Bartramia longicauda* 26–30 cm
A strange, rather plover-like wader with tall upright posture; often holds long slender neck vertically. Head is small for bulk of bird, yet looks large, rounded, while neck looks rather thin. Unlike most waders, the Upland Sandpiper prefers drier grassland – the prairies of N America and similar environs. Usual habitat open grassland or prairies, commonly on farmland, pasture, stubble or airfields; likely to use similar habitats on migration. Behaviour, posture and feeding actions are more like a plover; runs and stops abruptly to jab at small prey. If alarmed, has sandpiper-like bobbing action of rear end. Flight is swift with steady, easy wing beats; feet do not reach beyond long tail. Often perches on posts or power poles; holds wings up briefly after landing. Makes various sounds; flight call is a piping 'quip-eip-eip', and lower, liquid 'pull-ip' or 'quie-lip'. Single confirmed record near Sydney, NSW, 1848.

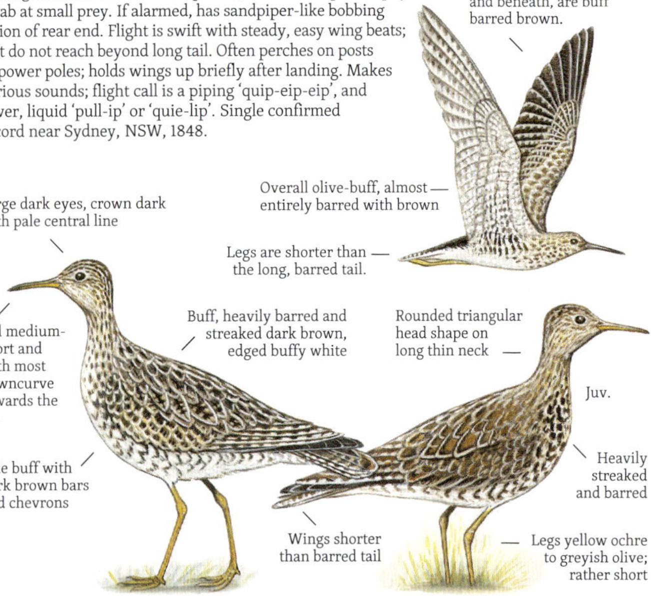

Blackish flight feathers; remainder of wings, above and beneath, are buff barred brown.

Overall olive-buff, almost entirely barred with brown

Large dark eyes, crown dark with pale central line

Legs are shorter than the long, barred tail.

Buff, heavily barred and streaked dark brown, edged buffy white

Rounded triangular head shape on long thin neck

Bill medium-short and with most downcurve towards the tip

Juv.

Pale buff with dark brown bars and chevrons

Heavily streaked and barred

Wings shorter than barred tail

Legs yellow ochre to greyish olive; rather short

## BUFF-BREASTED SANDPIPER & RUFF

**Buff-breasted Sandpiper** *Calidris subruficollis* 18–21 cm
An unusual, plover-like sandpiper that, like the Upland Sandpiper, is usually a bird of dry open grassland and pasture, sometimes in vicinity of swampy areas. In Aust., usually in short sparse grass, lawn, salt marsh or low samphire close to estuaries, lagoons, swamps; only rarely on mudflats. The upright stance when alert combines with the rounded head on a long, slender neck and high-stepping walk to make this species distinctive. May crouch in the grass with compact, hunched posture when feeding. Approached, may crouch, freeze or run. In flight, looks long winged; keeps a low, zigzag path. Usually silent; flight call a soft 'tchu' or 'prreei-t'. Rare; vagrant or perhaps regular visitor to SE and S Aust.

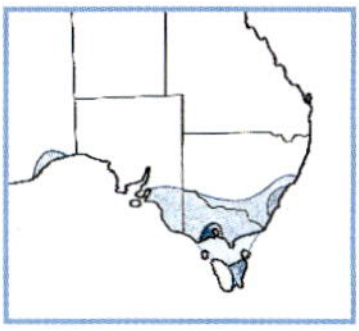

Only very faint, fine wing bars

Often stands tall; walks with head jerking, plover-like.

Dark-mottled trailing edge

Dark patch

Dark streaked

Thinner buff margins; plumage has more dark brown and rufous

Large dark eye in plain pale face

Wide golden buff margins to each feather

White underwing

Dark plain outer wing

Deepest buff is from face to breast.

Primaries are level with tail tip.

Juv.

Br.

Distinctive in overall buff tones, stronger about the face, sides of neck and breast

**Ruff** *Calidris pugnax* Ruff 26–32 cm; Reeve 20–25 cm
The name 'Reeve' is used for the female, 'Ruff' for the species or for the male, which is not only substantially larger, but has remarkable breeding plumage and displays. Males in breeding season develop large ear tufts and exaggerated, colourful ruffs about the neck and shoulders; colours vary; hardly any two birds are the same. Some have extensive black contrasting against faces of bare red skin; others have ruffs of snowy white, rufous or barred brown, and facial skin of ochre, orange or red. However, all this activity occurs in N hemisphere; breeding is in N Europe and Asia. Most arrive in Aust. in dull non-breeding plumage, but a few may arrive with remnants intact, or may start to acquire breeding colours before they leave in the autumn. Usual habitat is mud flats and sedges around fresh or saline lakes, estuaries, tidal pools. Usually silent or giving low grunts. An uncommon but quite regular visitor to Aust.

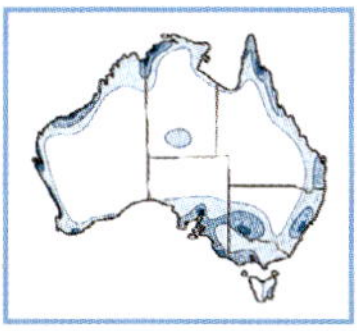

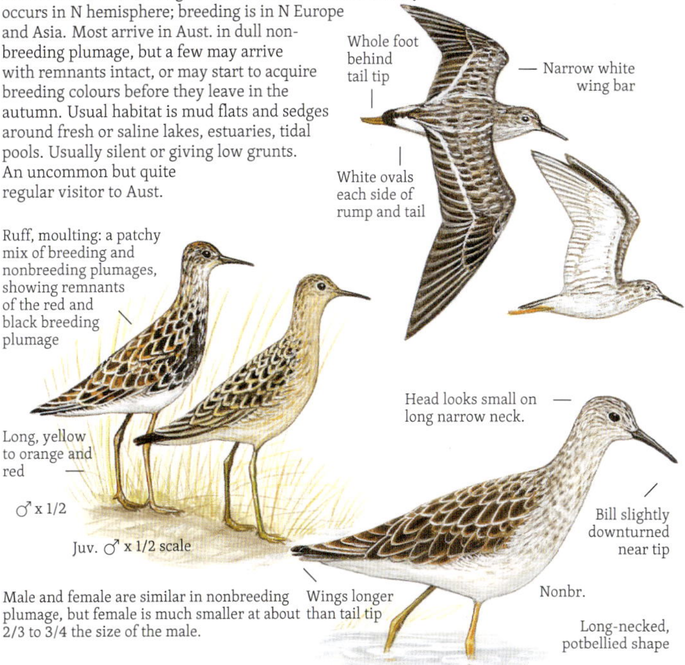

Whole foot behind tail tip

Narrow white wing bar

White ovals each side of rump and tail

Ruff, moulting: a patchy mix of breeding and nonbreeding plumages, showing remnants of the red and black breeding plumage

Head looks small on long narrow neck.

Long, yellow to orange and red

Bill slightly downturned near tip

♂ x 1/2

Juv. ♂ x 1/2 scale

Male and female are similar in nonbreeding plumage, but female is much smaller at about 2/3 to 3/4 the size of the male.

Wings longer than tail tip

Nonbr.

Long-necked, potbellied shape

135

# 9 PHALAROPES TO PLOVERS & PRATINCOLES

This group includes seven families. While most species of several of the families or genera can quite easily be named, others will almost certainly need the detailed entries in the main text. Some of these species that are unlikely to have even a tentative identification made here are not included on this page – only one or two examples of their genus are shown. Waders are a notoriously difficult group. It is helpful to join bird club wader-watching excursions with others who are familiar with these birds and always keen to assist beginners.

Willie Wagtail = 20 cm

## scolopacidae
### page 146

Red-necked Phalarope

Wilson's Phalarope

## jacanidae
### page 147

Comb-crested Jacana

## burhinidae
### page 150

Bush Stone-curlew

Beach Stone-curlew

## haematopodidae
### page 149

Pied Oystercatcher

## recurvirostridae
### page 151

Red-necked Avocet

## charadriidae
### page 138

Pacific Golden Plover — Nonbr.

Grey Plover — Nonbr.

Caspian Plover

Inland Dotterel

Little Ringed Plover — Nonbr., Br.

Black-fronted Dotterel

Double-banded Plover — Br., Nonbr.

Hooded Plover

Red-capped Plover

Red-kneed Dotterel

Lesser Sand Plover — Br., Nonbr.

Greater Sand Plover — Br., Nonbr.

Masked Lapwing

Oriental Plover — Br., Nonbr.

Banded Stilt

Pied Stilt

## glareolidae
### page 137

Oriental Pratincole

Australian Pratincole

# ORIENTAL & AUSTRALIAN PRATINCOLES

**Oriental Pratincole** *Glareola maldivarum* 23–24 cm
Elegant in flight; has long, pointed wings, deeply forked tail; hawks swallow-like for insects, especially around dawn and dusk, and frequently in large flocks. Often over wetlands where clouds of insects accumulate; at times around bush fires; otherwise generally on open plains, open areas around tidal flats, beaches, wetlands. During heat of day, rests on areas of flat, open land not far from water – paddocks, airfields, mudflats, roads. May squat low, well concealed by their colour on rough ground or in small depressions, even the impressions from hooves. Voice, compared with Australian Pratincole, a rather rough but not unpleasant 'chak-a-chak'; started by one or a few birds, taken up by flock. With sharper 'cha-rik' or 'krik' sounds intermixed, becomes an almost continuous confusion of 'char-rakichikiakak' noise. A migrant from India, SE Asia; arrives N Aust. with tropical wet season, Nov.–Jan.; dispersive, nomadic; flocks range widely seeking most productive sites. Leave Aust. to return to SE Asia between Feb. and April.

**Australian Pratincole** *Stiltia isabella* 22–24 cm
Graceful, slender pratincole; stands tall and erect on long legs. Hawks for insects swallow-like on backswept, finely pointed wings. Long-legged, often dashes about on ground after insects and other small prey; bobs head. Frequents treeless and sparsely wooded plains, grasslands, areas of sparse vegetation like claypans, gibberstone; never far from water of lagoon or livestock dam. Sandy rufous plumage matches bare ground typical of habitats. Call is an undulating, pleasant, cheery whistle; as flight call, a varied 'whit-WEIT', 'WEET-whit-whit-whit' and 'wirr-ie-WEIT'; as alarm calls, similar but sharp, urgent, loud. Flying birds' calls attract attention before they are seen; also often call in thunderstorms. Widespread, in places common through N and NE; occurrence varies, unpredictable.

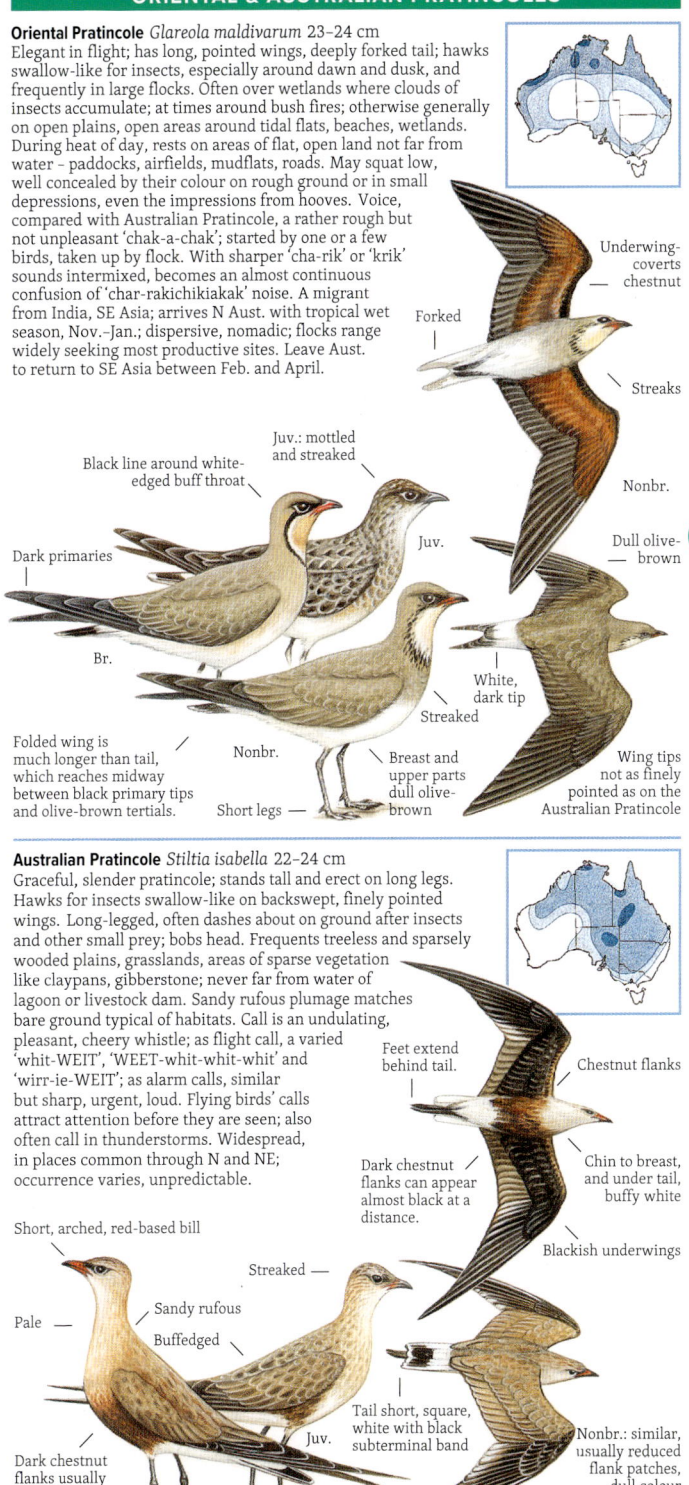

137

# GOLDEN PLOVERS

**Pacific Golden Plover** *Pluvialis fulva* 24–26 cm
Slender, upright, shy, wary; feeds in typical stop-start plover manner. Mainly coastal habitats; usually in small parties or quite large flocks on estuaries, intertidal mudflats, beaches, reefs, salt marshes, offshore islands; only rarely far inland. Usually silent on ground; calls in alarm, at take-off, in flight and when landing. In flight, a high, clear, musical 'tiu-ee' or 'tchu-ett', the second note higher. Similar calls, more subdued, when feeding or resting birds are put to flight and on coming in to land. In alarm, calls are more urgent, sharper. Common migrant from Arctic; arrives Aug.–Sep.; disperses to suitable habitat around coast; leaves Apr.–May when some will have moulted into breeding plumage. A few remain through southern winter.

Previously a race of the American Golden Plover, *Pluvialis dominica*.

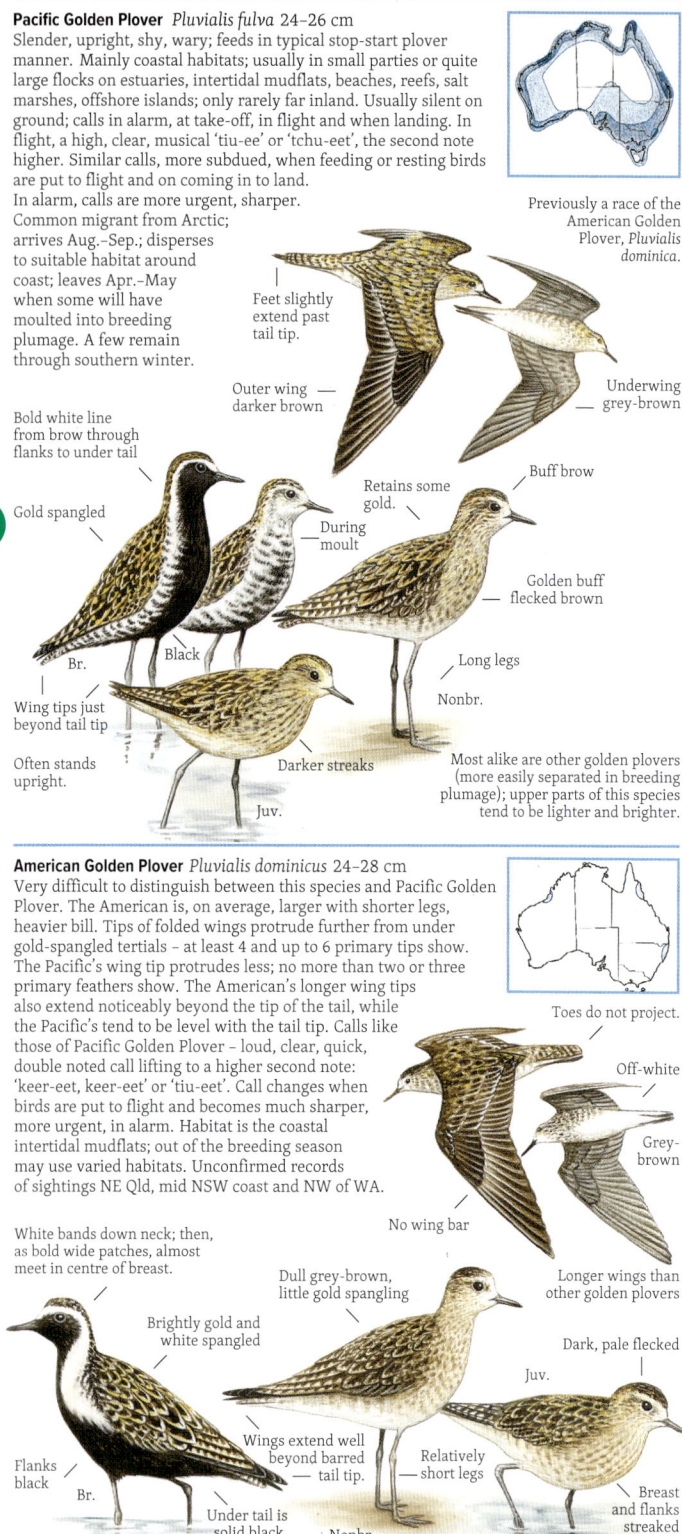

Most alike are other golden plovers (more easily separated in breeding plumage); upper parts of this species tend to be lighter and brighter.

**American Golden Plover** *Pluvialis dominicus* 24–28 cm
Very difficult to distinguish between this species and Pacific Golden Plover. The American is, on average, larger with shorter legs, heavier bill. Tips of folded wings protrude further from under gold-spangled tertials – at least 4 and up to 6 primary tips show. The Pacific's wing tip protrudes less; no more than two or three primary feathers show. The American's longer wing tips also extend noticeably beyond the tip of the tail, while the Pacific's tend to be level with the tail tip. Calls like those of Pacific Golden Plover – loud, clear, quick, double noted call lifting to a higher second note: 'keer-eet, keer-eet' or 'tiu-eet'. Call changes when birds are put to flight and becomes much sharper, more urgent, in alarm. Habitat is the coastal intertidal mudflats; out of the breeding season may use varied habitats. Unconfirmed records of sightings NE Qld, mid NSW coast and NW of WA.

## EURASIAN GOLDEN & GREY PLOVERS

**Eurasian Golden Plover** *Pluvialis apricaria* 26–28 cm
Larger than the other golden plovers; slightly smaller than the Grey Plover. Within usual range tends to be very alert, shy, easily put to flight, when it may circle warily for some time before landing. In N hemisphere, uses permanent grassland, farmland, intertidal mudflats, salt marsh, lake shores. Two races – northern (*altifrons*) from N Scandinavia into Arctic Russia, and southern (*apricaria*), from S Scandinavia to Ireland. If this species is eventually added to the Aust. bird list, it would seem that the northern race would be the more likely – that population is more highly migratory. Calls may differ in Aust. if species is recorded; a clear, liquid or yodelled flight call, 'tloo-ee'; clear and loud enough to be heard before the bird is seen. Common in usual range; no accepted records for Aust.; several possible sightings.

**Grey Plover** *Pluvialis squatarola* 28–30 cm
Medium-sized; large head and heavy bill; long legs; often stands hunched, body horizontal and head pulled down, but also stands tall, alert. Solitary or in small groups, at times with golden plovers. Shy, tends to stay far out on shallows or flats. Powerful, swift, graceful flight, loose flocks in irregular lines. Typical plover behaviour – stop-start running, pecking, often out into shallows; probes into sand or mud, occasionally pasture, for molluscs, marine worms, crustaceans. Habitat coastal, usually marine shores of estuaries or lagoons on broad, open mudflats, sandy bars or beaches, rock platforms and reef flats of rocky coasts; inland but near the coast on margins of salt lakes and swamps. Flocks often silent; may flush without calling; distinctive flight call – loud, triple noted whistle, plaintive, undulating 'whee-oo-eeir, whie-oo-eeir'. Common migrant. Occurs almost all parts of coast; common W and S coasts, abundant in Kimberley, scarce in Tas.

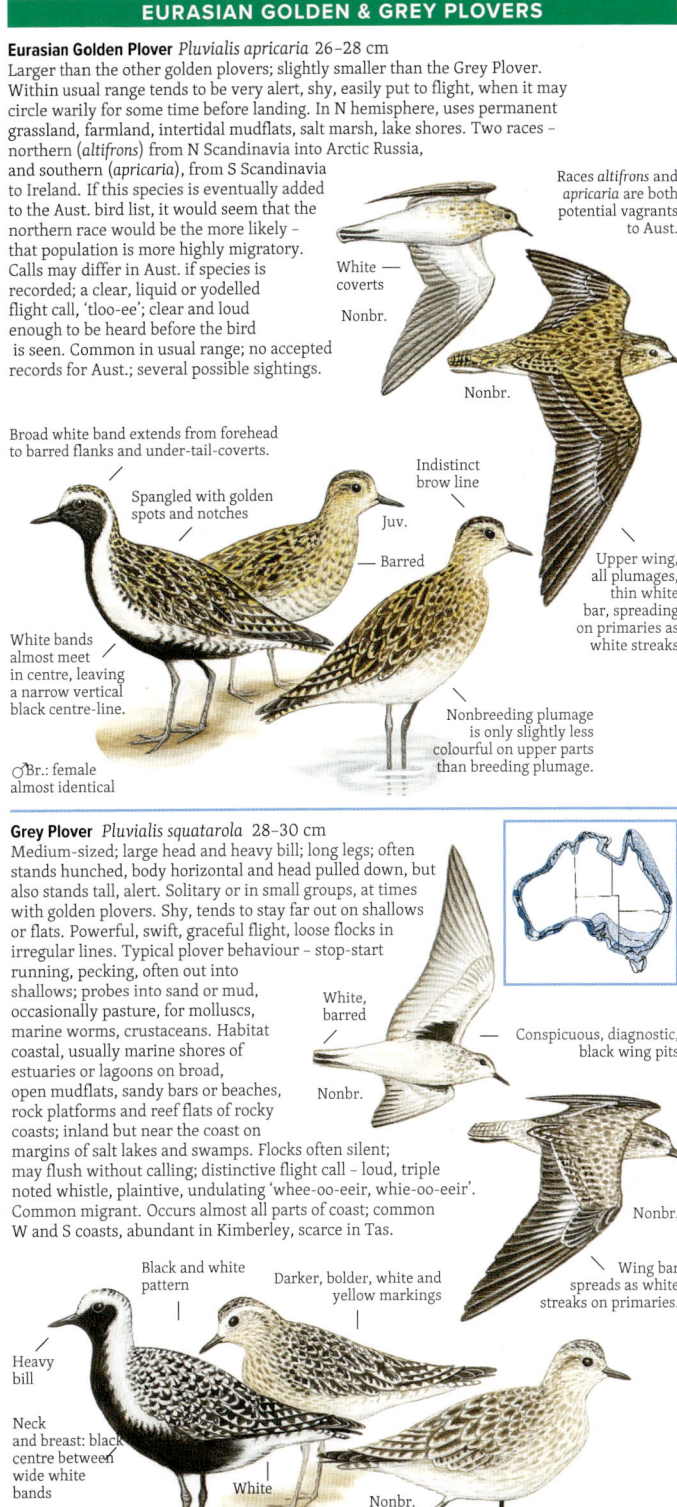

139

# RED-KNEED DOTTEREL & HOODED PLOVER

**Red-kneed Dotterel** *Erythrogonys cinctus* 17–19 cm
A medium-sized, long-legged, rather plump plover; holds body level or with breast low, tail high. Often stands motionless, then bobs head and dashes forward at prey. Uses well-vegetated freshwater wetlands – swamps, lakes, billabongs, interior claypans, sewage ponds, overflows from bores, windmills – working the shallows among emergent vegetation. Probes around muddy shoreline; wades and sometimes swims while feeding.
In flight, birds twist and turn; travel fast with shallow, flickering wing beats. Flight and alarm call a rather sharp, double 'tet-tet' or 'chit-chit'; gives a musical 'prit-prit-pri-t' trill on being put to flight. Resident; nomadic in response to rainfall; common.

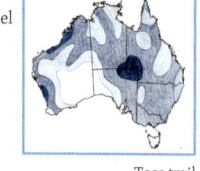

Bill red, tipped black
Toes trail behind tail tip.
White coverts and trailing edge
Wide white trailing edge
Black or brown hood does not extend down to chin or throat.
Black primaries
Back of neck black; lacks encircling white collar.
Imm. plumage shows beginning of breast band and flank streaks.
Flanks grade from black to chestnut.
Wide breast band
Juv.
Legs down to 'knee' red; lower leg is blue-grey.
At all ages has sharp edge between white throat and black or brown hood.

**Hooded Plover** *Thinornis cucullatus* 19–23 cm
A small plover with black hood and bars that disrupt the shape, making it inconspicuous when still. Uses sandy beaches of ocean, estuaries, coastal lakes and inland salt lakes. Usually in pairs or small parties; dashes about near the water's edge, pecking, bobbing, often darting down the beach as waves recede. In flight, shows boldly patterned wings. Has a husky, very abrupt, almost barking 'kep, kep, kep' or 'kue, kue, kue' at intervals of 1 or 2 seconds; also a similar, soft, flute-like 'quiep, quiep' and a higher piping call. Endemic species vulnerable to human disturbance of beaches, consequently declining in heavily populated locations.

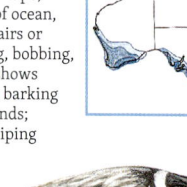

Black hood
White collar continuous around back of neck
White sides to rump and tail
Bright red eye-ring
Light brownish-grey
Under-wing white, dark-tipped
Bold white wing bar spreads wide on primaries.
Thin black hind neck collar forks onto breast.
Legs orange
Collar and side of breast band smudgy grey-brown
Juv.

**Black-fronted Dotterel**
**Black-fronted:** eggs are laid among riverbed stones.

Red-kneed Dotterel
Hooded Plover

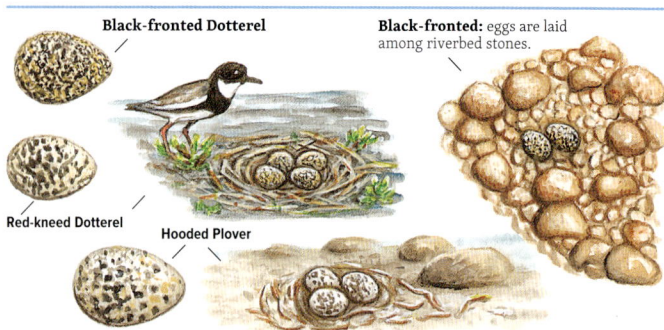

## BLACK-FRONTED DOTTEREL & RINGED PLOVERS

**Black-fronted Dotterel** *Elseyornis melanops* 16–18 cm
Small, slender plover, distinctive plumage pattern. Usually on freshwater wetlands, shallow, muddy bottomed swamps, billabongs, lake margins, temporary claypan pools; only rarely on saline coastal waters, tidal mudflats or other shoreline sites. Gregarious, gatherings from a few to many birds. When foraging, holds body horizontal; bobs head; runs on twinkling legs; stops abruptly; pecks; runs again. Contact call a regular, sharp 'tip-tip-tip-' at intervals of a second or two; becomes much louder, sharper in alarm. Common and widespread; sedentary, dispersive.

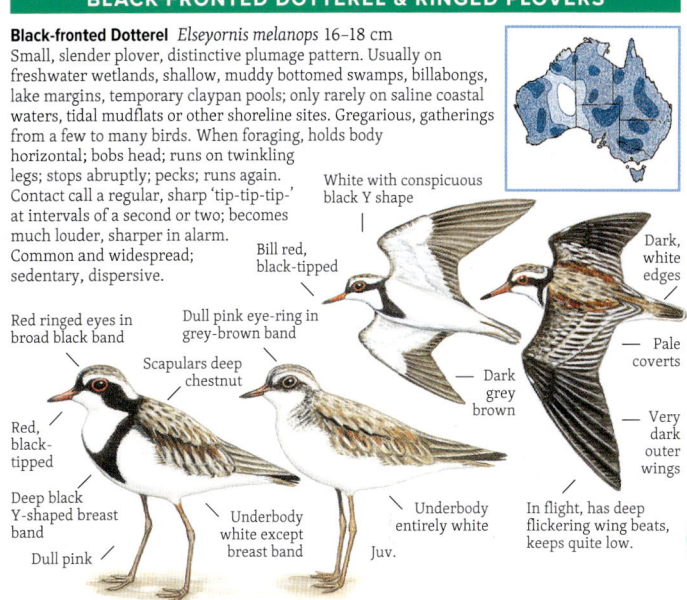

**Ringed Plover** *Charadrius hiaticula* 18–20 cm
Small robust wader; sightings in Aust. only of solitary birds on tidal mudflats and sand of estuaries, salt marsh, saltpans, sewage ponds. Has a wary, quick, scurrying action when foraging; pauses and stands tall. Glides to land; holds wings up on alighting. Breast band reduced or broken. Has a distinctive contact call, mellow or fluty, two notes, second higher, 'too-lee, too-lee'; also a low 'too-weep'; sharper versions in alarm. Rare migrant or accidental visitor; sightings quite widespread – Qld, NSW, Vic., SA, NT.

**Little Ringed Plover** *Charadrius dubius* 14–17 cm
Small, slender plover; busy, erratic, wary. Inhabitant of muddy edges or mudflats of tidal or freshwater wetlands including estuaries, lakes, lagoons, dams, ponds. Voice is a descending, clear 'peeeooo', usually as a flight call – contrasts with the rising call of the Ringed Plover; in alarm, an abrupt, rapidly repeated 'pip-pip!, pip!, pip-pip'. Most likely to be confused with the Ringed Plover, which lacks prominent eye-ring, has an obvious wing bar, an orange bill, and a different call. Rare but regular visitor.

## SMALL PLOVERS

**Red-capped Plover** *Charadrius ruficapillus* 14–16 cm 
Busy, gregarious: rushes along water's edge, stops abruptly, bobs head nervously, darts forward. Alone or in flocks, at times of hundreds of birds. Coastal – sheltered estuaries, salt marsh lagoons; also inland on salty edges of waterways, brackish pools, claypans. Greatest numbers occur occasionally on inland salt lakes. Flocks fly fast, erratically, displaying dark then light surfaces. Call, as contact, faint 'wit, wit, wit' or trilled 'twitwit twitwit' or'trrrrrit'. Gives a frantic chirring in distraction display; as alarm call a plaintive 'twink'. Common.

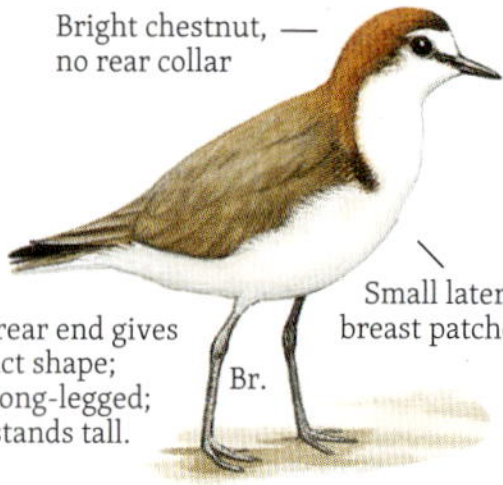

Bright chestnut, no rear collar

White

Bright chestnut, no rear collar

Pale upper parts

Blackish

Short rear end gives compact shape; looks long-legged; often stands tall.

Small lateral breast patches

Br.

White coverts

Narrow wing bar expands as white streaks on very dark primaries.

Grey-brown flight feathers

Dull chestnut-brown

Small side patches

Brownish

'Broken wing act' to lead predator away

Nonbr.

Juv.

Nest is a shallow depression scraped in beach sand, usually beside a small plant or tidal debris. Eggs are well camouflaged, but often lost to gulls or other predators, or beach traffic or disturbance.

---

**Kentish Plover** *Charadrius alexandrinus* 15–17 cm
A small, widespread plover, slightly larger than Red-capped Plover. Behaviour typical of small plovers. There are about 6 subspecies in all zoogeographical regions except Aust. American races are known as 'Snowy Plover' and are very pale. Flight call is a sharp but not loud 'twit' or 'wit' and rougher 'pirrr'. Habitat within usual range includes, most commonly, marine beaches, tidal flats. Rarely inland. In Aust., the record is from coastal mangrove mudflats. Rare vagrant; one verified occurrence on NT coast, Darwin, 1988.

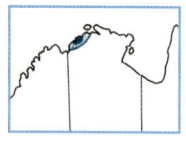

Sole Aust. record probably N Asian race *dealbatus*

Like Red-capped Plover; slightly larger. Comparisons are between these species.

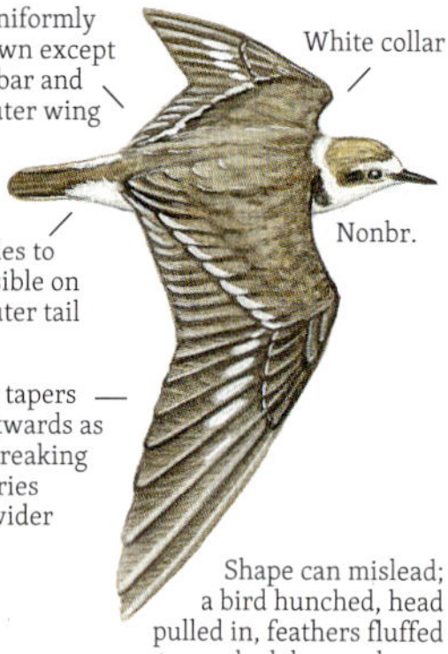

Longer brow line

Almost uniformly olive-brown except for wing bar and darker outer wing

White collar

Black through eye and on forehead

Nonbr.

Br.

Larger, darker patches at sides of breast

White sides to rump; visible on spread outer tail

White collar continues narrow but unbroken around back of neck.

Wing bar tapers wider outwards as broken streaking on primaries spreads wider

Nonbr.

Black or grey or olive-grey, long

Shape can mislead; a bird hunched, head pulled in, feathers fluffed out, may look large, plump. When active and alert, plumage is tight, sleek; neck outstretched, standing tall, it will look quite different.

## DOUBLE-BANDED & LESSER SAND PLOVERS

**Double-banded Plover** *Charadrius bicinctus* 18–21 cm 
On tidal mudflats, beaches, exposed reefs, salt marshes, freshwater wetlands, inland salt lakes, short grass of golf courses, airfields. Usually in groups or flocks. Behaviour typical of small plovers: abrupt stop-start, run-pause movements. Tends to stand more upright than other small plovers. Call is a loud 'pit', usually high pitched, but varied – sometimes a double 'tink-tink'. Flocks in flight maintain contact with constant musical tinkling of the many individual 'tink' and 'chip' calls. Widespread, common migrant to SE Aust., autumn to spring, with scattered records from as far as NE Qld and W coast of WA.

Breeds in NZ, winters in Aust. Most seen here are in dull grey-brown nonbreeding plumage.

White sides to rump; visible on spread tail

Dark outer wing, narrow wing bar. Underwing white

The only plover in Aust. region with two breast bands or two tabs each side on nonbreeding. More upright stance than most small plovers.

Some breeding females may be almost as strongly coloured as the males; others are like nonbreeding.

Black mask

Bands of black and of chestnut

Mask dull brown

No white collar

♀ Br.

Dull brown

Some birds have brown band rather than chestnut.

♂ Br.

Legs quite long, olive-yellow or olive-grey

Nonbr., both sexes: black and chestnut replaced by dull brown

Brown bands, often broken centrally; lower may be no more than a smudged tab of dull brown.

---

**Lesser Sand Plover** *Charadrius mongolus* 18–21 cm 
Small plover, may be seen on intertidal sandflats and mudflats, beaches, estuary mudflats and sandbars, reef flats. Some remain through winter in N. In both bright breeding and dull nonbreeding plumages while in Aust. Many moult to breeding colour before departing in autumn. Call differs slightly from that of Greater Sand Plover; usual call a quick, hard 'chrik' and 'chrik-it' on taking flight. As flight call, soft, musical 'tirrrit-tirrrit tritt'. Nonbreeding migrant Sep.–Apr.; abundant Qld, uncommon elsewhere.

Numerous races. Most likely to reach Aust. are ● *mongolus*, from interior of E Russia, and ● *atrifrons* from Tibet and Himalayas.

Similar to Large Sand Plover, but smaller and with rounded head, shorter legs and bill

Bright chestnut cap

White forehead

White sides of rump and tail show when tail is spread.

Black line

Black

No white collar

Bright light chestnut

● ♂ Breeding

● ♂ Br.

Narrow wing bar, spreads wider on primaries

● ● ♀ Black of male replaced by brown, but may still have chestnut on breast band.

Rufous, black replaced by brown, but same pattern

Underwing coverts white

Brown breast tabs

Buff tint to brown of side tabs

♂ Br.

Nonbr.

Juv.

Juv.: buff-brown breast patches and buff fringing to tertials and scapulars

In level flight, with feet drawn up under tail, toe tips are about level with tail tip.

# GREATER SAND & CASPIAN PLOVERS

**Greater Sand Plover** *Charadrius leschenaultii* 22–25 cm
Head looks large. Relaxes in hunched, horizontal stance. Alert, is more upright, neck extended. Gregarious, often in mixed flocks on coastal, intertidal mudflats and sandbanks of sheltered bays and estuaries, sandy cays of coral reefs, reef platforms, less often coastal salt marsh, brackish and, very rarely, freshwater wetlands. Often rather quiet, calling from flocks feeding and in flight – a short rippling 'drrit' or 'treet', extended as 'trrrri-trrrri-trrri-trrri', so rapid individual notes are not discernible, but at times becoming a rattling sound, almost unique. Most alike are Lesser Sand Plover, Oriental Plover. Regular migrant, Aug.–May; most common in N.

● *stegmanni* from NE Siberia most likely

Has a proportionately large head, rather angular-rectangular in shape, especially at rear of crown. Common head-down posture of resting bird makes head seem even larger than when active.

Chestnut nape; usually some grey at centre of crown

White may have fine black edge

● Has white patch on forehead where some other races mostly or entirely black.

Brown breast band; may have break at centre.

Light chestnut

White

Light grey, darker at wing tips

Large, heavy, black bill

Br.

Long

Wing tips slightly longer than tail

Olive-grey, straw or grey-brown

Nonbr.

Feet show behind tail.

Dark

White wing bar spreads widest on inner primaries.

Female in breeding plumage is like male, but black often replaced by dark brown. Juvenile has buff margined upper parts.

---

**Caspian Plover** *Charadrius asiaticus* 18–20 cm
Although different from the Oriental Plover in breeding plumage, these two are almost identical in other plumages and have at times been combined as one species. The Caspian breeds around the Caspian Sea in W Asia. Usual habitat differs substantially from similar plovers: much drier inland habitats, grassland, lake-edge mud, temporary flood water. Most migrate to S and E Africa where they winter inland on dry plains, like the Pine Creek area of inland NT where Australia's sole specimen was shot in 1896. Behaviour is similar to the Oriental Plover. Usual call is a loud, sharp 'tyip' or 'tyik' and, more rapidly, a rattling 'tip-tp-tp-tptptp'; also a loud, shrill, whistled 'kwheeeit'. Very rare vagrant or accidental; few sightings, but some confirmed; most reports from coastal NT and NE Qld.

Nonbr.

Distinctive only in breeding plumage, otherwise difficult to separate from more common Oriental Plover.

Brown rump and tail

Feet extend slightly beyond tail tip.

Narrow wing bar leads out to flash of white on inner primaries.

Long, attenuated rear body

Light chestnut

Dark border

Head pattern as when breeding, but dull brown.

♂ Br.

Mottled grey-brown

Long, olive or grey

Under-wing-coverts white

Br.

Female breeding: breast-band is mostly brown or has a few scattered chestnut feathers; head more like nonbreeding.

Nonbr.

Juv.: similar to nonbreeding, but with buff tint to breast band; upper surface feathers are dark centred with rufous or buff fringes.

## ORIENTAL PLOVER & INLAND DOTTEREL

**Oriental Plover** *Charadrius veredus* 21–25 cm 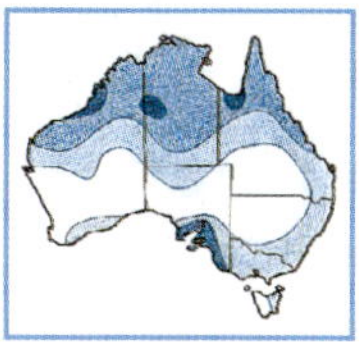
Long-legged plover of semi-arid regions: open grassland, claypans or gibberstone plains. Less often on marine sites such as tidal mudflats typically used by other plovers. Also, occasionally, where dense vegetation of spinifex, heath or similar has recently been burnt. On migration in N Aust., gathers in flocks on open, thinly vegetated, grassland; these flocks reach hundreds, even thousands, of birds. Often forages at night, roosts by day with other waders on beaches or mudflats. May be seen in company with Inland Dotterels or pratincoles. Foraging is typical – stop-start, running, bobbing down to peck at prey. Tends to be rather wary, bobbing head when alarmed; runs rapidly. If put to flight, has powerful action; fast and high while twisting and turning erratically, the many birds of the flock seem always to change direction in unison. Flight call is a soft, very quick 'tik' or 'tink' repeated irregularly, especially among birds of a flock; also has a sharp, rapid trill. Common annual migrant, Sep.–Nov.

Very pale-headed; brown of crown and through eye pale, soft-edged

White underparts a strong contrast between dark brown wings

Br.

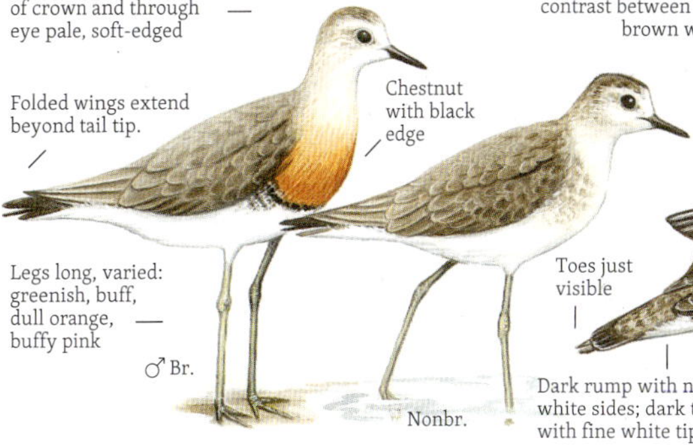

Folded wings extend beyond tail tip.

Chestnut with black edge

Nonbr.

Legs long, varied: greenish, buff, dull orange, buffy pink

Toes just visible

♂ Br.

Nonbr.

Dark rump with narrow white sides; dark tail with fine white tip

Breeding female has grey-brown breast band; may have some chestnut intermixed.

Only very fine wing bar with no white streaks to bases of inner primaries

Long, dark, pointed wings

**Inland Dotterel** *Peltohyas australis* 19–22 cm 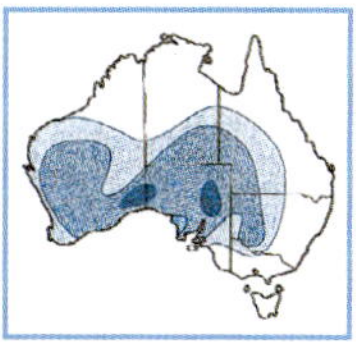
Unique wader adapted to arid habitats: arid plains, open flat country with sparse vegetation – gibber plains, claypans, gravelly flats, open bare clay soils. At times seen near water of ephemeral creeks on claypans, but can be far from any water. Gregarious, small to large flocks; mainly nocturnal; runs to escape – takes flight as a last resort, showing long wings, buff-brown patterning. Quiet outside breeding season unless disturbed. Then, as contact by birds separated from flock, a sharp, metallic 'kwik' or 'quoik' given as a single or double-noted call, or as a longer sequence with notes at half-second intervals. Calls persist while flock is scattered, running; has been heard from flocks feeding at dusk. Endemic species, probably nomadic or locally migratory.

Chestnut flanks

Unusual in having vertical dark line through each eye and meeting as a band across the crown

Y-shaped breast-band

Faint streak through eye

Deep buff

Darker

Twin encircling black bands

Wing tips are about level with tail tip.

White

Chestnut flanks

Juv.

Long legs; has upright alert stance.

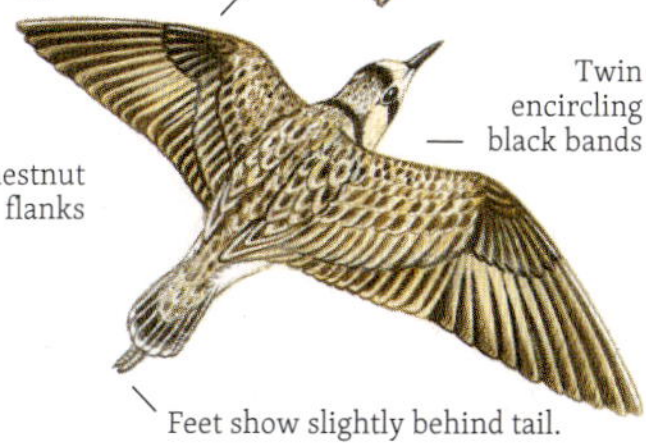

Feet show slightly behind tail.

145

## PHALAROPES; PHEASANT-TAILED JACANA

**Red-necked Phalarope** *Phalaropus lobatus* 18–19 cm
Gregarious; congregates on seas rich in plankton; occasionally is blown ashore or shelters from gales on coastal wetlands. Often swims – floats very high, lightly, on water – or wades; infrequently comes to land. Feeds by pecking at small creatures on water's surface. Male smaller than female and paler in breeding plumage. Call, in flight, a soft 'chek' and 'chwik'. Most alike is Grey Phalarope in nonbreeding plumage, but has a heavier build, eye-patch is square-ended rather than slender and downturned, and upper parts usually plainer, paler grey. Recordings mainly coastal, but also inland on brackish, saline or fresh pools and lagoons and their muddy margins. Spends nonbreeding period at sea. Rare vagrant.

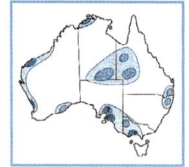

Breeding females are more brightly coloured than males.

**Grey Phalarope** *Phalaropus fulicaria* 20–22 cm
A large, plump phalarope, appears to have long slender neck, small head, bold white wing bar, extensive white under wing. This species and the Red-necked both spend much time at sea while escaping the N winter. Strays or shelters from time to time on lagoons or other protected coastal waters. Aust. records are from coastal sites, typically brackish and saline lagoons surrounded with salt marsh or wet grassland with exposed mud margins. Swims quite high in water, perhaps with tail held higher than the Red-necked, and, like it, often spins on the water surface, pecking at surrounding floating items. In flight, erratic over short distances, or has a distinctive, side to side 'jinking'. Usually quiet, but, as flight and contact call, a sharp 'whit' or 'twit'. In alarm, a more musical 'zhwhit'. Rare vagrant or migrant; few confirmed records; most sightings from SE and SW Aust.

**Pheasant-tailed Jacana**
*Hydrophasianus chirurgus* 35 cm
Larger than Comb-crested Jacana, and without any wattle rising above crown. Has obvious white wings, above and beneath; gold and black line from brow to shoulder. Breeding: very long dark tail. Nearest usual occurrence in Java, Timor and NG. An extremely rare vagrant to Aust. Several sightings, Pilbara and Kimberley regions of WA.

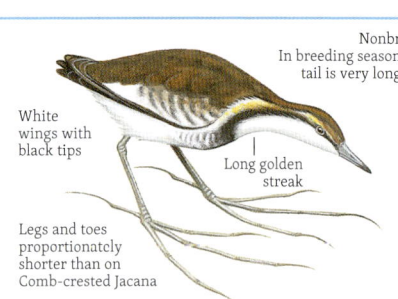

# WILSON'S PHALAROPE; COMB-CRESTED JACANA

**Wilson's Phalarope** *Phalaropus tricolor* 22–24 cm
Often with other waders – very active, forages in shallows, swinging bill through water, but less aquatic than other phalaropes. Forages less in deeper water, more often darting about in shallows and around waterline. Runs erratically; stalks insects. Uses marine situations in sheltered tidal estuaries, lagoons; more often around swamps, shallows of lakes. Usually silent, occasionally a soft, nasal, grunted 'aarrgh'. Migrant from Arctic, usually winters S to seas off W Africa and S America. In Aust., extremely rare, uncertain vagrant: several sight records in SE Aust.

Plumage of breeding male has darker brown and dull orange replacing the female's lighter, brighter grey and rufous.

Broad black eye stripe continues down neck.

Legs yellow on nonbreeding; extend behind in flight.

White fringes on fresh plumage

Grey from crown to upper back

Pale grey

Rufous

Dark eye streak on white face

Faint white wing bar

Legs dark when breeding

♀ Br.

Nonbr.

Toes lobed; unlikely to be visible in field

**Comb-crested Jacana** *Irediparra gallinacea* 20–27 cm
Highly adapted to a life entirely on tropical lagoons. Use lakes, swamps and dams where there are waterlilies or other extensive cover of floating vegetation. Not only feeds but nests on the floating lily leaves, building a flimsy nest of fine green water-plant stems, largely supported by the lilies. Chicks, already with long toes at hatching, are quickly able to avoid predators such as raptors and water pythons that congregate at wetlands. Small chicks are often carried tucked up under the wings. Calls quite often and regularly with diverse sounds. Some are typically given in confrontation with other Jacanas; others are twittering, chittering, piping notes, usually given while in flight or standing tall in upright, alert pose. Sequences may blend from one to the other. Also has a sharp, nasal alarm calling and searching for small water creatures and insects. Locally abundant; sedentary on large, rich, undamaged wetlands in remote or protected far N coastal Aust. Increasingly uncommon to scarce towards the SE extremity of its range.

Also known as Lotusbird.

Extremely long toes enable these birds to walk on lily leaves. The slender claws are enormously elongated, especially that of the hind toe.

In flight, long legs and toes trail conspicuously, far behind.

Under wing is entirely black.

Outer wing and all secondaries brownish-black

Coverts mid olive-brown

Tail well short of folded wings

Dark olive-brown with lighter iridescent highlights

Deep pink, rather glossy surfaced, fleshy wattle

Small, pale wattle and comb; rufous crown

Golden-buff

Deep rufous

White, no buff

Juv.

Legs and feet olive-green, olive-grey or dark grey.

## LAPWINGS

**Masked Lapwing** *Vanellus miles* 35–39 cm
A large, conspicuous, noisy and often aggressive plover; well known in those parts of Aust. where it occurs. Habitat varied, but typically open, short-grassed sites, both natural and modified, often beside water of swamps, lagoons and salt marshes. Noisy early in nesting season; strongly defends nest site; dives at intruders, at times striking with wing spurs. In pairs or small family groups during breeding season; at other times in large flocks that may number several hundred. When feeding, stalks slowly, deliberately, body horizontal, dipping and stabbing at prey; when alert, stands upright. Voice is loud, penetrating. As alarm and in threat, noisy 'karrek-karrak-karrak'. If the young are approached, the noise lifts in pitch and tempo, 'karri-karrik-karrik' and 'kek-kekkekekeke'. Common breeding resident of N and E Aust.

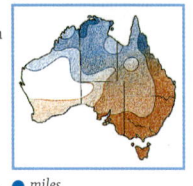

● *miles*
● *novaehollandiae*

Races overlap; many with intermediate size and varying extent of yellow wattles and black.

● **Masked Lapwing**
Larger wattles, small black cap, neck entirely white

Small black cap
Large yellow facial wattles
Small wattles, black hind neck
● Black hind neck
White with wide black band
Spur
Spur
Long red legs
● **'Spur-winged Plover'**
Former species, now race

Flight feathers black, wing tips broad
Fringed buff
Juv.

Nest is a shallow scrape in ground, normally 3 to 4 eggs.

---

**Banded Lapwing** *Vanellus tricolor* 25–29 cm
Uses open country with short grass, low, sparse shrubbery, open acacia or eucalypt woodland where trees are sparsely scattered. Often on agricultural land – ploughed paddocks, grazed pasture, emergent crops or mown grass such as golf courses. Strongly territorial when nesting; at other times highly gregarious. If disturbed, quick to take flight with abrupt jerky wing beats. Usually a grating, strident, resonant, very abrupt 'kerr-kerr-kerr-kerr-' given in rapid-fire bursts. Also a higher, sharper 'quirrrk-quirrrk-quirrk', at times with a ringing effect. Endemic species; common and widespread across S Aust.; vagrant, nomadic or dispersive.

Lapwings are named for their flight; the broad-ended wings flap with a distinctive, hesitant action.

Rump and tail white except for broad black band near the tip

Black cap with white eye-stripe; yellow eye-ring and bill; small red wattle over bill

Plain grey-brown

Black

Wing bar an obvious white diagonal across each wing

Juv.

Dark brown, buff margins

Bold, heavy, black vertical line from bill down to lower breast, from front a deep, rounded U shape enclosing white throat

Legs pink to pinkish grey

Under-wing-coverts and axillaries are white in contrast against black wing tips.

# OYSTERCATCHERS

**Pied Oystercatcher** *Haematopus longirostris* 42–50 cm
Coastal: beaches and mudflats of inlets, bays, ocean beaches and offshore islets; less often rocky coasts, headlands. Takes oysters and other marine life along water-line. Flight is fast and direct with rapid, shallow wing beats; keeps along the beach or out to nearby reefs and islets. Solitary, in pairs or in small family groups. Calls are high, not harsh or sharp, but ringing, mellow, resonant, 'quip-quip-quip-quip-quip', very even in pitch. In flight, 'quip-a-peep, quipapeep, quipapeep'. In alarm, loud, sharp, high-pitched 'kervee-curvee-curvee'. Common; but vulnerable to disturbance; becoming uncommon in parts of SE.

- In alert, upright posture
- White point or hook up between black of breast and of wing
- Flight feathers dark, coverts white
- White bar narrows to a fine point.
- Sharp edge to black breast
- White edge of wing stripe may be visible; much larger on SIPO, below.
- Legs medium long, stout, pink
- Brown eye
- Dark brown with orange-buff tips and notches
- Black tail, white rump
- Juv.
- Stands, forages, in hunched posture, bill downward. When alert, stands tall, neck extended.
- Black upper parts with white wing bars tapering inwards to a point, narrowest near the body, and not including any of the trailing edge, which is entirely black, unlike SIPO

**South Island Pied Oystercatcher (SIPO)** *Haematopus finschi* 42–48 cm
Vagrant to Aust. coast.

- Bill longer, finer, orange-red
- Habitat in Aust., sandy beaches, where it may pass unnoticed among Pied Oystercatchers
- White
- White wing bar, broad near body; includes part of the trailing edge.
- Legs 1/3 shorter, making SIPO look dumpy, short-legged, especially when seen in company with Pied Oystercatchers.
- White stripe on folded wing larger than on Pied
- White of rump and lower back form a sharp white triangle between wings.

**Sooty Oystercatcher** *Haematopus fuliginosus* 40–52 cm
Large, stocky oystercatcher; habitat purely marine. Usually on rocky shorelines, high rocky islets, boulders below cliffs, wave-cut platforms and reefs. Also visits and forages on sandy beaches and coves between rocky headlands. Solitary, pairs or flocks. When disturbed, walks away rapidly; flies if close pressed – low to waves, straight ahead or circling back 50 m or so. High, clear, piping calls; sharper, more piercing and quicker than the Pied; usually 'kier-kier-kier-kier-kier…'. May develop into a double-noted 'kwi-keer, kwikeer, kwikeer'; often given in flight. Also gives loud whistling calls before taking flight, and piercing calls if an intruder approaches nest. Sedentary; generally uncommon, scarce on disturbed coastlines, common on parts of N coasts.

- ● *fuliginosus*
- ● *ophthalmicus*: NW, N and NE coasts
- Scarlet eye-ring and iris
- Scarlet-pink bill
- Shorter than legs of Pied Oystercatcher; deep crimson-pink
- Stocky build; plumage entirely sooty black
- Overall black
- Nonbr.: legs, eyes, bill, dull colour
- ● Smaller eye-ring, shorter bill
- ● Has wider eye-ring, heavier bill

149

## STONE-CURLEWS

**Bush Stone-curlew** *Burhinus grallarius* 55–60 cm
The wailing call is one of the most characteristic sounds of the bush at night. By day the secretive, camouflaged birds, though large, are hard to see; they freeze flat on the ground or skulk away. Calls are eerie, a drawn out, mournful 'wee-ier, wee-ieer, whee-ieeer, whee-ieer-loo'; each call rises, strengthening, faster, building to a climax, then trails away. At night, the spine-tingling sounds carry far across lonely bush and paddocks. Generally unmistakeable; nearest perhaps is the Beach Stone-curlew. Habitat usually open woodland, lightly timbered country, mallee and mulga – anywhere with groundcover of small sparse shrubs, grass or litter of twigs. Avoids dense forest, closed canopy habitats. Common across N and NE; uncommon to rare in SE, SW.

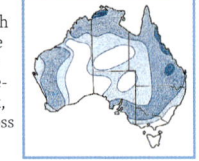

Intermediate, grey-brown form

Tall with long legs. Large head with big yellow eyes, cryptic plumage

Large, baleful yellow eye

Long legs trail far behind tail tip; pale yellow, olive or grey.

Dark tips to feathers

Dark stripe under eye, down side of neck

White across outer primaries

Dark tips to outer feathers

Rufous morph: overall warmer colour tones

Patterned and dark streaked with pale band across wing

Grey morph

Conspicuous white spot

Protective display over young

Nest is a scrape on the ground; camouflaged eggs lie on bare ground.

**Beach Stone-curlew** *Esacus magnirostris* 54–56 cm
Confined to the marine tidal zone – mudflats, mangroves, sandy, stony and rocky shores, sheltered or exposed to ocean breakers. Less strictly nocturnal than the Bush Stone-curlew; forages with slow, deliberate, heron-like actions. Usually wary; if disturbed flies ahead showing distinctive wing pattern. As alarm call is a quick 'chwip' repeated at intervals of about a second, perhaps also as 'chwip-chwip'. Territorial calls are given at night, like those of the Bush Stone-curlew, but harsher and at higher pitch, a 'weer-liew' repeated about eight times, each higher and faster. Considered vulnerable due to coastal disturbance, only secure in remote parts of N coast.

Feet trail only slightly beyond tail tip.

Thick white brow, strongly downcurved

Large, very heavy bill, also evident on chicks; yellow at base

Long black and white horizontal wing bar

Dark

Short white bar

Heavy bill evident on chicks

A large bird of unusual appearance: heavy horizontal body, tall neck, deep heavy bill, strongly patterned face, short legs

## STILTS & RED-NECKED AVOCET

**Banded Stilt** *Cladorhynchus leucocephalus* 35–43 cm
Unique endemic wader, most likely to be seen on ocean beaches after dispersal from breeding. On salt lakes of coast and inland: in large flocks on temporary flooded saltpan lakes, also on marine beaches of estuaries and intertidal flats. Calls, from flock, musical resonant yapping; varied, deeper, more mellow than Pied Stilt. Individual notes range from abrupt 'ohk' to slightly longer 'chowk' and 'chowk-ok'. In a large assembly, intermingled calls of hundreds of birds become a pleasant babble of 'ohk-chowk-ok-chok-ohk', a mixture of high and low, loud and soft. Common in parts of range; disperses widely.

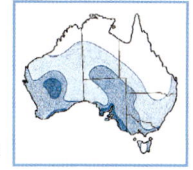

**Pied Stilt** *Himantopus leucocephalus* 33–37 cm
Slender, elegant, gregarious wader with incredibly long legs. Struts gracefully through shallows; bobs head and calls when alarmed. Flies swiftly; long legs trail; wings beat strongly. Inhabitant of shallow wetlands – interior claypans, flooded paddocks, salt lakes. Call is a regular high, nasal, yapping with slight variations of pitch and strength, like a toy trumpet: 'ap, ak, ap-ap, ak, ap, ap, ak-ap'. Higher pitch and more nasal than the mellow notes of the Banded Stilt; slight variations from individuals of group. Widespread; common; absent from sandy deserts; dispersive or nomadic.

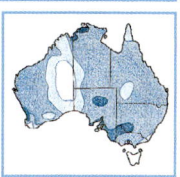

Aust. stilts are race *leucocephalus* of this widespread species.

**Red-necked Avocet** *Recurvirostra novaehollandiae* 40–48 cm
Unique to Aust. A graceful, colourful, large wader with long, slender, strongly upcurved bill. Uses both salt and freshwater wetlands. Large numbers tend to congregate on large, shallow salt lakes, particularly as salinity increases through evaporation; then feeds on brine shrimp. Also use shallow freshwater swamps and lakes, temporary flood waters, claypans, dams, saltworks. Feeds by wading through shallow water; sweeps the submerged bill from side to side. Almost always in large flocks or groups, often with stilts. Often silent; usually calls when disturbed – a yapping similar to that of the Pied Stilt, but less abrupt, higher and more metallic or nasal, 'aik, airk airk, airk, aik, aik' with slight variations. Quite common over most of continent, but scarce near N and NE coasts, vagrant to Tas.

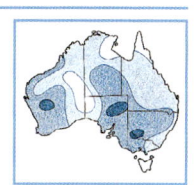

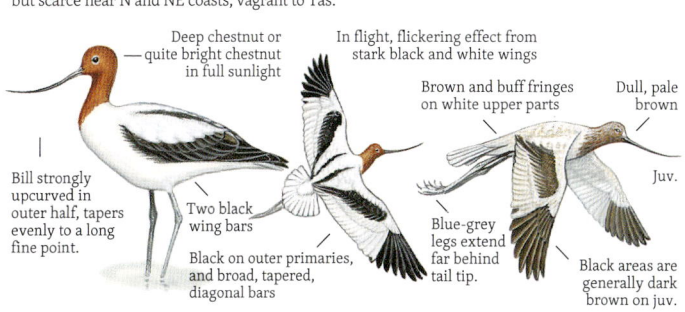

# 10 SKUAS, JAEGERS, GULLS, TERNS & NODDIES

This group contains but one family, the Laridae, commonly known as the gull family, for these birds of coasts and waterways are so familiar worldwide. The gulls, genus *Larus*, represent a middle ground between the very large predatory skuas and jaegers and the mostly slender, lightly built terns. In the guide chart below, most of the gull and tern species are included, but only one representative each of skua and jaeger. These, however, are shown on the opposite page, along with others of the family. These are the large species, and also those with very dark plumage across the upper surface of the wings at least. Also gathered together on page 153 are eleven species of smaller gulls and terns that have some form of distinctive wing markings; many of the other species have rather plain wings, not so helpful in identification.

# SKUAS, JAEGERS, GULLS, TERNS & NODDIES

Silver Gull = 40 cm

laridae (cont.)
page 166

Whiskered Tern

White-winged Black Tern

Black Tern

Brown Noddy

Black Noddy

Lesser Noddy

Grey Ternlet

White Tern

Adults and immatures of skuas, jaegers, gulls and noddies that have very dark wings in common. Medium to large species; not to scale.

Great Skua, Pomarine Jaeger, Pacific Gull, Imm., South Polar Skua, Kelp Gull, Long-tailed Jaeger, Arctic Jaeger, Imm., Imm., Lesser Black-backed Gull, Imm., Black-tailed Gull, Brown Noddy, Lesser Noddy

Gulls and terns with distinctive markings on upper wings, breeding or nonbreeding; plain-winged species not shown. Not to scale.

Silver Gull, Little Tern, Black-headed Gull, Fairy Tern, Laughing Gull, Bridled Tern, Sooty Tern, Sabine's Gull, Whiskered Tern, Franklin's Gull, White-winged Black Tern

153

## SKUAS & ARCTIC JAEGER

**Brown Skua** *Stercorarius antarcticus* 51–66 cm
Usually keeps to the open oceans, but occasionally found over inshore waters. Primary habitat is open ocean and continental shelf slopes, less often inshore waters, rarely bays or estuaries. Powerful flight; uses its great agility and acceleration in aerial pursuit of other seabirds for their food. Silent except in breeding colonies. Regular but uncommon winter visitor to S Aust. seas.

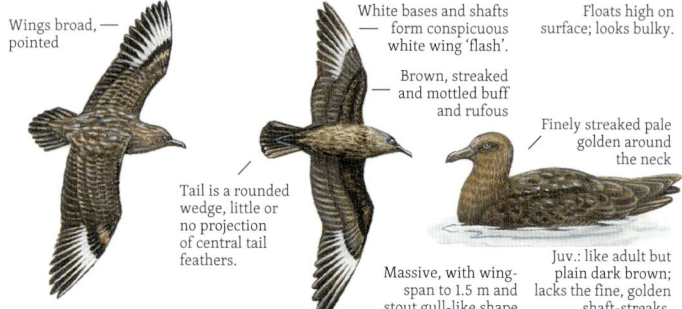

**South Polar Skua** *Stercorarius maccormicki* 51–54 cm
Large gull-like predator; forages over open sea; plummets from heights up to 6 m to take prey on or just beneath the surface. Silent at sea; squeals and screams at Antarctic breeding grounds. Usual habitat is the islands and ice shelf of Antarctica. Usually offshore, only occasionally close inshore, rarely in bays or harbours. Few records confirmed; scattered sightings around the southern coastline.

**Arctic Jaeger** *Stercorarius parasiticus* 41–46 cm
Medium-sized with slender bill, long narrow wings. Aggressive aerial pirate: attacks and robs other seabirds of their catch. Silent away from the breeding grounds except for some high, nasal squealing at others of its kind when squabbling over food. In nonbreeding months, subtropical and sub-Antarctic seas, inshore waters, shallower waters of the continental shelf and bays, estuaries and harbours. Arrives Aust. waters Oct.–Nov.; departs Apr.–May.

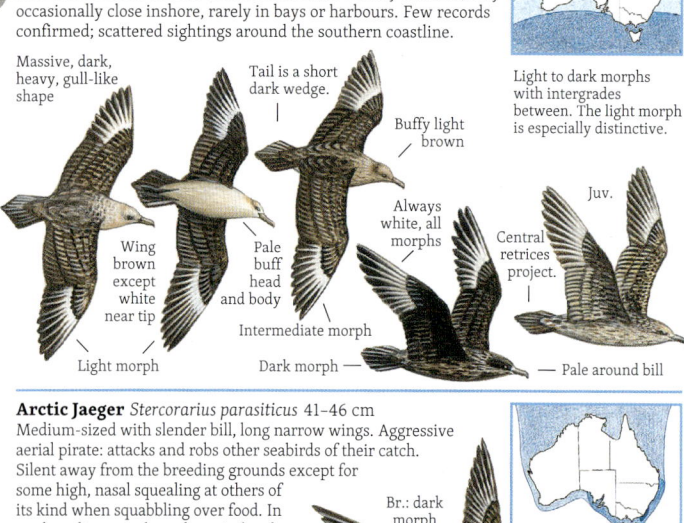

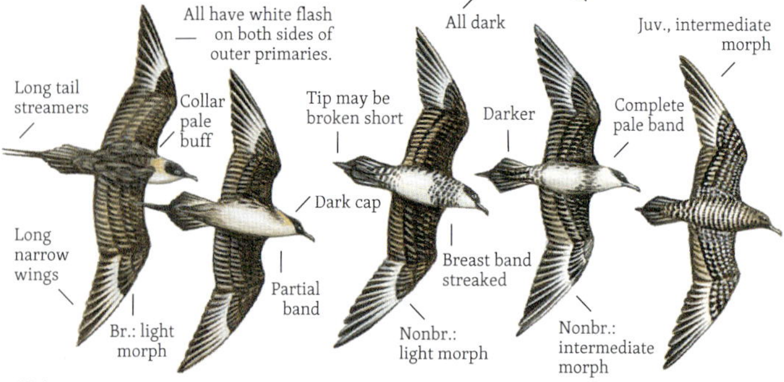

# JAEGERS

## Pomarine Jaeger *Stercorarius pomarinus* 66–78 cm

A more heavily built jaeger, rather large-headed, barrel-chested. In flight, gives impression of more power than other jaegers; deep wing beats like large skua or gull. Makes direct attacks on chosen seabird victim, flying with great agility through dives and tight turns until the harassed bird, which may be up to small albatross size, gives up its catch. During migration, wanders tropical and subtropical seas; occurs in Aust. coastal waters Oct.–May. Most common along edge of continental shelf, scarce closer inshore, only rarely seen in harbours and estuaries. There are light and dark morphs; the light form is thought to outnumber the dark by a ratio of about 20:1. At sea, calls when competing for scraps around fishing trawlers. Uncommon, nonbreeding migrant; recorded most frequently off Qld coast. This is the most common species of jaeger off NSW, and there are many records around most of the coast – Vic., northern Tas., SA, WA and NT.

## Long-tailed Jaeger *Stercorarius longicaudus* 50–55 cm

In flight, lighter, more elegant and tern-like than other jaegers; the buoyancy imparted by undulations of body with each wing beat. Patrols low, almost stalls, then often dips to surface. Robs other seabirds, harassing with quick, agile flight; but often feeds tern-like at sea surface. Adults polymorphic, but almost all are light phase. Dark morph is very rare; the intermediate, if it exists, is extremely rare. Apparently intermediate plumaged juveniles become light adults, as do light juveniles. Rarely if ever call away from breeding grounds. In nonbreeding months wanders oceans; tends to use edge of continental shelf where most Aust. sightings occur, rarely inshore waters. Regular but uncommon visitor.

Juveniles occur in light, dark and intermediate morphs, all barred, especially on under-tail- and under-wing-coverts.

# WHITE-HEADED GULLS

**Pacific Gull** *Larus pacificus* 50–66 cm
Very large gull, coastal habitat: nominate race of E Aust. more often uses sheltered beaches; those in W also seen on coasts exposed to open ocean. Often on islands, occasionally inland rivers, coastal lakes. Patrols coastal seas and shoreline in heavy, lumbering flight interspersed with glides. Seen in flocks, small groups, alone. Two races: nominate in SE Aust. and *georgii* in SA and WA. In alarm, gives a mournful 'ohk-ohk', or, shorter and sharper with greater anxiety, 'owk-owk-owk', barked 'ok-ok-ok' flying over intruders. Long, peevish, thin, piping 'airrk, airrk'. Endemic to Aust.; common on W and S coasts; scarce parts of SE where it is being displaced by the Kelp Gull.

● *pacificus*
● *georgii*

Page headings 'White-headed' and 'Black-headed' describe adults; if head patchy black or brown, check under both headings for other plumages.

Massive gull, adult's wings mostly black, body white; juv. blotchy brown

Usually black band

Developing white rump, black band

Whiter than juvenile

Darker, nearer to black of adult

Brown, paler tip

Juv.

Mottled brown-white

Imm., yr 3

White head

Massive, pinkish base

Dark brown, paler on face

White eye

Red eye

Massive, yellow, red-tipped bill

Black

Juv.

Breeding: slightly brighter colours of bill, orange legs and orbital ring compared with nonbreeding.

White wing bar

Folded, plain dark primaries extend far beyond tip of black-banded tail.

Immatures are recognisable up to the third or fourth year, but exact definition of age is complicated by individual variation.

**Kelp Gull** *Larus dominicanus* 50–62 cm
Slightly smaller than the endemic Pacific Gull; bill not as massive. Combination of black wings, all-white tail, white-tipped primaries and wider white trailing edge to wings makes identification of adult birds easy. Habitat in Aust. is almost entirely coastal; usually within sight of land; rarely more than 10 km offshore. Uses the more protected parts of coast – estuaries, bays, mudflats, offshore islands, also on wetland near coast. Call common throughout year. Starts softly, then loud notes: 'huh, huh, ee-arh-har-har-har-har' or 'yho-yho-yho-'; usually around 15 notes in series. Not recorded for Aust. until 1943; increasing; now common in parts of E coast and breeds in WA, NSW and Tas.

No black

Wide white edge

Thin white edge

Pale, mottled brown

Head brown with pale streaks

Dark tip

Partly barred

White 'mirror' on first primary; white tips to outer six

Imm., yr 2

Juv.

Upper parts dark-brown-fringed dull off-white

White

Black

Juv.

White primary tips

Off-white, streaked brown

Juv.

Red on lower tip

Legs bright yellow when breeding, at other times dull yellowish to grey-green

Bold white crescents

All white tail, no band

156

## WHITE-HEADED GULLS

### Black-tailed Gull *Larus crassirostris* 45–48 cm
Grey-backed, long-winged, short-legged, medium-sized gull. Stands slightly taller than the common Silver Gull, even though the legs are relatively short. Flight is similar to that of Silver Gull, but slower wing action. Call is a hard bass 'kaou, kaou' or 'yaou'; also plaintive mewing. Within its normal range, this gull is found on sandy or rocky coasts, inshore waters and islands, estuaries, harbours. Occasional vagrants to Aust.: sightings at Darwin, NT; Port Phillip Bay, Vic.; WA.

### Lesser Black-backed Gull *Larus fuscus* 50–60 cm
A very large gull that may occur in Aust. waters. This species migrates from Iceland, N Europe and N Russia, but not usually as far as Aust. Habitat in normal range includes coasts, inland and inshore waters, estuaries, lagoons. Occurrence on Aust. seas is as yet unproven, but the number of claimed sightings suggests that an acceptably confirmed sighting may yet occur.

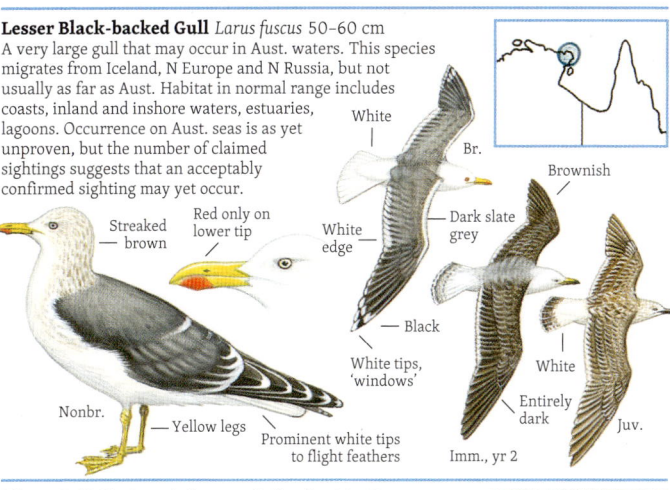

### Silver Gull *Chroicocephalus novaehollandiae* 38–42 cm
Well known, often tame; noisy, bold, gregarious, scavenging; often in large flocks. Habitat diverse, from the surf and cliffs of ocean coasts to offshore islands, inland rivers, lakes, temporary floodwaters, cultivation, ponds, coastal towns, rubbish dumps. Flight direct, unhurried. Diverse calls, especially in flocks. Includes an aggressive, rough 'karrgh-karrgh-karrgh', deeper, guttural 'korrr', higher 'kwee-arrgh', wheezy sounds from juveniles. Widespread, opportunistic, adaptable: abundant in S, less common in N, sporadic inland.

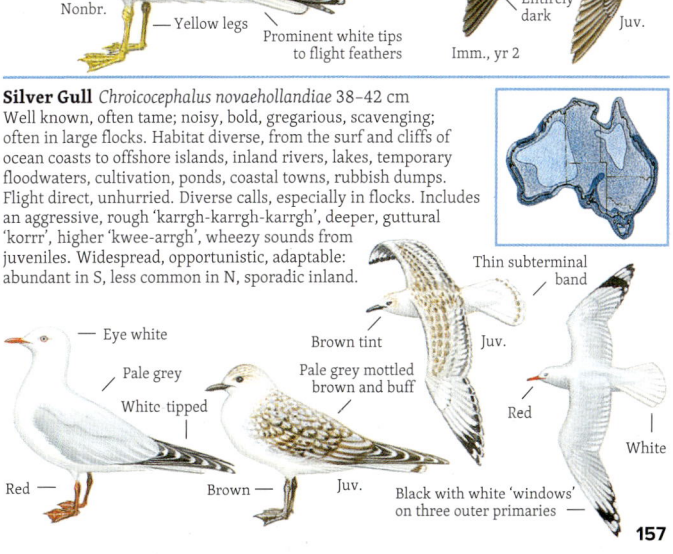

# BLACK-HEADED GULLS

**Franklin's Gull** *Leucophaeus pipixcan* 32–35 cm
A small distinctive gull whose breeding grounds are prairie marshes of N America. On migration uses both coastal and inland habitats. Along main routes south occurs in very large flocks. In Aust. sightings have been on sandy beaches, mudflats, rocky spits and near coastal wetland, all on sheltered parts of the coast in estuaries and bays. Perched, has a rather low, hunched horizontal posture. In flight, is agile, light and buoyant with paddling beats of slightly rounded wings. Feeds by dipping and snatching at surface. Silent outside colonies, where call is a soft 'krruk'. Widespread sightings, although few verified and accepted; rare vagrant, early spring to autumn.

Immature has nape half-hood, but less intensely black than on nonbreeding adult.

Imm. and juv. have black legs, adults red

Large white tips to primaries

Legs are shorter than those of Laughing Gull and, in breeding plumage, hood comes lower down front of neck.

Outer wing dark — No wing tip spots

White spot or 'mirror'

Nonbr. has diagnostic half-hood over nape.

White trailing edge

Mid blue-grey

Black band across tips of primaries

Hood includes nape, throat, upper fore-neck.

Some have a faint pink flush on the white plumage.

**Laughing Gull** *Leucophaeus atricilla* 36–40 cm
Distinctive in breeding plumage, with fine white crescents above and below eye, conspicuous on the black heads. While only occasional solitary vagrants have been sighted in Aust., this gull in its usual range is gregarious. Typical habitat is coastal – could be likely in any situation favoured by the Silver Gull. Scavenges in harbours, along tide line, over coastal reaches of rivers; follows fishing boats. Flight is slower – deeper, more leisurely wingbeats – than Silver Gull. Readily and skilfully snatches food from sea surface. On land, stands tall, upright on long legs when alert, or more horizontal when relaxed. The call is a strident, high 'ca-ha', often prolonged to a frenetic laughing: 'ha-ha-ha-haah-haah'. Records from Tas., S coast WA, SE Vic. and Cairns, NE Qld.

A small, slender gull. The sleek appearance is emphasised by long wings that, when folded, extend well beyond the tail tip.

Dusky markings through eye, around nape

Spots small

Bill is long, heavy, with downcurved top edge

Legs red to black breeding, greyish nonbreeding

Wings pointed, long

Dark / White

Hood over nape and throat, but not to neck

White trailing edge

Dark, nape to eye

Large, heavy

Pale, dusky

Narrow white trailing edge

Outer wing plain — dark brown

**Sabine's Gull** *Xema sabini* 27–31 cm

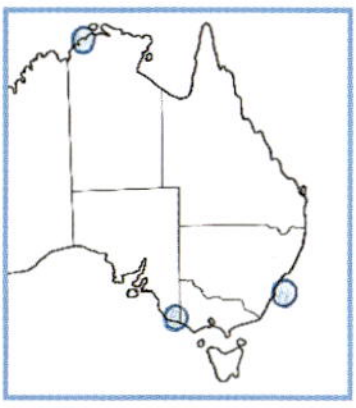

Small, slender bodied, long pointed wings, distinctive upper wing pattern. The tail has a shallow fork, unique in Aust. region. This species differs from most other gulls in retaining juvenile plumage until its arrival in southern wintering areas. Breeds arctic parts of Russia, Alaska, Canada. Very few travel as far S. as Aust., where it occurs as solitary vagrants. In the nonbreeding season is mainly pelagic, favouring deep water along edge of continental shelves, but also occasionally comes over shallower shelf waters. Flight is tern-like, light and buoyant, deep wing beats, little gliding. Feeds tern-like, picking up items from water or land. The call is a harsh grating. Sightings are very widely spaced in NT, NSW and SA; all occurring autumn to winter months.

**Black-headed Gull** *Chroicocephalus ridibundus* 36–42 cm

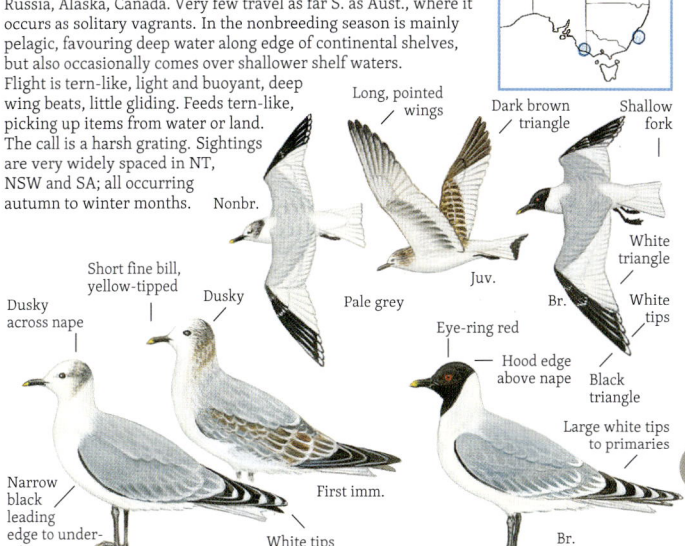

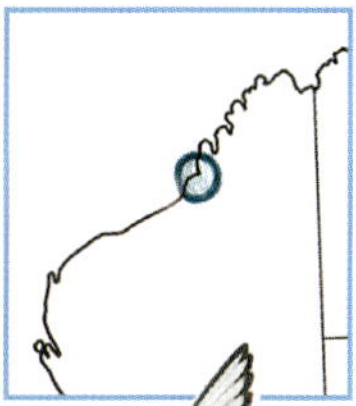

A small, slender gull with thin, long, pointed wings, rapid wing-beats; general appearance like Silver Gull, but slightly smaller. Diagnostic wing pattern of white triangle is conspicuous between black margins of wing tips; hood usually looks near black at a distance. Habitat in usual range is varied: shallow wetland, coast and inshore water, bays, estuaries, salt marsh, wet grassland and ploughed fields. In Aust., recorded at sewage ponds. Vocal throughout the year: varied screams; harsh, quite high pitched scolding 'karrgh' and 'kraaak'; various other calls – a drawn out 'kreeoo' and 'kekk' in alarm. Abundant, widespread, gregarious species; breeds Iceland, Europe, through Asia to Japan, China, Kamchatka Peninsula. Migrates to Africa, SE Asia; reaches Indonesia, NG, and, rarely, N Aust. A confirmed sighting in breeding plumage at Broome.

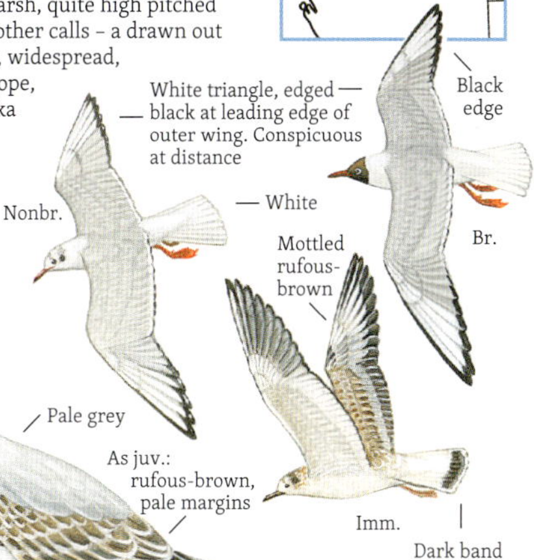


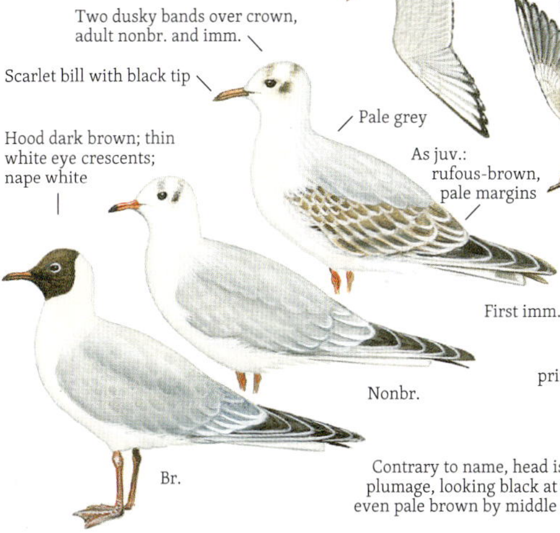


Black tip only on outer primaries; no white windows

Contrary to name, head is very dark brown in fresh plumage, looking black at a distance; fades to mid or even pale brown by middle of N hemisphere summer.

## GREATER CRESTED & LESSER CRESTED TERNS

**Greater Crested Tern** *Thalasseus bergii* 43–48 cm
This common, familiar large tern of bays and harbours often roosts on boats and jetties. Habitat includes ocean beaches, offshore islands and out over deeper pelagic waters, inshore on estuaries, bays, harbours, coastal lagoons, inland on major rivers, occasionally on saline lakes, salt ponds near coast. Flight is powerful, swift, long pointed wings angled back, deep beats. May be in mixed flocks with other terns and gulls. Forages by plunging from several metres to take prey just under the surface. Noisy, especially in breeding colonies, at night and in dawn flights. Intruders near nests or chicks cause very noisy reaction from colony. Alarm call an abrupt 'wep'. Common advertising call a raucous 'graaak' or 'kirrak'. Widespread and common around Aust. coast; breeding resident.

Large, black capped tern with wide white gap between black forehead and base of bill

Lemon or citrine yellow, unlike the warmer orange-yellow or orange of the Lesser Crested Tern

Lesser Crested, below, has orange tone to straight bill and much less white on forehead.

*Labels:* Silvery mid grey — Forked tail — Nonbr. — White under-wing-coverts — Br. — Solid black only at rear, on crest — Long scythe-like dark grey wings — Nonbr. — Black nape — Mottled coverts make paler panel. — Juv. — White gap — Bill slightly down-curved — Br. — Mottled brown — Juv. — Legs black, all ages, plumages

**Lesser Crested Tern** *Thalasseus bengalensis* 38–43 cm
Closely resembles the Greater Crested Tern, but is smaller, more slender, paler grey. In flight, shows shorter but slim wings giving more delicate, compact jiz. If the two species are seen together the size difference is obvious. Voice like Greater Crested but less harsh: a grating 'kik-kerek' and ratchetting 'gr-a-a-a-k', each 'a' hard, metallic, abrupt, also a slower, harsh 'grruk-uk-uk'. The long, shaggy crests of the two species of crested tern separate them from all other Aust. terns; the Caspian has a much shorter crest and massive red bill. Frequents coastal seas using shores of sandy beaches, coral cays, exposed reefs, islands; on parts of coast uses mudflats of estuaries, creek channels. Breeding resident, most common in NE Qld, breeding on reef islands S to Capricorn Group. Most colonies have a few to several hundred pairs; a few have above a thousand pairs. Also breeds NW WA, Adele Is. and Bedout Is.

*Labels:* Silvery grey — Dark edge — White — Black cap — Br. — Orange, straight — Crested cap only over nape, crown streaked — Pale — Forked — Black cap down almost to bill — Juv. — Bill colour: nonbr. and juv. similar, only slightly less colourful than breeding — Juv. — White — Short, black — Br. — Light pearly grey — Primaries silvery grey; may have darker outer primaries.

## CASPIAN & GULL-BILLED TERNS

**Caspian Tern** *Hydroprogne caspia* 48–54 cm
Flight like that of large gulls – powerful, deep, regular, wing beats, yet swift and graceful on long, slender, backswept wings. Patrols surf line and inshore waters, bill typically downward; often pauses, turns or briefly hovers heavily before plunging into water. Habitat usually coastal: prefers sheltered estuaries, inlets, bays, harbours, lagoons with muddy or sandy shores. Also extends well inland on fresh or salt lakes, temporary floodwater, large rivers, reservoirs, sewage ponds. Usual calls, often given while hovering overhead, are a loud, deep, rasping, abrupt 'owgk, owgk' or 'kowk, kowk, kowk' at irregular intervals and, varied slightly, 'urgk, urgk', or, becoming agitated, a rapid 'urgk-urk-uk-uk-uk-'. Also drawn out, harsh, aggressive sounding 'ar-ar-rr-rk' or 'kraark' and higher, squealed 'ai-air-arrk'. Breeding resident species. Usually sedentary, although those using temporary wetland of interior then disperse; such movements cause numbers to fluctuate seasonally.

**Australian Gull-billed Tern**
*Gelochelidon macrotarsa* 36–42 cm
At a distance, looks like a large white gull but has a bulkier body, broader wings, and is more rounded than is typical of terns. Close up, the heavy gull-like bill is noticeable. Occurs across every continent as an inland species, only rarely over the ocean. Unusual in nesting on inland waters, fresh or saline. Often uses temporary water on mudflats or claypans, saltpans, salt marsh, open flood plains in arid regions where heavy rain has caused extensive shallow flooding. Out of the breeding season, seems to prefer lagoons and salt marshes near the coast. Breeds in colonies on small islands of shallow inland waters. Call is a hoarse, nasal, quavering 'ar-ark, ar-ark, ar-arrk'. Nomadic; highly dispersive.

# WHITE-FRONTED & BLACK-NAPED TERNS

**White-fronted Tern** *Sterna striata* 35–43 cm
The black cap does not touch the bill, a fairly narrow but obvious white band separating the bill and cap; this is the 'White Front' of the name. Habitat marine, usually exposed ocean coast – beach, rocky headland, offshore islet; less often in sheltered estuaries, harbours. When disturbed or in flight, gives a squeaky, rasping 'ki-erk', beginning sharply, then vibrating, still high; the whole call quite quick. Also gives a more abrupt 'kiek' as a contact call and an angry scream at aerial predators. From a colony or flock, a confusion of sharp squeaks intermingle with lower, but not deep, rasping sounds. Most breed in NZ, visiting SE Aust. in substantial numbers (more juveniles than adults) through winter. There is a small breeding colony in SE Bass Str.

**Black-naped Tern** *Sterna sumatrana* 30–32 cm
Distinctive, a very pale tern with a conspicuous black band around the nape. Although pale grey above, it can look white in strong light. Direct, fast flight; short, quick wing beats; often skims low across the water or swoops to snatch small fish from just beneath the surface. Tends to rest and roost on sandspits, beaches, rocks at water's edge; forages both inside barrier reefs and over open seas beyond the outer reef. Habitat includes offshore coral cays, reef islets, rarely continental coasts, estuaries, harbours. Call is a sharp, scratchy, abrupt 'chaik-chaik, chaik, chaik' at irregular intervals; also harsh scolding sounds, 'karrk-karrk-'. Resident species; most common along Great Barrier Reef.

On adult, back of nape tapers forward to terminate in point ahead of eye; similar on immatures, though with streaked crown; still recognisable in the indistinct speckled brown of juveniles.

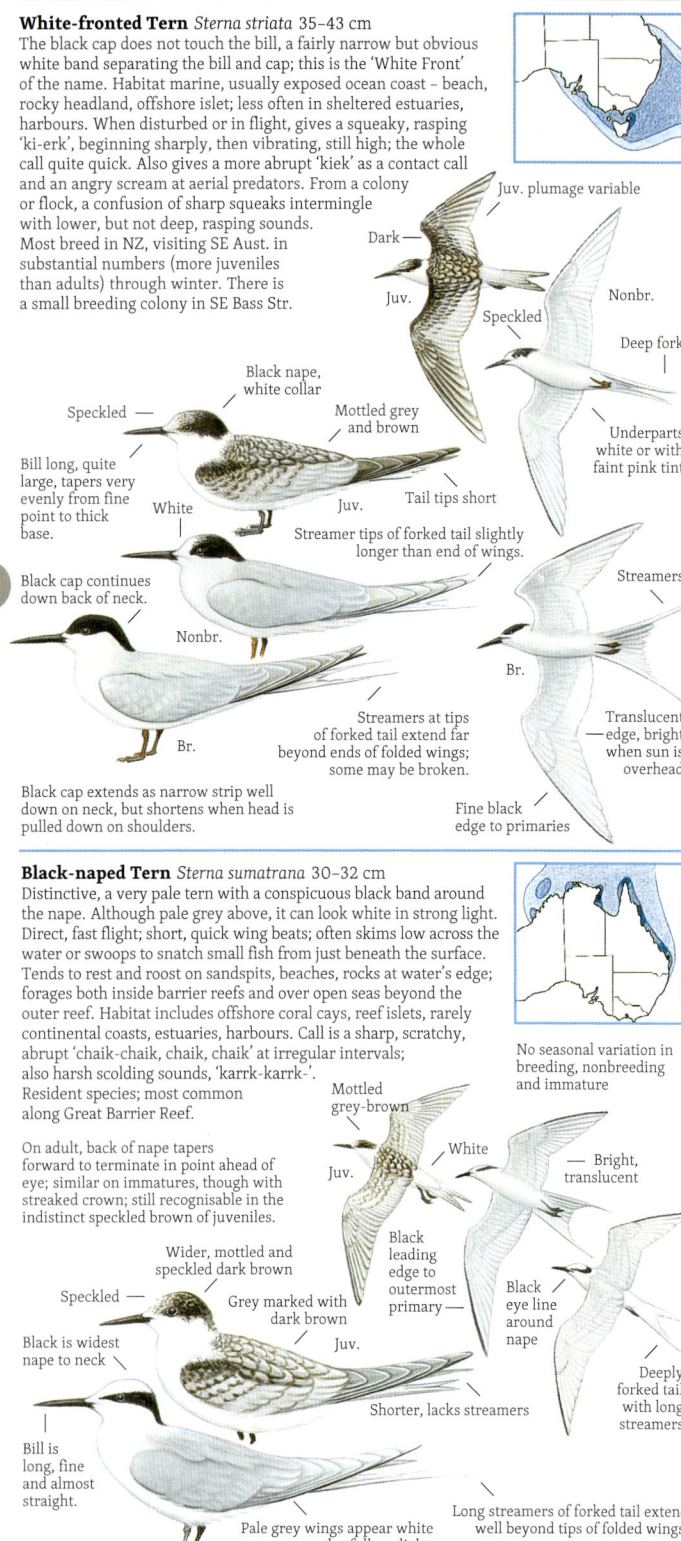

162

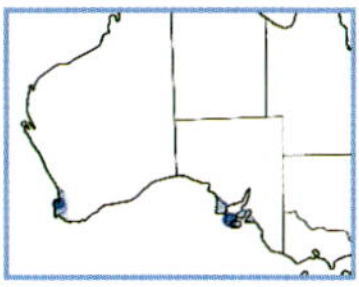

# ANTARCTIC & ROSEATE TERNS

**Antarctic Tern** *Sterna vittata* 32–36 cm
Larger and looking bulkier than Arctic and Common Terns, the Antarctic Tern is in breeding plumage through the summer, Oct.–Apr., when those northern terns are in nonbreeding plumage. Habitat usually the sub-Antarctic seas, coasts and islands, some moving over oceans out of breeding season. Most feeding is over shallow inshore waters. Flight like White-fronted Tern – steady, direct, regular, deep wing beats. Patrols some 5–10 m above the sea; hovers then plunges; also shallow dips to surface. Call is a high 'chirr-chirr-chirrah'; a rattling alarm call and a scolding, squawking noise when nest site threatened. Widespread in southern oceans, usually N as far as S tip of NZ. Rare vagrant to Aust. waters where there have been two specimens: one SW WA; one near Kangaroo Is., SA.

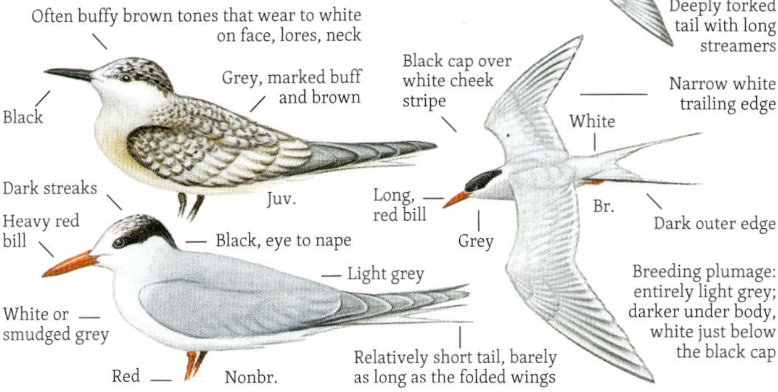

**Roseate Tern** *Sterna dougallii* 31–38 cm
Smallest and palest of the 'commic' terns (that is, all similar to the Common Tern). Has a delicate appearance in flight; wing beats are shallow; flight is graceful. Pale upper parts look almost white, and underparts are gleaming white. The faint pink blush for which the bird is named is evident only while breeding. Flight is direct and fast, not as wavering or buoyant as other commic terns. Usual call is a rising 'ch-vriek' or clear 'chu-ick'; also gives a deep, grating 'krarrk'. Habitat is marine, coastal, often coral reefs, foraging over reef, lagoons and surrounds. Usually keeps away from mainland shore, but may use shallow water just 100 m or so offshore. Common N Aust., breeds on islands of NE and W; range expanding southward.

163

## COMMON & ARCTIC TERNS

**Common Tern** *Sterna hirundo* 32-37 cm
One of most common terns of N hemisphere, uncommon in Aust. This species was thought to be quite rare here, no doubt largely due to difficulties of recognition when among other very similar terns. First noticed in Aust. in 1944, and, for many years, thought to be a rare visitor; now reported more often and in greater numbers. Race *hirundo* breeds from N America and Europe to W Russia; race *longipennis*, E Siberia. Habitat is marine, typically well offshore, but occasionally coastal waters, bays, estuaries, ocean beaches. Call in flight is a repeated 'kik-kik-kik' or 'kirrik-kirrik'; also gives angry 'krek, krek'; 'kwarrr' in conflict. On N and E coasts, black-billed race moderately common; SW WA, red-billed nominate race is a rare vagrant that also, although very rarely, reaches SA and Vic.

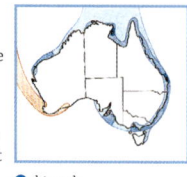

● *hirundo*
● *longipennis*

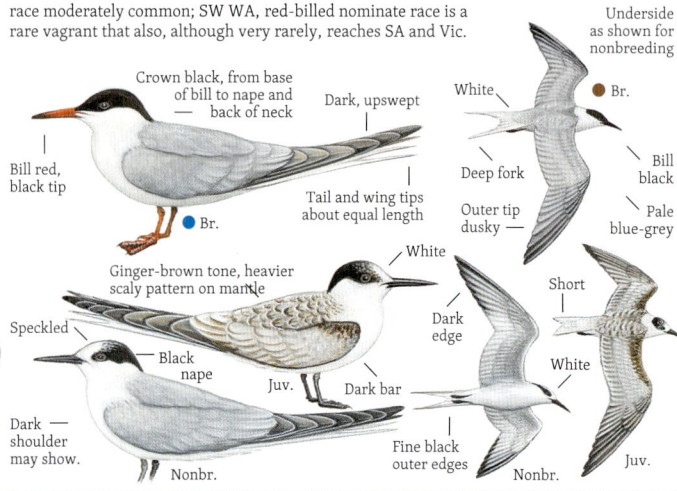

**Arctic Tern** *Sterna paradisaea* 32-37 cm
Famed for its remarkable migration: each small bird travels from breeding grounds on continents close to the North Pole and escapes the extreme cold by spending the northern winter in the Antarctic polar summer. Most seem either to follow the W coast of the Americas or go down the coast of Europe and the W coast of Africa. While these routes are far from Aust., some are sighted around the S and E coasts. The Arctic Tern is very like the Common, but with a longer tail and shorter bill, legs and neck, and differs in details of the various plumages, specifically the outer wing, the greyer underbody and the greater extent of translucent flight feathers along the rear edge of the wing. Flight graceful, light and buoyant; shallow wing beats. In Aust., most have been sighted resting on estuarine sand-spits or beaches, reefs, jetties. Call is a shrill 'tcheek' and 'tree-teeair'; sharp 'kee-kiear' calls, sharper than Common Tern. Regular migrant but in very small numbers.

# LITTLE & FAIRY TERNS

**Little Tern** *Sterna albifrons* 20–28 cm 
A very small tern. Tends to live and feed over shallower coastal waters – the estuaries, lagoons and channels around river and harbour entrances, and along the shallows close inshore. Attracted to sandbars, estuaries and river channels. Gregarious, usually in small groups or flocks, often congregating in much larger flocks at favoured roosting islets. In flight, quick fluttering wing beats, often hovers briefly, then uplifts the wings to allow a sudden drop closer to the water before a final plunge. Gives a short, sharp high 'kiep' or 'queeik', and slightly harsh 'krik-kriek'. Widespread around warmer seas of all continents. In Aust., breeds on E and NE coasts, Tas. to Gulf of Carpentaria; declining in SE; rare in SW.

In Aust., race *sinensis* has:
(a) a resident population with yellow bill when breeding;
(b) migrants from N Hemisphere, nonbreeding in Aust., all have black bills.

Saunder's Tern, *S. saundersi*, is a possible vagrant.

**Fairy Tern** *Sterna nereis* 22–27 cm
A very small tern, only slightly larger than the Little Tern and very similar looking, especially in plumages other than breeding. Compared with the larger terns, flight of these species is more fluttering with faster wing beats, more agility, hovering and sudden changes of direction. Both patrol quite low, often stopping to hover, bills downward, then plunging to take small fish from close beneath the surface. Habitat essentially marine: sheltered coasts, bays, inlets, estuaries, coastal lagoons, ocean beaches – rarely out to sea or even out of sight of land. Also wetland near the coast, including salt ponds, lakes. Like Little Tern, favours sites with sand spits and small sand islets in river-mouth channels. Feeding is usually over shallow waters, close inshore, or over barely submerged sandbars, but also on seaward side of reefs and islands. Call is an abrupt, nasal 'arrak, arrik, arrik'; also has fast excited 'krik-krikrikkrik'. Generally uncommon; rare where beaches are disturbed.

Differs from the Little Tern in being slightly larger and more solid. Appears deeper chested in silhouette, and with proportionately larger head and bill, shorter, thicker legs, together with small differences in plumage stages.

**Whiskered Tern** *Chlidonias hybridus* 23–25 cm

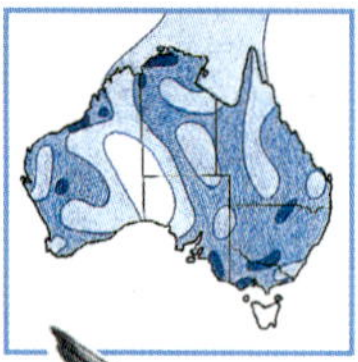

A distinctive marsh tern; in breeding plumage the dark grey underparts blend to a white band across the cheeks, the 'whiskers' in bold contrast against the lower edge of the black cap. In its habitat preferences, being an inland species, uses shallow freshwater wetland, permanent or temporary floodwater, interior claypans, irrigated pasture; only occasionally estuaries or similar marine wetland. Usually in flocks patrolling back and forth over wetlands with steady wing beats; straight direct flight at height 5–10 m, often dipping, shallow diving to take any small aquatic creature. The call is a harsh, rather hoarse, rasping 'kierch' or 'kerrrik', and sharper 'krik' or 'krit-ik'. Breeding species; common.

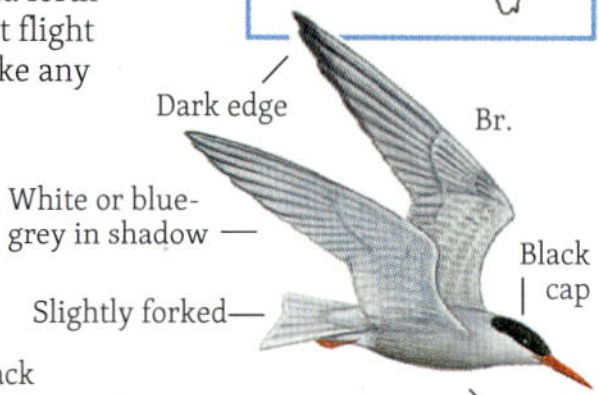


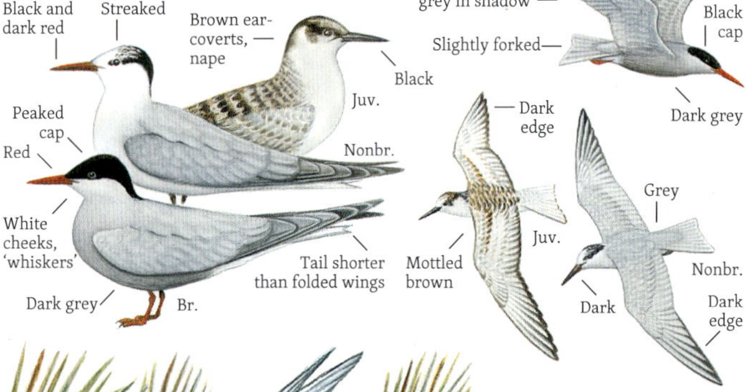


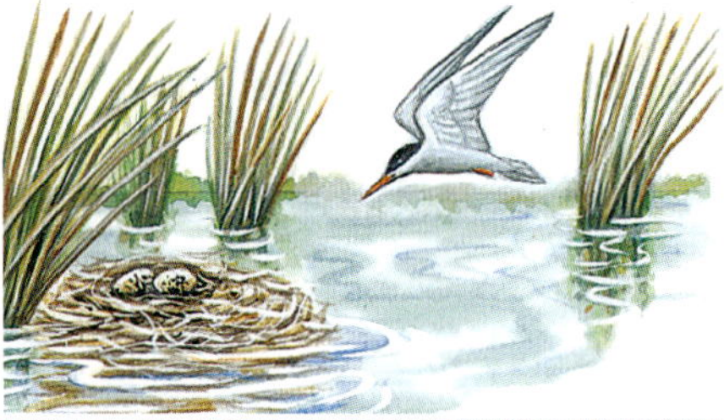

Nests in colonies, often large, situated in shallow, often temporary, waters of inland swamps. Nest is a rough platform of vegetation built on samphire or other low shrubs standing in shallow water. Parents share the nest building and incubation. Vigorously and noisily defend the colony by diving at intruders.

**White Tern** *Gygis alba* 30–33 cm

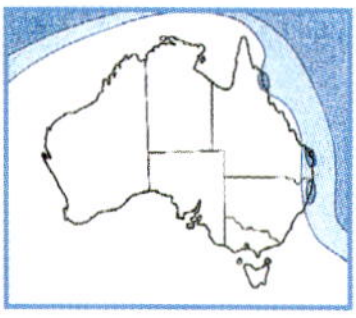

Oceanic, often seen in small groups close to island colonies; further out to sea, in pairs or alone. Flight can be buoyant and fluttering or swift and darting. Travels fast with deep, effortless wing beats, or circles and flutters, forked tail spread, square-tipped when fully expanded. To feed, it hovers, dips, and flutters over the water, picking up prey, only rarely plunging below the surface. Inquisitive and tame, may come close overhead. Call is a muffled rasping or grinding sequence that gradually fades: 'arrag-arrag-arrug-urrug-urrug'. A tern of subtropical and tropical seas; usually keeps far from land, feeding over open ocean except in breeding season – then forages over lagoons and reefs around islands. Occasional visitor to E Aust. Breeds Lord Howe and Norfolk islands.

Flight feathers and tail are translucent; overhead, sunlight passes through, giving brilliant white, especially seen against deep blue of sky.

Undersides of wings and other shadowed surfaces take on some reflected tones of sea or foliage, appear faintly blue-grey.

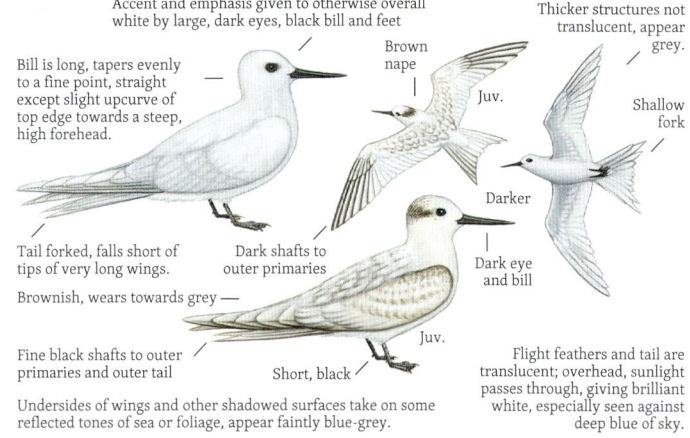

# WHITE-WINGED BLACK & BLACK TERNS; GREY TERNLET

**White-winged Black Tern** *Chlidonias leucopterus* 20–23 cm
A small marsh tern, unmistakeable in distinctive breeding plumage; most alike are Black Tern and Whiskered Tern. Habitat includes marine and freshwater coastal wetland, the latter including river pools, billabongs and inundated floodplains. Tidal habitats are typically estuaries, lagoons, harbours. Gregarious, usually in flocks; feeds noisily, patrolling with erratic, buoyant flight, frequently dipping to surface. Call is a sharp 'krik', a hoarse, deepish 'keerek', and fast 'kiek-kiek-kiek'. Regular migrant to Aust.; quite common across northern coast, further S. rare, erratic occurrence.

**Black Tern** *Chlidonias niger* 22–24 cm
Easily identified in breeding plumage; black head and body, dusky grey wings, pale grey rump, tail and under-wing-coverts. Nonbreeding and juveniles have a largish cap over crown, ear-coverts and nape; dark patches at sides of breast; white collar, slightly darker inner-wing leading edge. Habitat includes freshwater lakes, swamps, floodwater, sheltered coast. In flight, slim, long, pointed wings; long slightly forked tail; long slender bill. Flight erratic, low, frequently dipping to touch the surface, hovering briefly; occasional shallow surface diving. Call is a sharp, high but weak 'kik-kik-kik-'; in alarm, a sharper, louder 'tweek, tweek'. Similar are White-winged Black Tern, Whiskered Tern in nonbreeding and juvenile plumages, but smaller cap, no breast patches on the side. Rare vagrant; several accepted records, few other sightings.

**Grey Ternlet** *Procelsterna albivitta* 25–30 cm
A small 'noddy' plumaged in soft shades of grey. Graceful, buoyant flight with strong wing thrusts. Habitat is typically inshore waters, along coasts, around islands; often forages over calmer waters in lee of cliffs or reefs; occasionally much further out over pelagic waters. Roosts and loafs on ledges of island cliffs. Usually forages in flocks; a confusion of wheeling, fluttering birds low above the waves, dropping to snatch at small surface creatures while pattering the surface with webbed feet. Gives rapid, sharp purr, almost a cat-like 'meiow', rising in pitch and frequency to a high squeak, 'qar-ar-ar-ar-air-aiik', a quick rattle that lasts from half to one second. A central Pacific species, Australia's mid E coast the western extremity of its range. Rare vagrant, some after storms in W Pacific.

Two morphs, the light grey and a rare dark grey, in which white is replaced by pale grey and mid grey of wings by a darker grey

## SOOTY & BRIDLED TERNS

**Sooty Tern** *Onychoprion fuscata* 33–36 cm + tail streamers 8–10 cm
Black and white oceanic tern, feeds in mixed flocks, usually far out from land. Habitat oceanic: tropical and subtropical Indian and Pacific Oceans. Frequents and feeds over pelagic waters far from land, returns to land only rarely to breed or escape rough weather. Occasionally seen close inshore on coast or inland after gales at sea. Rarely alights on water; powerful flight, deep wing beats; wheels and soars, dives close to water, skimming the surface for small fish and crustaceans; some nocturnal species also taken. Does not often call at sea, but noisy in colonies with almost constant yapping calls, rather like a small dog, 'ker-waka-wak', from which comes the alternative common name, 'Wide-awake'. As alarm, a harsh, drawn out 'kraaak'. In colonies the incessant cacophony of hundreds of 'wak-awak' and similar calls intermingle so that it is difficult to single out the call of any individual bird. Breeding species off coasts of WA and Qld, occasional sightings further S; some colonies number thousands.

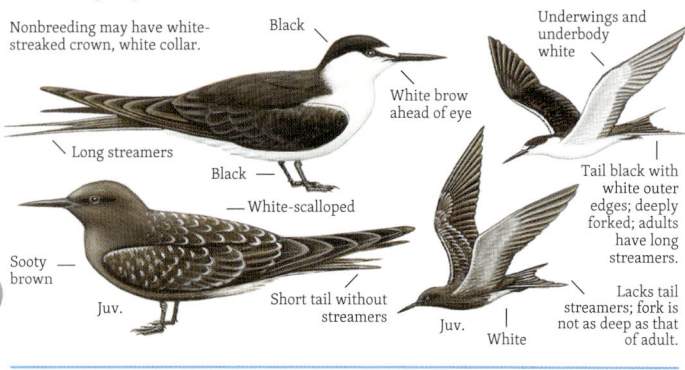

**Bridled Tern** *Onychoprion anaethetus* 30–32 cm
Superficially similar to the Sooty Tern, but distinctive in many details. In distance, lacks the strongly contrasting, crisp black and white pattern of the Sooty; instead has brown upper parts with dulled, faintly grey-tinted whites of underbody and under-wing-coverts. Has voice like yapping of a small dog – similar to the yap of a Black-winged Stilt – very abrupt, nasal 'ak, ak-ak, ak-ak-aak' or 'wep-wep, wep-wep-wup', and harsher, scolding 'airrrgh, arrrrgh'. Noisy if nesting colony is approached; much calling at night in breeding season. Habitat covers tropical and subtropical seas, often far from land. Usually forages on open seas, but frequents breeding islands, reefs and, occasionally, inshore waters. More likely to be seen from land than is Sooty Tern. Common NW to NE Australian coasts; rare to absent in SE.

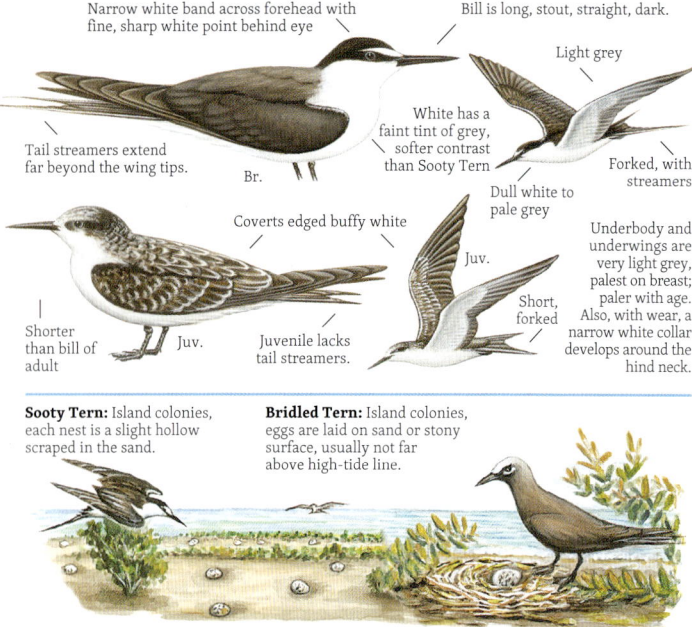

**Sooty Tern:** Island colonies, each nest is a slight hollow scraped in the sand.

**Bridled Tern:** Island colonies, eggs are laid on sand or stony surface, usually not far above high-tide line.

## NODDIES

### Brown Noddy *Anous stolidus* 40–45 cm
Large brown noddy with ashy white cap; in flight, paler grey under wing, two-toned brown on upper wing. Gregarious – often dense flocks attack shoals of fish, usually far out at sea. Large body, slower wing beats give heavier jiz than other noddies. Voice is grating or guttural, 'urk-urk-rrk-rrk-rrkrrk'; calls combine as constant, murmuring background noise in colonies. In alarm or threat, harsh, squawked 'kar-rrark', 'kraaaa'. Plain brown head, fine pale fringes to upper-wing coverts. Juveniles similar to those of Sooty and Bridled Terns. Breeding habitat: coastal waters near island colonies. Otherwise oceanic. Breeding resident, islands off WA and Qld.

Largest noddy: white forehead, ashy grey crown merging to dark brown body

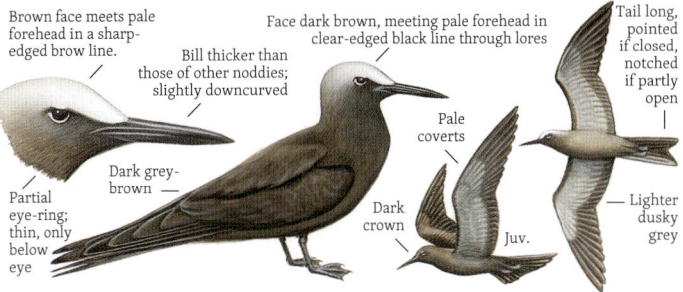

Brown face meets pale forehead in a sharp-edged brow line.
Partial eye-ring; thin, only below eye
Dark grey-brown
Bill thicker than those of other noddies; slightly downcurved
Face dark brown, meeting pale forehead in clear-edged black line through lores
Dark crown
Pale coverts
Tail long, pointed if closed, notched if partly open
Juv.
Lighter dusky grey

### Black Noddy *Anous minutus* 35–40 cm
Medium-sized, slender; all brownish black except silvery white cap and pale central tail; reflective highlights along rear of underwing. Habitat in Aust., islands, offshore reefs of Qld and Kimberley coast, surrounding seas and reefs. Usually in flocks over schools of fish; hovers before dropping to the surface. Flight lighter, faster wing beats, more fluttering than Brown Noddy. Voice is harsh, fast, rattling, 'ak-ar-r-r-r-k', 'ak-ak-ak-akak', 'ak-aairak-ak-air-ark'. Breeding resident; common near colonies; casual visitor further S.

Lores black, in stark contrast to the silvery white cap
Partial eye-ring
Brownish black
Whiter cap in greater contrast with black plumage than on other noddies
Long, slender, almost straight
Compared with Common Noddy, above, Black Noddy is smaller, more slender, darker sooty black; has finer, straighter bill.
Glossy highlights
Brown-black, no pale bar
Brownish black
Pale
Juv.: as adult, but larger white cap onto nape; faint pale edges on upper parts

### Lesser Noddy *Anous tenuirostris* 30–34 cm
A delicately plumaged, graceful noddy confined to the Indian Ocean. In flight, more like Black Noddy than larger Brown Noddy. Makes a purring sound, churring or soft rattling; similar to, but softer than, the sounds of other noddies; a colony of thousands described as a 'vast purring chorus'. Also has louder rattle of alarm. Habitat in Aust. centred on low, flat coral islands of the Abrolhos Group, with reefs and lagoons with dense, low, fringing mangroves. Forages around islands, reefs and lagoons, and further out to sea. Rare in Aust. except that the species is locally abundant at and near the W coast breeding colonies; probably resident.

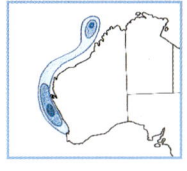

Partial eye-ring above and beneath eye
Light grey of crown blends softly, evenly, into blue-grey neck and darker body.
Tip of shallowly forked tail level with ends of folded wings
Juv.: whiter cap with sharper transition to nape; pale markings on mantle and scapulars
Almost white, pearly grey forehead; dark oval in front of eye
Long, fine, slightly downcurved
Lores pale grey, softly merging into forehead and lower face
Long, narrow wings
Shallow fork, tip rounded when spread

169

# 11 DOVES & PIGEONS

This group is one of the most widely recognised bird families. The family likeness is evident in all species: there is a common pigeon character of shape, movement, behaviour and voice that defines the family Columbidae. As with many other families, the range of sizes is great and is not readily apparent unless all are shown to the same scale, as here. The family's general tolerance of new environs has allowed many feral species to establish in Australia. Most are included in the main text, but not in this basic quick identifier.

Magpie = 40 cm

## columbidae
page 171

- White-headed Pigeon
- Crested Pigeon
- Bar-shouldered Dove
- Elegant Imperial-Pigeon
- Brown Cuckoo-Dove
- Spinifex Pigeon
- Partridge Pigeon
- Collared Imperial-Pigeon
- Pacific Emerald Dove
- Squatter Pigeon
- Wonga Pigeon
- Common Bronzewing
- Chestnut-quilled Rock-Pigeon
- Black-banded Fruit-Dove
- Torresian Imperial-Pigeon
- Brush Bronzewing
- White-quilled Rock-Pigeon
- Wompoo Fruit-Dove
- Topknot Pigeon
- Flock Bronzewing
- Diamond Dove
- Superb Fruit-Dove
- Peaceful Dove
- Rose-crowned Fruit-Dove

## COLLARED IMPERIAL-PIGEON

**Collared Imperial-Pigeon** *Ducula mullerii* 40–43 cm
Diagnostic bold black neck band. A NG species, poorly known. Juvenile not yet described. Flight strong, fast and direct. Individual birds or small flocks cross the narrow channel to Boigu Is., an Australian territory in Torres Str., to forage in canopy of fruiting trees, thus a potential vagrant to Cape York or NT. Call not well known, but includes a rising sequence of about five cooing notes. Habitat: dense lowland rainforest, swamp forest or mangroves, usually near rivers or lakes. Feeds in fruiting trees in small parties, pairs or alone; usually seen only when flying across open space of waterways or above the forest canopy.

- Cinnamon-rufous cap
- Chin, throat, face, nape collar, white
- Upper mantle glossy maroon or wine red
- Black collar, broadest at nape and unbroken around the neck
- Back, rump, wings, grey
- Mauve-pink
- White band
- Rufous-pink under tail
- Feet deep pink

## BANDED FRUIT-DOVE; BLACK & WHITE PIGEONS

**Black-banded Fruit-Dove** *Ptilinopus alligator* 33–35 cm
Identified by obvious black breast band. Occurs in small groups, pairs or alone in pockets of rainforest; feeds on figs or other fruits early and late; shelters in dense foliage through heat of day. Shy and often well concealed in foliage; never far from water – visits waterholes at least once a day; sudden take off causes loud wing clap. Flight powerful, clearly audible, whistling wing beats. Its 'advertising' call is a deep booming 'coo' repeated 6–8 times at intervals of several seconds. Inhabits rugged sandstone escarpment among gorges and cliffs in patches of monsoon forest with fruiting trees. Scarce; restricted but secure.

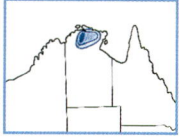

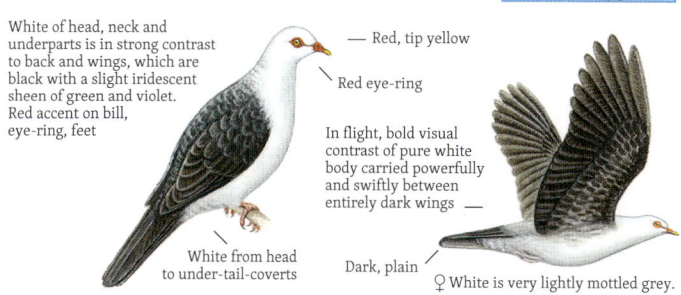

**White-headed Pigeon** *Columba leucomela* 38–40 cm
Large; entire head and underparts white; back and wings black. One of the shyest and wariest of pigeons. Often the first sign of its presence is a loud clatter of wings as a flock panics into flight close overhead. Inhabits rainforest, favouring edges and regrowth, cleared land with abundant Camphor Laurel, suburban gardens. Call a soft, high-low 'whOO-wuk, whOO-wuk'; the first half is loud, emphasis on the resonant, musical 'OO', the second half much softer and fading away. Scarce, but common NE NSW; dispersive.

**Torresian Imperial-Pigeon** *Ducula spilorrhoa* 38–42 cm
Large pigeon, conspicuous in big flocks commuting from offshore roosting and nesting islands to rainforest, adjacent eucalypts and mangroves where it feeds. When it travels, flight is direct and swift with powerful, steady uninterrupted wing beats. But, when feeding in foliage of fruiting trees or shrubs, species is shy and wary; usually then in small parties. Call is usually a double noted 'ook!-whuuu'; the first part abrupt, the second drawn out, moaning, fading. In display, 'coo-whoo-hoo': soft then loud, with bowing movements. Sedentary or migratory; common, probably increasing in Qld.

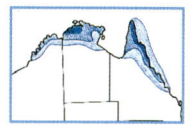

Juv. (not shown): light grey instead of white; buff tint to underbody; black bars fainter

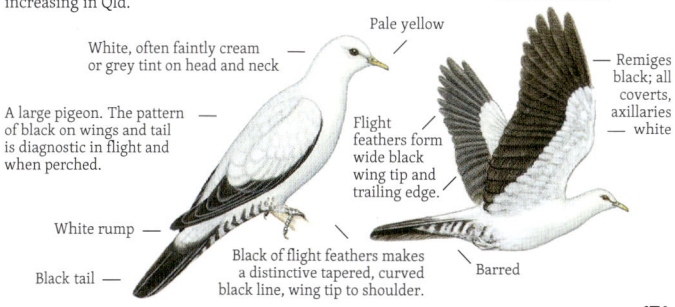

171

## ELEGANT IMPERIAL-PIGEON; SMALL FRUIT-DOVES

### Elegant Imperial-Pigeon *Ducula concinna* 44–48 cm
A quite recent vagrant recorded at Darwin added this species to the Aust. bird list, but it could equally likely be encountered elsewhere along the N coast. In its usual range on tropical islands, the species inhabits the canopy of monsoon forest, forest edges and lightly wooded cultivated land. The solitary bird in suburban Darwin was using remnant rainforest species, especially the dense canopy of Banyan Fig. Voice a deep, gruff 'urr-wooo' and a shorter, abrupt, almost barked version – the two intermixed or sometimes given by two birds. Extremely rare vagrant to N Aust.; usually nomadic.

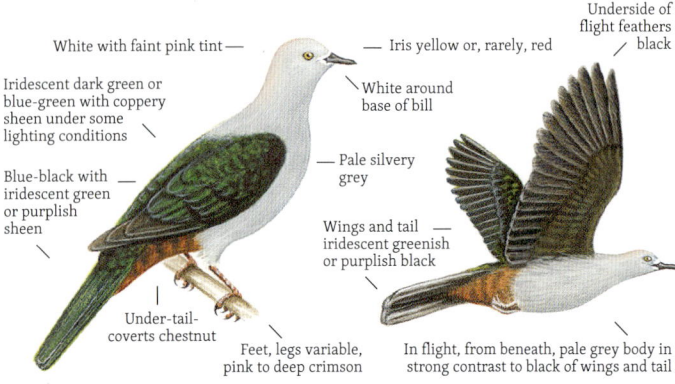

### Superb Fruit-Dove *Ptilinopus superbus* 22–24 cm
A small fruit-dove. Male is resplendent in bright colours, yet is often difficult to sight. Small parties and pairs feed high in fruiting rain-forest trees. Flight, with whistling sound from wings, is fast, direct, through canopy spaces. Inhabitant of rainforest and similar closed forest – monsoon forest, regrowth, lantana thickets, woodland adjoining rainforest at all altitudes. Forages mostly within rainforest, usually high, but lower when shrubbery carries fruits. Gives a series of mellow, musical calls, beginning softly, slowly, rising in pitch and volume until a loud and clear 'whoop, whoop, whooop'; also low 'oom', in steady sequence. Quite common in N of its range; considered endangered in NSW.

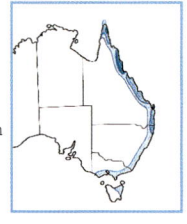

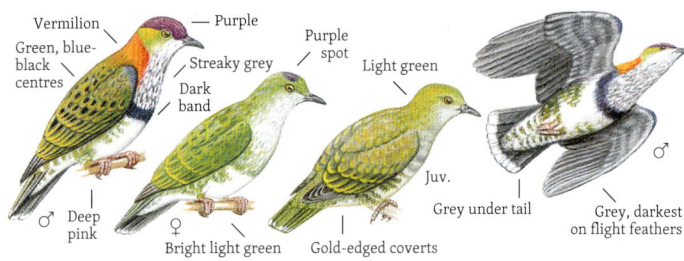

### Rose-crowned Fruit-Dove *Ptilinopus regina* 22–24 cm
Similar to Superb, but different enough for adult males to be easily separated. Inhabits rainforest, monsoon forest, vine scrub, mangroves, swampy woodland. Noisy, with loud calls given in long series: 'whup-whooo, whp-whooo, whp-whooo', the first part low and abrupt, followed by a drawn out 'whooo'. Also as 'whu-whoo, whu-whoo-whuk-whukwuk...', becoming faster and fading away. Quite common; migratory, nomadic, dispersive.

● ewingii
● regina

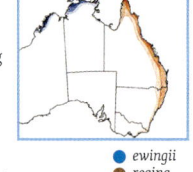

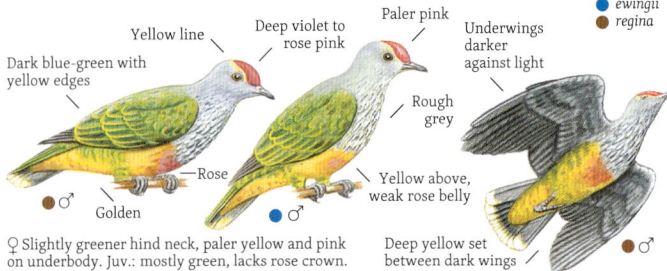

♀ Slightly greener hind neck, paler yellow and pink on underbody. Juv.: mostly green, lacks rose crown.

## WOMPOO, BROWN & PACIFIC EMERALD DOVES

**Wompoo Fruit-Dove** *Ptilinopus magnificus* 38–48 cm
Despite its size, is often hard to find in the rainforest canopy. Can be revealed by sounds of falling fruit or by calls. Inhabits tropical and subtropical rainforest, monsoon and closed gallery forest, more rarely temperate rainforest of SE Qld and NSW, and wet eucalypt forest near rainforest. Call powerful, deep, reverberating, as in the name 'wom-poo', but the first part is like the 'plonk' of a rock dropped in deep water: 'g'lonk-ooo', or 'wollocka-roo'. Nomadic, wanders in search of fruiting trees. Quite abundant in NE Qld, declining in S - rare in NE of NSW, almost extinct S of Sydney.

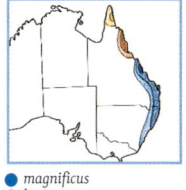

- ● *magnificus*
- ● *keri*
- ● *assimilis*

● Slightly darker green, breast more maroon than purple
● Breast paler purplish maroon

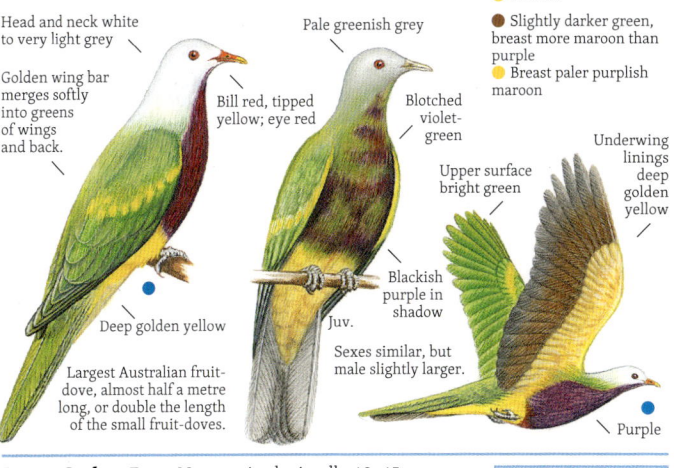

Head and neck white to very light grey

Golden wing bar merges softly into greens of wings and back.

Deep golden yellow

Largest Australian fruit-dove, almost half a metre long, or double the length of the small fruit-doves.

Pale greenish grey

Bill red, tipped yellow; eye red

Blotched violet-green

Blackish purple in shadow

Juv.

Sexes similar, but male slightly larger.

Upper surface bright green

Underwing linings deep golden yellow

Purple

**Brown Cuckoo-Dove** *Macropygia phasianella* 40–45 cm
Typically a bird of forest edges, seen along tracks, in clearings and regrowth. In small parties or pairs, feeds in the vegetation or on the ground. Inhabits rainforest, wet eucalypt forest, brigalow scrub, regrowth, thickets of lantana, wild tobacco. Has a loud, almost ringing call – from afar sounds like a single, clear, rising 'waaalk'. Common in N, rare in SE.

Male has slight iridescence on neck and mantle, reflecting tints of violet or green depending on the light direction.

♂

Rufous-cinnamon

General impression: very long bodied, slender, small-headed

Deep rufous-brown

Rufous cap

Scalloped

♀

Juv.

Long, tapering tail

Subspecies uncertain: perhaps none, possibly *quinkan*, NE Cape York, and *phasianella*. Considerable clinal variation

Bright rufous cap

Fine dark barring and scalloping to throat, neck

**Pacific Emerald Dove** *Chalcophaps longirostris* 23–27 cm
Small, plump dove, typically seen in pairs, rarely in larger groups. Forages on ground or feeds on seeds or fruits in trees. Lives in rainforest, monsoon forest, wet eucalypt forest, melaleuca woodland, lantana thickets, regrowth scrub along creeks. Call is a series: begins with a very low purring that gradually rises higher: 'p-ur-r-r-oom, prooom'. Abundant in N and NE, declining in SE of its range.

Iridescent emerald green wings and back, underparts pinkish cinnamon

Pinkish or violet-tinted brown

Green, edged cinnamon

White shoulder

Pink

Juv.

Dark band

Rufous, barred brown

● *longirostris*
● *chrysochlora*

Chestnut or cinnamon

## TOPKNOT & WONGA PIGEONS; ROCK DOVE

**Topknot Pigeon** *Lopholaimus antarcticus* 40–45 cm
The unique backswept crest makes the Topknot easily identifiable close up; even far off, its silhouette is distinctive. It is also recognised by its flight: far and fast across open country in search of rainforest trees with ripening fruit. Flocks of thousands were common before extensive clearing and uncontrolled hunting. They wander the forests in smaller numbers now. Flight is powerful, fast and straight; birds swoop in long glides, wings flat, from ridge top to valley; or circle high over forests. Habitat includes rainforest – remnant and regrowth – and nearby eucalypt forest. Forages high in canopy, occasionally in undergrowth, but not on the ground. Usually silent, but has low, resonant grunt: 'whug, whug, whug'. Feeding or fighting, short sharp screeches from the flock. Nomadic; common only in N.

**Wonga Pigeon** *Leucosarcia melanoleuca* 35–40 cm
A large, plump, grey and white pigeon with distinctive markings. Takes off with loud wing claps; on landing, may lift tail showing black flecked under-tail-coverts. Glides on flat wings with occasional quick, flicking wing flaps. Found in rainforest, eucalypt forest and woodland, brigalow and tea-tree thickets, vine scrub. Call is a long series of notes, quite high, rapid, very penetrating: 'whoik-whoik-' repeatedly, at constant pitch, becoming monotonous. Sedentary; uncommon.

**Rock Dove** *Columba livia* 33–35 cm
Also known as domestic pigeon, Rock Dove and Homing Pigeon. Widespread, introduced. Familiar as domesticated cage and racing bird, and in feral colonies. Diverse cultivated varieties in many different plumage patterns and colours. Call is soft, rather musical cooing. Lives in built environs and surrounding country. Introduced soon after European settlement; now widespread, mainly around towns and farms.

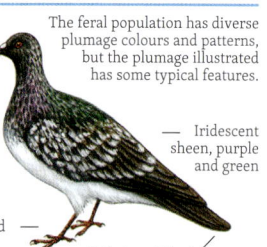

## PEACEFUL & DIAMOND DOVES; FERAL DOVES

### Peaceful Dove *Geopelia placida* 20–24 cm
Small dove; forages on ground; roosts, rests and nests in shrubs or trees. Undulating flight gives glimpses of rufous wing linings, dark flight feathers and pale tips to dark outer tail. Usually in open forest, woodland, tall shrubs and watercourse tree lines through arid country. Call is a pleasant, musical, lilting, loud and clear 'coo-wi-ook, coo-wiook, coo-iook, cooiook'. Also a throaty but musical 'quo-r-r-r-r'. Very common in better watered regions of N; sedentary.

In flight, rufous under-wing-coverts and grey flight feathers show clearly.

Long, tapering tail

Fine dark cross-barring

Eye-ring, iris, cere, blue-grey

White chin

Scalloped black and white

Faint tint of pink

Feet deep pink

● *placida*
● *clelandi*: plumage overall slightly warmer brownish grey

Juv.: like adults, but barring less distinct

### Diamond Dove *Geopelia cuneata* 20–24 cm
Usually small parties feed on the ground, running fast if disturbed, or flying on whistling wings. Flight swift, undulating. Widespread in dry areas – grassy woodland, semi-arid grassland, spinifex, mulga and similar dry scrub – but never far from water. Call: two versions given, each of four notes, all clear, pleasant, musical. One has the four notes alternating low-high, low-high, 'coor-cooo, coor-cooo'; in the other, the four notes are quite level in pitch, 'coo-cooo, coo-cooo'. Also gives a drawn out 'quorrr'. Common, but fluctuating populations that move to coastal SE when interior is in drought.

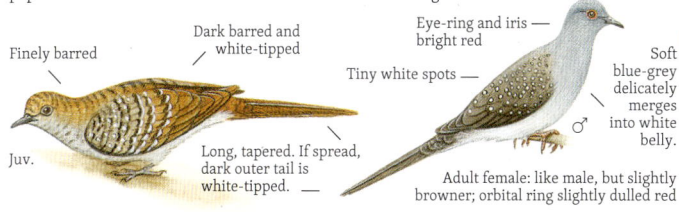

Finely barred

Dark barred and white-tipped

Eye-ring and iris bright red

Tiny white spots

Soft blue-grey delicately merges into white belly.

Juv.

Long, tapered. If spread, dark outer tail is white-tipped.

♂

Adult female: like male, but slightly browner; orbital ring slightly dulled red

### Laughing Dove *Spilopelia senegalensis* 26 cm
Small introduced dove of suburbs, country towns, parks, schoolyards, rail yards, farms; rarely in natural bushland. Voice is a laughing, bubbling, mellow sequence like 'did-you-see-a-cuckooo', the last word loudest, clearest: 'quook-kuk-a-kuk-KUKooo'. Introduced, Perth about 1898; now around most SW towns and farms.

Head and neck pale brown with violet-pink tint

Usually has brownish spotting, but not on juv.

Blue-grey shoulders are a diagnostic feature, both with wings folded and in flight.

In flight, white outer tail feathers show.

### Spotted Dove *Spilopelia chinensis* 30–32 cm
Feral species. Two subspecies introduced between 1860 and 1920; they have since interbred in some localities. Call is a high, triple noted, musical cooing: 'cook-oo-ook', 'coo-coo-crooo' with variations. Habitat is mostly suburban parks and gardens, remnant vegetation, farms, plantations. Rarely in natural bush. Common, and spreading to country towns.

Head pale grey

Neck patch black with white spots

Pale pinkish brown

Feet pink

Central tail feathers are brown, hiding the white-tipped outer tail feathers that are so conspicuous in flight.

White

### Barbary Dove *Streptopelia roseogrisea* 29–30 cm
Domesticated form of either the African Collared-Dove, *S. roseogrisea*, or the Eurasian Collared-Dove, *S. decaocto*. The adult is distinctive: it is sandy or creamy buff with a clearly marked black half-collar around the hind neck. Juveniles lack the half-collar. This race is unlike any other dove in Aust.: the only similar sized species, the Spotted Turtle-Dove, is much darker, while the black half-collar of the adult Barbary is diagnostic. Birds feed on the ground. Occasional feral colonies have occurred in the suburbs of Sydney, Perth and Darwin, and there have been isolated aviary escapes elsewhere.

## SPINIFEX & CRESTED PIGEONS; BAR-SHOULDERED DOVE

**Spinifex Pigeon** *Geophaps plumifera* 20–24 cm
Unique little reddish, plump, upright, red-crested, ground-feeding pigeon - unmistakeable. Small parties live on rocky ground, blending with red rock and gold spinifex. Runs erratically, stopping to stand tall, alert; flushes abruptly, quail-like, with loud whirr of wings; glides away on stiff downcurved wings, giving a brief glimpse of rufous back and wings. Usually in spinifex of rocky ranges, gorges, less often in sandy country; never far from water. Voice soft, husky, gives a throaty yet musical 'coo, coo-roo!' with final 'roo!' uplifting and clearer, the sequence repeated for long periods. Sedentary; common.

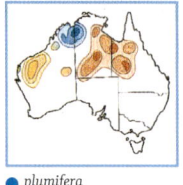

● *plumifera*
● *leucogaster*
● *ferruginea*

Finely tapered, upright crest, sandy buff to cinnamon, similar all races

Colour of races tends to become warmer westwards, from sandy buffy brown in W Qld through cinnamon to rufous in Pilbara, WA.

Face pattern same for all races

Nominate white-bellied race, western NT to NE Kimberley

Bare red skin around red eye, all races

Reduced white on belly

Short, rounded wings

Fine white-over-black lines, all races

Variable extent of white, medium to large area

Entirely red-bellied; brightest rufous of the three races

Juv.: similar but paler; orbital skin khaki; iris pale yellow

---

**Crested Pigeon** *Ocyphaps lophotes* 31–35 cm
Usually in small flocks; feeds on ground. In flight accelerates with vigorous bursts of whistling wing beats interspersed with fast, direct glides with wings stiff and flat; on landing, tips forward, long tail lifting vertically. Inhabits open woodland, acacia scrub, farmland, roadside tree lines, homesteads and yards, always near water. Avoids dense, wet coastal habitats. The call is a rather musical 'whoo' starts softly, lifts up and strengthens, ending abruptly, 'whoo,-whoo,-whoo-'; also a very quick 'woop!' Very widespread and common; range expanding as heavier vegetation is cleared.

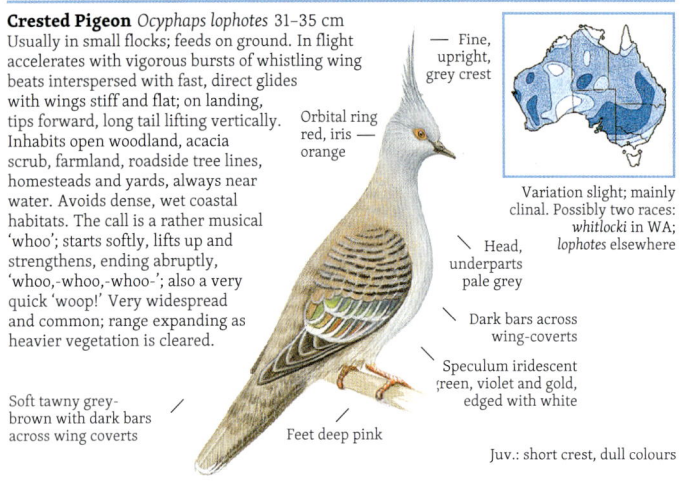

Fine, upright, grey crest

Orbital ring red, iris orange

Variation slight; mainly clinal. Possibly two races: *whitlocki* in WA; *lophotes* elsewhere

Head, underparts pale grey

Dark bars across wing-coverts

Speculum iridescent green, violet and gold, edged with white

Soft tawny grey-brown with dark bars across wing coverts

Feet deep pink

Juv.: short crest, dull colours

---

**Bar-shouldered Dove** *Geopelia humeralis* 27–30 cm
Medium sized, long tailed. Feeds on open ground near cover; roosts and nests in shrubs or trees. Habitat is usually near wetland, especially inland, woodland, forest, monsoon and vine scrub, mangroves, brigalow scrub and spinifex.
Call is a double noted, cheery 'cookaw-cookor', 'cookaw-cookor'; first part is higher, stronger and clearer, the second slightly lower, fading. Also, a long bubbling, laughing, descending 'cook-aw-cookaw-cookaw-cookaw-'. Common.

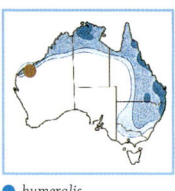

Coppery rufous with dark barring

Plumage and bare eye-ring blue-grey; eye reddish

● Grey-brown, barred;
● Paler

● *humeralis*
● *headlandi*

Other races have been described across N Aust., but generally considered clinal rather than race status.

● Blue-grey, not barred
● Paler, with pinkish tint

Feet deep pink or pinkish brown

In flight, underwing chestnut with dark trailing edge

# BRONZEWINGS

## Flock Bronzewing *Phaps histrionica* 27–31 cm

Nomadic inhabitant of grasslands of N interior; formerly in immense flocks, still occurs in thousands. Dispersed on reddish ground, feeding birds are inconspicuous. Morning and evening visits to water can be spectacular, with tight flocks crowding the water's edge then bursting up with a muffled roar of wings, displaying fanned, white-tipped tails. Flight is fast and direct with deep, strong action of pointed, backswept wings. Lives on open plains among clumped grasses, small shrubs with open spaces. No audible calls at water or in flight. Birds in feeding flocks give a soft, low, drawn out moaning at even pitch and volume; in display a 'wook' call. Common, but of erratic occurrence.

Bold pattern of rufous and dark grey

Female face pattern like male, usually with much less black

Black head, white around bill, white line encircling ear-coverts

Deep blue-grey

Tail grey, white tip

In distant flocks the rufous and blue-grey helps identify the species.

Primaries white-tipped

Mottled grey and cinnamon

Nape to tail-coverts and inner wing-coverts rufous-brown

Juv.: like paler faced females; muted colours, pale scallops on breast, back, rump, wing-coverts

## Common Bronzewing *Phaps chalcoptera* 30–36 cm

Solitary or in small parties; feeds on ground. Habitat is diverse, most of continent: coastal forest, woodland, arid scrub, mallee, heath, alpine woodland, farmland. Call is a carrying, mournful, slow, resonant and deeply vibrating 'whooo'; begins softly, is carried through at very even, low pitch, dropping slightly to finish; repeated at several second intervals. The display call is very different – a short, soft 'whoo-, hoo-hoo'. Endemic; locally nomadic; common.

Pale grey

Buff

White cheek stripe curves back and wider, down towards blue sided neck.

Underwing bronze-rufous

Less blue

Blue of neck blends to pinkish breast.

Coverts dark olive-brown with multicoloured iridescence

Deep, dull pink

Blue-grey, tipped white

## Brush Bronzewing *Phaps elegans* 25–33 cm

More secretive and wary than Common Bronzewing; flushes suddenly with loud wing clapping. Uses habitats with dense groundcover dominated by shrubs, scattered trees of banksia, leptospermum, casuarina. Call higher, shorter, faster than Common Bronzewing's, a repeated 'hoo, hoo, hoo, hoo', each call slightly higher, quicker. Uncommon to rare.

Rufous under wing

● *elegans*
● *occidentalis*: SW of WA

● Forehead may be paler buff, back more olive.

Cinnamon crown

Chestnut

Rich deep chestnut

Flight fast, direct; may give glimpse of chestnut shoulders, rufous under wings.

Soft grey blue, face to neck and entire underbody

177

## SQUATTER & PARTRIDGE PIGEONS

**Squatter Pigeon** *Geohaps scripta* 26–30 cm
A heavily built, ground-dwelling pigeon, usually in small flocks or pairs; responds to disturbance by 'freezing' or running, darting erratically among grass tussocks. If pushed too closely, a flock will burst into flight with a clapping of wings, birds scattering widely to trees or back to ground. Flight fast – flapping interspersed with swift glides on stiff, downcurved wings. Uses open grassy woodland on sandy soils interspersed with low gravelly ridges. These poorer soils have a more open, shorter grass cover that allows birds to move more easily, run faster than does the densely matted grass of heavier blacksoil country. These pigeons are never far from water. Voice is low, murmuring, seems to come from deep within, muffled or throaty, yet rather musical; sequence is often varied – fast, slow, slow or slow, fast, fast – 'coo-coo, coore, coore, coo-coo, cooor'. Also higher, uplifting 'coo-hooop!' Widespread NE Aust.; uncommon to locally common; rare in S of range; sedentary, partly nomadic.

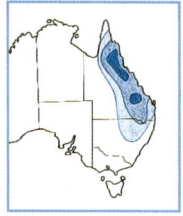

● *scripta*
● *peninsulae*: Cape York

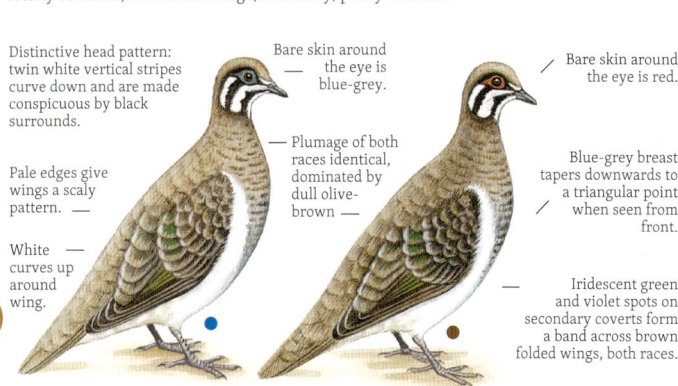

Distinctive head pattern: twin white vertical stripes curve down and are made conspicuous by black surrounds.

Pale edges give wings a scaly pattern.

White curves up around wing.

Bare skin around the eye is blue-grey.

Plumage of both races identical, dominated by dull olive-brown

Bare skin around the eye is red.

Blue-grey breast tapers downwards to a triangular point when seen from front.

Iridescent green and violet spots on secondary coverts form a band across brown folded wings, both races.

**Partridge Pigeon** *Geohaps smithii* 25–28 cm
This species is the NW replacement for the Squatter Pigeon of NE Aust. and shares its choice of habitat. Both are terrestrial, feeding, nesting and usually roosting on the ground, colours blending well with environs. Will squat or stand motionless if approached, then runs erratically through the grass or leaps into flight with loudly whirring wings. Other behaviour shared by both species includes a breeding season that extends throughout the year with maximum activity in the midwinter dry season. Like the Squatter, this is a gregarious species: usually in small parties, occasionally forms flocks of many hundreds. Calls similar to those of Squatter, but slightly more husky and querulous, slow, varying in strength – some strong, some softly fading and very low, often just single notes or a series more widely spaced, 'khwoor, khwuoor, kwoo-kwoo-kwooor, khwuoor', or higher, clearer 'oo-poo-poor' in alarm. Uses open tropical woodland, often on sandy, stony ground with short, rather sparse grass. Locally common; sedentary.

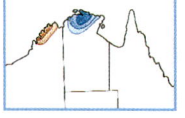

● *smithii*
● *blaauwi*
● Identifying features same as described for nominate race, except bare facial skin yellow rather than red

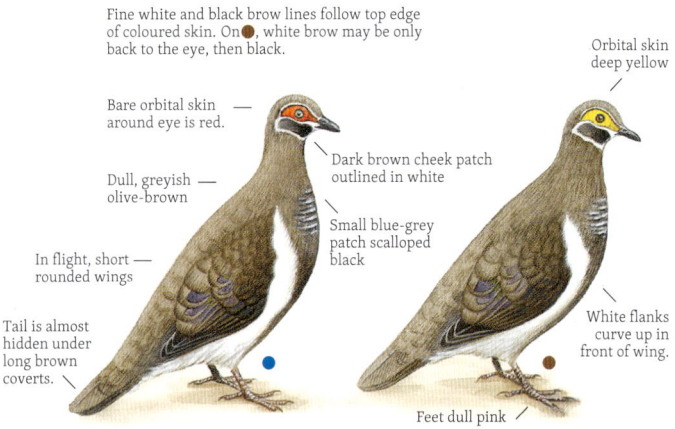

Fine white and black brow lines follow top edge of coloured skin. On ●, white brow may be only back to the eye, then black.

Bare orbital skin around eye is red.

Dull, greyish olive-brown

In flight, short rounded wings

Tail is almost hidden under long brown coverts.

Dark brown cheek patch outlined in white

Small blue-grey patch scalloped black

Orbital skin deep yellow

White flanks curve up in front of wing.

Feet dull pink

## ROCK-PIGEONS

**White-quilled Rock-Pigeon** *Petrophassa albipennis* 29 cm
The droop-winged stance gives a low, hump-backed, short-legged shape. With the attenuated rear end and tail, and the dark, scaly looking plumage, this and the Chestnut-quilled Rock-Pigeon appear almost reptilian. Lives in small flocks, pairs or alone in sandstone escarpment country; usually seen perched on huge boulders or ledges of cliffs where the birds run with agility. The pigeons launch themselves into flight, gliding down to search for seed among boulders or out on grassy, open woodland nearby. If startled, they take off with a loud clapping of wings, flying up to the sanctuary of the cliffs where they characteristically run a few metres across the rock after landing, sometimes hiding in rock crevices. Call starts with rough, low, grating 'grr-'; rises to an abrupt, high, sharp '-oook', repeated at regular intervals, 'grr-oook, grr-oook'. Also has a softer, more husky or throaty 'coo, car-ook'. Inhabitant of rugged sandstone ranges, gullies, gorges. Locally common in suitable habitat; sedentary.

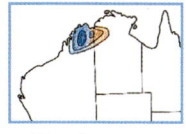

- *albipennis*
- *boothi*
- Overall lighter cinnamon-brown with little or no white on outer wing
- Overall darker dusky brown with large white panels on outer wing

Rock-Pigeons are large-bodied, small-headed, ground-dwelling birds: two species, both plumaged dark brown with small identifying differences.

White line under eye, down-turned at ear-coverts

Attenuated rear body: back, rump and tail-coverts have dark, scaly feather pattern.

Tail long, rounded

Larger white 'quill', the white base of outer primaries

Head looks small

Rounded, relatively short wings, fold far short of tail tip.

In flight shows little, if any, white on outer wing

Smaller white 'quill', the white base of outer primaries

Plumage lighter brown; white panel on primaries very small

Juv. (both races): almost identical to adults

**Chestnut-quilled Rock-Pigeon** *Petrophassa rufipennis* 31 cm 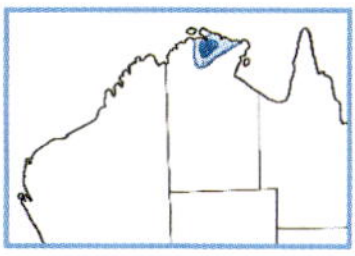
The Chestnut-quilled, like the White-quilled, is specialised for a restricted type of environment – the cliffs and gorges of rugged sandstone escarpments. The very limited habitat would always restrict the total population, while the intervening flat country breaks these sedentary rock-pigeons into separated populations, facilitating the evolution of two species, and of races among the widespread populations. In flight, behaviour and use of habitats, this species is like the White-quilled. Call is loud, musical, rapid 'cook-ar-ook', from which comes the Aboriginal name Kukarook. The display call is a low, grinding, grating 'owrrgh-owrrgh-owrrgh', uttered at the bottom of each bow. Neither species is found far from water, usually pools in gorges. Restricted range, but locally common.

Bright chestnut across the outer wing

When in flight, the reddish or white wing patches of both species are conspicuous, but vanish as wings fold on landing so that these dark birds appear to vanish. Combined with their short run at landing, this could confuse any pursuing predator.

Thin but prominent line under eye, back around ear-coverts

Sooty, brown, pale-edged, 'scaly' feathers, with long-rumped shape, give reptilian looks.

Sooty black, speckled silvery grey

Entirely sooty brown without lighter or darker bands or tips

Orange-chestnut streak on folded wing.

# 12 COCKATOOS

The families Cacatuidae and Psittacidae hold Australia's 54 species of the Order Psittaciformes (the world's total is some 330 to 356 species). The 14 species of cockatoo are usually large, stocky and crested, and are spread across most habitats from rainforest to arid. Some are entirely arboreal in their foraging, others terrestrial, while some use both resources. All species nest in hollows, requiring large old trees to provide big enough hollows: some species are thus threatened by loss of old-growth trees.

Magpie = 40 cm

## cacatuidae
### page 183

- Palm Cockatoo
- Baudin's Black-Cockatoo (Long-billed)
- Red-tailed Black-Cockatoo
- Carnaby's Black-Cockatoo (Short-billed)
- Western Corella
- Long-billed Corella
- Pink Cockatoo
- Glossy Black-Cockatoo
- Gang-gang Cockatoo
- Sulphur-crested Cockatoo
- Yellow-tailed Black-Cockatoo
- Galah
- Little Corella
- Cockatiel

# LORIKEETS & PARROTS

Australian parrots, family Psittacidae, have evolved diversely in size and colour, as well as in ecology: species exploit almost every habitat from alpine to tropical rainforest, and from beaches to central deserts. The parrots range in size from the very large, long-tailed Australian King-Parrot and the heavily built Eclectus to the tiny Double-eyed Fig-Parrot and Little Lorikeet. Most resources are used: nectar, seeds and larvae of wood-boring insects.

Budgerigar = 20 cm

## psittacidae
**page 189**

Rainbow Lorikeet

Eclectus Parrot

Superb Parrot

Scaly-breasted Lorikeet

Regent Parrot

Varied Lorikeet

Red-cheeked Parrot

Princess Parrot

Green Rosella

Musk Lorikeet

Double-eyed Fig-Parrot

Crimson Rosella

Little Lorikeet

Australian King-Parrot

'Yellow Rosella'

Purple-crowned Lorikeet

Red-winged Parrot

'Adelaide Rosella'

# PARROTS

# PALM COCKATOO & RED-TAILED BLACK-COCKATOO

**Palm Cockatoo** *Probisciger aterrimus* 55–65 cm
Largest Australian cockatoo; found singly, in pairs or small parties. Confined to lowland rainforest and its margins, and into adjoining eucalypt and swamp woodland. Rarely ventures far from rainforest habitat; roosts high. Each morning groups congregate, display, disperse to feeding trees. Likely to be seen in flight over breaks in forest canopy; big broad wings have steady, deep action; glides on downcurved wings. Feeds on seeds, nuts, berries; able to crack extremely hard nuts of pandanus palms. Has call of two notes; low then rising through a rasping beginning to a sharp, abrupt screech: 'aar-rraiik!'. Common in restricted range.

Nest hole typically a wide deep hollow in the trunk of a rainforest tree

Crest feathers long, back-curved

Bare facial skin, varies, orange-pink to scarlet

Bill huge, dark grey; female has smaller face patch and bill.

Wings almost square-tipped

Black without any colour panels

From beneath, entirely black except red face

Juv.: pinkish grey facial skin and yellow-edged feathers on underparts and wing linings

**Red-tailed Black-Cockatoo** *Calyptorhynchus banksii* 50–65 cm
A magnificent, conspicuous, noisy cockatoo. Travels to water morning and evening in loud, often very large flocks. Male has bulky black crest overhanging heavy bill; unbarred scarlet panels in tail. Female is spotted and edged yellow; tail yellow to orange, barred black; immature similar. Contact call, usually in flight, is a grating, metallic 'karraak', 'karrark' and 'airrk'; some calls hard and grating, others husky or squeaky, whistling 'kreeeeik'. Similar is the Glossy Black-Cockatoo, but smaller, browner, very short, low crest. Habitats are diverse: tall wet mountain forests of SE to open tropical forests of far N, and tree-lined rivers of interior. Nomadic, migratory; common in N, SW; rare, threatened in SE.

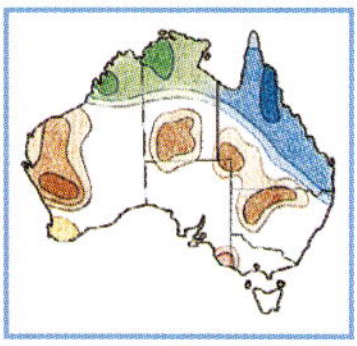

● *banksii*: largest, tail and wing long; bill large; female tail orange-yellow
● *macrorhynchus*: large; bill massive, large notch; female tail pale yellow
● *samueli* and
● *naso*: both have smaller bodies, bills and crests
● *graptogyne*: smallest, female has orange-red tail.

Most differences between races are slight; noticeable are overall size, tail colour, bill size and shape.

Crest down

● Bill very large, heavy, top mandible notched

Massive helmet-like crest overhangs far forward of the heavy grey bill.

Female and imm. spotted and barred yellow ♀

Tails: ●● yellow, some with slight orange
● pale yellow, some partly orange
● deeper orange, touch of red
● reddish orange, partly red

Brilliant translucent red panels

♂

● Tail orange-red, brilliant against sun
● Slightly more orange. Males of other three races similar in plumage to ●

♂

Flight appears lazy, buoyant, languid; often glides, lands with tail spread, displaying colour.

183

## GLOSSY & YELLOW-TAILED BLACK-COCKATOOS

**Glossy Black-Cockatoo** *Calyptorhynchus lathami* 46–50 cm
Highly specialised; reliant on casuarina seeds and requires habitats that include these trees. Inhabits forest and woodland with abundant casuarina trees. Not as conspicuous as the Red-tailed; usually in smaller flocks. Groups, commonly of up to ten birds, spend most of the day quietly feeding in the foliage of casuarina trees, the only sound being the busy clicking of bills as they demolish the hard, woody seed capsules. Call weaker, higher and more wheezy than that of the Red-tailed. Varied, rather soft, wheezing and grating 'kee-aiirrk', 'airr-riiek', 'airrk', 'airrek', 'arr-errk'. Uncommon, perhaps declining; limited by remaining habitat.

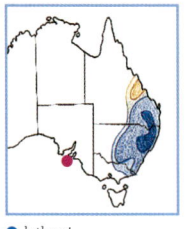

● *lathami*
● *erebus*
● *halmaturinus*

The three subspecies differ mainly in size and shape of bill:
● medium to large bill;
● disproportionately small bill;
● substantially larger bill.

Crest small, low, usually flat, following head contour

Bill dark, rounded, massive

Plumage brownish-black, not obviously glossy

Red tail panels

Brilliant red translucent panels when tail is spread

Low flat crown; very small crest at front, both sexes, usually folded down

Flight appears lazy, buoyant, languid; flocks drift along, flying high; often glide and land with tails spread, displaying red or orange.

Variable irregular patches of yellow around neck and face; male may have a few sparse yellow spots.

---

**Yellow-tailed Black-Cockatoo** *Zanda funerea* 58–65 cm Flight buoyant, effortlessly wheeling among the treetops with slow, deep wing beats and floating glides. In pairs or small parties; forms large flocks in winter. Occupies diverse habitats: coastal, inland and alpine, eucalypt forest, woodland, rainforest. Feeds largely on seeds of native trees and shrubs including eucalypts, banksias, hakeas, xanthorrhea. Plantation pines are worked for both seeds and wood-boring insects. Has carrying, wailing calls, 'why-eeela, weee-la'; in alarm, harsh screeches. Grinding noises when feeding. Nomadic or locally migratory; moderately common.

● *funerea*
● *xanthanotus*
● *whiteae*

Yellow ear patch
Red eye-ring
Bill dark
Yellow scalloping

Yellow-tailed and Short-billed species have similar wide, short-tipped bills.

● Smaller, browner. Has heavier yellow scalloping on underparts; male lacks black speckling on yellow panels of tail, female more lightly speckled on yellow of tail

Eye-ring grey, bill pale

Light speckling

● Tail panels flecked black on yellow
● Panels plain mid to pale yellow

Imm.: browner plumage with wider pale yellowish margins; bill pale, eye-ring grey

Dense black speckling on yellow

Translucent yellow panels glow brightly when backlit, sunlight passing through brightly

## WHITE-TAILED BLACK-COCKATOOS

### Baudin's Black-Cockatoo (Long-billed)
*Zanda baudinii* 55–60 cm

Lives in forest, woodland, farm trees. Feeds mainly on seed from large woody capsules of marri, a common SW eucalypt. Strips bark from dead trees in search of wood-boring insects; can damage fruit crops. Except for bill, calls and behaviour, almost identical to the Short-billed. Both white-tailed species give a call briefly described as the 'Wy-lah' call. This species has a shorter call than the Short-billed, less of the long, drawn out, wheezy wailing, but a rather more clear, whistled sound, rising clear and high then falling away. The calls are not harsh: 'wiee-ier', 'whyie-ier' or 'whyie-rrk'. Also deep, querulous corella-like sounds. Locally nomadic; secure while sufficient old-growth forests remain.

Bill has a long, fine point, narrow in front view. Enables extraction of seed from large, hard marri capsules, and cuts into bark and dead wood for larvae of wood-boring insects.

Both Short-billed and Long-billed Black-Cockatoos have long tails with obvious white panels.

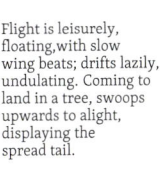

Female has pale grey eye-ring and bill, and white ear patch.

Male has red eye-ring, black bill, dusky white ear patch.

Flight is leisurely, floating, with slow wing beats; drifts lazily, undulating. Coming to land in a tree, swoops upwards to alight, displaying the spread tail.

Although the ranges of the two white-tailed species overlap, the Long-billed tends to be further to the SW in heavier, wetter forest. Both may be seen in small or very large, noisy flocks.

Both white-tailed species have the white ear patch.

Long tip is often tucked down among feathers.

### Carnaby's Black-Cockatoo (Short-billed)
*Zanda latirostris* 55–60 cm

Closely related to (and previously a race of) the Yellow-tailed; similar flight and call. Prefers habitat of forest, woodland, heath, farms; feeds on banksias, hakeas, dryandras – often on ground; also exploits pine plantations. Often in large wandering flocks. Voice is a loud, querulous, high, drawn out, wailing, wheezy, complaining 'ai-whiieer-la' or 'wy-ieeer-la'; the emphasis on the extended middle wail. With considerable experience, the calls can be helpful in separating the two WA white-tailed black-cockatoos. Declining due to clearing and resultant loss of food plants and scarcity of large hollow nest trees; local migratory and nomadic movements.

Feeding behaviour helps separate the WA white-tailed species. The Long-billed is adapted to extract marri eucalypt seed; the Short-billed more often feeds on pines, grevilleas, banksias and other Proteaceae, and more often searches on the ground for fallen seed.

Carnaby's Black-Cockatoo (Short-billed) is more closely related to the Yellow-tailed than to the Long-billed, above. The differences between the two white-tailed species are more significant.

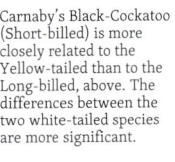

Large white tail panels

Pale-edged, scalloped pattern

Both Short-billed (Carnaby's) and Long-billed (Baudin's) Black-Cockatoos have long tails with obvious white panels.

Yellow-tailed and Short-billed species have similar broad, short-tipped bills.

## GANG-GANG COCKATOO & COCKATIEL

**Gang-gang Cockatoo** *Callocephalon fimbriatum* 33–36 cm
Distinctive dark-grey cockatoo with wispy, fine-plumed crest. Bulky head and shoulders taper to short tail. Uses dense, tall, wet forests of mountains and gullies, and alpine woodland. Once settled in a tree and feeding on seeds, nuts and berries, Gang-gangs can be hard to see: when feeding or resting in dense foliage, they remain silent except for soft growling sounds. Sometimes reveal their presence by dropping debris as they rip seed capsules apart, but remarkably placid and tolerant of close approach. Long wings and deep powerful wing beats make the Gang-gang a strong flier; often travels tree to tree by swooping low then rising steeply to land. Contact call is a drawn out, creaky rasp, 'gr-raer-iriek!', beginning as a rough croak, ending with a squeak like a rusty hinge. Common in prime habitat of S NSW and NE Vic., but becoming rare where habitat is degraded.

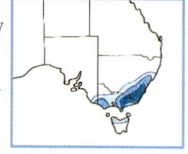

**Cockatiel** *Nymphicus hollandicus* 31–33 cm
Graceful small cockatoo with slender silhouette. Widespread in open country, woodland, scrub or grassland with scattered trees near water. Flight swift on slender, flickering, backswept wings, long tail trailing. Travels direct and without the undulations typical of parrots; small flocks keep together with precision formation flying. The Cockatiel's classification is still uncertain, it being the sole member of its genus. Until quite recently, it was placed in the parrot family. Evidence now suggests it is a cockatoo, and it is placed in that family, to which it has anatomical similarities and behavioural links in breeding rituals and the begging calls of the young. Biochemical comparisons support this classification. Has a distinctive, loud, clear, flight call, a slightly husky, mellow, rolling yet sharply penetrating 'whee-it, wheeit' or 'querr-eel', which gives the name 'Quarrion'. Common, abundant in some northern inland regions; highly nomadic – follows rain into drier regions or moves towards coast during inland droughts.

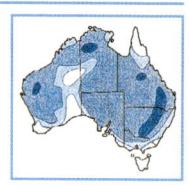

# CORELLAS

**Long-billed Corella** *Cacatua tenuirostris* 38–41 cm
A white cockatoo with crimson or salmon-pink splashed about face and neck. Savannah woodland, open forest, grassland with scattered or watercourse trees, roadside and paddock trees; never far from water. Usually in pairs or alone in spring breeding season; at other times in small to large flocks. Usual contact call in flight is a wavering, nasal, falsetto, triple-noted 'ar-aer-ek, ar-aer-aerk', which, from a flock, becomes a raucous din. When disturbed, very loud, harsh screeches.

Perth; introduced

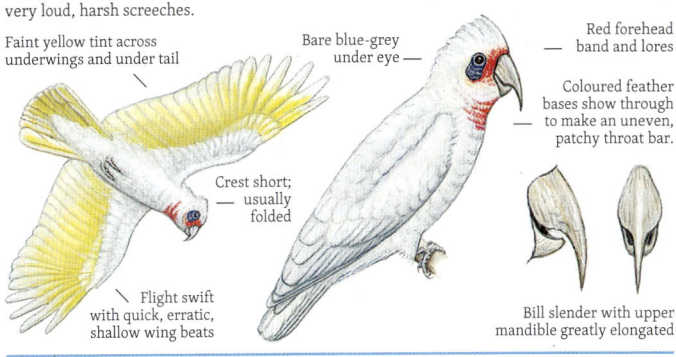

Faint yellow tint across underwings and under tail

Bare blue-grey under eye

Red forehead band and lores

Coloured feather bases show through to make an uneven, patchy throat bar.

Crest short; usually folded

Flight swift with quick, erratic, shallow wing beats

Bill slender with upper mandible greatly elongated

**Little Corella** *Cacatua sanguinea* 36–39 cm
Large flocks use tree-lined watercourses and adjacent plains; savannah woodland, mulga, mallee. Feeds in flocks on ground; congregates on trees to strip the leaves. Very noisy in flight and perched. Call varies – a harsh, resonant, nasal, brassy 'air-er-ek' from sharp to guttural – 'aier-ek, aier-rr-k, aer-rk, errk, urrk, aiirk'; also loud screeches. Flocks of thousands create a crescendo of screeches and hollow croaks. In breeding season flocks are smaller, pairs keep closer to their nest sites. Generally abundant.

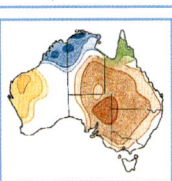

● *sanguinea*: largest body; little or no pink at lores; eye-ring pale grey
● *normantoni*: smallest, pinkish orange lores
● *gymnopsis*: deeper red lores, darker blue-grey eye-ring
● *westralensis*: longer wings, tail; larger red loral spot

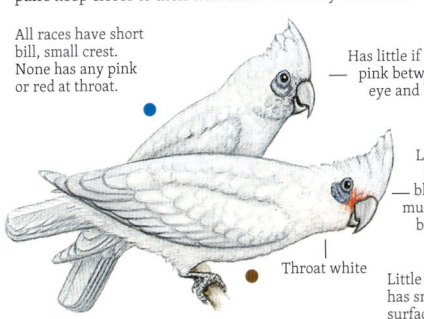

All races have short bill, small crest. None has any pink or red at throat.

Has little if any pink between eye and bill

Lores red, eye-ring blue-grey, much wider below eye

Throat white

Little Corella bill has smooth surface; short tip.

**Western Corella** *Cacatua pastinator* 36–39 cm
Habitat is open forest, woodland, farmland with abundant trees in shelter belts, road reserves, paddocks of SW WA. Prior to settlement there was a wide gap between northern and southern races. With clearing of wheat belt woodland and forest, the N race has extended further S while the S race has contracted southwards to the more heavily forested SW corner. Calls are like those of the Little Corella. N race common; S race restricted, uncommon.

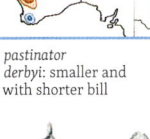

● *pastinator*
● *derbyi*: smaller and with shorter bill

Uses its long bill to dig for roots and corms. The white plumage is then often stained rust or brown.

Taller crest

Crimson between eye and bill

Crimson at throat

Upper wing entirely white; underwing pale yellow

Western Corella bill: elongated, gradual curvature, not as long and fine as the bill of the Long-billed.

# SULPHUR-CRESTED COCKATOO & GALAH

**Sulphur-crested Cockatoo** *Cacatua galerita* 45–50 cm
A big, noisy white cockatoo with yellow crest. Habitat is diverse, ranging from high rainfall forests – eucalypt, rainforest, coastal mangroves – to semi-arid inland regions, watercourse trees and partly cleared farmland. In S Aust. forms huge flocks with regular roosts and midday shelters on tree-lined watercourses, from where the birds fly out to open country to feed on the ground. In tropics, flocks are small; the birds are more arboreal, feeding on seeds, berries and flowers of trees and shrubs. Has stiff-winged, irregular flap and glide flight. Makes loud, raucous, unpleasant screeches, usually an intermix of harsh and sharp sounds, varying from deep, grinding and guttural to powerful, piercing screeches of ear-splitting intensity: 'airrrik, aarrrk, ahrk, aieirrk, aieirieik!'. Sedentary; common to abundant. Introduced to Perth area.

- *galerita*
- *fitzroyi*: reduced or paler yellow in plumage, bluish eye-ring, little if any yellow on ear-coverts

High, forward curving, deep yellow crest

- Yellow feathers of crest curve forward
- Crest feathers straighter

Bill dark grey to black

- Ear-coverts yellow
- Ear coverts white

Pale yellow tint

- Bare whitish skin around eye
- Bare skin pale blue, some white

Male: iris dark brown
Female: iris red-brown to dark brown

Crest folds almost flat, usually with slight upwards curve.

- Nominate race, Cape York to Tas.
- NT and Kimberley

Characteristic flight pattern of brief glides interspersed with bursts of shallow, rapid wing beats. Travels high, then drops suddenly to tree-top level in spiralling glide; displays crest at landing.

**Galah** *Eolophus roseicapilla* 35–38 cm
Familiar pink and grey cockatoo. In flight, deep, abruptly varied wing beats; steady and direct or a wild, erratic, crazy route across the sky. Often big flocks wheel in unison, showing massed pink then grey. Habitat covers diverse open country: open woodland, sparsely vegetated interior, coastal areas opened by clearing. Feeds on ground or low shrubs in small parties or flocks. Voice rather harsh, metallic and abrupt, yet not unpleasant: 'chirrink-chirrink, chirrink-chirrink' and variations 'chzink-chzink', 'czink-czink-czink-czink'; in alarm, harsh, scolding, rasping screeches. Abundant across most of range; increased with clearing of heavier forests.

- *roseicapilla*
- *albiceps*
- *kuhli*: smaller; pink eye-ring; crest only on front of crown, shorter and very pale pink

- Crest larger, pale pink blending to deeper colour on neck; when folded makes head of this race appear larger, bird seems of heavier build.

Grey or white 'warty' eye-ring

All races: upper body and wings light grey, wing tips darker; underparts mostly pink

♀ Eye red
♂ Eye brown

Rump and tail very pale grey

- Crest very pale pink, almost white, sharply distinct from darker hind-neck
- Deep pink eye-ring
- Usually has deeper pink body and very pale crown.

# PINK COCKATOO TO SCALY-BREASTED LORIKEET

**Pink Cockatoo** *Lophochroa leadbeateri* 35–40 cm
In pairs, small family parties and, rarely, large flocks; at times intermingles with Galahs or Little Corellas. Uses open, sparsely timbered grassland, farmland with well-treed paddocks, mulga and similar open scrub, open mallee country, callitris and casuarina country, watercourse trees. Never far from water. Contact call, given frequently while in flight, is a thin, querulous, drawn out, wavering screech, the sound stuttering or undulating, hoarse and scratchy, 'ar-ai-ar-a-ar-iagh, ai-ra-a-iagh'. Sedentary; uncommon.

● *leadbeateri*
● *mollis*

● Crest scarlet and yellow

On landing, uplifted wings display deep pink, usually only glimpsed in flight.

● Crest scarlet

Pale to mid-pink

Male has brown eyes; female has deep red, and often a wider yellow band in the crest.

Dorsal pearly white

Underwing deep salmon pink with white margins

Flies with irregular, shallow, almost fluttering wing beats interspersed with glides on downcurved wings.

**Rainbow Lorikeet** *Trichoglossus moluccanus* 26–31cm
Distinctive flight silhouette; flocks dart, twist and wheel among trees, or fly straight and direct above the forest canopy. Diverse habitats: rainforest, eucalypt forest, woodland, farmland, paperbark woods and heath, mangroves. Feeds on fruit, nectar, blossoms, seeds, berries and orchard fruit. In flight, gives frequent, quite pleasant, softly rasping or vibrating musical screeches; softer mellow chattering and subdued screeches while feeding; quiet twittering while resting. Common in N and E, uncommon in S of range, extremely rare vagrant to Tas; introduced in Perth suburbs.

● *moluccanus*
● *rubritorquis*

● Shows some N to S variation, but all of this race have the lemon-green nape collar.

Bill and eye bright red

Yellow bar along wing

Streaky blue-violet

Collar orange-red

Bright leaf green

Red, sides yellow

Blue-violet

Tail long, finely tapered

● 'Red-collared Lorikeet' has deep orange-red breast, wing linings and hind neck collar.

**Scaly-breasted Lorikeet** *Trichoglossus chlorolepidotus* 23 cm
A plain green parrot with spectacular underwing colours revealed in flight. Uses most vegetation types along lowlands of E coast where there are flowering trees, eucalypt forest, woodland, heath-woodland, paperbarks. Gives sharp, short, clear screeches, rather like the Rainbow Lorikeet, but usually sharper. The 'chewip' call varies from high and sharp to mellow or slightly rasping. Also gives scolding 'charr!' and soft chatterings. Common; nomadic.

A hint of the scarlet under-wings shows beneath bend of shoulder

Scarlet, vermilion and dark grey underwing

Upper parts entirely bright green

Bill, eye, bright red

Bill and eye brown

Imm.

Yellow edges

'Scaly' pattern

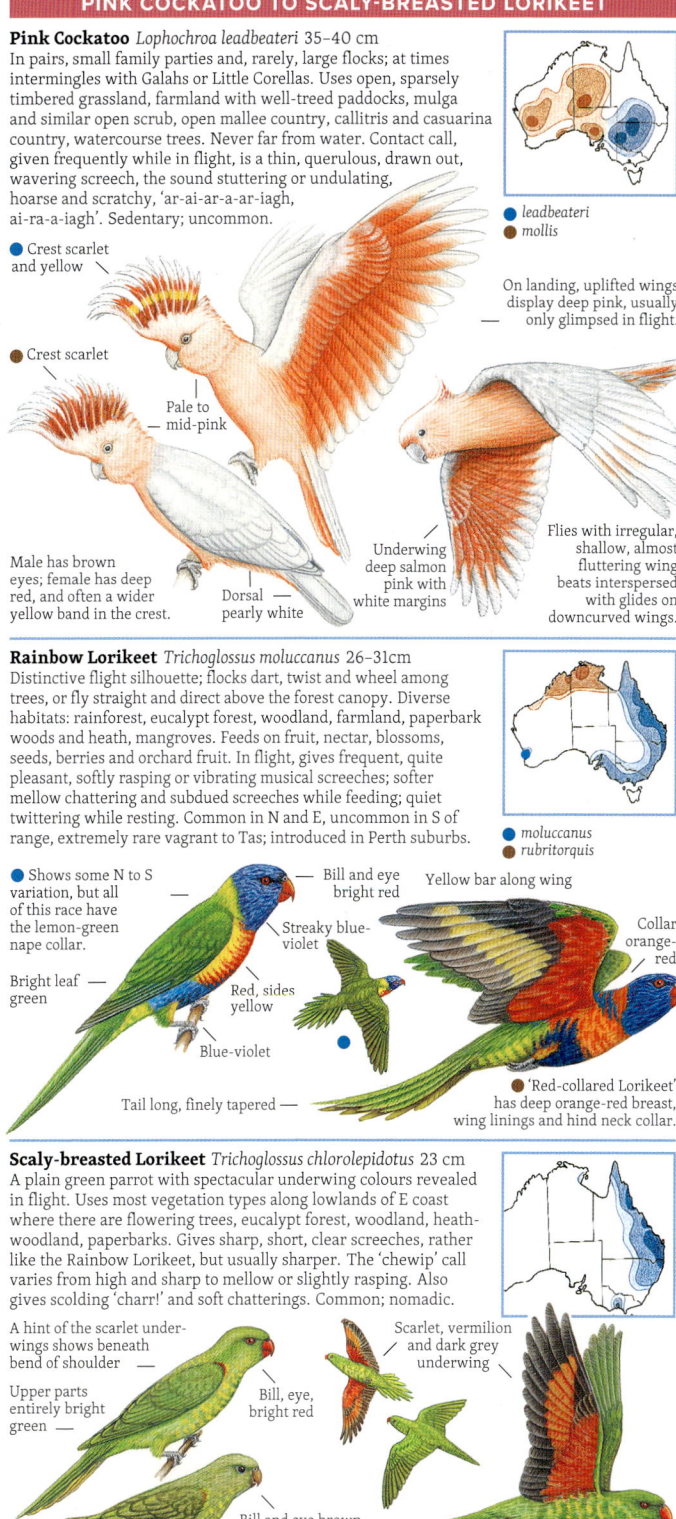

189

## VARIED, LITTLE & PURPLE-CROWNED LORIKEETS

**Varied Lorikeet** *Psitteuteles versicolor* 18–20 cm
Small parties or large flocks feed on flowers or fruit: favours nectar of bloodwoods and melaleucas. Flocks dart in tight formation among trees in swift, direct flight. Found in tropical forest and eucalypt or melaleuca woodland wherever trees or lower shrubs are flowering; often in heavier vegetation near watercourses. Contact call, given almost constantly in flight, is a thin, shrill, metallic screech, noticeably higher pitched but not as loud as the calls of the Rainbow Lorikeet. While feeding, chatters busily – quick, sharp little screeches intermingling with softer scoldings. Moderately common; abundant in tropical NT.

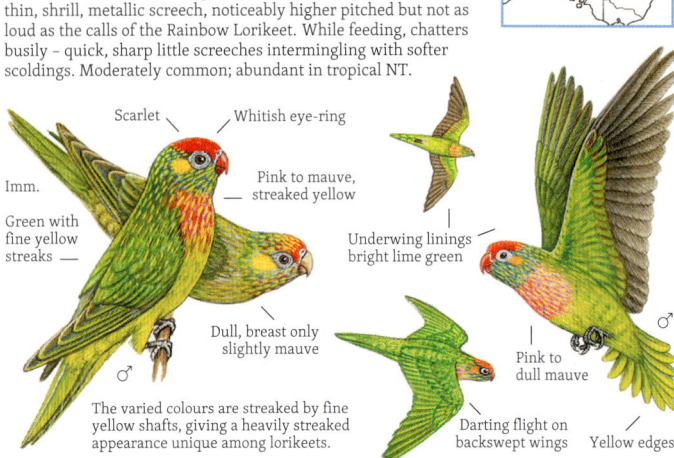

**Little Lorikeet** *Parvipsitta pusilla* 15–16 cm
Flight swift, direct; rapidly whirring small wings give a fleeting glimpse of yellowish green underwing linings; widely fanned tail reveals orange-red colour that is usually hidden. Uses forest, woodland; favours open country – trees along watercourses and paddock trees. Feeds on nectar, pollen, fruits, berries, seeds. Gives sharp, short screeches and warbling 'zrrit, zrit' or 'chirrit, chrrit'. Gives almost constant, pleasant, subdued, chatter-like screeches with intermingled brief, harsh scoldings while feeding. Nomadic; most common in N and E, uncommon in S; vagrant to Tas.

**Purple-crowned Lorikeet** *Parvipsitta porphyrocephala* 16 cm
Almost continuous sharp, high screeches attract attention as pairs or flocks dash through treetops, sunlight at times catching crimson beneath their wings. The green plumage of these tiny parrots allows them to merge into foliage as they climb among leaves and flowers in their search for nectar and pollen. When eucalypts are in flower, noisy flocks may be heard, and are easily seen in low mallees, but hard to sight in high crowns of forest trees. Calls almost continuously in flight, a high, slightly metallic or vibrating 'tziet, tziet'. When feeding in treetops, soft chattering mixes with louder flight calls. Moderately common; nomadic; erratic, locally abundant to rare.

# MUSK LORIKEET & DOUBLE-EYED FIG-PARROT

**Musk Lorikeet** *Glossopsitta concinna* 21–23 cm
Predominantly bright leaf-green with splashes of bright crimson, yellow and blue, but not easily seen dashing across the sky in screeching flocks or climbing among dense foliage and flowers of tall eucalypts. Often intermingles with Purple-crowned or Little Lorikeets, the Musk obviously larger. Flight very swift, direct with audible whirr of rapidly vibrating wings. Attracted to woodland, open forest, mallee, cleared country with trees along watercourses and roads. In flight, gives a shrill metallic screech as contact call. Feeds in treetops with much soft chattering, the varied notes intermingling, sharp, husky and querulous. Common to uncommon; locally abundant where flowering is heavy. The Musk Lorikeet is nomadic, perhaps more seasonal and predictable than other lorikeets in its search for nectar.

Crown dark turquoise-blue, nape and upper back olive-brown

Forehead through eye to ear-coverts and bill, bright crimson

Diving away, shows colours that are usually hidden when perched: dark brown flight feathers, yellow-green wing linings and golden-pink of spread tail

Upper parts green with bronze tone on nape and mantle

Tail yellow, merging to red at base

Yellow patches on flanks, largely hidden by wings when perched

Female has much less blue on crown; immature has dull reds, brown bill.

Flight swift, body tapering to pointed tail, backswept wings

## Double-eyed Fig-Parrot
*Cyclopsitta diophthalma* 13–15 cm
Extremely small, short-tailed parrot of tropical rainforest canopy, scurrying mouse-like along branches and among foliage. Core habitat is rainforest, from low riverine monsoon forest types of Cape York to lowland and high-altitude tall rainforest of NE Qld. Ventures into fringing woodland or other habitats when moving between forest. Flight is swift, direct, slightly undulating, with short bursts of wing beats; usually travels above the canopy. Contact call, a sharp, penetrating 'tseit-tseit-tseit-', usually in flight and on landing; alarm is a shrill screech. Races in separated tracts of rainforest have at times been described as species. Two N races are secure.

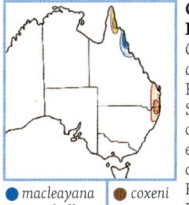

● *macleayana*  ● *coxeni*
● *marshalli*

Smallest Aust. parrot, once considered three separate subspecies.

## Coxen's Fig-Parrot
*Cyclopsitta coxeni* 13–15 cm
Formerly the S race, *coxeni*, of the Double-eyed, sometimes called the 'Blue-Browed Fig Parrot', this tiny SE Qld and NE NSW rainforest parrot is critically endangered, having lost much of its lowland rainforest habitat. Fewer than 100 may remain. Some consider it likely extinct.

### 'Marshall's Fig-Parrot'

Red edge to innermost wing-coverts, both races

Race with largest red cheek patch, but small blue brow line

Very narrow blue edge under tail

Blue forehead; dull white cheek patch with blue border

Yellow edges to breast, yellow flanks hidden under folded wing (both races)

Outer edges of flight feathers blue (both races)

Blue

Yellow edges to breast, yellow flanks largely hidden under folded wing.

Like Double-Eyed races, has pale yellow bands on base of flight feathers.

### 'Red-browed' or 'Blue-faced' Fig-Parrot

Cheeks red, underlined blue; eye-ring blue

Red

Face dull white, edged with blue

Underwing (both races and Coxen's, both sexes) has pale yellow bands across base of flight feathers and greater coverts.

Diving away, Coxen's and Double-Eyed races show stumpy tail, yellow wing bars.

## ECLECTUS & RED-CHEEKED PARROTS

**Eclectus Parrot** *Eclectus roratus* 40–43 cm
Magnificent bulky parrot; sexes vastly different, both spectacular. Male looks his best when in flight or displaying underwings and flanks. Usually in rainforest, but ventures into adjoining eucalypt woodland to reach other areas of rainforest. Very noisy, obvious; small flocks roost in tall rainforest trees. In morning, small groups, occasionally large numbers, noisily move out to feed on big fruiting trees. Eclectus travel high above the canopy. Heavy, slow and direct, purposeful, deep strokes, wings not lifted above shoulder height, the action interspersed with glides on down-bowed wings. Call is a harsh screech with the recurrent strongly rolled 'r' giving heavily vibrating throaty roughness: 'arrrk-arrrk-arrrk' and 'airrk'. While feeding, an occasional wailing cry or soft mellow whistle. Quite common in optimum rainforest; also in Solomons, NG, Indonesia.

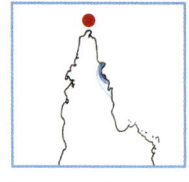

● *macgillivrayi*
● *polychloros*: much smaller; male slightly deeper green, towards viridian rather than the yellowish-green of the Cape York male

Male brilliant green with red eye and yellow-orange bill

Female has crimson head to upper breast; eyes yellow, bill black.

Glimpses of scarlet flanks

Dark flight feathers

Crimson thighs, flanks and under-tail-coverts

Brilliant scarlet wing linings; intense blue leading edge

In flight, the brilliant red and green of the male's plumage is no less spectacular than the crimson and deep blue-violet of the female.

---

**Red-cheeked Parrot** *Geoffroyus geoffroyi* 21–24 cm
Small bright parrot of Cape York's dense tropical rainforest and margins of adjoining woodland. Uses canopy foliage; feeds on seeds, fruits, berries, usually in pairs or small flocks. Hard to see high in the leaves, but noisy, which helps locate where birds are feeding. May be glimpsed fluttering high in the branches, dropping fruit and debris. The flight is distinctive, swift and direct, swerving and twisting so that bright blue of underwings shows; wing beats are shallow and rapid, without gliding. Calls are loud, metallic, piercing 'airk, airk, airk' and long rapid series, even pitch and strength: 'haik-haik-haik-haik'. Common within its very limited range.

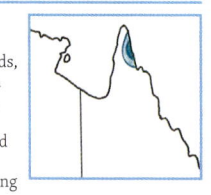

Crown to nape intense blue-violet

Cheeks rose-red, extending to forehead and ear-coverts

Underbody bright leaf-green, lightly mottled dark green

Bronzed green

Olive-brown

Green

Greenish yellow

Imm.
Brown to grey-brown

Under-wing-coverts bright light blue

Dark blue-grey

Males are slow to gain their mature red-cheeked plumage. Initially green-headed, then brown like the female, they finally reach full male colours after two years.

## KING & RED-WINGED PARROTS

**Australian King-Parrot** *Alisterus scapularis* 42–44 cm
In pairs, small flocks; flight direct, heavy, with full, regular wing beats. In the wild, very wary; flies far if disturbed. Keeps to heavier coastal and mountain forest in breeding season, including rainforest, eucalypt forest, palm forest, dense river-edge forest and closely adjoining eucalypt woodland. After breeding, King-Parrots wander further to lowland, farmland, shelter-belts, parks, gardens. In flight gives abrupt, sharp, clear 'krassiek', 'k-wiek', 'chriek' or 'charrak' with slight variations; or rapid 'chrak-chrak-chrak-'. Alarm call is a harsh, metallic, screeched 'karrark!'. Males give a sequence of high whistles, 'chreip, peeip, peip', while perched. Sedentary, dispersive; usually common where habitat remains.

- ● *scapularis*
- ● *alisterus*: intergrade form, doubtful if distinct
- ● *minor*: smaller, otherwise not distinguished by plumage or other feature

Long, dark tail
Flight fast, direct, on backswept wings
Entire head, neck and underbody bright scarlet
Intense emerald-viridian back and wings with pale turquoise band
Scarlet, dark-tipped
Bright green, but dark against light, in shade
An impressive sight, diving away, showing bright colours
Dull green, red tinge
Scarlet
♂
Lower back to tail-coverts, deep blue
♀
Plumage colours intense; may seem darker if backlit or in shade.
Long tail black with green or blue sheen

**Red-winged Parrot** *Aprosmictus erythropterus* 31–32 cm
An unmistakeable parrot clad in brilliant green; crimson wings set in dark surrounds add visual impact. Occupies open eucalypt forest and woodland, usually with open grassy understorey; also fringes of rainforest or monsoon forest; further inland, mulga, brigalow, casuarina and callitris; mangroves along the far N coast. Typically seen in small parties or pairs feeding on fruits, seeds, nectar and insects in foliage of trees and shrubs; rarely on ground. Tend to be wary, easily put to flight; rise with much calling. Flight call a sharp, metallic 'chrrik-chrrik, chrrik-chrrik' or 'crillik, crillik'; in distance sounds like a Budgerigar chattering. Alarm call is a series of harsher, more abrupt screeches, 'chak-chak-chak, chrak-k-kak'. Nomadic, dispersive; common through most of its range.

- ● *erythropterus*
- ● *coccineopterus*: with broad band of hybrids or a cline on Cape York Peninsula. Size appears to be the only difference, and subspecies status perhaps not warranted.

Scarlet shoulder patch set in black against the body's bright yellowish green creates a spectacular plumage.
Upper-tail-coverts greenish yellow
Identified by unique flight action, even in the distance or when no colours are visible
Lower back mid-azure blue
Eye and bill orange-red
♂
Spectacular in flight with deep downward beats displaying scarlet inner-wing panels
♀
Green, edged and tipped yellow
♂
In flight, has unique erratic wing beat with a slight pause at the bottom of each deep stroke, giving a jerky rowing action and a buoyant, weaving path.

## REGENT, PRINCESS & SUPERB PARROTS

**Regent Parrot** *Polytelis anthopeplus* 38–40 cm
Colourful, gregarious, long-tailed; flies swiftly on backswept wings. Eastern population inhabits floodplain woodland, mainly of river gums, with big old trees for nesting hollows. WA race uses open forest and woodland, appears to be increasing in range and numbers, while eastern race may be less secure. Voice is a distinctive, mellow, rolling, deep 'quarrak-quarrak-quarrark'; scolds and chatters. Moderately common; nomadic.

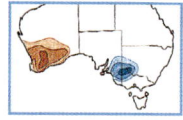

● *anthopeplus*
● *monarchoides*: has slightly brighter colours.

**Princess Parrot** *Polytelis alexandrae* 35–45 cm
One of Australia's most beautiful parrots; inhabits arid regions with sparse trees, eucalypts, casuarinas, acacias, spinifex, also vicinity of salt lakes with succulents and saltbush groundcover. May be seen in pairs or small parties, travelling with slightly irregular wing beats and undulating flight; drops slowly to the ground with fluttering, almost hovering, wing action. Much time is spent on the ground searching for seed; spinifex seed is probably a major part of the diet – observers have noted that it is seldom found far from spinifex. Rather quiet; gives a loud, unmusical 'kee-ahrk-carruk' occasionally. Highly nomadic; may appear after an absence of years, even decades, after flooding rain when fresh vegetation covers part of their range.

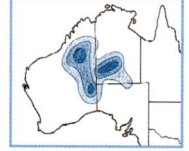

**Superb Parrot** *Polytelis swainsonii* 37–42 cm
Usually seen in small parties or flocks. River red gum, box and similar forest, river-edge forest, nearby mallee, native cypress, farmlands. Varied calls: commonly a strong, penetrating, rather rough yet musical 'querr-ieek, querrieek' with final 'ieek' loud and sharp; or 'krak-karrark'. Other calls include a sharp, penetrating, whistled 'whiek, whiek, whiek' at regular intervals, and harsh, deep, scolding, 'quarrarrk'. Migrates from SW Riverina to central-north NSW along Namoi and Macquarie rivers for winter. Has declined with habitat loss; species now common only locally, mainly in a few protected habitats.

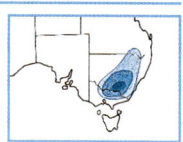

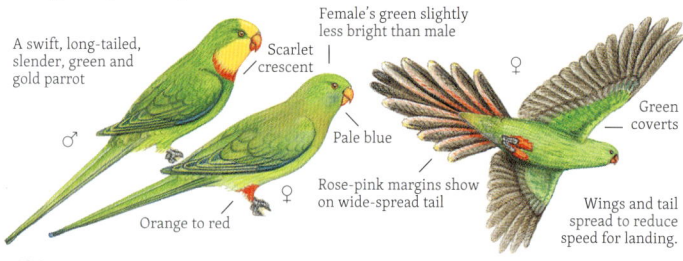

# GREEN & CRIMSON ROSELLAS

**Green Rosella** *Platycercus caledonicus* 32–38 cm 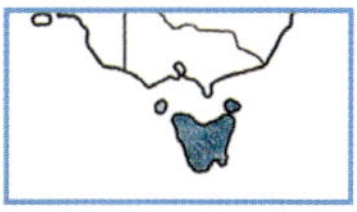
In most Tasmanian habitats except treeless moorland and farmland; common in heavy rainfall districts, dense forest and mountains. Usually seen in flocks or small parties, feeding on the ground, or in foliage of trees. Their flight is strong, swift, with only slight undulations; bursts of quick wing beats are broken by brief glides. In flight, often draw attention with a distinctive 'kzink' or 'kussink'. Often also a very high, ringing call, three piercing whistles, second note highest: 'whee-whieit-whee'. Also a harsh, sharp, metallic 'k-ziek, kziek-kziek'. From pairs, parties or flocks, typical rosella chattering, but rather harsh. Abundant; locally nomadic.

Black with green feather margins

Colours not as bright as on adult

In flight, sunlit yellow is bright against blues and greens of sky and forest.

Bright red band

Black with blue margins

Blue cheek patch

Juvenile is greener overall.

Central feathers bronze-green; outers blue, edged white

Greenish yellow

**Crimson Rosella** *Platycercus elegans* 32–37 cm
Plumaged bright-red, orange, yellow and blue, the parrots within *elegans* show greater colour difference than is usual in one species. Now combined with the Crimson as a single species, the 'Adelaide Rosella' and 'Yellow Rosella' are subspecies, but so distinctly different in appearance that these names continue in common use. All are boldly coloured and conspicuous. The combined habitat of the subspecies includes heavy wet forest, inland river belt trees, mallee and scrubby eucalypt country. The call is a clear, ringing 'k-teee-tip, k-tee-tip', the central 'teee' loud, high and clear; variations include 'k-tee-it-tip', 'kteeeit-tip' and 'tip-teee'. All races common.

Crimson

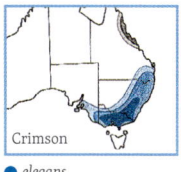

● *elegans*
● *nigrescens*: like *elegans* but smaller and darker; juv. much darker

Entirely in crimson and deep blue, a dramatic simple pattern

Olive-green

Juv.

Long tail tipped pale blue

Deep blue-violet

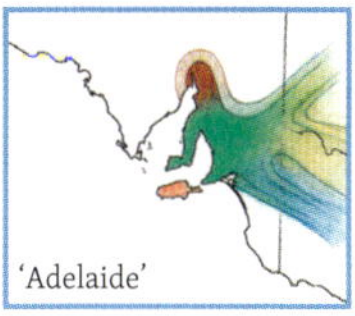

Patchy crimson and green

Crimson rump

Green

Outer web of primaries blue-violet

Tail blue

Black centred, broadly edged yellow

Patchy crimson and green

Varies, orange to yellow.

'Adelaide'

● *subadelaidae*
● *fleurieuensis*
● *elegans*
● *flaveolus*
● *melanoptera*: larger, darker; Kangaroo Is.
Colours intergrade and mingle in this zone where races meet.

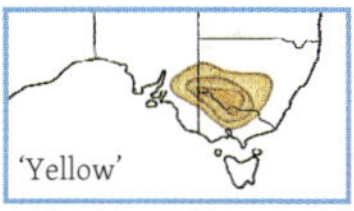

Red frontal band

Tail tips white

Bend of wing blue

Outer webs of primaries blue

'Yellow'

● *flaveolus*: full range

# PALE-HEADED & NORTHERN ROSELLAS

**Pale-headed Rosella** *Platycercus adscitus* 28–32 cm
Feeds in pairs or small flocks on ground or among foliage; flight undulating; obvious pale head. Habitat usually grassy woodland, farmland with scattered trees, lines of trees along watercourses and roads, dry scrubby ridges, but prefers lowlands rather than the highest parts of ranges. Call in flight is an abrupt 'czik-czik-, czik-czik-'. From trees gives high but soft, thin and slightly tremulous 'fee-e-fee-e-fe-e' or 'fwe-we-we-wee'. Abundant, sedentary; has benefited from thinning of heavier forests.

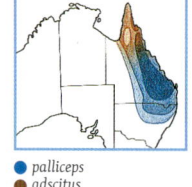

- ● *palliceps*
- ● *adscitus*

**Pale-Headed Rosella**

Deeper golden yellow compared with the paler, narrower feather edging of the blue-cheeked nominate race

Cheeks white with yellow tint

Blue panel in blackish wing

Blue extends to upper breast or throat.

Blue

Red

Upper-wing- and under-wing-coverts deep blue

Central tail feathers blue with slight green tint towards coverts; outer tail blue, tipped white

Tail green, gradually blue towards the tip

**'Blue-cheeked Rosella'**

Females: both races, like males, but with underwing stripe

Cheek patch white with wide blue-violet lower edge

Yellow

**Northern Rosella** *Platycercus venustus* 28–30 cm
Usual habitat grassy open forest and woodland, especially where dominated by eucalypt or melaleuca trees; further inland, where woodland gives way to more open country, the tree belt lining watercourses. Also visits coastal-swamp forest and mangroves to feed. Usually small parties of five or ten birds feed on ground in cooler morning and evening, or shelter in leafy treetops during heat of day, where they may attract attention with their sharp calls or typical rosella chatter accompanied by much tail wagging. The flight is undulating, often swooping close to ground, then turning steeply upwards, tail and wings spread, to land. Calls high, very clear and ringing; includes very rapid, high, piercing 'whit-whit-whit-whit-', so fast that ten or twenty notes are given in two or three seconds at an unwavering, even pitch. Also quieter, more husky, mellow 'chak, chak-chak, chakchakchak'. Sedentary; uncommon, but not scarce in most suitable habitat; population appears to be declining.

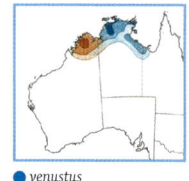

- ● *venustus*
- ● *hillii*

Three species, Eastern, Pale-headed and Northern Rosellas, are sometimes grouped as the 'White-cheeked Rosella Complex'.

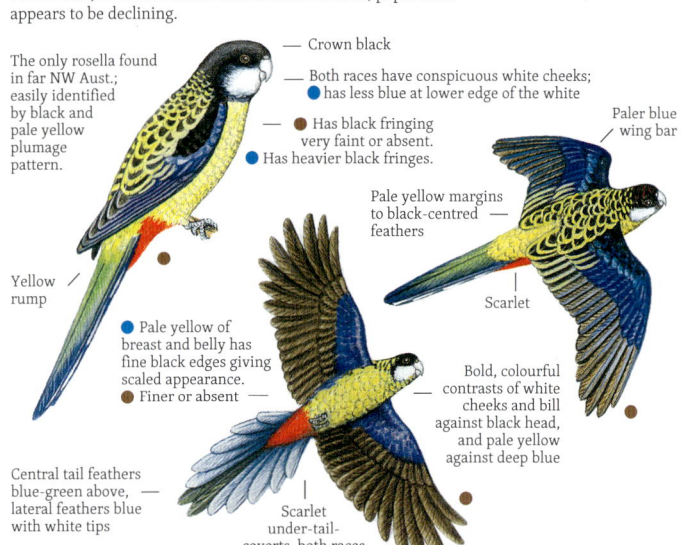

The only rosella found in far NW Aust.; easily identified by black and pale yellow plumage pattern.

Crown black

Both races have conspicuous white cheeks; ● has less blue at lower edge of the white

● Has black fringing very faint or absent.
● Has heavier black fringes.

Paler blue wing bar

Pale yellow margins to black-centred feathers

Yellow rump

Scarlet

● Pale yellow of breast and belly has fine black edges giving scaled appearance.
● Finer or absent

Bold, colourful contrasts of white cheeks and bill against black head, and pale yellow against deep blue

Central tail feathers blue-green above, lateral feathers blue with white tips

Scarlet under-tail-coverts, both races

# WESTERN & EASTERN ROSELLAS

**Western Rosella** *Platycercus icterotis* 25–28 cm
An unobtrusive bird; usually in pairs or small family parties rather than in flocks. Diverse habitats, from tall wet karri to dry woodland and mallee well inland towards the Nullarbor. More common in woodland of salmon gum and wandoo, and farmland with scattered trees; less common in heavy wet karri and jarrah; scarce on most of sandy SW coastal plain. Feeds quietly on the ground or in foliage. The flight is lighter, more fluttering than other rosellas, with only slight undulations. Less wary and timid, and less aggressive in aviaries. Two races, with transition between gradual rather than clearly defined and shown by the spread of red down the back. Calls are clear, musical, high-pitched, ringing 'quink-quink-quink-quink' and slightly softer 'whip-a-wheee'. Generally quite common; probably gaining from land clearing, provision of water.

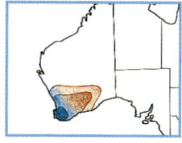

- ● *icterotis*
- ● *xanthogenys*

The only rosella species with yellow cheeks. The White-cheeked Complex has three species and the Blue-cheeked has two.

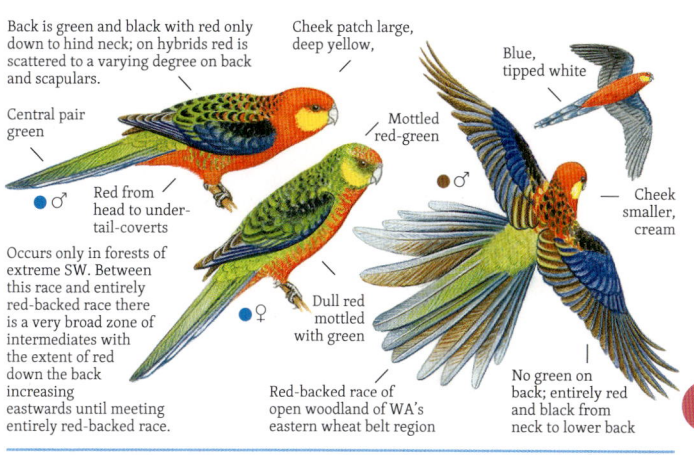

Back is green and black with red only down to hind neck; on hybrids red is scattered to a varying degree on back and scapulars.

Central pair green

Red from head to under-tail-coverts

Occurs only in forests of extreme SW. Between this race and entirely red-backed race there is a very broad zone of intermediates with the extent of red down the back increasing eastwards until meeting entirely red-backed race.

Cheek patch large, deep yellow,

Mottled red-green

Dull red mottled with green

Red-backed race of open woodland of WA's eastern wheat belt region

Blue, tipped white

Cheek smaller, cream

No green on back; entirely red and black from neck to lower back

**Eastern Rosella** *Platycercus eximius* 29–33 cm
The brightly coloured Eastern Rosella is familiar in small flocks around towns and along roadsides; feeds mainly on the ground. Inhabits woodland with scattered trees, but usually with grassy groundcover, also farmland, watercourse trees, crops, parks, gardens; generally more open environs than Crimson Rosella, and usually below 1200 m altitude. Flight is undulating when flying tree to tree – dives down, swoops up – but travels high and level over longer distances. In flight, gives a brisk, sharp, clear, rapid 'quink-quink, quink-quink' and even more rapid 'whit-whit-whit-whit-'. Slower, much more drawn out, is a high, clear, ringing 'pee-pt-eee'; there is also much rapid, confused chattering within groups. Abundant through most of its range where suitable habitat exists; has benefited from partial clearing into areas of heavy forest.

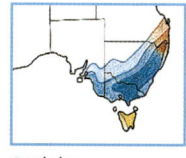

- ● *eximius*
- ● *elecica*: larger, more extensive yellow in plumage
- ● *diemenensis*: largest race, significantly larger than nominate, this including bill, tail, tarsus, both sexes; red a deeper crimson, white cheek patch larger

- ● Black back feathers of nominate race have narrow yellow margins.
- ● 'Golden-mantled Rosella' has wider deep yellow margins.

Bright red

Cheeks white

Tails relatively broad, not very long

Scarlet

Rump light, bright green

Deep blue

Black

Yellow

Central tail feathers blue-green, laterals blue, tipped white

Female and immatures tend to have slightly greener, finer edgings to the black feathers of the back.

# AUSTRALIAN RINGNECK

**Australian Ringneck** *Barnardius zonarius* 34–38 cm
At present all these parrots with yellow 'ringneck' hind collars are races of a single species, the Australian Ringneck. Travels with strong, undulating flight, swooping low then up to land in a tree, long tail spread. Often feed on the ground on native seeds and plants, or on spilled grain in paddocks or on roadsides. Previously they have been grouped as separate species with long-established common names still in use. The races of Australian Ringneck differ not just in appearance but in calls and habitat. However, they interbreed in the transitional zones.

- 🔵 *zonarius*
- 🟢 *semitorquatus*
- 🟤 *barnardi*
- 🟡 *macgillivrayi*
- 🔴 *parkeri*

● **'Port Lincoln Parrot':** Drier regions – woodland, mallee, mulga, spinifex, especially tree-lined watercourses, roadside and farm trees. Call is a ringing double-noted 'klingit, klingit, klingit' and rapid, sharply ringing 'kling-kling-kling-'. Much noisy chattering in groups. Common in E; abundant in WA.

● **'Twenty-eight Parrot':** Named for its triple-noted call, more mellow than sharply ringing, 'teu-wit, teoo', or 'twen-ty-eight'. Other calls are more like those of the Port Lincoln. Confined to heavier forest of the wet SW of WA, the jarrah, karri, wandoo and marri.

The Australian Ringneck species is made up of dark-headed W races and paler-headed E races, sometimes grouped as two species. All have the yellow collar. The races have at times been listed as separate full species, and many common names are still in widespread use as a convenient way of describing the present races.

● **'Mallee Ringneck':** Feeds both in trees and on the ground. Habitats include mallee, mulga, eucalypt woodland, river gums along watercourses, farmland with scattered trees. In flight, brief bursts of deep wing beats interspersed with short glides that give an undulating motion. Calls in flight, a high, ringing 'kling-kling-kling'; much varied chattering while feeding in foliage. Sedentary or locally nomadic; common.

● **'Cloncurry Ringneck':** Confined to arid, hilly, open woodland and grassland of the Selwyn Ranges, especially around large eucalypts along watercourses. This race appears to be holding its numbers near Cloncurry and around the upper reaches of the Diamantina R.

● **'Innamincka Ringneck':** Confined to upper Cooper Creek. *Not pictured.

# RED-CAPPED & SWIFT PARROTS

**Red-capped Parrot** *Purpureicephalus spurius* 34–37 cm
Colourful almost to the point of being gaudy. Mature male has areas of solid, bright, deep colour: crimson, yellow, deep purplish blue. Likely to be seen in almost any habitat: forested areas of marri, karri and jarrah forests; woodland of marri, wandoo and banksia; mallee; heath; farmland with remnants of any of this vegetation. Flight is swift, undulating; bursts of wing beats are interspersed with glides; travels level rather than swooping low to the ground like many other parrots. In flight gives a distinctive, rolling 'kchurrrink!'. From trees, abrupt, rough 'chrrek!' and rapid 'chirek-achek'. Sedentary or local wanderings; common.

Spectacular large parrot, unique colour pattern: strong, bold, bright colours; bill has long, fine upper mandible.

Cap bright red

Deep yellow

Deep purplish blue

Thighs to under-tail-coverts bright red

The rump is bright, slightly greenish yellow; conspicuous as the birds fly up and away (when the colours of head and breast may not be visible).

The range of this parrot approximately matches that of the eucalypt known as marri. Using its long, fine bill tip, the Red-cap can extract seed from the marri's large, hard, woody seed capsules, an ability shared with the Carnaby's Black-Cockatoo (Long-billed), although neither species is limited to this food source. Other parrot species must cut away the top half of the marri capsule.

Red of crown is dull, may be streaked green.

Small red band

Red with flecks of green

Faint wing stripe, females only

Central pair bronze green

Bill long, hooked

Imm. is less obviously a Red-cap: has red forehead and dull brownish violet breast.

The female can be almost as colourful as the male.

Imm.: red with flecks of green

**Swift Parrot** *Lathamus discolor* 23–26 cm
Aptly named for its swift flight. Flocks dart across the sky or weave fast and low among trees with audibly whirring wings and clear sharp calls. Seeks forest and woodland with flowering trees. Often with lorikeets; feeds on nectar, scale insects, fruits. Calls differ from those of lorikeets; lack their screeching, scratchy, harsh sounds; have pleasant, musical, high notes, sharp but clear. In flight, contact call is 'chi-wit, chiwit, chiwit'. Also greatly varied chatters and trills mixed with deeper, mellow sounds and clear, sharp notes. Common.

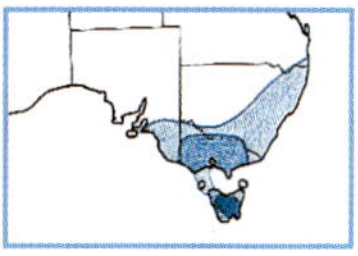

Deep emerald green

Crown dark blue

Forehead to throat, crimson

Scarlet of inner edges of tertials

Bend of wing scarlet

Tail pointed, unusual dusky maroon

Spectacular in flight with scarlet under-wing-coverts, dark flight feathers, bright green body

Scarlet

Sexes similar, female slightly duller, has creamy underwing bar. Juvenile is like female, but dark iris, paler red on throat and under tail.

Scarlet under tail and spots along flanks

## RED-RUMPED, MULGA & PARADISE PARROTS

**Red-rumped Parrot** *Psephotus haematonotus* 26–28 cm
Found in pairs to large flocks, feeding on the ground or on flowering trees or foliage. On open grassy and lightly timbered plains, timbered watercourses, mallee, farmland not far from water. Flight swift, slightly undulating; often travels high. Calls sharp, metallic, scratchy, squeaky, abrupt: 'chwie-chwiep, chwie-chwiep–'; long squeaky 'chwieee'. Also lower, rather husky, harsh 'chwier-querrk', harsh scolding 'querrk, querrk' intermixed with fine squeaks. Sedentary, rarely dispersive; common.

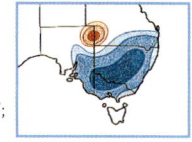

● *haematonotus*
● *caeruleus*: plumage paler, female more grey than olive

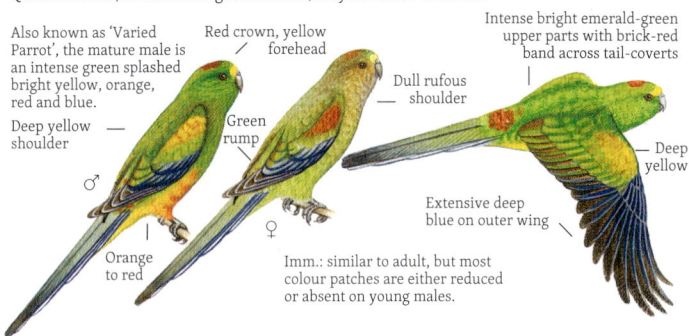

**Mulga Parrot** *Psephotellus varius* 26–30 cm
Undulating low flight that displays intense colours of male; generally quiet and unobtrusive. Found across a wide expanse of semi-arid regions in mulga, mallee, saltbush, usually with scattered small trees, lines of big trees along watercourses, drier farmland; never far from water. Flight call is a distinctive, slightly husky yet still sharp and strong 'zwit-zwit, zwit-zwit' or 'chwit-chwit'; from trees, a very rapid, sharp 'wit-wit-witwitwit' and softer, fast chattering. Quite common; abundant in good seasons, may be scarce in others.

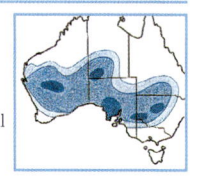

**Paradise Parrot** *Psephotus pulcherrimus* 27–30 cm
One of the most beautiful parrots; lived in small family parties or pairs; fed on ground. Habitat was the open grassy woodland and scrubby grassland of broad river valleys and plains where termite mounds are common. Records indicate nesting in spring to early summer. The nest was a tunnel dug into a large but rather low and rounded termite mound, unlike the tall mounds used by Golden-shouldered and Hooded Parrots. Call a series of soft, rather musical whistles; in alarm, a sharp metallic call. Almost certainly extinct.

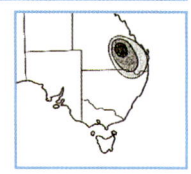

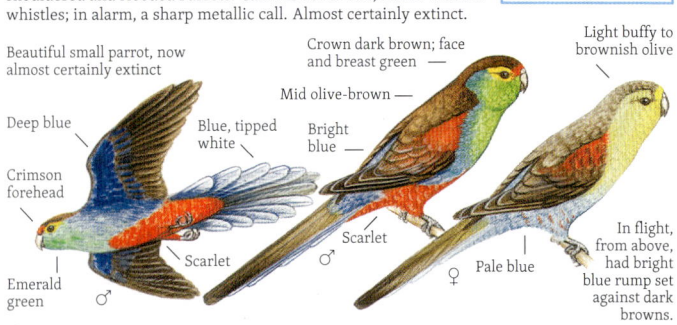

## GOLDEN-SHOULDERED & HOODED PARROTS

**Golden-shouldered Parrot** *Psephotellus chrysopterygius* 26 cm
A rare and beautiful parrot, inhabitant of tropical grassy woodland where the vegetation is of scattered eucalypts and paperbarks, and occasionally river-edge mangroves, and where large termite mounds abound. Feeds in small parties or pairs, usually on the ground. Not timid: flies to nearby trees; soon returns when disturbance ceases. Flight swift with only slight undulation. During heat of day rests in shady foliage; visits waterholes early in the morning, occasionally during day. Flight call is a sharp, scratchy, metallic yet musical, pleasant 'chwit, chwit' and 'chirrit, chirrit'; also sharp, quiet, varied chattering from treetops. Silent on ground. Endangered, probably by altered habitat, changes in burning and regeneration.

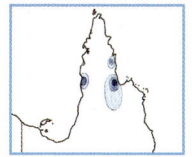

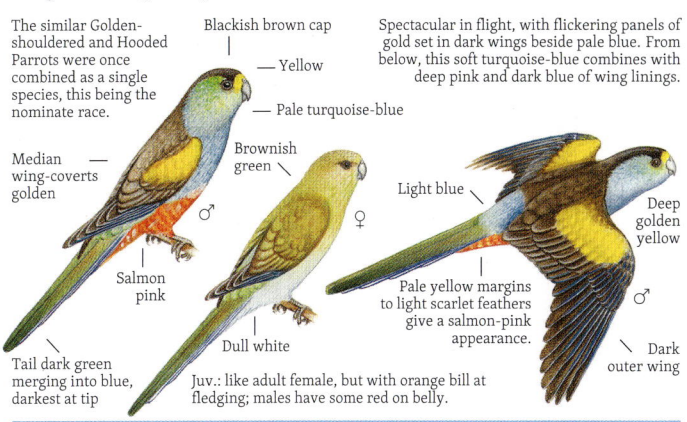

**Hooded Parrot** *Psephotellus dissimilis* 25–26 cm
Bold strong colours make the male spectacular, and even more so in flight; golden wing panels flicker with the action of outspread dark wings. Travels swiftly, slightly undulating. Not timid, flies up to nearby trees if flushed; often feeds along roadsides. Habitat is dry savannah woodland and open forest with grass or spinifex, termite mounds, plains or stony ridges not far from water. Call is a sharp, very high, thin 'chseit, chseit' in flight and sharp, metallic 'chsink, chsink'; slightly harsh or scolding 'charrak, charrak'. Quiet, squeaky, budgie-like chatter in trees. Sedentary; uncommon, localised.

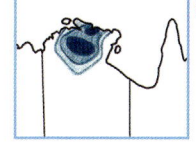

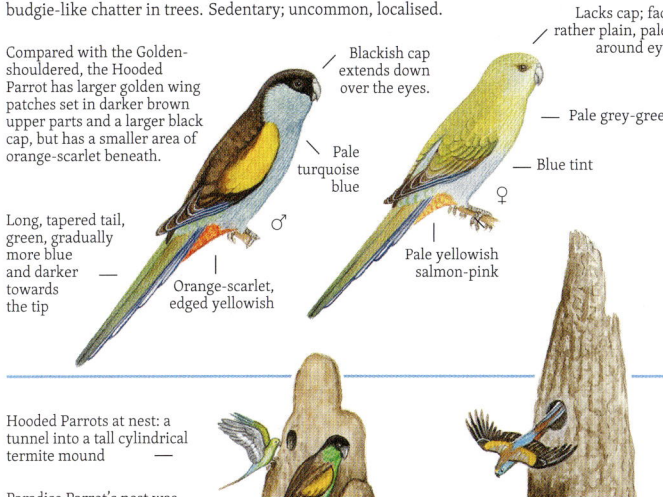

## BLUE-WINGED, ELEGANT & ORANGE-BELLIED PARROTS

**Blue-winged Parrot** *Neophema chrysostoma* 20–22 cm
Usually in pairs or small flocks, foraging on ground, early and late in the day. Forest to alpine grassland, also mulga, saltbush and coastal dunes. Flight is swift, direct with little undulation. Calls are very high, thin tinkling, in bursts, fast then slow, 'tsiwee-tsiwee-tsiweet, tsi-weet, tsi-weet', more like highest squeaks of a thornbill or fairy-wren than a parrot. Abundant in Tas.; common in Vic. and SE of SA; elsewhere uncommon to rare.

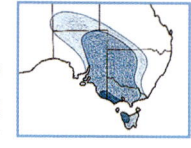

Crown dull golden, merging to dull golden olive-green on back, wing-coverts, rump

Displays extensive deep blue of both upper- and under-wing-coverts in flight.

Blue-green

Frontal band blue almost back to eyes

Lores and around eyes yellow

Light, bright golden yellow

This parrot has the largest, deepest blue shoulder and wing panels of all *Neophema* species.

No blue frontal band

Juv.

Imm. and female are slightly less colourful than male.

Female and juveniles have reduced blue, mottled green, on shoulders.

---

**Elegant Parrot** *Neophema elegans* 22–23 cm
In pairs or, out of the breeding months, in small to large flocks, feeding on the ground. Usual habitat is woodland, lightly timbered grassland, partly cleared farmland, margins of clearings in heavy forest, tree-lined watercourses, mallee, mulga. Call, in flight, is a sharp, squeaky 'chwit' or 'tzit'. Is usually silent while feeding, but may give occasional faint, sharp, squeaky twitters, like some of the squeakiest sounds from fairy-wrens. Usually only locally nomadic; generally common.

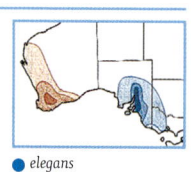

● *elegans*
● *carteri*: both sexes have shorter tail and the male has a smaller bill than male of *elegans*.

Head golden olive with two-tone blue frontal band. Olive-green on back, wings and rump

A rather plain golden olive parrot with rather narrow wing-edge of blue curving back from shoulder

Blue continues back slightly behind eye.

Similar to adult, but lacks blue forehead band.

Lower breast to under tail brightest yellow of plumage

♂

In flight from above or rear: pale blue tail, deep blue outer wing

● Juv.

Female (not shown) has dull colours compared with male, blue frontal band reduced.

---

**Orange-bellied Parrot** *Neophema chrysogaster* 20–22 cm
Most of day spent hidden under low vegetation; feeds on ground at dawn and dusk. Wary, easily flushed; flies high with loud calls; travels far before dropping again to ground. Breeds in summer in SW Tas. on buttongrass and swampy sedgeland plains. Winters S coast of Vic. and SE of SA, then using tidal flats, salt marsh and heath, islets, pasture close to shore. In alarm, makes rapid, buzzing 'zzt-zzt-zzt-zzt'; also tinkling contact calls in flight. Migratory; endangered, possibly still declining. One of Australia's rarest birds.

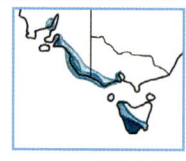

Has orange belly patch, but this may be difficult to see, especially when parrots are feeding on the ground.

'Grass' green

Flight feathers brown, outer margins blue

Frontal band plain blue; does not extend behind eye.

Yellowish-green, face to breast

Shoulder deep blue

Unique orange patch

♂

Bright yellow

♀ Similar, dull colours, less blue, much less orange

In flight, shows blue under-wing-coverts, blue secondaries.

Juvenile lacks blue forehead band.

Slightly dulled greens

Orange patch is smaller, much paler.

Imm.

All the *Neophema* parrots tend to sit with breast feathers fluffed out to overlap and partly conceal bend of wing, thus hiding some blue.

# SCARLET-CHESTED, TURQUOISE & ROCK PARROTS

**Scarlet-chested Parrot** *Neophema splendida* 18–21 cm 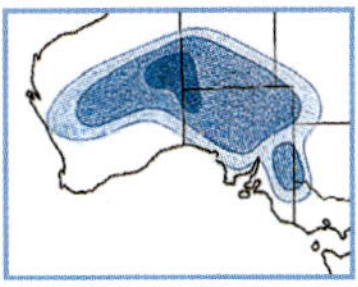
Although colourful, these parrots are inconspicuous in their natural habitat: open woodland of eucalypt, she-oak, mulga with spinifex and saltbush. They feed on the ground; fly low, keeping close to cover. In pairs or small parties, rarely large flocks. May remain hidden unless accidentally flushed by close approach. The flight is swift and rather erratic or fluttering. Calls are soft, mellow, abrupt, chattering, twittering 'tooweet', 'chwit', 'chweet'. Nomadic; irruptive; scarce but at times locally common.

Quiet and unobtrusive; easily overlooked even though so brightly coloured

Deep cobalt blue

Upper parts rich bright green

Scarlet

♂

Bright yellow

Wings and tail spread as 'air-brakes' on landing.

Flight feathers edged deep blue

Wing-coverts pale blue

Extensive light cobalt blue

♀

Underside of outer tail feathers deep golden yellow

Juv.: like female, but bill and cere orange; males have darker blues.

**Turquoise Parrot** *Neophema pulchella* 19–21 cm
Feeds inconspicuously on ground in woodland and open grassland, both natural and partly cleared. Flight is swift, erratic, fluttering wing action, brief glides on downcurved wings. The tail is spread on take off and landing, displaying extensive yellow. Calls are high, weak, musical, tinkling, 'tzeit-tzeit, tzeit-tzeit'; faint high twittering when feeding or at water. Rare, locally common; semi-nomadic.

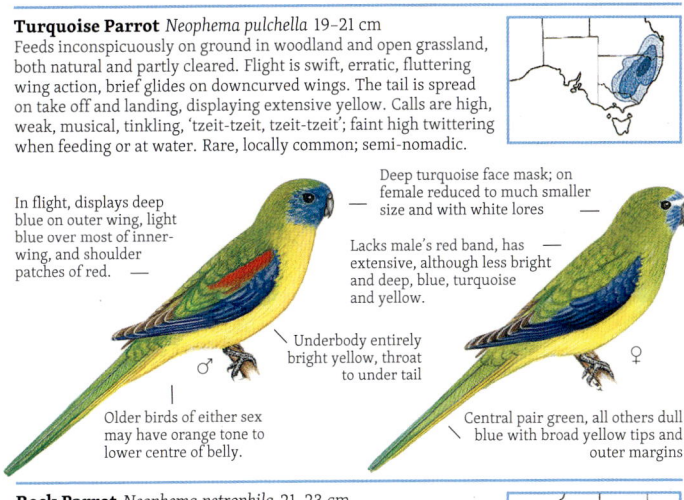

In flight, displays deep blue on outer wing, light blue over most of inner-wing, and shoulder patches of red.

Deep turquoise face mask; on female reduced to much smaller size and with white lores

Lacks male's red band, has extensive, although less bright and deep, blue, turquoise and yellow.

Underbody entirely bright yellow, throat to under tail

♂

Older birds of either sex may have orange tone to lower centre of belly.

♀

Central pair green, all others dull blue with broad yellow tips and outer margins

**Rock Parrot** *Neophema petrophila* 21–23 cm
Usually not seen unless flying or flushed, or located by tinkling calls. In flocks, at times quite large, on offshore islands along coastline and in dunes and heath, usually within sound of the sea. Rarely far inland. If flushed from dense low cover, birds dash away with a flurry of quick wing beats and burst of calls. Erratic flight, jerky wing beats. Sharp, thin, tinkling calls often given in flight: 'tsee, tzit-tseit'. Sedentary; common.

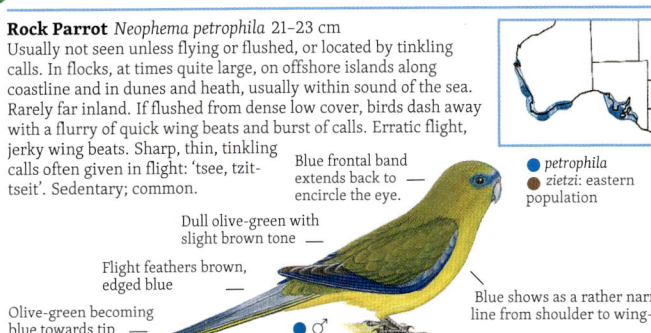

Blue frontal band extends back to encircle the eye.

● *petrophila*
● *zietzi*: eastern population

Dull olive-green with slight brown tone

Flight feathers brown, edged blue

Blue shows as a rather narrow line from shoulder to wing-tip.

Olive-green becoming blue towards tip

♂

203

# BLUE BONNETS & BOURKE'S PARROT

## Eastern Blue Bonnet
*Northiella haematogaster* 27–34 cm
Usually feeds on the ground in pairs or small parties; often in shade by roadside seeking seeds, flying to nearby tree if disturbed. Inhabits mostly open country: lightly timbered grassland, mulga, mallee, she-oak, watercourses and paddock trees. Roosts quietly in foliage by day. Flight undulating, erratic, dropping low on leaving a tree, swooping up to another, tail spread to show white-tipped blue outer feathers. Flight contact call is harsh, nasal with metallic, scratchy quality, abrupt, 'chrak, chrak': shorter, faster in alarm; sometimes longer; soft whistles and chattering.

● *haematogaster*;
● *haematorrhous*;
● *pallescens*: paler, more yellow tint

● *narethae*

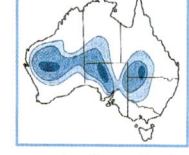

## Naretha Blue Bonnet
*Northiella narethae* 27–34 cm
A former race split from the Eastern, from which it is isolated on the Nullarbor with a slightly softer, muscial 'cloot-cloot' call.

**'Yellow-vented Blue Bonnet'** ●
- Forehead, face, cheeks and throat masked in deep violet-blue
- Rusty red wing-coverts; olive or yellow some other races
- Light red
- Bright light red
- Vent area bright light yellow

**'Red-vented Blue Bonnet'** ●
- All races and Naretha have entire outer wing and most of secondaries deep blue
- Leading edge deep blue
- Face masked deep blue
- Has largest red shoulder patch
- Rump olive-orange
- Red
- Thighs and vent area entirely feathered red

Forehead light turquoise-blue, obviously paler than rest of blue on the head
Lores down to throat all very deep violet-blue
Thighs feathered pale yellow

Three races — formerly four until Naretha was considered a distinct species — sufficiently distinctive to have acquired common names.

**'Pallid Blue Bonnet'** ●
From the Lake Eyre Basin, *pallescens*, or 'Pallid Blue-bonnet' is a pale form of the nominate race.

## Bourke's Parrot *Neopsephotus bourkii* 19–22 cm
In the soft dim twilight just after the sun has dipped below the horizon, these parrots come cautiously to water. On the ground, they blend with reddish earth at the edge of the pool. Their flight is direct, swift, low with audibly whirring wings, showing pink and blue underparts, tail prominently tipped white if spread. During the day when birds feed under mulga, their plumage colours blend well with greys and browns of weathered dead limbs and shadowy ground. Lives in mulga and similar acacia scrub, tree-lined watercourses, semi-arid woodland. Flight call is not very high, scratchy or harsh, but is mellow, pleasantly musical yet quite sharp, penetrating and clear: 'chiew-eet, chw-iet, chew-iet'. Silent or soft musical twittering. Shrill double noted alarm call. Nomadic; locally fairly common, but big seasonal fluctuations.

# BUDGERIGAR; GROUND & NIGHT PARROTS

**Budgerigar** *Melopsittacus undulatus* 17–20 cm
In flight, warbling, musical 'tirrrit', 'tir-rit, tirit'; on taking off, a rapid, rasping or scolding 'tzzit-tzit-tzit, tzt-zt-zt', fading away. Habitat of spinifex, saltbush and grassland, plains and ranges, either treeless or with scattered trees or shrubs along watercourses and stands of open eucalypt woodland scattered through open country, providing nesting hollows and water from creek beds. Budgerigars can survive for some time without water but are rarely found far from it; fast-flying flocks can quickly cover considerable distances to water. Highly nomadic, following heavy rains to congregate where lush grass ensures abundant seed, quickly raising several broods of up to eight young. Abundant.

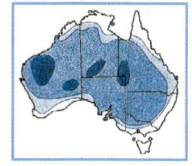

**Eastern Ground Parrot** *Pezoporus wallicus* 28–30 cm
**Western Ground Parrot** *Pezoporus flaviventris*
A shy, elusive bird, usually not seen unless flushed, or at twilight when calls are heard and birds can be seen fluttering low over heath of coasts, ranges, drier ridges in swamps or, rarely, grassland near heath. In Tas., the stronghold of this species, it favours buttongrass plains. Calls at dawn and dusk, a series of piercing, ringing, resonating whistles, rising in steps, each note flowing on almost unbroken, but abruptly higher than the preceding. Also lower notes at more even pitch. Cheerful budgie-like warbling and sharp, rapid trills. Uncommon to rare; endangered in many parts by habitat loss in excessively frequent fires. Western Ground Parrot may now only survive east of Esperance.

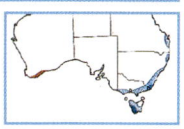

● *wallicus*
● *flaviventris*: previously a race, now full species

**Night Parrot** *Pezoporus occidentalis* 22–24 cm
Nocturnal; hides in dense spinifex by day, drinks after dark, then feeds among spinifex. Often appears associated with spinifex or found among samphire bushes on margins of salt lakes. Old records of calls are of a sharp squeak when flushed; contact call in flight a short, sharp whistle, several times, rapidly. A low, drawn out, double-noted whistle when coming to water. In alarm, a harsh, almost croaking sound. Rare, thought probably extinct until a dried specimen was discovered on a Qld roadside in 1990.

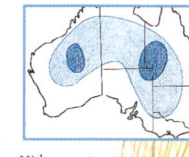

# 13 CUCKOOS, OWLS, FROGMOUTHS & NIGHTJARS

This family group contains cuckoos, Pheasant Coucal, hawk-owls, masked owls, frogmouths, nightjars and owlet-nightjar. These families are not closely related, being from three different orders. Within each family, size varies greatly. As for most other families, knowledge of the relative sizes helps bird recognition. In the main text, the scale usually differs from page to page to provide detail. Useful size comparison can be made here, where all birds are at a common scale.

Willie Wagtail = 20 cm

## cuculidae
page 207

- Oriental Cuckoo
- Pacific Koel
- Pallid Cuckoo
- Brush Cuckoo
- Channel-billed Cuckoo
- Chestnut-breasted Cuckoo
- Fan-tailed Cuckoo
- Black-eared
- Horsfield's Bronze-
- Shining Bronze-
- Little Bronze-
- Gould's Bronze-

## centropodidae
page 213

- Pheasant Coucal

## strigidae
page 214

- Powerful Owl
- Rufous Owl
- Barking Owl
- Australian Boobook
- Brown Hawk-Owl (not illustrated)

## tytonidae
page 216

- Greater Sooty Owl
- Eastern Grass Owl
- Lesser Sooty Owl
- Australian Masked Owl
- Eastern Barn Owl

## podargidae
page 218

- Tawny Frogmouth
- Papuan Frogmouth
- Marbled Frogmouth

## caprimulgidae
page 220

- White-throated Nightjar
- Spotted Nightjar
- Large-tailed Nightjar

## aegothelidae
page 221

- Australian Owlet-nightjar

# ORIENTAL & PALLID CUCKOOS

## Oriental Cuckoo *Cuculus optatus* 28–33 cm

Large cuckoo, underparts boldly barred; long pointed wings; swift, slightly undulating, dashing flight like a small falcon. Often solitary, dispersed; some congregate at points of departure for the return migration. Always shy and elusive, quiet; slips away unobtrusively if approached. Hunts from perch, darting out to take insects from foliage and ground. Usually silent; may give up to six whistled notes, 'kee-kee-kee', and harsh 'graak-graak-gak-ak-ak'. Inhabitant of rainforest margins, monsoon forest, vine scrub, riverine thickets, wetter, densely canopied eucalypt forest, paperbark swamp, mangroves. Typically in denser vegetation with more closed canopy than is usual for the Pallid Cuckoo. Migrates from Eurasia as far S as Indonesia, NG and N Aust. Some remain through Aust. winter. Uncommon.

Those visiting Aust. are mostly *optatus* from NE Asia, but possibly also slightly smaller nominate race, *saturatus*, of S Asia.

Large cuckoo, boldly barred. On females, the grey upper breast and throat may have a slight buff tint.

Upper parts are grey with a slight iridescence, at times giving a bronzed gloss.

Yellow eye-ring and base of bill

Bright rufous, barred brown

Plain, pale slaty grey

Barred

Orange feet

White tips to underside of tail

Rare rufous 'hepatic' coloration, females only: almost entirely barred

Pointed, backswept wings and swift flight give a falcon-like action jiz.

Entirely barred beneath

Obvious white tips, dull bars or spots

Flight feathers dark grey-brown, barred buff or whitish on primaries; coverts whitish, barred dark grey

## Pallid Cuckoo *Cacomantis pallidus* 28–34 cm

Large, long tailed, superficially falcon-like in flight, but feeds on large insects from ground and foliage. Eats hairy caterpillars avoided by most birds. Typically inhabits open country; avoids dense, closed vegetation types. In spring, the male advertises with repeated series of slightly husky, mellow whistles on a rising scale, 'quip-peer-peer-peeer-peeeer-peer-', 'quip-quip-pip-pip-pieeer'. Female responds with a long, drawn out, husky, sharp 'queeeep!'. Migratory in S; elsewhere only attracts attention by calls in spring; common.

Sexes differ: male light and dark grey morphs and intermediates; females have light rufous and deep rufous morphs.

Yellow eye-ring

From rear, pale patch on nape

Upper parts grey-brown. Large spots along covert margins on browner sub-adult; few and faint on greyer, plainer, paler adult

Pale morph

Juveniles follow adults in their respective morph colours.

Females of dark rufous morph have rufous, barred breast; upper parts are rufous mottled dark brown.

Notched white

First plumage is seen on large nestling or recently fledged juveniles. Upper parts are heavily blotched, streaked dark brown and white.

Juv.

Buffy white angular notchings to edges of feathers of wings, back, tail

Pale grey wing linings

## BRUSH & CHESTNUT-BREASTED CUCKOOS

**Brush Cuckoo** *Cacomantis variolosus* 22–26 cm
Takes its name from use of 'brush' in rainforest. Inhabits dense, often closed canopy habitats that include rainforest, monsoon and gallery forest, stream thickets, paperbark swamp, eucalypt forest and woodland. There it tends to sit silent and motionless for long periods, occasionally darting out to take caterpillars or other insects from vegetation: can be hard to sight unless it moves. Only in the breeding season does the species become conspicuous; males are then very active and noisy, chasing other males, pursuing females. The flight is swift on pointed, backswept wings, slightly undulating and showing pale wing bars. Gives a long series of descending, clear, mellow yet piercing calls, 'feee-ip, feeeip, feeeip-'. A ringing, metallic 'pee-whip-ee, pee-whip-ee' may be either slow, gently rising and quite mellow, or rapid, rising in pitch, ever quicker and higher, until continuous – an excited, frenzied, piercing sound. Common across N Aust.; uncommon migrant to SE Aust. Some winter on northern islands.

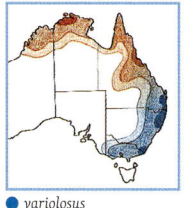

● *variolosus*
● *dumetorum*: smaller; very slightly paler above and beneath; wide Qld intergrade zone between races

**Chestnut-breasted Cuckoo** *Cacomantis castaneiventris* 24 cm
Secretive bird of understorey and mid levels of tropical rainforests; watches from perch then drops to ground, or flutters around foliage to take insects or other small prey. Habitat typically dense, closed canopy in rainforest, monsoon forest, mangroves, river-edge thickets. Sits still for long periods; inconspicuous unless calling. Nominate race in Aust. and NG; two other races in NG. Voice is a wavering, rattling, musical trill that is sharply penetrating if close; similar to descending trill of Fan-tailed Cuckoo; some similarity to call of Yellow-billed Kingfisher. Also a mournful, husky, 'wheeer-wheeer-'. Uncommon; sedentary or migratory.

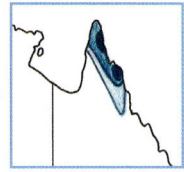

Race *castaneiventris* is in both NG and Aust.; other races occur in NG.

# FAN-TAILED & BLACK-EARED CUCKOOS

**Fan-tailed Cuckoo** *Cacomantis flabelliformis* 25–27 cm
Adopts a rather steeply upright posture, often keeping to a perch for long periods while calling. Habitat includes wet eucalypt forest, open forest and rainforest margin, but typically less dense than habitats preferred by Brush and Chestnut-breasted Cuckoos. Flight is undulating; elevates tail on landing. Hunts from perch; waits patiently, dropping to take insects from ground, foliage. Call is a descending, rattling, mellow, musical trill, quite loud if close; similar to trill of Chestnut-breasted, which undulates rather than descends and may be slightly sharper. Also a musical, almost level, vibrating 'pheeweer' and high, thin, whistled 'fweeit'. Common; may be sedentary or locally migratory.

The presence of a breeding species of cuckoo is signalled by finding its egg in a host species' nest. Cuckoo eggs may match colour of host eggs, but are often larger. Those in darker, hooded nests need not match host eggs so well. Host species names are those in *italics*.

In flight, Fan-tailed Cuckoo is similar to Chestnut-breasted, but with the rich rufous-chestnut of that species replaced by much paler cinnamon-buff.

**Black-eared Cuckoo** *Chrysococcyx osculans* 19–21 cm
Solitary, inconspicuous; hunts from perch in shrub or tree, dropping to ground to take insects, including hairy caterpillars that are disliked by most other birds. Occurs across most of the Australian mainland, avoiding only the wet, heavily forested east coast and similar SW corner of WA. Present in most drier habitat types: open woodland, mulga and mallee, sparsely vegetated arid country with spinifex, grassland or salt marsh, widely scattered trees and shrubs, lines of vegetation along watercourses. Usually quiet, but through the breeding season males call from a prominent open perch, giving a piercing, drawn out, slightly descending whistle repeated at regular intervals, each call of the same length and each at the same penetrating high pitch as the previous: 'feeeieuw, feeeieuw, feeeiew'. Also a loud but less piercing 'fee-ew-it, fee-ew-it' or 'fee-ew-eer'. Occasionally an abrupt, quieter 'feeit'. Migrates into SE and SW for the summer, only rarely reaching Tas. Movements in central Aust. are uncertain. Present across N Aust. throughout the year; probably some travel to NG. Generally uncommon, but is easily overlooked and probably common at times.

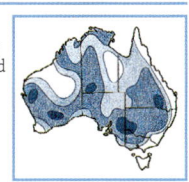

## BRONZE-CUCKOOS

**Horsfield's Bronze-Cuckoo** *Chrysococcyx basalis* 14–17 cm
A common cuckoo throughout Aust. in almost all habitats except the densest and wettest vegetation types. Usual habitat includes open forest, woodland, roadside trees and farm shelter belts. Hunts from a perch; darts to ground to take insects. Flight swift; glimpse of pale underwing bar. Has a sharp, piercing, descending whistle, 'tsieeew, tsieeew, tsieeew'; the initial 'tsie' an extremely high, thin squeak of sound, then descending and strengthening through the final 'eew'. Also a quick, cheery chirrup like that of a Budgerigar. Common migrant to S.

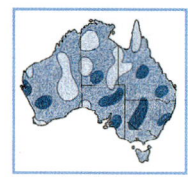

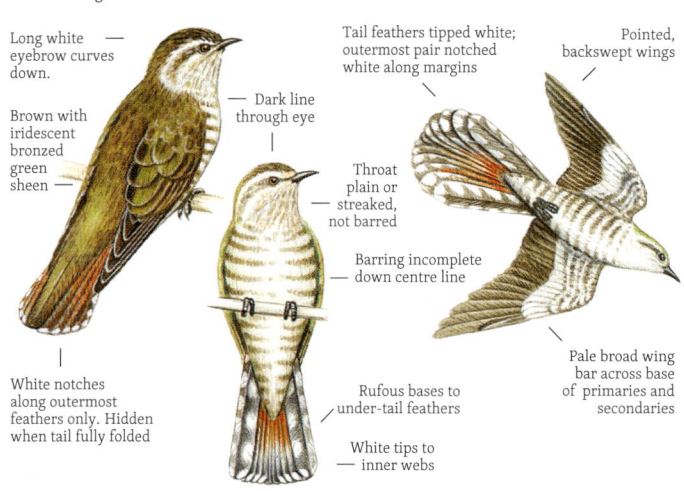

Long white eyebrow curves down.

Brown with iridescent bronzed green sheen

Dark line through eye

Tail feathers tipped white; outermost pair notched white along margins

Pointed, backswept wings

Throat plain or streaked, not barred

Barring incomplete down centre line

White notches along outermost feathers only. Hidden when tail fully folded

Rufous bases to under-tail feathers

White tips to inner webs

Pale broad wing bar across base of primaries and secondaries

**Little Bronze-Cuckoo** *Chrysococcyx minutillus* 14–15 cm
Small, inconspicuous in dense habitats unless calling. Varied habitats: woodland and fringes of heavier closed vegetation – rainforest edge, tropical monsoon forest, paperbark swamp, mangroves, lush gardens – often near water. Hunts from perch; sallies out to take insects on wing as well as from ground and foliage. Call is a long, descending, rippling trill, loud, clear and ringing; also gives series of 4 to 6 descending notes like a slower, drawn out version of the trill, 'tiew-tiew-tiew-tiew-'. Call is almost identical to that of Gould's Bronze-Cuckoo, which appears to continue longer, up to 8 notes in a sequence. In N of range, sedentary; race *barnardi* migratory from SE to NE Aust. Common.

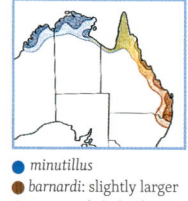

- 🔵 *minutillus*
- 🟤 *barnardi*: slightly larger
- 🟡 Intergrade hybrid zone between *minutillus* and Gould's (but perhaps not *barnardi*, which is further S in breeding season)

The Little Bronze-Cuckoo and Gould's (opposite) are closely alike; Gould's may be a race of this species rather than a separate full species.

🔵 Green has a very slight rufous sheen.
🟤 Green has little if any rufous overtone; female plumage similar.

Male: red eye-ring and iris

Green of plumage, has a less rufous tone than has Gould's.

🔵 Appears to prefer more open woodland vegetation.
🟤 Denser habitats, especially vine thickets, riverine rainforest and mangroves

🔵 ♂

No rufous tint to breast; short dark scallops

Eye-ring pale creamy fawn, iris brownish

Underside of tail rufous except for outer tail feathers

Bars broken, irregular, but do meet across centre

♀

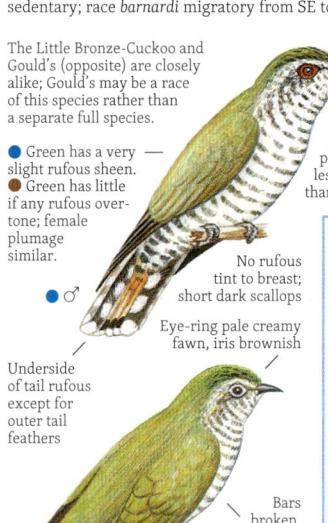

Horsfield's, Shining, Scarlet Robin, Yellow-rumped Thornbill, Gould's, Little, Large-billed Warbler

Some cuckoo eggs with nests and eggs of usual host species. Identification of these eggs reveals that a cuckoo is a breeding species at a site. Eggs tend to match those of the host, but larger.

## BRONZE-CUCKOOS

### Shining Bronze-Cuckoo *Chrysococcyx lucidus* 16–18 cm
Hunts from perch; darts out to take insects from foliage and ground. Uses mid to upper strata of wet dense rainforest, eucalypt forest, woodland. Voice includes several different call sequences, all very high pitched, metallic. As a very long series, piercingly high, at even pitch: 'wheee-wheee-wheee...'. Another call begins high and sharp, ends wavering downwards: 'phee-ieer, phee-ierr, phee-ier'. Race *plagosus* breeds Aust.; *lucidus*, nominate race, breeds in NZ, passes through E coast of Aust. to and from wintering grounds on islands to N of Aust. Common.

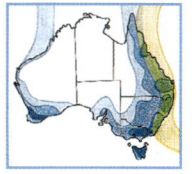

● *plagosus*
● *lucidus*

● Breeding migrant to SW and coastal E Aust.
● Visits E Aust; breeds NZ.

Crown and back iridescent metallic green

Bronzed, glossy iridescent green; crown coppery bronze

From front, white specks up forehead

Broad pale bar across primaries and secondaries

White tips hidden when tail is folded

Olive-brown bars, unbroken, up to chin

In Aust., most are ●, 'Golden Bronze-Cuckoo'. Nominate race is ● 'Shining Bronze-Cuckoo', an uncommon transient visitor to E coast.

Imm.: upper parts grey-brown; underbody dull white, faintly barred

### Gould's Bronze-Cuckoo *Chrysococcyx russatus* 14–15 cm
Very small, more rufous than any other bronze-cuckoo. Habitat includes rainforest, monsoon and dense river-edge forest, mangroves, paperbark swamp. Compared with the Little Bronze, this species or race is more inclined to use heavier closed-forest habitats in the NE Qld part of its range. Hunts insects in flight, on foliage or on the ground. Call is like Little Bronze, a descending trill; also a slower, descending 'tiew-tiew-', but may extend to 8 notes rather than the 6 maximum notes of the Little Bronze. Sedentary; common.

**Variation:** Taxonomy of bronze-cuckoos, genus *Chrysococcyx*, is confused. Gould's and Little Bronze-Cuckoos have been considered separate species, but the differences are slight and interbreeding occurs. Little and Gould's may be the same species. Gould's Bronze would then be race *russatus* of *C. minutillus*.

Compared even with the more rufous-toned race *minutillus* of the Little Bronze-Cuckoo, the plumage of Gould's Bronze-Cuckoo has a more obvious rufous suffusion over its green.

Rufous tint to white of upper breast; Little Bronze lacks this rufous cast to the white.

Eye-ring pale grey to tan

Rufous tint on breast, both sexes

♂

Green plumage has strongly bronzed iridescence, wing-covert margins are bronze.

Scallops broken, rather than the long, neat bars of Shining Bronze

Tail has a more intense and more extensive rufous tone.

In flight, both this and the Little Bronze show a pale bar across the underside of primaries and secondaries.

♀

# PACIFIC KOEL & LONG-TAILED CUCKOO

**Pacific Koel** *Eudynamys orientalis* 40–46 cm
A large parasitic cuckoo well known in its range for its carrying 'koo-eel' call. Breeding males often call for long periods, day and night, advertising their territories. The male's blue-black iridescent plumage differs substantially from the female, immature and juvenile plumages. Habitat usually rainforest, monsoon forest and dense wet eucalypt forest (especially margins), leafy trees of river edges, farmland, woodland and gardens. The Pacific Koel usually keeps to closed-canopy forest margins and vegetation mainly comprised of dense leafy trees where it can be difficult to see unless calling from a perch unscreened by intervening foliage. May be in pairs, small groups or alone. Usually wary and elusive, although males display noisily and chase females in the breeding season. Feeds on native fruit, and, now, cultivated fruit. Male has a mellow, rather musical, ringing, 'quow-eel, quow-eel, quowee, quowee'; initially the 'quo' quite deep, the final 'ee' lifting in pitch. The sequence of 5–10 calls starts slowly with pauses of several seconds between notes, but becomes quicker and higher. Has a similar but more rapid sequence: 'quowil-quoil-quoil-quoi-quoi'. Female gives a series of four or so shrieking whistles: 'quieek-quieek-quieek-quieek'. Tends to be silent outside the breeding season. Most likely to be confused with the Spangled Drongo, but its outcurving tail feathers are distinctive.

Occurs widely from India through SE Asia to S China, Malaysia, Indonesia, NG and, as a summer migrant, into N and E Aust.

● *cyanocephala*
● *subcyanocephala*

**Variation:** Many subspecies (17 or 18) from India to NG, China, Solomon Is. and Aust., where there are two. Race *cyanocephala* occurs down the E coast. The smaller race, *subcyanocephala*, occurs across N Aust. It also differs in having less streaking on the head.

Pacific Koel selects hosts that build open cup nests: Magpie-lark, miners, Figbirds, and others of similar size.

*Magpie-lark*
*Koel*

Very large cuckoo, male black with iridescent blue-green sheen

Eyes bright red

Crown black

Buffy, streaky, black-edged throat

Immature female is like adult female, but eye may remain brown into second summer.

Blackish-brown, spotted white

Crown rufous, streaked

Dark line through eye

White to deep buff, barred

Pale form: this is the form that is similar to the NZ Long-tailed Cuckoo.

♂

Dark, heavily spotted

Wings are often held 'sagging' low to each side of tail.

♀ Dark form

Tail long, straight sided, round tipped ♀

Long tail, rounded, barred buffy white and purplish brown

Long, round-tipped tail, black with blue-green sheen

♂

Adult females occur in two forms: the dark form has a black cap and blackish-brown and white-spotted upper parts. The pale form has a rufous crown, chestnut replaces dark brown on the back. Many intermediate forms. The NZ breeding Long-tailed Cuckoo (below) is most like the pale form of this species.

♀ Dark form

Juv.: like pale female, but with off-white, dark centred cap and brown eye

**Long-tailed Cuckoo** *Eudynamys taitensis* 40–42 cm
No confirmed records for Australia, but normal range comes close. Breeds in NZ, perhaps Norfolk Is.; migrates N to Polynesia and islands NE of NG. Reports of sightings in Aust. mostly from NT, near Darwin, Mt Todd, Maningrida; also Qld at Cato Is.

A very large cuckoo, almost same size as female of the Pacific Koel; tail is proportionately longer. In shape, colours and plumage pattern, the adult and juvenile are quite like the pale-type female Pacific Koel. However, the adult differs in having upper parts boldly barred rufous-brown. The upper parts of the Long-tailed are mostly barred cinnamon and brown, whereas the female Pacific Koel is darker brown, entirely and heavily spotted white, except the rufous-streaked head and white-barred tail. The juvenile Long-tailed is more like the female Pacific Koel: upper parts are dark brown spotted white. But, like all the Long-tailed Cuckoos, it has longitudinal streaking of the underbody, unlike the female pale Pacific Koel's fine, transverse, wavy barring.

# CHANNEL-BILLED CUCKOO & PHEASANT COUCAL

**Channel-billed Cuckoo** *Scythrops novaehollandiae* 58–65 cm
A huge cuckoo, largest of all parasitic birds, with massive bill giving more the appearance of a hornbill. Inhabitant of rainforest, monsoon and eucalypt forest, woodland, river-edge thickets, swamp woodland. Flight is strong, wings pointed and at times partly backswept, giving a rather long bodied, falcon-like silhouette. Has strong, deep wing beats and often travels high, conspicuously. Channel-billed Cuckoos feed mostly on fruit, but also take insects, especially large insects such as locusts. During breeding season they are usually in pairs or solitary; at other times they may be in small flocks. Voice is loud, raucous with maniacal crowing and squawking: 'awrrk, aworrk, oirrk, oik-oik-oik'. Variations – 'aiirrk-aiirrk-ark-ark-urk' and crow-like 'arrk, arrk, arrk, airrk, urrgk urrgk'. Quite common, but uncommon in Kimberley. Departs Jan.–Feb. from SE, Mar.–Apr. from N Aust.

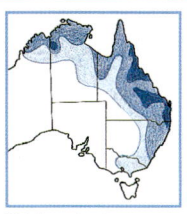

This is a migratory species; breeds in Aust. in spring and summer and migrates to NG, Indonesia and other northern islands for the winter.

In flight, this species looks like a flying cross; with huge bill it is over half a metre long and the wingspan is near a metre.

So large a cuckoo requires an almost equally large host species. The Channel-billed Cuckoo lays its eggs into nests of ravens, crows, currawongs, Australian Magpies, Sparrowhawk and Brown Falcon.

**Pheasant Coucal** *Centropus phasianinus* 60–75 cm
A large, superficially pheasant-like member of the cuckoo family. Unlike other Australian cuckoos, it is a ground-dweller, not arboreal, inhabiting tall grass and other dense groundcover of river floodplains, tropical woodland, watercourses, canefields, lantana thickets, roadside and riverbank undergrowth, mangrove fringes, swampy heath. Its barred and intricately patterned plumage matches the colours and detail of ground debris – leaf litter, grass and twigs. With short, rounded wings its flight is weak, typically moving higher into shrubbery or trees in a series of flapping leaps, eventually returning to ground in a steep glide, terminating in a rather clumsy plunge into the grass. It does, however, fly quite high and far at times. The Pheasant Coucal also differs from other Australian cuckoos in building its own nest. Call is a long, rapid series of resonant notes; initially slow, accelerating, slowing, then fading. At other times, begins abruptly, rapid notes, sustained, then falling away: 'coop, coop, cook-kook-kook-kook'. Sometimes accompanied by a second bird at different pitch. Also a metallic, tapping 'chak-chok, chowk, chowgk'. Sedentary; common.

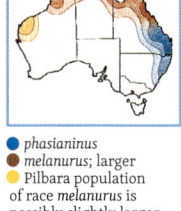

● *phasianinus*
● *melanurus*; larger
● Pilbara population of race *melanurus* is possibly slightly larger, no apparent difference in plumage; have in past been listed as separate race, *highami*.

## RUFOUS OWL, AUSTRALIAN BOOBOOK & BROWN HAWK-OWL

**Rufous Owl** *Ninox rufa* 45–55 cm
Large rufous-fronted owl, most likely to be seen roosting in dense foliage of a rainforest tree. Hides in dense vegetation of rainforest, river-edge gallery forest, monsoon scrub, swamp, mangroves; hunts through adjoining eucalypt woodland. A powerful hunter, its prey is often large: Brush Turkey, Frogmouth, Scrubfowl, Sugar Glider. Usual call a slow, deep 'whooo-hoo'; the first 'whoo' drawn out, the second note shorter, same or slightly lower pitch. When both sexes call, male's hoots are deeper; courting male may give up to seven rapid hoots. Probably rare in most parts of its wide range.

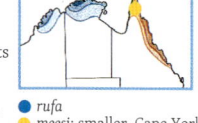

● *rufa*
● *meesi*; smaller, Cape York
● *queenslandica*; darker

The darker races of NE Aust have most of the face blackish-brown; the paler NW race has smaller dark eye patches.

Alert postures; relaxed, head is low, blending 'neckless' to shoulders; appears more broadly, heavily built.

White downy head and front

♂

Underparts light rufous-buff densely barred with rich rufous-brown

Female smaller, usually darker

Back and wings mottled dark brown

Juv.

Heavily barred white

**Australian Boobook** *Ninox boobook* 25–35 cm
Small brown owl with a dark patch around each eye, most conspicuous on the paler western-inland race, obscure on dark races. Habitat diverse: almost anywhere with trees, especially open eucalypt forest and woodland. Rare in dense rainforest, except for NE Qld race *lurida*. Preys mostly on insects and arthropods, but small birds and mouse-sized mammals are also taken; insects are sometimes caught in flight. Roosts by day in dense foliage. The voice is a mellow, musical double hoot, 'kook-kook' or 'book-book' at varied pitch; the second note is lower, throaty, vibrating in deepest calls. Common; sedentary.

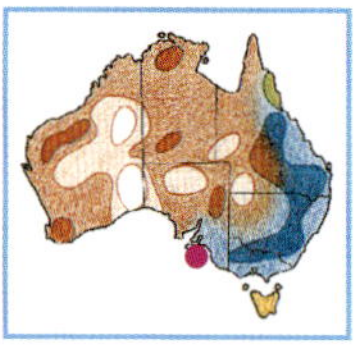

● *boobook*
● *ocellata*
● *leucopsis*
● *lurida*
● *halmaturina*

Dark brown

Barred, all races

Smaller, darker reddish

Pale, goggle-like rims around dark eye patches; eyes usually pale

Upper parts dark brown spotted white and rufous

Large white X shape with centre between eyes

Streaked rufous
Streaked buff

Juv.

Paler race of interior and W regions

**Brown Hawk-Owl** *Ninox scutulata* 30 cm
Golden eyes emphasised by surrounding very dark brown hood over head and down onto the upper breast.

Dark red-brown
Tail dark-barred

Head and face dark brown without typical owl facial disc or eye-patches; dull white between eyes

Solid brown at throat; begins to break into streaks from breast downwards.

The Brown Hawk-Owl has been described as looking like a rather dumpy, heavy-bodied hawk in flight. This species is a winter migrant from SE Asia, S to Indonesia, and, rarely, NW Aust. Several records: 1973, 1991; these were of race *japonica*.

## POWERFUL & BARKING OWLS

### Powerful Owl *Ninox strenua* 60–65 cm
Difficult to detect on its daylight roost except for its use of limbs that, although shaded by a dense overhead canopy, are usually clear of concealing low foliage. Uses eucalypt forest, preferring tall wet forest of ranges where the territories centre on densely vegetated gullies. Also found marginally in lower or drier forest that holds both prey and large hollows. Male's territorial call is a clear, almost ringing, carrying, single or double 'whoo'; begins softly and rises clear and loud; may be heard a kilometre away. When a pair is calling, the male's call is deeper; female's higher hooting seems to be a response or contact call. Various totally different sounds are given near the nest hollow. Sedentary; uncommon.

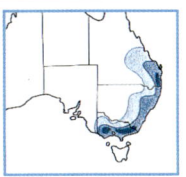

Downy white head and underparts with brown streaked crown and dark around the eyes

Dark grey-brown, flecked white

A huge owl, Australia's largest, with powerful talons, dark disruptive colours and a pattern that blends into the forest background.

Dark surrounds enhance large golden eyes.

Rufous-brown heavily barred white

Underparts white with dark grey, broadly V-shaped chevrons

Underparts downy white with sparse dark streaks

Yellow feet, huge claws, feathered legs

Juv. Juveniles have very short tails when they leave the nest; they remain for months with the parents.

Prey includes possums, gliders, roosting birds and some terrestrial animals, usually rabbits and small marsupials.

### Barking Owl *Ninox connivens* 35–45 cm
Medium-sized owl superficially resembling Australian Boobook, but larger, taller, more upright with brilliant yellow eyes. Typically in open country with stands of trees, tree-lined watercourses, paperbark swamp, N and NW Aust. Presence may be revealed by barking calls. Daytime roost may be betrayed by small birds' noisy harassment. Call is a rapid 'wook-wook', clear, carrying, pleasant, rather musical, especially when a pair is calling; male slightly deeper, female responds in a higher, clearer note. Close by, a soft, gruff 'arr' may be heard, leading in to the loud, clear barks, becoming 'arr-wook-wook'. Can be a brief flurry of calls, a fast, excited, higher 'wook-wook,-wook-wook-'; at other times the calls die away to a soft, low, relaxed 'wuf-wuf, wuf-wuf'. Common NW and N, but now uncommon in SW and SE. Sedentary.

● *connivens*
● *peninsularis*

Crown grey to grey-brown

Fine black brow lines slope up and out from inner corner of each eye

Underparts downy with softer dark streaking

Broken, untidy vertical streaking

Larger, paler in southern Aust.

Tail softly barred

Darker, slightly smaller northern form. There is a gradual, clinal size increase southwards; also becomes paler until merges into southern form.

● Juv.

## BARN & SOOTY OWLS

**Eastern Barn Owl** *Tyto javanica* 30–40 cm
Best known of all owls almost throughout the world; named for its preference elsewhere for roosting and nesting in barns or church steeples; in Aust., usually in large, deep hollows of trees. Often seen briefly in flight, caught in car headlights, when it appears starkly white, or perched on a roadside fence post.
Inhabitant of diverse open country: woodland, grassland, farmland; roosts in trees, caves, buildings, crops. Voice is a husky, often wavering screech, 'skeeaiirr', often while in flight and more often in breeding season. Nomadic, irruptive and varying; locally rare to common.

In flight, underside of wings and body almost entirely white; looks brilliant white when caught in headlights flying across road or on roadside perch.

Legs moderately long, but much shorter than those of Grass Owl; lightly feathered, almost to the foot

Feet and talons are quite lightly built.

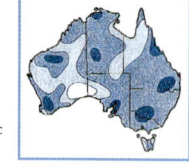

Heart-shaped facial disc

Softly intermixed buff, chestnut and grey, finely spotted

Juv.: fluffy white, mostly while in nest hollow

Underparts white with fine darker spots heavier on female

In flight, the trailing feet reach tail tip; Grass Owl's feet extend well beyond tail tip.

**Lesser Sooty Owl** *Tyto multipunctata* 30–38 cm
The smaller of two very similar sooty grey owls. Inhabitant of dense mountain rainforest and adjoining wet eucalypt forest. The Lesser Sooty is very much smaller – females weigh about 540 g compared with about 875 g for the female Greater Sooty – and has obvious differences such as a barred rather than a spotted breast. Sexual dimorphism is much less pronounced in the Lesser Sooty. Hunts low in rainforest; roosts in hollows or large epiphytes. Gives a piercing, downward whistle or shriek. Sedentary; quite commonly seen in optimum habitat.

Sooty dark grey with large silvery white spots and overall dense, fine white speckling that makes plumage appear paler

Very little size difference between sexes

Silvery facial disc is broadly oval; streaks radiate out from eyes.

Breast is silvery grey with fine, dark, broken barring fading to silvery white towards under-tail area.

In flight, underwing is very pale grey with darker grey barring.

Often hunts in darkness, dropping onto prey from perches low in rainforest.

**Greater Sooty Owl** *Tyto tenebricosa* 40–50 cm
A large, heavily built, powerful, dark grey-black owl; large-headed appearance. Extremely secretive, seldom seen. Legs and feet are huge with long, powerful talons; can take quite large prey, including the Sugar Glider, Ringtail Possum, Southern Greater Glider, Long-nosed Potoroo, rats, rabbits. Roosts and nests in high hollows, caves. Contact call, often given in flight, is a long, descending, vibrating screech; from a distance sounds more whistle than screech. Other screeches are almost uniform in pitch, similar to the Eastern Barn Owl's screech but far louder, more powerful, harsh and metallic. Found in tall wet eucalypt forest of coastal ranges, especially steep, heavily vegetated gullies, often E to SE aspect. Territorial, sedentary, rarely seen; probably not uncommon in rugged, optimum habitat.

Nest is a very high hollow or cave.

Huge black eyes set in very dark centres of wide facial disc.

Facial disc varies, brownish buff tint or almost white; chestnut 'tear' marks

Back and wings are sparsely spotted white with little of the very fine speckling of the Lesser Sooty.

Underparts vary from pale grey to deep rufous-buff, softly spotted white. Females darker

Broad wings fold short to give stubby rear body shape.

Massive, fully feathered legs, powerful feet and long talons can kill animals to the size of the Southern Greater Glider, Potoroo or rabbit.

## MASKED & EASTERN GRASS OWLS

**Australian Masked Owl** *Tyto novaehollandiae* 35–57 cm
Larger, more heavily built version of the Eastern Barn Owl with thicker, more powerful legs, feet and talons; usually darker in colour. Four, possibly five, races, each with colour morphs. Nominate race has light, dark and intermediate morphs, inhabits much of mainland Aust. except tropical N, where race *kimberli* is smaller with pale and intermediate morphs (some see Cape York individuals as a separate race, *galei*). Palest is white beneath, pale grey above, the palest of all barn owls in Aust. In Tas., race *castanops* is usually dark, but some are intermediate. Females of Tas. race are, at 1.25 kg, some 50% heavier than mainland females, and double the weight of males of race *kimberli*. Gives a drawn out, harsh, unwavering screech, stronger and deeper than Eastern Barn Owl's; also cackling, rattling shrieks and raspings. Roosts and nests in heavy forest; hunts over open woodland and farmland. Uncommon to rare, but common in Tas.

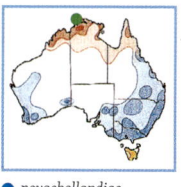

- 🔵 *novaehollandiae*
- 🟤 *kimberli*
- 🟡 *castanops*
- 🟢 *melvillensis*

Mask broad, rounded, with darker border

Intermediate morph has rufous tint on face and breast.

● (Not shown) Small; upper parts pale buff, light grey markings

Wine-red tone to facial disc

Chestnut and buff, heavily patterned with dark brown or black

Dusky rufous

🟡 ♀ Dark morph

Pale morph has white underparts.

Some are white with pale buffy grey upper parts; few if any dark morphs.

Feathered legs, powerful feet

Female is always larger, often much darker, than male.

Tasmanian race of Australian Masked Owl is largest, females to 57 cm.

**Eastern Grass Owl** *Tyto longimembris* 32–38 cm
Hunts over grass and other very low vegetation, flying slowly, low, back and forth, dropping down to snatch small prey. At times, perhaps when food is scarce, will hunt in the day. Otherwise, daylight hours are spent in 'squats', cave-like spaces under dense vegetation. Flushed from cover in daylight, the owl bursts up in panic, glides slowly before dropping again to cover. Seen from below against overhead sun or bright sky, the shadowed underparts can appear dark, but the flight feathers of wings and tail are brightly translucent, and barring is emphasised. At night, in car or spotlight beams, shows white beneath like Eastern Barn Owl, or rufous-buff. (Eastern Barn Owls also sometimes roost on ground.) Gives a harsh, rasping screech similar to that of Eastern Barn Owl; also low, soft screeches, trills. Lives on open, often swampy country with dense low groundcover of tall grass, heaths, cane grass, sugarcane, margins of mangroves. Nomadic, irruptive; usually rare, locally common.

Often hunts near dusk: slow, leisurely, flapping flight interspersed with slow glides, long legs trailing or dangling low ready to snatch prey from the ground.

In flight, feet trail well beyond tail tip. At take-off and landing, long legs dangle far below the body.

Crown dark brown

In flight, a blunt-faced, large-headed jiz typical of barn owl family

Back patchy brown, buff and rufous, unlike the pale grey of the Eastern Barn Owl

♂

Beneath: males whitish, females light to mid rufous-buff

Tail barred

Legs long, sparsely feathered tarsus

Nests on ground under dense grass.

# TAWNY FROGMOUTH

**Tawny Frogmouth** *Podargus strigoides* 34–52 cm
Probably the best known Australian nocturnal bird; occasionally seen in camouflage pose on an exposed limb, stiffly posed to mimic a broken branch. The streaked and mottled plumage looks like old, weathered wood or bark, the bill and bristles resemble the jagged end of a broken branch, and the untidy white spots and dark streaks are like lichens and sap stains on old timber. Yellow eyes look through narrowed slits; the head turns almost imperceptibly to follow an intruder's movements. Habitat mostly open woodland and eucalypt forest, and wide variety of other vegetation types; least often seen in treeless desert, rainforest and wet, dense eucalypt forest. Hunts mostly ground-dwelling creatures: large insects, spiders, frogs, small mammals and ground birds. Watches and listens from a low perch, gliding down on broad, silent wings to take prey on ground. Call is a muffled, low, resonant, rapid 'ooo-oom-oom-oom'; from males, a similarly rapid 'ar-oom, aroom, aroom-aroom'. Sedentary; common.

● *strigoides*
● *brachypterus*
● *phalaenoides*

If disturbed in daylight will imitate a broken limb – bill uplifted, feathers tight against stiffly outstretched body, colour and pattern like bark or weathered wood. Watches through eyes narrowed to fine slits.

Bristles extend past bill tip.

Eyes pale to deep yellow. If seen in lights at night, is transformed: eyes wide, relaxed.

Long dark streaks

Male always in streaked greys or grey-browns; all races. Grey morph females similar

Rufous birds, and the rare E coast chestnut phase, are always females.

Females have both grey and rufous morphs; males always grey. Both sexes darker in humid coastal regions

Three races, grey both sexes; females of two races may be of grey or rufous morph.
● Large; males grey; females grey or chestnut
● Medium-sized, paler, appears not to have a rufous morph. In SW of WA, in moister habitat, is slightly larger and darker, like ●
● Smaller, males paler grey; females grey or rufous

Frogmouth nests are rough, flimsy, untidy collections of sticks in a fork 2–15 m above ground. Above: Tawny Frogmouth

The Ant Plant of NE Qld makes a convenient nest platform for Marbled and Papuan Frogmouths.

Rufous tint

All flight feathers barred

Nestling

## PAPUAN & MARBLED FROGMOUTHS

**Papuan Frogmouth** *Podargus papuensis* 50–60 cm
At 60 cm in length, this is a huge frogmouth; its massive head adds to the bulky appearance, but its size is exaggerated by loose bulky feathering. Although nearly the size of a Rufous Owl, it is only one-third the weight. With small, thin, weak feet and slow flight, these frogmouths are no match for big owls, often becoming their prey. Most frequent the margins of dense closed forest such as rainforest and monsoon forest; seems to avoid the interior of the forest, preferring to hunt out through more open tropical woodland; but using denser vegetation for daytime concealment. Roost sites tend to have a dense overhead canopy and, at times, as in mangroves, are over water. Call is a soft, low, steady 'orrm-orrm-orrm-orrm', slower and deeper than calls of either smaller frogmouth, and with a slight but discernible rough or harsh quality. Also has a descending sequence of bubbly sounds, often ending with a 'clak', perhaps a snapping of the bill, and a harsh, long, scolding, rasping scream: 'haaarrrrck!'. The screech is often intermixed with harsh, loud, unpleasant frog-croak noises. Common in N; scarce in S.

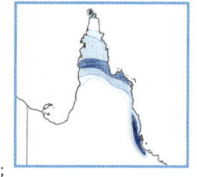

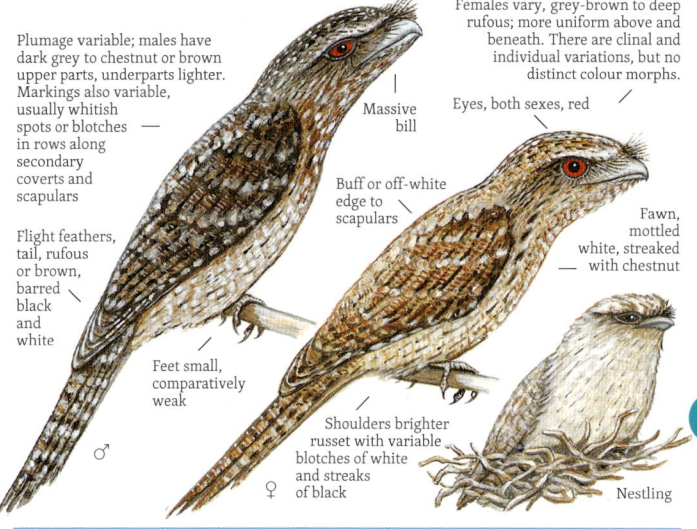

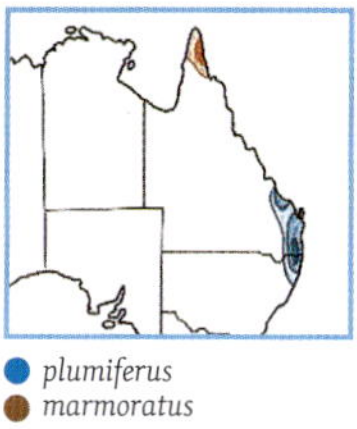

**Marbled Frogmouth** *Podargus ocellatus* 37–46 cm
Beautifully plumaged frogmouth with deep orange or reddish eyes and extensive mottling or marbling of white, grey and brown that gives the look of rough bark, complete with pale patches of lichens, fissuring and weathering. Hunts from low branches, dropping down to take small creatures from the ground. Race *marmoratus* of NE Qld lives in river-edge gallery forest, monsoon forest, rainforest, adjoining eucalypt or melaleuca forest and woodland; race *plumiferus* lives in patches of subtropical lowland rainforest at altitudes up to about 800 m. Call is usually a double-noted, rapid, musical 'whooo-whoop, whooo-whoop', the second part uplifted; a guttural, growling 'quorr-orr, quorr-orr', which is deep, throaty, vibrating – a rapid, descending bubbling sound. Is probably uncommon but secure on Cape York; habitat of *plumiferus* on mid-E coast is greatly reduced; race is now apparently rare, perhaps endangered.

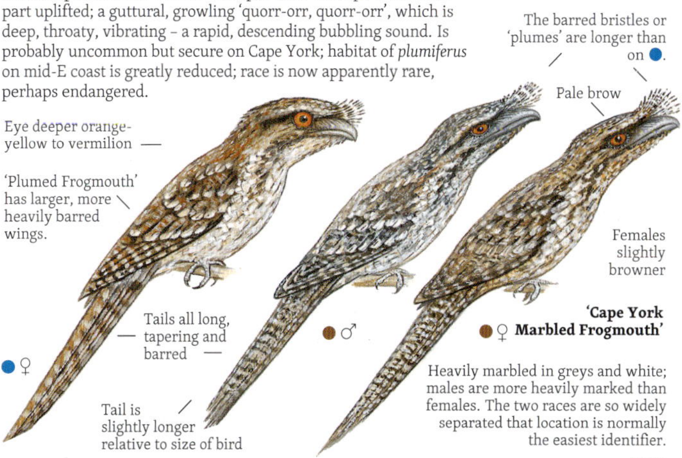

● *plumiferus*
● *marmoratus*

## NIGHTJARS

**Large-tailed Nightjar** *Caprimulgus macrurus* 25–28 cm
Tropical nightjar with prominent white on tips of tail and wing feathers. Habitat includes margins of rainforest, monsoon and vine forest, mangroves, and adjoining open tropical woodland. These nightjars are rarely deep in the forest, but roost and nest on the ground on the heavy layer of leaf litter near the edge or just outside closed forest. There, they are close to dense cover for daytime concealment, but have open woodland for night hunting nearby. Call, in distance, is a hollow 'quok-quok-quok-' or 'chop-chop-chop'; close at hand the call is a quick 'quorrok-quorrok-quorrok', usually, three to six calls, brief pause, then repeat. Each 'quorrok' is very abrupt. Common in undamaged habitat; vulnerable to stock damage in small patches of monsoon and similar vegetation across N Aust.

Long wings and light body give buoyant, quiet, fluttering, butterfly-like flight.

Twin thin white streaks

White spots across wing-coverts

Tropical northern nightjar unique in having large white patches in both the wing tips and outer tail.

Rufous face underlined by two white streaks

Wide white wing bar, narrower on female

Capture of flying insects aided by bristles

In flight, displays combination of white wing patches and very large white tips to the outer tail feathers.

Plumage pattern makes bird difficult to see.

Roosts and nests on leaf litter.

Folded wings much shorter than tail

**White-throated Nightjar** *Eurostopodus mystacalis* 32–37 cm
A large, dark nightjar that, if flushed, shows no large white patches on wing tips or tail. In headlights or spotlight beam its eyes return a fiery red reflection. Like other species, the disruptive plumage patterns and colours of this nightjar are almost impossible to detect on the ground among leaves, twigs and rocks. Usual habitat of forest, woodland, heath, often the more open, drier ridges among rocks and fallen timber. Nightjars are most active around dusk and dawn when their long winged, hawk-like silhouette may be seen against a darkening sky. The call is a long sequence of musical, bubbling sounds; begins with very slow notes that become more rapid, rising to a crescendo, then abruptly switching to a long cascade of bubbling notes: 'awwk, awwk, awk, awk-AWk-AWk-AWk-quok-quok-quok-quok-quok-', rather like the initial chuckle of a Laughing Kookaburra. Common.

Egg laid on leaf litter

In flight, identified by absence of white in tail together with small white wing spots in place of white bars

Erratic, stiff 'double-beat' wing action; long glides on uplifted wings

Juv.: overall rufous tone, dull compared with adult

Grey brow line, very dark face

Dark back

Tips of long wings almost reach end of tail

Undersurfaces buff, barred brown

Sexes alike

220

# SPOTTED NIGHTJAR & OWLET-NIGHTJAR

**Spotted Nightjar** *Eurostopodus argus* 29–32 cm
Most colourful of nightjars; a bird of drier and more open country where the ground is red or brown, and littered with stones, grey and brown twigs, and yellow, ochre and rufous dry leaves. The intricately patterned plumage mimics these colours so that a Spotted Nightjar roosting or brooding on the ground is almost invisible. Habitat encompasses eucalypt, acacia, mulga and other open woodland like mallee and spinifex; favours stony ridges. Hawks aerially for insects with fluttering, bat-like flight; most active just after dark, and again towards dawn. Has a beautiful call that carries far on still nights in open country. Starts slow, accelerating, becoming rapid and loud, then dies away in a series of bubbling sounds: 'coooor, cawoor, cawow-caWOW-caWOW-tooka-tooka-tooka...'. If flushed from nest by day, is likely to fly a short distance before dropping to ground. Quite common, especially in more remote regions; part migratory.

Extensively spotted and barred

Bold white streak

Underparts relatively plain — rufous

Ruby-red eye reflection in spotlight

Rufous collar

White around throat

In flight, identified by white wing patches, but no white patches in tail

Folded wings shorter than tip of tail

Juv.: extensive cinnamon-russet on upper parts

**Australian Owlet-nightjar** *Aegotheles cristatus* 21–25 cm
A tiny owl-like bird with eyes that seem huge. Captures insects in flight or on the ground at night. Its distinctive, pleasant churring, is a common night call of the Australian bush, especially in the arid interior where even the smallest stand of stunted trees seems to have its resident owlet-nightjars. At a distance, the call is a pleasant, vibrating, musical churring, ending slightly sharper. Closer, the churring in the first, loudest part of the call is like a rolling of r's that gives a throaty 'quar-rr-rgh-'. It then blends to a higher, sharper, but not as loud, '-a-kak' – the whole call an undulating 'quar-rr-rgh-a-kak, quar-rr-rgh-a-kak'. Also gives a single, sharp, penetrating, metallic 'aeiiirk!'. Calls at all times of year and can often be heard during the day, when it may be seen sunning at the entrance to its hollow. Flushes easily; darts to another hollow, seemingly familiar with all the hide-outs within its territory. Diverse habitats: coastal and mountain rainforest, eucalypt forest, woodland, tree-lined watercourses, mulga and spinifex with scattered clumps of trees. Common; widespread.

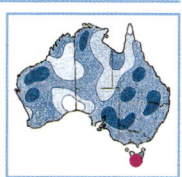

● *cristatus*
● *tasmanicus*

Tasmanian Owlet-nightjar is smaller, but no significant plumage difference. Colour morphs: light grey, dark grey, rufous-grey, and rufous

Hollow for daylight roosting, or as nest

Huge eyes, each set in a dark line from base of bill to back of crown. A third, finer, central dark line extends from bill to crown.

Tail and wings with softly speckled barring

Rufous-grey morph

Eyes, unlike those of true nightjars (above), do not glow brightly in a spotlight beam.

Most strongly rufous are those found in the Pilbara region and across arid parts of the interior; occurs in both sexes.

Soft, fringe-edged flight feathers give quiet, fluttering, moth-like flight.

Rufous morph

Juv.: similar to adult, but with less strongly marked face and shorter facial bristles and tail

# 14 SWIFTS, KINGFISHERS, BEE-EATER & DOLLARBIRD

The birds of this group are from five families. Of the swifts, Apodidae, Australia has a single breeding species, the Glossy Swiftlet. Others are visitors, regularly and common to extremely rare vagrants. Two kingfisher families occur in Australia: Alcedinidae, the plunge-fishing 'river kingfishers' have two species, while 'tree king-fishers', Halcyonidae, mostly dry-land hunters, have eight, possibly nine, including two kookaburras. The bee-eaters, Meropidae, and rollers, Coraciidae, are each represented by a single species.

Scarlet Robin = 12 cm

## apodidae
### page 223

Glossy Swiftlet

Australian Swiftlet

Uniform Swiftlet

White-throated Needletail

Fork-tailed Swift

House Swift

## halcyonidae
### page 227

Buff-breasted Paradise-Kingfisher

Common Paradise-Kingfisher

Forest Kingfisher

Red-backed Kingfisher

Blue-winged Kookaburra

Sacred Kingfisher

Torresian Kingfisher

Laughing Kookaburra

Yellow-billed Kingfisher

## alcedinidae
### page 226

Azure Kingfisher

Little Kingfisher

## meropidae
### page 231

Rainbow Bee-eater

## coraciidae
### page 231

Oriental Dollarbird

## QUICK INDEX WITH LINKS TO FAMILY

- (3) **Albatross** 30
- (23) Apostlebird 338
- (9) Avocet 151
- (19) **Babbler** 302
- (6) Baza 83
- (14) Bee-eater 231
- (20) Bellbird 313
- (5) Bittern 76
- (12) Black-Cockatoo 183
- (24) Blackbird 344
- (12) Blue Bonnet 204
- (21) Boatbill 314
- (13) Boobook 214
- (4) Booby 64
- (23) Bowerbird 339
- (17) Bristlebird 254
- (5) Brolga 80
- (13) Bronze-Cuckoo 209
- (11) Bronzewing 177
- (1) Brush-turkey 12
- (12) Budgerigar 205
- (7) Bush-Hen 106
- (7) Bustard 109
- (22) Butcherbird 330
- (8) Button-quail 111
- (1) **Cassowary** 11
- (23) Catbird 338
- (18) Chat 292
- (19) Chowchilla 302
- (22) Cicadabird 323
- (25) Cisticola 353
- (12) Cockatiel 186
- (12) Cockatoo 183
- (7) Coot 107
- (12) Corella 187
- (4) Cormorant 68
- (7) Corncrake 106
- (13) Coucal 213
- (7) Crake 102
- (5) Crane 80
- (23) Crow 336
- (13) Cuckoo 207
- (11) Cuckoo-Dove 173
- (22) Cuckoo-shrike 323
- (8) Curlew 118
- (22) Currawong 332
- (4) **Darter** 67
- (17) Diamondbird 252
- (3) Diving-Petrel 47
- (14) Dollarbird 231
- (9) Dotterel 140
- (11) Dove 171
- (8) Dowitcher 116
- (21) Drongo 317
- (2) Duck 16
- (8) Dunlin 126
- (6) **Eagle** 89
- (5) Egret 73
- (1) Emu 10
- (16) Emu-wren 244
- (16) **Fairy-wren** 239
- (6) Falcon 94
- (21) Fantail 320
- (17) Fernwren 260
- (17) Fieldwren 256
- (12) Fig-Parrot 191
- (22) Figbird 327
- (26) Finch 358
- (26) Firetail 364
- (21) Flycatcher 318
- (18) Friarbird 274
- (13) Frogmouth 218
- (4) Frigatebird 66
- (11) Fruit-Dove 171
- (3) Fulmar 37
- (12) **Galah** 188
- (4) Gannet 62
- (2) Garganey 21
- (17) Gerygone 261
- (3) Giant-Petrel 36
- (18) Gibberbird 293
- (8) Godwit 116
- (9) Golden Plover 138
- (25) Goldfinch 354
- (2) Goose 15
- (6) Goshawk 90
- (25) Grassbird 351
- (16) Grasswren 245
- (2) Grebe 24
- (25) Greenfinch 354
- (8) Greenshank 122
- (10) Gull 156
- (2) **Hardhead** 18
- (61) Harrier 86
- (3) Hawk-Owl 214
- (17) Heathwren 257
- (5) Heron 71
- (6) Hobby 94
- (18) Honeyeater 273
- (5) **Ibis** 78
- (11) Imperial Pigeon 170
- (5) **Jabiru** 81
- (9) Jacana 147
- (19) Jacky Winter 301
- (10) Jaeger 154
- (1) Junglefowl 12
- (6) **Kestrel** 94
- (12) King-Parrot 193
- (14) Kingfisher 226
- (6) Kite 84
- (8) Knot 126
- (13) Koel, Pacific 212
- (14) Kookaburra 230
- (9) **Lapwing** 148
- (19) Logrunner 302
- (12) Lorikeet 189
- (15) Lyrebird 233
- (22) **Magpie** 331
- (21) Magpie-lark 317
- (2) Mallard 19

# FAMILY GROUPS

| | |
|---|---|
| **1** | 10–13 |
| **2** | 14–27 |
| **3** | 28–59 |
| **4** | 60–69 |
| **5** | 70–81 |
| **6** | 82–101 |
| **7** | 102–109 |
| **8** | 110–135 |
| **9** | 136–151 |
| **10** | 152–169 |
| **11** | 170–179 |
| **12** | 180–205 |
| **13** | 206–221 |
| **14** | 222–231 |
| **15** | 232–237 |
| **16** | 238–249 |
| **17** | 250–269 |
| **18** | 270–293 |
| **19** | 294–303 |
| **20** | 304–313 |
| **21** | 314–321 |
| **22** | 322–333 |
| **23** | 334–341 |
| **24** | 342–349 |
| **25** | 350–357 |
| **26** | 358–365 |

Non-passerines
Passerines

## INDEX OF COMMON NAMES (CONT.)

Russet-tailed 343
Song 342
'Tin-tack' (see White-fronted Chat) 292
Treecreeper, Black-tailed 237
Brown 237
'Little' (see White-throated) 236
Red-browed 236
Rufous 237
White-browed 236
White-throated 236
Triller, Varied 323
White-winged 323
Tropicbird, Red-tailed 63
White-tailed 63
Turkey, Wild 12
Turnstone, Ruddy 134
Turtle-Dove, Laughing 175
Spotted 175
'Twelve Apostles' (see White-browed Babbler) 302

### W

Wagtail, Black-backed 356
Citrine 357
Eastern Yellow 357
Grey 357
White 356
Willie 320
Warbler, Arctic 350
Australian Reed- 351
Oriental Reed- 351
Speckled 257
Wattlebird, Little 272
Red 272
Western 273
Yellow 272
Wedgebill, Chiming 305
Chirruping 305

Weebill 260
Whimbrel 118
Whipbird, Eastern 305
Mallee (see Western Whipbird) 305
White-browed 305
'Western' (see White-browed) 305
Whistler, 'Brown' (see Grey) 309
Gilbert's 309
Golden 311
Grey 309
Mangrove Golden 310
Olive 309
Red-lored 309
Rufous 310
White-breasted 310
Western 311
Whistling-Duck, Plumed 16
Spotted 16
Wandering 16
White-eye, Ashy-bellied 348
'Capricorn' 348
Yellow 349
Whiteface, Banded 269
Chestnut-breasted 269
Southern 268
Willie Wagtail 320
Woodswallow, Black-faced 328
Dusky 328
Little 328
Masked 329
White-breasted 329
White-browed 329

### Y

Yellowlegs, Greater 121
Lesser 121

Albert's Lyrebird

## INDEX OF COMMON NAMES (CONT.)

ler's 50
sh-footed 51
ttering 54
at 53
tton's 55
tle 55
nx 54
nk-footed 52
ort-tailed 52
ooty 51
reaked 53
edge-tailed 50
luck, Australian 18
adjah 18
eler, Australasian 19
orthern 19
e-thrush, Arafura 312
ower's 312
rey 313
ufous' (see Arafura) 312
andstone 312
Vestern' (see Grey) 313
e-tit, 'Eastern' (see Crested) 311
rested 311
Jorthern' 311
Vestern' 311
fflewing' (see Black-faced
uckoo-shrike) 324
reye 348
lla, 'Black-capped' 308
Striated' 308
Varied' 308
White-headed' 308
White-winged' 308
a, Brown 154
South Polar 154
lark, Eurasian 354
pe, Australian Painted 115
Latham's 114
Pin-tailed 114
Swinhoe's 115
nglark, Brown 353
Rufous 352
arrow, Eurasian Tree 354
House 354
parrowhawk, Collared 90
pinebill, Eastern 290
Western 290
Spinetail, Silver-rumped 224
Spinifexbird 352
Spoonbill, Royal 79
Yellow-billed 79
Starling, Common 345
Metallic 345
Stilt, Banded 151
Pied 151
Stint, Little 128
Long-toed 129
Red-necked 128
Temminck's 129
Stone-curlew, Beach 150
Bush 150
Stork, Black-necked 81
Storm-Petrel, Black-bellied 57

Grey-backed 56
Leach's 58
Matsudaira's 59
Swinhoe's 58
Tristram's 58
White-bellied 57
White-faced 59
Wilson's 56
Sunbird, Olive-backed 292, 349
Swallow, Barn 346
Pacific 346
Red-rumped 347
Striated (see Red-rumped) 347
Welcome 346
White-backed 346
Swamphen, Australasian 108
Swan, Black 15
Mute 15
Swift, Fork-tailed 225
House 224
Swiftlet, Australian 223
Glossy 223
Uniform 224

### T
'Tang' (see White-fronted Chat) 292
Tattler, Grey-tailed 125
Wandering 125
Teal, Chestnut 20
Grey 21
Tern, Antarctic 163
Arctic 164
Australian Gull-billed 161
Black 167
Black-naped 162
Bridled 168
Caspian 161
Common 164
Fairy 165
Greater Crested 160
Lesser Crested 160
Little 165
Roseate 163
Sooty 168
Whiskered 166
White 166
White-fronted 162
White-winged Black 167
Ternlet, Grey 167
Thornbill, Brown 266
Buff-rumped 265
Chestnut-rumped 266
Inland 267
'Little' (see Yellow) 264
Mountain 268
Slaty-backed 267
Slender-billed 265
Striated 264
Tasmanian 268
'Varied' (see Buff-rumped) 265
Western 264
Yellow 264
Yellow-rumped 265
Thrush, Bassian 343
Blue Rock- 344

## INDEX OF COMMON NAMES (CONT.)

  Noisy 234
  Papuan 234
  Rainbow 234
Plains-wanderer 111
Plover, American Golden 138
  Caspian 144
  Double-banded 143
  Eurasian Golden 139
  Greater Sand 144
  Grey 139
  Hooded 140
  Kentish 142
  Lesser Sand 143
  Little Ringed 141
  Oriental 145
  Pacific Golden 138
  Red-capped 142
  Ringed 141
  'Snowy' (see Kentish) 142
  'Spur-winged' (see Masked Lapwing) 148
Pratincole, Australian 137
  Oriental 137
Prion, Antarctic 49
  Broad-billed 48
  Fairy 49
  Fulmar 48
  Salvin's 49
  Slender-billed 48
Pygmy-Goose, Cotton 20
  Green 20

### Q

Quail, Brown 13
  California 13
  King 13
  Stubble 13
  'Swamp' 13
Quail-thrush, Chestnut 306
  Chestnut-breasted 307
  Cinnamon 307
  Nullarbor 307
  Spotted 306
  Western 307
'Quarrion' (see Cockatiel) 186

### R

Rail, Buff-banded 103
  Chestnut 107
  Lewin's 105
Raven, Australian 336
  Forest 337
  Little 336
  'Relict' (see Forest) 337
Redshank, Common 120
  Spotted 120
Redthroat 260
Reed-Warbler, Australian 351
  Oriental 351
Reeve (and Ruff) 135
Riflebird, Magnificent 335
  Paradise 335
  Victoria's 335
Ringneck, Australian 198
Robin, Buff-sided (see White-browed) 296
  Dusky 298

  Eastern Yellow 299
  Flame 297
  Grey-headed 295
  Hooded 298
  Mangrove 298
  'Northern Yellow' (see Eastern Yellow) 299
  Pale-yellow 300
  Pink 296
  Red-capped 297
  Rose 296
  Scarlet 297
  Western Yellow 299
  White-breasted 299
  White-browed 296
  White-faced 300
Rock-Pigeon, Chestnut-quilled 179
  White-quilled 179
Rockwarbler 255
Rosella, 'Adelaide' 195
  'Blue-cheeked' (see Pale-headed) 196
  Crimson 195
  Eastern 197
  'Golden-mantled' (see Eastern) 197
  Green 195
  Northern 196
  Pale-headed 196
  Western 197
  'Yellow' 195
Ruff (and Reeve) 135

### S

Sanderling 131
Sandpiper, Baird's 131
  Broad-billed 127
  Buff-breasted 135
  Common 124
  Cox's 132
  Curlew 127
  Dunlin 126
  Green 123
  Marsh 123
  Pectoral 133
  Sharp-tailed 132
  Stilt 133
  Terek 124
  Upland 134
  Western 130
  White-rumped 130
  Wood 124
Scrub-bird, Noisy 235
  Rufous 235
Scrub-robin, Northern 295
  Southern 295
Scrubfowl, Orange-footed 12
Scrubtit 255
Scrubwren, Atherton 258
  'Buff-breasted' (see White-browed) 259
  Large-billed 258
  'Spotted' (see White-browed) 259
  Tasmanian 259
  Tropical 258
  White-browed 259
  Yellow-throated 259
Sea-Eagle, White-bellied 92
Shearwater, Audubon's 55

## INDEX OF COMMON NAMES (CONT.)

**P**

Paradise-Kingfisher, Buff-breasted 227
    Common 227
Pardalote, 'Black-headed' (see Striated) 253
    'Eastern Striated' (see Striated) 253
    Forty-spotted 252
    Red-browed 253
    Spotted 252
    Striated 253
    'Yellow-rumped' (see Spotted) 252
    'Yellow-tipped' (see Striated) 253
Parrot, Australian King- 193
    Australian Ringneck 198
    Blue Bonnet 198
    Blue-winged 202
    Bourke's 204
    'Cloncurry Ringneck' 198
    Double-eyed Fig- 191
    Eastern Ground 205
    Eclectus 192
    Elegant 202
    Golden-shouldered 201
    Hooded 201
    'Mallee Ringneck' 198
    Mulga 200
    'Multi-coloured' (see Mulga Parrot) 200
    Night 205
    Orange-bellied 202
    Paradise 200
    'Port Lincoln' 198
    Princess 194
    Red-capped 199
    Red-cheeked 192
    Red-rumped 200
    Red-winged 193
    Regent 194
    Rock 203
    Scarlet-chested 203
    Superb 194
    Swift 199
    Turquoise 203
    'Twenty-eight' 198
    'Varied' (see Mulga Parrot) 200
    Western Ground 205
Parrot-Finch, Blue-faced 365
Peafowl, Indian 12
Pelican, Australian 61
'Peewee' (see Magpie-lark) 317
Penguin, Adelie 25
    Chinstrap 25
    Erect-crested 27
    'Fairy' (see Little) 25
    Fiordland 27
    Gentoo 26
    King 26
    Little 25
    Macaroni 26
    Magellanic 25
    Southern Rockhopper 27
    Royal (see Macaroni) 26
    Snares 27
Petrel, Antarctic 37
    Atlantic 43
    Barau's 45
    Black 39
    Black-bellied Storm 57
    Black-winged 44
    Blue 46
    Bulwer's 43
    Cape 37
    Common Diving- 47
    Cook's 46
    Gould's 44
    Great-winged 38
    Grey 42
    Grey-backed Storm- 56
    Herald 40
    Jouanin's 43
    Juan Fernandez 46
    Kerguelen 39
    Kermadec 41
    Leach's Storm- 58
    Matsudaira's Storm- 59
    Mottled 44
    Northern Giant- 36
    Providence 38
    Snow 37
    Soft-plumaged 40
    South Georgian Diving- 47
    Southern Giant- 36
    Swinhoe's Storm- 58
    Tahiti 41
    Tristram's Storm- 58
    Westland 39
    White-bellied Storm- 57
    White-chinned 38
    White-faced Storm- 59
    White-headed 42
    White-necked 45
    Wilson's Storm- 56
Phalarope, Grey 146
    Red-necked 146
    Wilson's 147
Pheasant, Common 12
Pigeons
    Brush Bronzewing 177
    Chestnut-quilled Rock- 179
    Collared Imperial 170
    Common Bronzewing 177
    Crested 176
    Elegant Imperial 172
    Feral 174
    Flock Bronzewing 177
    Partridge 178
    Spinifex 176
    Squatter 178
    Topknot 174
    Torresian Imperial 171
    White-headed 171
    White-quilled Rock- 179
    Wonga 174
    'Yellow-eyed Imperial' (see Elegant Imperial) 172
Pilotbird 255
Pintail, Northern 21
Pipit, Australian 355
    Red-throated 355
Pitta, Blue-winged 235

## INDEX OF COMMON NAMES (CONT.)

White-streaked 286
White-throated 289
Yellow 280
Yellow-faced 278
Yellow-plumed 282
Yellow-spotted 278
Yellow-throated 281
Yellow-tinted 283
Yellow-tufted 281

**I**

Ibis, Australian White 78
    Glossy 78
    Straw-necked 78
Imperial-Pigeon, Collared 170
    Elegant 172
    Pied 171

**J**

'Jabiru' (see Black-necked Stork) 81
Jacana, Comb-crested 147
    Pheasant-tailed 146
Jacky Winter 301
Jaeger, Arctic 154
    Long-tailed 155
    Pomarine 155
Junglefowl, Red 12

**K**

Kestrel, Nankeen 94
King-Parrot, Australian 193
Kingfisher, Azure 226
    Buff-breasted Paradise- 227
    Common Paradise- 227
    Forest 229
    Little 226
    Red-backed 228
    Sacred 228
    Torresian 228
    Yellow-billed 229
Kite, Black 85
    Black-shouldered 84
    Brahminy 88
    Letter-winged 84
    Square-tailed 86
    Whistling 85
Knot, Great 126
    Red 126
Koel, Pacific 212
Kookaburra, Blue-winged 230
    Laughing 230

**L**

Lapwing, Banded 148
    Masked 148
Logrunner, Australian 302
Lorikeet, Little 190
    Musk 191
    Purple-crowned 190
    Rainbow 189
    'Red-collared' (see Rainbow) 189
    Scaly-breasted 189
    Varied 190
Lyrebird, Albert's 233
    Superb 233

**M**

Magpie, Australian 331
    'Black-backed' 331
    'Western' 331
Magpie-lark 317
Mallard, Northern 19
Malleefowl 12
Mannikin, Chestnut-breasted 359
    Nutmeg. See Munia, Scaly-breasted 360
    Pictorella 359
    Yellow-rumped 360
Manucode, Trumpet 334
Martin, Fairy 347
    Tree 347
Miner, Bell 276
    Black-eared 277
    Noisy 277
    Yellow-throated 277
Mistletoebird 292, 349
Monarch, Black-faced 316
    Black-winged 316
    Frilled 315
    Pied 315
    Spectacled 316
    White-eared 315
'Moonface' (see White-fronted Chat) 292
Moorhen, Dusky 106
'Morepork' (see Tawny Frogmouth) 218
Munia, Scaly-breasted 360
Myna, Common 344

**N**

Native-hen, Black-tailed 108
    Tasmanian 109
Needletail, White-throated 225
Nightjar, Australian Owlet- 221
    Large-tailed 220
    Spotted 221
    White-throated 220
Noddy, Black 169
    Brown 169
    Lesser 169

**O**

Oriole, Green 326
    Olive-backed 326
Osprey, Eastern 88
Ostrich 11
Owl, Australian Boobook 214
    Australian Masked 217
    Barking 215
    Brown Hawk- 214
    Eastern Barn 216
    Eastern Grass 217
    Greater Sooty 216
    Lesser Sooty 216
    Powerful 215
    'Red Boobook' (see Australian Boobook) 214
    Rufous 214
    'Tasmanian Masked' (see Australian Masked) 217
Owlet-nightjar, Australian 221
Oystercatcher, Pied 149
    Sooty 149
    South Island Pied 149

## INDEX OF COMMON NAMES (CONT.)

**G**

Galah 188
Gannet, Australasian 62
    Cape 62
Garganey 21
Gerygone, 'Black-throated' (see Fairy) 263
    Brown 262
    Dusky 261
    Fairy 263
    Green-backed 262
    Large-billed 261
    Mangrove 261
    Western 262
    White-throated 263
Giant-Petrel, Northern 36
    Southern 36
Gibberbird 293
Godwit, Bar-tailed 117
    Black-tailed 117
    Hudsonian 116
Golden Plover, American 138
    Eurasian 139
    Pacific 138
Goldfinch, European 354
Goose, Cape Barren 15
    Cotton Pygmy- 20
    Green Pygmy- 20
    Magpie 15
Goshawk, Brown 91
    Grey 90
    Red 91
Grassbird, Little 351
    Tawny 352
Grasswren, Black 245
    Carpentarian 246
    Dusky 248
    Eyrean 249
    Grey 246
    Kalkadoon 248
    Short-tailed 247
    Striated 247
    Thick-billed 249
    Western 249
    White-throated 245
Grebe, Australasian 24
    Great Crested 24
    Hoary-headed 24
    Little 24
Greenfinch, Common 354
Greenshank, Common 122
    Nordmann's 122
Gull, Black-headed 159
    Black-tailed 157
    Franklin's 158
    Kelp 156
    Laughing 158
    Lesser Black-backed 157
    Pacific 156
    Sabine's 159
    Silver 157

**H**

Hardhead 18
Harrier, Papuan 87
    Spotted 87
    Swamp 86

Hawk-Owl, Brown 214
Heathwren, Chestnut-rumped 257
    Shy 257
Heron, Great-billed 72
    Grey 72
    'Nankeen Night' 75
    Pied 71
    Rufous Night (see 'Nankeen Night Heron') 75
    Striated 76
    White-faced 73
    White-necked 71
Hobby, Australian 94
    Eurasian 95
Honey-Buzzard, Oriental 83
Honeyeater, Banded 290
    Bar-breasted 285
    Black 290
    Black-chinned 289
    Black-headed 289
    Blue-faced 275
    Bridled 279
    Brown 284
    Brown-backed 284
    Brown-headed 288
    Crescent 286
    Cryptic 278
    Dusky 291
    Eungella 279
    Fuscous 283
    'Golden-backed' (see Black-chinned) 289
    Graceful 278
    Green-backed 284
    Grey 284
    Grey-fronted 282
    Grey-headed 282
    'Helmeted' (see Yellow-tufted) 281
    Kimberley 279
    Lewin's 278
    Macleay's 276
    Mangrove 280
    New Holland 287
    Painted 287
    Pied 291
    Purple-gaped 282
    Red-headed 291
    Regent 275
    Rufous-banded 285
    Rufous-throated 285
    Scarlet 291
    Singing 280
    Spiny-cheeked 273
    Striped 273
    Strong-billed 288
    Tawny-breasted 276
    Tawny-crowned 285
    Varied 280
    White-cheeked 287
    White-eared 281
    White-fronted 286
    White-gaped 279
    White-lined 279
    White-naped 288
    White-plumed 283

## INDEX OF COMMON NAMES (CONT.)

    Peaceful 175
    Rose-crowned Fruit- 172
    Spotted 175
    Superb Fruit- 172
    Wompoo Fruit- 173
Dowitcher, Asian 116
    Short-billed 116
Drongo, Spangled 317
Ducks
    Australasian Shoveler 19
    Australian Shelduck 18
    Australian Wood 16
    Blue-billed 17
    Chestnut Teal 20
    Cotton Pygmy-Goose 20
    Freckled 17
    Garganey 21
    Green Pygmy-Goose 20
    Grey Teal 21
    Hardhead 18
    Northern Mallard 19
    Musk 17
    Northern Pintail 21
    Northern Shoveler 19
    Pacific Black 21
    Pink-eared 19
    Plumed Whistling- 16
    Radjah Shelduck 18
    Spotted Whistling- 16
    Wandering Whistling- 16
Dunlin 126

### E

Eagle, Gurney's 92
    Little 89
    Wedge-tailed 93
    White-bellied Sea- 92
Egret,
    Eastern Cattle 73
    Eastern Reef 73
    Great 74
    Intermediate 75
    Little 74
Emu 11
Emu-wren, Mallee 244
    Rufous-crowned 244
    Southern 244

### F

Fairy-wren, 'Black-backed Wren'
    (see Splendid) 241
    Blue-breasted 243
    'Lavender-flanked' (see Variegated) 242
    Lovely 243
    'Purple-backed' (see Variegated) 242
    Purple-crowned 240
    Red-backed 239
    Red-winged 243
    Splendid 241
    Superb 241
    'Turquoise Wren' (see Splendid) 241
    Variegated 242
    White-winged 240
Falcon, Black 96
    Brown 97
    Grey 96
    Peregrine 95
Fantail, Arafura 321
    Grey 321
    Mangrove 320
    Northern 320
    'Rufous' (See Arafura) 321
    'White-tailed' (see Grey) 321
    'Wood' (see Arafura) 321
Fernwren 260
Fieldwren, Rufous 256
    Striated 256
    'Western' (see Rufous) 256
Fig-Parrot, Double-eyed 191
Figbird, Australasian 327
    'Green Figbird' (see Australasian Figbird) 327
    'Yellow Figbird' (see Australasian Figbird) 327
Finch, Black-throated 361
    Blue-faced Parrot- 365
    Crimson 362
    Double-barred 358
    Gouldian 365
    Long-tailed 361
    Masked 361
    Painted 363
    Plum-headed 360
    Red-browed 362
    Star 363
    'White-bellied Crimson' (see Crimson) 362
    'White-eared' (see Masked) 361
    Zebra 359
Firetail, Beautiful 364
    Diamond 364
    Red-eared 364
Flycatcher, Blue-and-White 344
    Broad-billed 318
    Kimberley (See Lemon-bellied) 301
    Leaden 318
    Lemon-bellied 301
    Narcissus 344
    Paperbark 319
    Restless 319
    Satin 318
    Shining 319
    Yellow-legged 300
Friarbird, Helmeted 274
    Horned 274
    Little 275
    'Melville Island' (see Helmeted) 274
    Noisy 274
    'Sandstone' (see Helmeted) 274
    Silver-crowned 274
Frigatebird, Christmas Island 66
    Great 67
    Lesser 66
Frogmouth, Marbled 219
    Papuan 219
    'Plumed' (see Marbled) 219
    Tawny 218
Fruit-Dove, Black-banded 171
    Rose-crowned 172
    Superb 172
    Wompoo 173
Fulmar, Southern 37

## INDEX OF COMMON NAMES

Bronze-Cuckoo, 'Golden' (see Shining) 211
    Gould's 211
    Horsfield's 210
    Little 210
    Shining 211
Bronzewing, Brush 177
    Common 177
    Flock 177
Brush-turkey, Australian 12
Budgerigar 205
Bulbul, Red-whiskered 345
Bush-hen, Pale-vented 106
Bushlark, Horsfield's 355
Bustard, Australian 109
Butcherbird, Black 330
    Black-backed 330
    Grey 330
    Pied 331
    Silver-backed 330
    'Tasmanian' (see Grey) 330
Button-quail, Black-breasted 111
    Buff-breasted 112
    Chestnut-backed 112
    Little 113
    Painted 113
    Red-backed 112
    Red-chested 113
Buzzard, Black-breasted 93
    Oriental Honey- 83

### C

Cassowary, Southern 11
Catbird, Green 338
    Spotted 338
Chat, Crimson 292
    Orange 293
    White-fronted 292
    Yellow 293
Chough, White-winged 337
Chowchilla 302
Cicadabird, Common 323
Cisticola, Golden-headed 353
    Zitting 353
Cockatiel 186
Cockatoo, Baudin's Black- 185
    Carnaby's Black- 185
    Gang-gang 186
    Glossy Black- 184
    Long-billed Black- 185
    Palm 183
    Pink ('Major Mitchell's') 189
    Red-tailed Black- 183
    Short-billed Black- 185
    Sulphur-crested 188
    Yellow-tailed Black- 184
Coot, Eurasian 107
Corella, Little 187
    Long-billed 187
    Western 187
Cormorant, Australian Pied 68
    Black-faced 69
    Great 68
    Little Black 69
    Little Pied 69

Corncrake 106
Coucal, Pheasant 213
Crake (see also Corncrake 106)
    Australian Spotted 105
    Baillon's 105
    Red-legged 102
    Red-necked 103
    Spotless 104
    White-browed 104
Crane (see also Brolga 80)
    Sarus 80
Crow, House 336
    Little 336
    Torresian 337
Cuckoo (see also Common Koel 212)
    Black-eared 209
    Brush 208
    Channel-billed 213
    Chestnut-breasted 208
    Fan-tailed 209
    'Golden Bronze-' (see Shining Bronze-) 211
    Gould's Bronze- 211
    Horsfield's Bronze- 210
    Little Bronze- 210
    Long-tailed 212
    Oriental 207
    Pallid 207
    Shining Bronze- 211
Cuckoo-Dove, Brown 173
Cuckoo-shrike, Barred 325
    Black-faced 324
    Ground 325
    White-bellied 324
    'White-breasted' (see White-bellied) 324
    'Yellow-eyed' (see Barred) 325
Curlew, Eurasian 119
    Far Eastern 119
    Little 118
Currawong, Black 332
    'Black-winged' 333
    'Brown' 333
    'Clinker' 333
    'Clinking' 333
    Grey 333
    'Kangaroo Island' 333
    Pied 332

### D

Darter, Australian 67
'Diamondbird' (see Spotted Pardalote) 252
Diving-Petrel, Common 47
    'Kerguelen' (see Common) 47
    South Georgian 47
Dollarbird, Oriental 231
Dotterel, Black-fronted 141
    Inland 145
    Red-kneed 140
Dove, Banded Fruit- 171
    Barbary 175
    Bar-shouldered 176
    Brown Cuckoo- 173
    Diamond 175
    Laughing 175
    Pacific Emerald 173

## INDEX OF SCIENTIFIC NAMES (CONT.)

albigularis 300
*Tribonyx mortierii* 109
   *ventralis* 108
*Trichodere cockerelli* 286
*Trichoglossus*
   *chlorolepidotus* 189
   *moluccanus* 189
      *moluccanus* 189
      *rubritorquis* 189
*Tringa brevipes* 125
   *erythropus* 120
   *flavipes* 121
   *glareola* 124
   *guttifer* 122
   *incana* 125
   *melanoleuca* 121
   *nebularia* 122
   *ochropus* 123
   *stagnatilis* 123
   *totanus* 120
   *ussuriensis* 120
*Tryngites subruficollis* (See *Calidris subruficollis*) 135
*Turdus merula* 344
   *philomelos* 342
Turnicidae 110
*Turnix castanota* 112
   *maculosa* 112

*melanogaster* 111
*olivii* 112
*pyrrhothorax* 113
*varius* 113
   *scintillans* 113
   *varius* 113
   *velox* 113
*Tyto javanica* 216
   *longimembris* 217
   *multipunctata* 216
   *novaehollandiae* 217
      *castanops* 217
      *kimberli* 217
      *melvillensis* 217
      *novaehollandiae* 217
      *tenebricosa* 216
Tytonidae 206

**V**

*Vanellus miles* 148
   *miles* 148
   *novaehollandiae* 148
   *tricolor* 148

**X**

*Xanthotis flaviventer* 276
   *macleayanus* 276
*Xema sabini* 159
*Xenus cinereus* 124

**Z**

*Zanda baudinii* 185
   *funerea* 184
   *latirostris* 185
*Zapornia pusilla* 105
   *tabuensis* 104
*Zoothera dauma* 343
   *heinei* 343
   *lunulata* 343
      *cuneata* 343
      *halmaturina* 343
      *lunulata* 343, 342
*Zosterops citrinella* 348
   *albiventris* 348
   *lateralis* 348
      *chlorocephalus* 348
      *chloronotus* 348
      *cornwalli* 348
      *lateralis* 348
      *ochrochrous* 348
      *pinarochrous* 348
      *vegetus* 348
      *westernensis* 348
   *luteus* 349
      *balstoni* 349
      *luteus* 349

## INDEX OF COMMON NAMES

The names given in quotation marks are widely used names for a species or subspecies but may not be the preferred BLA common name for that species.

**A**

Albatross, Amsterdam 30
   Atlantic Yellow-nosed 32
   Black-browed 32
   Buller's 34
   'Chatham Island' (see Shy Albatross) 33
   Grey-headed 34
   'Grey-backed' (see Shy Albatross) 33
   Laysan 35
   Light-mantled Sooty 35
   Shy 33
   Sooty 35
   Southern Royal 31
   Wandering 30
Apostlebird 338
Arafura Fantail 321
Avocet, Red-necked 151

**B**

Babbler, Chestnut-crowned 303
   Grey-crowned 303
   Hall's 303
   'Red-breasted' (see Grey-crowned) 303
   White-browed 302
'Baldy-head' (see White-fronted Chat) 292
Baza, Pacific 83
Bee-eater, Rainbow 231
Bellbird, Crested 313
Bittern, Australian Little 76

   Australasian 77
   Black 77
   Yellow 76
Black-Cockatoo, Baudin's (Long-billed) 185
   Carnaby's (Short-billed) 185
   Glossy 184
   Red-tailed 183
   Yellow-tailed 184
Blackbird, Common 344
Blue Bonnet 204
   'Naretha' 204
   'Red-vented' 204
   'Yellow-vented' 204
'Blue Jay' (see Black-faced Cuckoo-shrike) 324
Boatbill, Yellow-breasted 314
Boobook, 'Red' (see Southern) 214
   Southern 214
Booby, Abbott's 65
   Brown 64
   Masked 65
   Red-footed 64
Bowerbird, Fawn-breasted 341
   Golden 339
   Great 341
   Regent 339
   Satin 339
   Spotted 340
   Tooth-billed 340
   Western 340
Bristlebird, Eastern 254
   Rufous 254
   Western 254
Brolga 80

## INDEX OF SCIENTIFIC NAMES (CONT.)

*Pycnoptilus floccosus* 255
    *floccosus* 255
    *sandlandi* 255
*Pyctonmotus jocosus* 345
*Pygoscelis adeliae* 25
    *antarcticus* 25
    *papua* 26
        *ellsworthii* 26
        *papua* 26
*Pyrrholaemus brunneus* 260
    *sagittata* 257

### Q
*Quoyornis georgiana* 299

### R
*Radjah radjah* 18
*Rallina fasciata* 102
    *tricolor* 103
*Rallus brachipus* 105
    *clelandi* 105
    *pectoralis* 105
*Ramsayornis fasciatus* 285
    *broomei* 285
    *fasciatus* 285
    *modestus* 284
*Recurvirostra novaehollandiae* 151
Recurvirostridae 136
*Rhapidura leucopygialis* 224
*Rhipidura dryas* 321
    *albiscapa* 321
        *albicauda* 321
        *albiscapa* 321
        *alisteri* 321
        *keasti* 321
        *preissi* 321
    *dryas* 321
    *leucophrys* 320
        *leucophrys* 320
        *melaleuca* 320
        *picata* 320
    *phasiana* 320
    *rufifrons* 321
        *intermedia* 321
        *rufifrons* 321
    *rufiventris* 320
        *gularis* 320
        *isura* 320
*Rostratula australis* 115
Rostratulidae 110

### S
*Scenopoeetes dentirostris* 340
Scolopacidae 110, 136
*Scythrops novaehollandiae* 213
*Sericornis beccarii* 258
    *dubius* 258
    *minimus beccari* 258
    *citreogularis* 259
        *cairnsi* 259
        *citreogularis* 259
        *intermedius* 259
    *frontalis* 259
        *ashbyi* 259
        *balstoni* 259
        *flindersi* 259
        *frontalis* 259
        *humilis* (see
            *S. humilis*) 259
        *laevigaster* 259
        *maculatus* 259
        *rosinae* 259
        *tregellasi* 259
        *tweedi* 259
    *humilis* 259

*keri* 258
*magnirostra* 258
    *howei* 258
    *magnirostra* 258
    *viridior* 258
*Sericulus chrysocephalus* 339
*Smicrornis brevirostris* 260
    *brevirostris* 260
    *flavescens* 260
    *occidentalis* 260
    *ochrogaster* 260
*Sphecotheres vieilloti* 327
    *ashbyi* 327
    *flaviventris* 327
    *vieilloti* 327
Spheniscidae 14
*Spheniscus magellanicus* 25
*Spilopelia chinensis* 175
    *roseogrisea* 175
    *senegalensis* 175
*Stagonopleura bella* 364
    *guttata* 364
    *oculata* 364
*Steganopus tricolor*. See
*Phalaropus tricolor* 147
*Stercorarius antarcticus* 154
    *longicaudus* 155
    *maccormicki* 154
    *parasiticus* 154
    *pomarinus* 155
*Sterna albifrons* 165
    *sinensis* 165
    *dougallii* 163
    *hirundo* 164
        *hirundo* 164
        *longipennis* 164
    *nereis* 165
    *paradisaea* 164
    *striata* 162
    *sumatrana* 162
    *vittata* 163
*Stictonetta naevosa* 17
*Stiltia isabella* 137
*Stipiturus malachurus* 244
    *halmaturinus* 244
    *hartogi* 244
    *intermedius* 244
    *littleri* 244
    *malachurus* 244
    *parimeda* 244
    *polionotum* 244
    *westernensis* 244
    *mallee* 244
    *ruficeps* 244
*Stizoptera bichenovii* 358
    *annulosa* 358
    *bichenovii* 358
*Stomiopera flava* 280
    *unicolor* 279
*Strepera fuliginosa* 332
    *colei* 332
    *fuliginosa* 332
    *parvior* 332
    *graculina* 332
        *ashbyi* 332
        *graculina* 332
        *magnirostris* 332
        *nebulosa* 332
        *robinsoni* 332
    *versicolor* 333
        *arguta* 333
        *halmaturina* 333
        *intermedia* 333

        *melanoptera* 333
        *plumbea* 333
        *versicolor* 333
*Streptopelia*. See *Spilopelia* 175
Strigidae 206
*Sturnus tristis*. See *Acridotheres tristis* 344
*Struthidea cinerea* 338
    *cinerea* 338
    *dalyi* 338
*Struthio camelus* 11
Struthionidae 10
Sturnidae 342
*Sturnus tristis* 344
*Sturnus vulgaris* 345
*Sugomel niger* 290
*Sula dactylatra* 65
    *leucogaster* 64
    *sula* 64
Sylviidae 350
*Syma torotoro* 229
*Symposiarchus trivirgatus* 316
    *albiventris* 316
    *gouldii* 316
    *melanorrhoa* 316

### T
*Tachybaptus*
    *novaehollandiae* 24
    *ruficollis* 24
*Tadorna radjah* (See *Radjah radjah*) 18
    *tadornoides* 18
*Taeniopygia bichenovii* (See *Stizoptera bichenovii*) 358
    *guttata* 359
*Tanysiptera galeata* 227
    *sylvia* 227
*Thalassachre bulleri* 34
    *bulleri* 34
    *platei* 34
    *cauta* 33
        *cauta* 33
        *eremita* 33
        *salvini* 33
    *carteri* 32
        *bassi* 32
        *chlororhynchos* 32
    *chrysostoma* 34
    *epomophora* 31
        *epomophora* 31
        *sanfordi* 31
    *melanophris* 32
        *impavida* 32
        *melanophris* 32
*Thalasseus bengalensis* 160
    *bergii* 160
*Thalassoica antarctica* 37
*Thinornis cucullatus* 140
*Threskiornis molucca* 78
    *spinicollis* 78
Threskiornithidae 70
*Todiramphus sordidus* 228
    *pilbara* 228
    *sordidus* 228
    *macleayii* 229
        *incinctus* 229
        *macleayii* 229
    *pyrrhopygius* 228
    *sanctus* 228
*Tregellasia capito* 300
    *capito* 300
    *nana* 300
    *leucops* 300

## INDEX OF SCIENTIFIC NAMES (CONT.)

*corniculatus* 274
   *corniculatus* 274
   *monachus* 274
*Philomachus pugnax* (See *Calidris pugnax*) 135
*Phoebastria immutabilis* 35
*Phoebetria fusca* 35
   *palpebrata* 35
*Phonygammus keraudrenii* 334
*Phylidonyris*
   *nigra* 287
      *gouldi* 287
      *nigra* 287
   *novaehollandiae* 287
      *campbelli* 287
      *canescens* 287
      *caudata* 287
      *longirostris* 287
      *novaehollandiae* 287
   *pyrrhopterus* 286
      *halmaturina* 286
      *pyrrhopterus* 286
*Phylloscopus borealis* 350
*Pitta erythrogaster* (See *Erythropitta macklotii*) 234
   *iris* 234
      *iris* 234
      *johnstoneiana* 234
   *moluccensis* 235
   *versicolor* 234
      *intermedia* 234
      *simillima* 234
      *versicolor* 234
*Pittidae* 232
*Platalea flavipes* 79
   *regia* 79
*Platycercus adscitus* 196
   *adscitus* 196
   *palliceps* 196
   *caledonicus* 195
   *elegans* 195
      *adelaidae* 195
      *elegans* 195
      *flaveolus* 195
      *fleurieuensis* 195
      *melanoptera* 195
      *nigrescens* 195
      *subadelaidae* 195
   *eximius* 197
      *diemenensis* 197
      *elecica* 197
      *eximius* 197
   *icterotis* 197
      *icterotis* 197
      *xanthogenys* 197
   *venustus* 196
      *hilli* 196
      *venustus* 196
*Plectorhyncha lanceolata* 273
*Plegadis falcinellus* 78
*Pluvialis apricaria* 139
   *altifrons* 139
   *apricaria* 139
   *dominica* 138
   *fulva* 138
   *squatarola* 139
*Podargidae* 206
*Podargus ocellatus* 219
   *marmoratus* 219
   *plumiferus* 219
   *papuensis* 219
   *strigoides* 218
      *brachypterus* 218

      *phalaenoides* 218
      *strigoides* 218
*Podiceps cristatus* 24
*Podicipedidae* 14
*Poecilodryas superciliosa* 296
   *cerviniventris* 296
   *superciliosa* 296
*Poephila acuticauda* 361
   *acuticauda* 361
   *hecki* 361
   *cincta* 361
      *atropygialis* 361
      *cincta* 361
      *nigrotecta* 361
   *personata* 361
      *leucotis* 361
      *personata* 361
*Poliocephalus*
   *poliocephalus* 24
*Poliolimnas cinereus* 104
*Polytelis alexandrae* 194
   *anthopeplus* 194
      *anthopeplus* 194
      *monarchoides* 194
   *swainsonii* 194
*Pomatostomidae* 294
*Pomatostomus halli* 303
   *ruficeps* 303
   *superciliosus* 302
      *ashbyi* 302
      *centralis* 302
      *gilgandra* 302
      *superciliosus* 302
   *temporalis* 303
      *rubeculus* 303
      *temporalis* 303
*Poodytes carteri* 352
   *gramineus* 351
      *goulburni* 351
      *gramineus* 351
      *thomasi* 351
*Porphyrio melanotus* 108
   *bellus* 108
*Porzana fluminea* 105
   *pusilla.* (See *Zapornia pusilla*) 105
   *tabuensis* (See *Zapornia tabuensis*) 104
*Prionodura newtoniana* 339
*Probisciger aterrimus* 183
*Procellaria aequinoctialis* 38
   *cinerea* 42
   *parkinsoni* 39
   *westlandica* 39
*Procellariidae* 28
*Procelsterna albivitta* 167
*Psephotellus chrysopterygius* 201
   *dissimilis* 201
   *varius* 200
*Psephotus*
   *chrysoptergius* (See *Psephotellus chyrsoptergius*) 201
   *dissimilis* (See *Psephotellus dissimilis*) 201
   *haematonotus* 200
      *caeruleus* 200
   *pulcherrimus* 200
   *varius* (See *Psephotellus varius*) 200
*Pseudobulweria rostrata* 41
*Psittacidae* 181
*Psitteuteles versicolor* 190
*Psophodes cristatus* 305

   *nigrogularis* 305
      *lashmari* 305
      *leucogaster* 305
      *nigrogularis* 305
      *oberon* 305
   *occidentalis* 305
   *olivaceus* 305
      *lateralis* 305
      *olivaceus* 305
*Pterodroma baraui* 45
   *cervicalis* 45
   *cookii* 46
   *externa* 45
   *incerta* 43
   *heraldica* 40
   *inexpectata* 44
   *lessonii* 42
   *leucoptera* 44
      *brevipes* 44
      *caledonicus* 44
      *leucoptera* 44
   *macroptera* 38
      *gouldi* 38
      *macroptera* 38
   *mollis* 40
   *neglecta* 41
      *neglecta* 41
      *nigripennis* 44
   *solandri* 38
*Ptilinopus alligator* 171
   *magnificus* 173
      *assimilis* 173
      *keri* 173
      *magnificus* 173
   *regina* 172
      *ewingii* 172
      *regina* 172
   *superbus* 172
*Ptilonorhynchus guttatus* (See *Chlamydera guttata*) 340
   *cerviniventris* (See *Chlamydera cerviniventris*) 341
   *maculatus* (See *Chlamydera maculata*) 340
   *nuchalis* (See *Chlamydera nuchalis*) 341
   *violaceus* 339
      *minor* 339
      *violaceus* 339
*Ptiloris magnificus* 335
   *paradiseus* 335
   *victoriae* 335
*Ptilotula flavescens* 283
   *flavescens* 283
   *melvillensis* 283
   *fuscus* 282
      *fuscus* 283
      *subgermanus* 283
   *keartlandi* 282
   *ornata* 282
   *plumula* 282
      *graingeri* 282
      *planasi* 282
      *plumula* 282
*Puffinus assimilis* 55
   *assimilis* 55
   *tunneyi* 55
   *gavia* 54
   *huttoni* 55
   *lherminieri* 55
   *puffinus* 54
*Purnella albifrons* 286
*Purpureicephalus spurius* 199

# INDEX OF SCIENTIFIC NAMES (CONT.)

*ruficauda*. (See *Bathilda ruficauda*) 363
*temporalis* 362
    *loftyi* 362
    *minor* 362
    *temporalis* 362
*Neophema chrysogaster* 202
    *chrysostoma* 202
    *elegans* 202
        *carteri* 202
        *elegans* 202
    *petrophila* 203
        *zietzi* 203
    *pulchella* 203
    *splendida* 203
*Neopsephotus bourkii* 204
*Neositttidae* 304
*Nesoptilotis flavicollis* 281
    *leucotis* 281
        *leucotis* 281
        *novaenorciae* 281
        *thomasi* 281
*Nettapus coromandelianus* 20
    *pulchellus* 20
*Ninox connivens* 215
    *connivens* 215
    *peninsularis* 215
    *boobook* 214
        *boobook* 214
        *halmaturina* 214
        *leucopsis* 214
        *lurida* 214
        *ocellata* 214
    *rufa* 214
        *meesi* 214
        *queenslandica* 214
        *rufa* 214
    *scutulata* 214
    *strenua* 215
*Northiella haematogaster* 204
    *haematogaster* 204
    *haematorrhous* 204
    *pallescens* 204
    *narethae* 204
*Numenius arquata* 119
    *arquata* 119
    *orientalis* 119
    *madagascariensis* 119
    *minutus* 118
    *phaeopus* 118
        *hudsonicus* 118
        *variegatus* 118
*Nycticorax caledonicus* 75
*Nymphicus hollandicus* 186

## O

*Oceanites oceanicus* 56
    *exasperatus* 56
    *oceanicus* 56
*Ocyphaps lophotes* 176
    *lophotes* 176
    *whitlocki* 176
Odontophoridae 10
*Onychoprion anaethetus* 168
    *fuscata* 168
*Oreoica gutturalis* 313
    *gutturalis* 313
    *pallescens* 313
*Oreoscopus gutturalis* 260
*Origma solitaria* 255
Oriolidae 322
*Oriolus flavocinctus* 326
    *flavocinctus* 326
    *flavotinctus* 326
    *kingii* 326
    *tiwi* 326
    *sagittatus* 326
        *affinis* 326
        *grisescens* 326
        *sagittatus* 326
Orthonychidae 294
*Orthonyx spaldingii* 302
    *melasmenus* 302
    *spaldingii* 302
    *temminckii* 302
*Oxyura australis* 17

## P

*Pachycephala fuliginosa* 311
    *inornata* 309
        *gilberti* 309
        *inornata* 309
    *lanioides* 310
        *carnarvoni* 310
        *fretorum* 310
        *lanioides* 310
    *melanura* 310
        *melanura* 310
        *robusta* 310
        *spinicauda* 310
    *olivacea* 309
        *apatetes* 309
        *bathychroa* 309
        *hesperus* 309
        *macphersoniana* 309
        *olivacea* 309
    *pectoralis* 311
        *fuliginosa* 311
        *glaucura* 311
        *pectoralis* 311
        *queenslandica* 311
        *youngi* 311
    *rufiventris* 310
        *falcata* 310
        *minor* 310
        *pallida* 310
        *rufiventris* 310
    *rufogularis* 309
    *simplex* 309
        *peninsulae* 309
        *simplex* 309
Pachycephalidae 304
*Pachyptila belcheri* 48
    *crassirostris* 48
    *desolata* 49
    *salvini* 49
    *turtur* 49
    *vittata* 48
*Pagodroma nivea* 37
    *confusa* 37
    *nivea* 37
*Pandion cristatus* 88
*Papasula abbotti* 65
Paradisaeidae 334
Pardalotinae 250
*Pardalotus punctatus* 252
    *millitaris* 252
    *punctatus* 252
    *xanthopygus* 252
    *quadragintus* 252
    *rubricatus* 253
        *rubricatus* 253
        *yorki* 253
    *striatus* 253
        *melanocephalus* 253
        *melvillensis* 253
        *ornatus* 253
        *striatus* 253
        *substriatus* 253
        *uropygialis* 253
*Parvipsitta porphyrocephala* 190
    *pusilla* 190
*Passer domesticus* 354
    *montanus* 354
Passeridae 350
*Pavo cristatus* 12
Pedionomidae 110
*Pedionomus torquatus* 111
*Pelagodroma marina* 59
    *dulciae* 59
*Pelecanoides georgicus* 47
    *urinatrix* 47
        *exsul* 47
        *urinatrix* 47
Pelecanoididae 29
*Pelecanus conspicillatus* 61
*Peltohyas australis* 145
*Peneothello pulverulenta* 298
    *alligator* 298
    *cinereiceps* 298
    *leucura* 298
*Pernis ptilorhynchus* 83
*Petrochelidon ariel* 347
    *nigricans* 347
        *neglecta* 347
        *nigricans* 347
*Petroica goodenovii* 297
    *multicolor* 297
        *boodang* 297
        *campbelli* 297
        *leggii* 297
    *phoenicea* 297
    *rodinogaster* 296
        *inexpectata* 296
        *rodinogaster* 296
    *rosea* 296
Petroicidae 294
*Petrophassa albipennis* 179
    *albipennis* 179
    *boothi* 179
    *rufipennis* 179
*Pezoporus flaviventris* 205
    *occidentalis* 205
    *wallicus* 205
        *wallicus* 205
*Phaethon lepturus* 63
    *rubricauda* 63
*Phalacrocorax carbo* 68
    *fuscescens* 69
    *sulcirostris* 69
    *varius* 68
*Phalaropus fulicaria* 146
    *lobatus* 146
    *tricolor* 147
*Phaps chalcoptera* 177
    *elegans* 177
        *elegans* 177
        *occidentalis* 177
    *histrionica* 177
Phasianidae 10
*Phasianus colchicus* 12
*Philemon argenticeps* 274
    *argenticeps* 274
    *kempi* 274
    *buceroides* 274
        *ammitophila* 274
        *gordoni* 274
        *yorki* 274
    *citreogularis* 275
        *citreogularis* 275
        *sordidus* 275

# INDEX OF SCIENTIFIC NAMES (CONT.)

*Lichmera indistincta* 284
    *indistincta* 284
    *melvillensis* 284
    *ocularis* 284
*Limicola falcinellus.* (See *Calidris falcinellus*) 127
*Limnodromus griseus* 116
    *semipalmatus* 116
*Limosa haemastica* 116
    *lapponica* 117
    *limosa* 117
*Lonchura castaneothorax* 359
    *flaviprymna* 360
    *punctulata* 360
*Lophochroa leadbeateri* 189
    *leadbeateri* 189
    *mollis* 189
*Lophoictinia isura* 86
*Lopholaimus antarcticus* 174
*Lugensa brevirostris* 39

## M

*Machaerirhynchus*
    *flaviventer* 314
        *flaviventer* 314
        *secundus* 314
*Macronectes giganteus* 36
    *halli* 36
*Macropygia phasianella* 173
    *quinkan* 173
*Malacorhynchus membranaceus* 19
Maluridae 238
Malurinae 238
*Malurus amabilis* 243
    *assimilis* 242
        *assimilis* 242
        *bernieri* 242
        *dulcis* 242
        *rogersi* 242
    *coronatus* 240
        *coronatus* 240
        *macgillivrayi* 240
    *cyaneus* 241
        *ashbyi* 241
        *cyaneus* 241
        *cyanochlamys* 241
        *elizabethae* 241
        *leggei* 241
        *samueli* 241
    *elegans* 243
    *lamberti* 242
    *leucopterus* 240
        *edouardi* 240
        *leuconotus* 240
        *leucopterus* 240
    *melanocephalus* 239
        *cruentatus* 239
        *melanocephalus* 239
    *pulcherrimus* 243
    *splendens* 241
        *callainus* 241
        *emmottorum* 241
        *melanotus* 241
        *splendens* 241
*Manorina flavigula* 277
    *flavigula* 277
    *lutea* 277
    *melvillensis* 277
    *obscura* 277
    *wayensis* 277
    *melanocephala* 277
        *leachi* 277
        *lepidota* 277

        *melanocephala* 277
    *melanophrys* 276
    *melanotis* 277
*Megalurus gramineus.* See *Poodytes gramineus* 351
    *timoriensis.* (See *Cincloramphus timoriensis*) 352
Megapodiidae 10
*Megapodius reinwardt* 12
    *carbonarius* 12
    *castanotus* 12
    *tumulus* 12
    *yorki* 12
*Melanodryas cucullata* 298
    *cucullata* 298
    *melvillensis* 298
    *picata* 298
    *westralensis* 298
    *vittata* 298
        *kingii* 298
        *vittata* 298
*Meleagris gallapavo* 12
*Meliphaga albilineata* 279
    *albilineata* 279
    *fordiana* 279
    *gracilis* 278
        *gracilis* 278
        *imitatrix* 278
    *lewinii* 278
        *amphochlora* 278
        *lewinii* 278
        *mab* 278
    *notata* 278
        *mixta* 278
        *notata* 278
Meliphagidae 270
*Melithreptus affinis* 289
    *albogularis* 289
        *albogularis* 289
        *inopinatus* 289
    *brevirostris* 288
        *brevirostris* 288
        *leucogenys* 288
        *magnirostris* 288
        *pallidiceps* 288
        *wombeyi* 288
    *gularis* 289
        *gularis* 289
        *laetior* 289
    *lunatus* 288
        *chloropsis* 288
        *lunatus* 288
        *validirostris* 288
*Melloria quoyi* 330
    *jardini* 330
    *rufescens* 330
    *spaldingi* 330
*Melopsittacus undulatus* 205
*Menura alberti* 233
    *novaehollandiae* 233
        *edwardi* 233
        *novaehollandiae* 233
        *victoriae* 233
Menuridae 232
Meropidae 222
*Merops ornatus* 231
*Microcarbo melanoleucos* 69
*Microeca fascinans* 301
    *assimilis* 301
    *fascinans* 301
    *pallida* 301
    *flavigaster* 301

    *flavigaster* 301
    *flavissima* 301
    *laetissima* 301
    *tormenti* 301
    *griseoceps* (See *Kempiella griseoceps*) 300
*Micropalama himantopus* (See *Calidris himantopus*) 133
*Milvus migrans* 85
*Mirafra javanica* 355
    *athertonensis* 355
    *forrestii* 355
    *halli* 355
    *horsfieldii* 355
    *melvillensis* 355
    *rufescens* 355
    *secunda* 355
    *sodergergi* 355
    *woodwardi* 355
*Monarcha frater* 316
    *melanopsis* 316
        *melanopsis* 316
*Monticola solitarius* 344
    *philippensis* 344
    *solitarius* 344
*Morus capensis* 62
    *serrator* 62
*Motacilla alba* 356
    *baicalensis* 356
    *leucopsis* 356
    *lugens* (See *M. lugens*) 356
    *ocularis* 356
    *cinerea* 357
    *citreola* 357
    *tschutschensis* 357
        *lutea* 357
        *macronyx* 357
        *simillima* 357
        *taivana* 357
        *tschutschensis* 357
    *lugens* 356
Motacillidae 350
Muscicapidae 342
*Myiagra alecto* 319
    *melvillensis* 319
    *tormenti* 319
    *wardelli* 319
    *cyanoleuca* 318
    *inquieta* 319
        *inquieta* 319
        *nana* 319
    *rubecula* 318
        *concinna* 318
        *okyri* 318
        *rubecula* 318
        *yorki* 318
    *ruficollis* 318
*Myzomela erythrocephala* 291
    *erythrocephala* 291
    *infuscata* 291
    *obscura* 291
        *harterti* 291
        *obscura* 291
    *sanguinolenta* 291

## N

*Nectarinia jugularis.* (See *Cinnyris jugularis*) 292, 349
Nectariniidae 342
*Neochmia modesta* (See *Aidemosyne modesta*) 360
    *phaeton* 362
        *evangelinae* 362
        *phaeton* 362

## INDEX OF SCIENTIFIC NAMES (CONT.)

subniger 96
Falconidae 82
*Falcunculus frontatus* 311
    *leucogaster* 311
    *whitei* 311
*Ficedula narcissina* 344
*Fregata andrewsi* 66
    *ariel* 66
    *minor* 67
        *minor* 67
        *palmerstoni* 67
*Fregetta grallaria* 57
    *tropica* 57
        *melanoleuca* 57
Fringillidae 350
*Fulica atra* 107
    *australis* 107
*Fulmaris glacialoides* 37

### G

*Gallinago hardwickii* 114
    *megala* 115
    *stenura* 114
*Gallinula tenebrosa* 106
*Gallirallus philippensis* (See
*Hypotaenidia philippensis*) 103
*Gallus gallus* 12
*Garrodia nereis* 56
*Gavicalis fasciogularis* 280
    *versicolor* 280
    *virescens* 280
        *cooperi* 280
        *forresti* 280
        *sonorus* 280
        *virescens* 280
*Geoffroyus geoffroyi* 192
*Gelochelidon macrotarsa* 161
*Geopelia cuneata* 175
    *humeralis* 176
        *headlandi* 176
    *striata* 175
        *placida* 175
        *clelandi* 175
*Geophaps plumifera* 176
    *ferruginea* 176
    *leucogaster* 176
    *plumifera* 176
    *scripta* 178
        *peninsulae* 178
        *scripta* 178
    *smithii* 178
        *blaauwi* 178
        *smithii* 178
*Gerygone chloronota* 262
    *chloronotus* 262
    *darwini* 262
    *fusca* 262
        *exsul* 262
        *fusca* 262
        *mungi* 262
    *levigaster* 261
        *cantator* 261
        *levigaster* 261
    *magnirostris* 261
        *cairnsensis* 261
        *magnirostris* 261
    *mouki* 262
        *amalia* 262
        *mouki* 262
        *richmondi* 262
    *olivacea* 263
        *cinerascens* 263
        *olivacea* 263
        *rogersi* 263
    *palpebrosa* 263
        *flavida* 263
        *palpebrosa* 263
        *personata* 263
    *tenebrosa* 261
        *tenebrosa* 261
        *christophori* 261
*Glareola maldivarum* 137
Glareolidae 136
*Glossopsitta concinna* 191
    *porphyrocephala.* (See
*Parvipsitta porphyrocephala*) 190
    *pusilla* (See *Parvipsitta
pusilla*) 190
*Glycichaera fallax* 284
*Glyciphila melanops* 285
    *melanops* 285
    *chelidonia* 285
*Grallina cyanoleuca* 317
    *cyanoleuca* 317
    *neglecta* 317
*Grantiella picta* 287
Gruidae 70
*Grus* (See *Antigone*) 80
*Gygis alba* 166
*Gymnorhina tibicen* 331
    *dorsalis* 331
    *eylandtensis* 331
    *hypoleuca* 331
    *longirostris* 331
    *telonocua* 331
    *terraereginae* 331
    *tibicen* 331
    *tyrannica* 331

### H

Haematopodidae 136
*Haematopus finshchi* 149
    *fuliginosus* 149
        *fuliginosus* 149
        *ophthalmicus* 149
    *longirostris* 149
Halcyonidae 222
*Haliaeetus leucogaster* 92
*Haliastur indus* 88
    *sphenurus* 85
*Halobaena caerulea* 46
*Hamirostra melanosternon* 93
*Heteromunia pectoralis* 359
*Heteromyias cinereifrons* 295
*Hieraaetus morphnoides* 89
*Himantopus leucocephalus* 151
    *leucocephalus* 151
*Hirundapus caudacutus* 225
Hirundinidae 342
*Hirundo neoxena* 346
    *carteri* 346
    *neoxena* 346
    *rustica* 346
        *gutturalis* 346
        *tahitica* 346
        *frontalis* 346
*Hydrobates leucorhoa* 58
    *matsudairae* 59
    *monorhis* 58
    *tristami* 58
Hydrobatidae 29
*Hydroprogne caspia* 161
*Hydrophasianus chirurgus* 146
*Hylacola cauta* 257
    *pyrrhopygia* 257
*Hypotaenidia philippensis* 103
    *mellori* 103
    *tounelieri* 103

### I

*Irediparra gallinacea* 147
*Ixobrychus dubius* 76
    *flavicollis* 77
    *sinensis* 76

### J

Jacanidae 136

### K

*Kempiella griseoceps* 300
    *kempi* 300

### L

*Lalage leucomela* 323
    *leucomela* 323
    *macrura* 323
    *rufiventris* 323
    *yorki* 323
    *tricolor* 323
Laridae 152
*Larus crassirostris* 157
    *dominicanus* 156
    *fuscus* 157
    *pacificus* 156
        *georgii* 156
        *pacificus* 156
*Lathamus discolor* 199
*Leipoa ocellata* 12
*Leucophaeus atricilla* 158
    *pipixcan* 158
*Leucosarcia melanoleuca* 174
*Lewinia pectoralis* 105
*Lichenostomus chrysops.* (See
*Caligavis chrysops*) 278
    *cratitius* 282
        *cratitius* 282
        *occidentalis* 282
    *fasciogularis* (See *Gavicalis
fasciogularis*) 280
    *flavescens* (See *Ptilotula
flavescens*) 283
    *flavicollis* (See *Nesoptilotis
flavicollis*) 281
    *flavus* (See *Stomiopera flava*)
280
    *frenatus* (See *Bolemoreus
frenatus*) 279
    *fuscus* (See *Ptilotula fuscus*)
283
    *hindwoodi* (See *Bolemoreus
hindwoodi*) 279
    *keartlandi* (See *Ptilotula
keartlandi*) 282
    *leucotis.* (See *Nesoptilotis
leucotis*) 281
    *melanops* 281
        *cassidix* 281
        *melanops* 281
        *meltoni* 281
    *ornatus* (See *Ptilotula
ornata*) 282
    *penicillatus* 283
        *calconi* 283
        *carteri* 283
        *leilavalensis* 283
        *penicillatus* 283
    *plumulus.* (See *Ptilotula
plumula* 282)
    *unicolor* (See *Stomiopera
unicolor* 279)
    *versicolor.* (See *Gavicalis
versicolor* 280)
    *virescens* (See *Gavicalis
virescens* 280)

## INDEX OF SCIENTIFIC NAMES (CONT.)

*murchisonii* 313
*pallescens* 313
*rufiventris* 313
*strigata* 313
*superciliosus* 313
*whitei* 313
*megarhyncha* 312
*parvula* 325
*rufogaster* 312
*woodwardi* 312
*Columba leucomela* 171
*livia* 174
Columbidae 170
*Conopophila albogularis* 285
*rufogularis* 285
*whitei* 284
Coraciidae 222
*Coracina lineata* 325
*lineata* 325
*maxima* 325
*novaehollandiae* 324
*melanops* 324
*novaehollandiae* 324
*subpallida* 324
*papuensis* 324
*apsleyi* 324
*artamoides* 324
*hypoleuca* 324
*oriomo* 324
*robusta* 324
*tenuirostris* (See *Edolisoma tenuirostre*) 323
Corcoracidae 334
*Corcorax*
*melanorhamphos* 337
*melanorhamphos* 337
*whiteae* 337
*Cormobates leucophaea* 236
*grisescens* 236
*intermedius* 236
*leucophaeus* 236
*metastasis* 236
*minor* 236
Corvidae 334
*Corvus bennetti* 336
*coronoides* 336
*coronoides* 336
*perplexus* 336
*mellori* 336
*orru* 337
*cecilae* 337
*orru* 337
*splendens* 336
*tasmanicus* 337
*boreus* 337
*tasmanicus* 337
*Coturnix colletti* 13
*victoriae* 13
*pectoralis* 13
*ypsilophora* 13
*australis* 13
*ypsilophora* 13
*Cracticus argenteus* 330
*mentalis* 330
*nigrogularis* 331
*nigrogularis* 331
*picatus* 331
*quoyi* (See *Melloria quoyi*) 330
*torquatus* 330
*cinereus* 330
*colletti* 330
*leucopterus* 330
*torquatus* 330
*Crex crex* 106
*Cuculus optatus* 207
Cuculidae 206
*Cyanoptila cyanomelana* 344
*Cyclopsitta coxeni* 191
*diophthalma* 191
*macleayana* 191
*marshalli* 191
*Cygnus atratus* 15
*olor* 15

### D

*Dacelo leachii* 230
*cervina* 230
*kempi* 230
*leachii* 230
*occidentalis* 230
*novaeguineae* 230
*minor* 230
*novaeguineae* 230
*Daphoenositta chrysoptera* 308
*chrysoptera* 308
*leocochepala* 308
*leucoptera* 308
*pileata* 308
*striata* 308
*Daption capense* 37
*australe* 37
*capense* 37
*Dasyornis brachypterus* 254
*monoides* 254
*broadbenti* 254
*broadbenti* 254
*caryochrous* 254
*littoralis* 254
*longirostris* 254
Dasyornithinae 250
*Dendrocygna arcuata* 16
*eytoni* 16
*guttata* 16
Dicaeidae 342
*Dicaeum*
*hirundinaceum* 292, 349
Dicruridae 314
*Dicrurus bracteatus* 317
*atrabectus* 317
*baileyi* 317
*bracteatus* 317
*carbonarius* 317
*Diomedea amsterdamensis* 30
*epomophora* 31
*epomophora* 31
*sanfordi* 31
*exulans* 30
*chionoptera* 30
*exulans* 30
*gibsoni* 30
Diomedeidae 28
*Dromaius novaehollandiae* 11
*novaehollandiae* 11
*rothschildi* 11
*Drymodes brunneopygia* 295
*superciliaris* 295
*colcloughi* 295
*superciliaris* 295
*Ducula spilorrhoa* 171
*concinna* 172
*mullerii* 170

### E

*Eclectus roratus* 192
*macgillivrayi* 192
*polychloros* 192
*Edolisoma tenuirostre* 323

*melvillensis* 323
*tenuirostre* 323
*Egretta garzetta* 74
*novaehollandiae* 73
*picata* 71
*sacra* 73
*Elanus axillaris* 84
*scriptus* 84
*Elseyornis melanops* 141
*Emblema pictum* 363
*Entomyzon cyanotis* 275
*albipennis* 275
*cyanotis* 275
*griseigularis* 275
*Eolophus roseicapilla* 188
*albiceps* 188
*kuhli* 188
*roseicapilla* 188
*Eopsaltria australis* 299
*australis* 299
*chrysorrhoa* 299
*georgiana.* (See *Quoyornis georgiana*) 299
*griseogularis* 299
*griseogularis* 299
*rosinae* 299
*Ephippiorhynchus asiaticus* 81
*Epthianura albifrons* 292
*aurifrons* 293
*crocea* 293
*crocea* 293
*macgregori* 293
*tunneyi* 293
*tricolor* 292
*Eremiornis carteri.* (See *Poodytes carteri*) 352
*Erythrogonys cinctus* 140
*Erythropitta macklotii* 234
*Erythrotriorchis radiatus* 91
*Erythrura gouldiae* (See *Chloebia gouldiae*) 365
*Esacus giganteus* 150
Estrildidae 358
*Eudynamys orientalis* 212
*cyanocephala* 212
*subcyanocephala* 212
*taitensis* 212
*Eudyptes chrysocome* 27
*chrysocome* 27
*filholi* 27
*moseleyi* 27
*chrysolophus* 26
*chrysolophus* 26
*schlegeli* 26
*pachyrhynchus* 27
*robustus* 27
*sclateri* 27
*Eudyptula minor* 25
*Eulabeornis*
*castaneoventris* 107
*Eurostopodus argus* 221
*mystacalis* 220
*Eurystomus orientalis* 231
*Excalfactoria chinensis* 13

### F

*Falco berigora* 97
*cenchroides* 94
*hypoleucos* 96
*longipennis* 94
*longipennis* 94
*murchisonianus* 94
*peregrinus* 95
*subbuteo* 95

## INDEX OF SCIENTIFIC NAMES

*hindwoodi* 279
*Botaurus poiciloptilus* 77
*Bubulcus coromandus* 73
*Bulweria bulwerii* 43
    *fallax* 43
Burhinidae 136
*Burhinus grallarius* 150
*Butorides striatus* 76
    *macrorhynchus* 76
    *stagnatilis* 76

### C

*Cacatua galerita* 188
    *fitzroyi* 188
    *galerita* 188
    *pastinator* 187
        *derbyi* 187
        *pastinator* 187
    *sanguinea* 187
        *gymnopsis* 187
        *normantoni* 187
        *sanguinea* 187
        *westralensis* 187
    *tenuirostris* 187
Cacatuidae 180
*Cacomantis castaneiventris* 208
    *castaneiventris* 208
    *flabelliformis* 209
    *pallidus* 207
    *variolosus* 208
        *dumetorum* 208
        *variolosus* 208
*Calamanthus campestris* 256
    *campestris* 256
    *hartogi* 256
    *isabellinus* 256
    *montanellus* 256
    *rubiginosus* 256
    *wayensis* 256
    *winiam* 256
    *cauta* (See *Hylacola cauta*) 257
        *halmaturina* 257
        *macrorhyncha* 257
        *whitlocki* 257
    *fuliginosus* 256
        *albiloris* 256
        *bourneorum* 256
        *diemenensis* 256
        *fuliginosus* 256
    *pyrrhopygia* (See *Hylacola pyrrhopygia*) 257
        *parkeri* 257
        *pedleri* 257
        *pyrrhopygius* 257
    *montanellus* 256
*Calidris acuminata* 132
    *alba* 131
    *alpina* 126
        *pacifica* 126
        *sakhalina* 126
    *bairdii* 131
    *canutus* 126
    *falcinellus* 127
    *ferruginea* 127
    *fuscicollis* 130
    *himantopus* 133
    *mauri* 130
    *melanotos* 133
    *minuta* 128
    *paramelanotos* 132
    *pugnax* 135
    *ruficollis* 128
    *subminuta* 129
    *subruficollis* 135

    *temminckii* 129
    *tenuirostris* 126
*Caligavis chrysops* 278
*Callipepla californica* 13
*Callocephalon fimbriatum* 186
*Calonectris leucomelas* 53
*Calyptorhynchus banksii* 183
    *banksii* 183
    *graptogyne* 183
    *macrorhynchus* 183
    *naso* 183
    *samueli* 183
    *baudinii*. (See *Zanda baudinii*) 185
    *funereus* (See *Zanda funerea*) 184
        *xanthanotus* 184
        *whiteae* 184
    *lathami* 184
        *erebus* 184
        *halmaturinus* 184
        *lathami* 184
    *latirostris* (See *Zanda latirostris*) 185
Campephagidae 322
Caprimulgidae 206
*Caprimulgus macrurus* 220
*Carduelis carduelis* 354
Casuariidae 10
*Casuarius casuarius* 11
*Carterornis leucotis* 315
*Cecropis daurica* 347
    *daurica japonica* 347
    *striolata* 347
Centropodidae 206
*Centropus phasianinus* 213
    *highami* 213
    *melanurus* 213
    *phasianinus* 213
*Cereopsis novaehollandiae* 15
    *grisea* 15
    *novaehollandiae* 15
*Certhionyx variegatus* 291
*Ceyx azureus* 226
    *azurea* 226
    *ruficollaris* 226
    *diemenensis* 226
    *pusillus* 226
        *halli* 226
        *ramsayi* 226
*Chalcites* (See *Chrysococcyx*)
*Chalcophaps longirostris* 173
    *chrysochlora* 173
Charadriidae 136
*Charadrius alexandrinus* 142
    *dealbatus* 142
    *asiaticus* 144
    *australis* (See *Peltohyas australis*) 145
    *bicinctus* 143
    *dubius* 141
    *hiaticula* 141
    *leschenaultii* 144
        *stegmanni* 144
    *mongolus* 143
        *atrifrons* 143
        *mongolus* 143
    *ruficapillus* 142
    *veredus* 145
*Chenonetta jubata* 16
*Cheramoeca leucosterna* 346
*Chlamydera cerviniventris* 341
    *guttata* 340

    *carteri* 340
    *guttata* 340
    *maculata* 340
    *nuchalis* 341
        *nuchalis* 341
        *orientalis* 341
*Chlidonias hybridus* 166
    *leucopterus* 167
    *niger* 167
*Chloebia gouldiae* 365
*Chloris chloris* 354
*Chroicocephalus novaehollandiae* 157
    *ridibundus* 159
*Chrysococcyx basalis* 210
    *lucidus* 211
    *minutillus* 210
        *barnardi* 210
        *minutillus* 210
    *osculans* 209
    *russatus* 211
Ciconiidae 70
*Cincloramphus cruralis* 353
    *mathewsi* 352
    *timoriensis* 352
*Cinclosoma alisteri* 307
    *castaneothorax* 307
        *castaneothorax* 307
        *marginatum* 307
    *castanotum* 306
        *clarum* 306
    *cinnamomeum* 307
    *punctatum* 306
        *anachoreta* 306
        *dovei* 306
        *punctatum* 306
Cinclosomatidae 304
*Cinnyris jugularis* 292, 349
*Circus approximans* 86
    *assimilis* 87
    *spilonotus* 87
*Cissomela pectoralis* 290
*Cisticola exilis* 353
    *alexandrae* 353
    *diminuta* 353
    *exilis* 353
    *lineocapilla* 353
    *juncidis* 353
        *laveryi* 353
        *leanyeri* 353
        *normani* 353
*Cladorhynchus leucocephalus* 151
Climacteridae 232
*Climacteris affinis* 236
    *affinis* 236
    *superciliosa* 236
    *erythrops* 236
    *melanurus* 237
        *melanurus* 237
        *wellsi* 237
    *picumnus* 237
        *melanota* 237
        *picumnus* 237
    *rufus* 237
*Collocalia esculenta* 223
    *esculens* 223
*Colluricincla boweri* 312
    *harmonica* 313
        *brunnea* 313
        *halmaturina* 313
        *harmonica* 313
        *kolichisi* 313

# INDEX OF SCIENTIFIC NAMES

## A

Acanthagenys rufogularis 273
Acanthiza apicalis 267
    albiventris 267
    apicalis 267
    cinerascens 267
    whitlocki 267
  chrysorrhoa 265
  ewingii 268
    ewingii 268
    rufifrons 268
  inornata 264
  iredalei 265
    hedleyi 265
    iredalei 265
    rosinae 265
  katherina 268
  lineata 264
    alberti 264
    clelandi 264
    lineata 264
    whitei 264
  nana 264
    flava 264
    modesta 264
    nana 264
  pusilla 266
    archibaldi 266
    dawsonensis 266
    diemenensis 266
    pusilla 266
    zietzi 266
  reguloides 265
    australis 265
    nesa 265
    reguloides 265
    squamata 265
  robustirostris 267
  uropygialis 266
Acanthizinae 250
Acanthorhynchus superciliosus 290
  tenuirostris 290
    cairnsensis 290
    dubius 290
    halmaturinus 290
    tenuirostris 290
Acanthornis magnus 255
  greenianus 255
  magnus 255
Accipiter cirrocephalus 90
  fasciatus 91
    didimus 91
    fasciatus 91
  novaehollandiae 90
Accipitridae 82
Acridotheres tristis 344
Acrocephalus australis 351
  australis 351
  gouldi 351
  orientalus 351
Actitis hypoleucos 124
Aegotheles cristatus 221
  cristatus 221
  tasmanicus 221
Aegothelidae 206
Aerodramus terraereginae 223
  chillagoensis 223
  terraereginae 223
  vanikorensis 224
Aidemosyne modesta 360
Ailuroedus crassirostris 338
  maculosus 338
    joanae 338

Alauda arvensis 354
Alaudidae 350
Alcedinidae 222
Alectura lathami 12
  lathami 12
  purpureicollis 12
Alisterus scapularis 193
  alisterus 193
  minor 193
  scapularis 193
Amaurornis moluccana 106
  ruficrissa 106
Amblyornis newtonianus. See
  Prionodura newtoniana 339
Amytornis ballarae 248
  barbatus 246
    barbatus 246
    diamantina 246
  dorotheae 246
  goyderi 249
  housei 245
  merrotsyi 247
  purnelli 248
  striatus 247
    rowleyi 247
    striatus 247
    whitei 247
  modestus 249
    modestus 249
    myall 249
    textilis 249
  woodwardi 245
Amytornithinae 238
Anas acuta 21
  castanea 20
  clypeata 19
  gracilis 21
  platyrhynchos 19
  querquedula 21
  rhynchotis 19
  superciliosa 21
Anatidae 14
Anhinga novaehollandiae 67
Anous minutus 169
  stolidus 169
  tenuirostris 169
Anseranas semipalmata 15
Anseranatidae 14
Antigone antigone 80
  rubicunda 80
Anthochaera carunculata 272
  carunculata 272
  clelandi 272
  woodwardi 272
  chrysoptera 272
    chrysoptera 272
    halmaturina 272
    tasmanica 272
  lunulata 273
  paradoxa 272
    kingi 272
    paradoxa 272
  phrygia 275
Anthus australis 355
  bilbali 355
  bistriatus 355
  rogersi 355
  cervinus 355
Aphelocephala leucopsis 268
  castaneiventris 268
  leucopsis 268
  whitei 268
  nigricincta 269

  pectoralis 269
Aplonis metallica 345
Apodidae 222
Aprosmictus erythropterus 193
  coccineopterus 193
  erythropterus 193
Aptenodytes patagonicus 26
Apus nipalensis 224
  pacificus 225
Aquila audax 93
  audax 93
  fleayi 93
  gurneyi 92
Ardea alba 74
  cinerea 72
  ibis. See Bubulcus
  coromandus 73
  intermedia 75
  pacifica 71
  sumatrana 72
Ardeidae 70
Ardenna bulleri 50
  carneipes 51
  creatopus 52
  gravis 53
  grisea 51
  pacifica 50
  tenuirostris 52
Ardeotis australis 109
Arenaria interpres 134
Arses kaupi 315
  kaupi 315
  terraereginae 315
  lorealis 315
Artamidae 322
Artamus cinereus 328
  cinereus 328
  dealbatus 328
  melanops 328
  normani 328
  cyanopterus 328
    cyanopterus 328
    perthi 328
  leucorynchus 329
  minor 328
    derbyi 328
    minor 328
  personatus 329
  superciliosus 329
Ashbyia lovensis 293
Atrichornis clamosus 235
  rufescens 235
    ferrieri 235
    rufescens 235
Atrichornithidae 232
Amaurornis cinerea 104
Aviceda subcristata 83
Aythya australis 18

## B

Barnardius zonarius 198
  barnardi 198
  macgillivrayi 198
  parkeri 198
  semitorquatus 198
  zonarius 198
Bartramia longicauda 134
Bathilda ruficauda 363
  clarescens 363
  ruficauda 363
  subclarescens 363
Biziura lobata 17
Bolemoreus frenatus 279

## GLOSSARY

See also the Illustrated Glossary on page 9.

**Accidental:** Straying beyond usual range or migratory path

**Adult:** A bird in its final plumage excluding seasonal changes

**Breeding plumage:** Acquired for duration of courting and nesting

**Cheek:** Area of plumage or skin under the eye from gape of bill back to ear-coverts

**Chevrons:** Broad V shaped markings on plumage; arrowheads

**Cline:** A gradual or graded series of small changes in plumage or some other characteristic across a geographical area

**Colonial:** Birds roosting together; nests close together, from touching to just a few metres apart

**Cosmopolitan:** Occurring through most of the world

**Diagnostic:** Feature so distinct that it is a conclusive diagnosis of identity

**Endemic:** Found only in a locality, region or continent

**Eye-ring:** A row of tiny feathers or area of skin forming a distinct ring encircling the eye

**Facial disc:** The pale facial feathers, heart-shaped or rounded, defined by a dark rim, that is typical of many owls and, less obviously, harriers

**Frontal shield:** A fleshy or horny rectangular or oval patch lacking feathers on the forehead area

**Gape:** Corner of mouth, a fleshy rim, often yellow, and most conspicuous on young birds

**Genus:** A group of closely related species, the genus name being the first of the two names given each species, and capitalised

**Gregarious:** In groups or flocks; suggests some degree of social interaction among members

**Hackles:** Longer, pointed, stiff looking feathers of neck or throat

**Immature:** Yet to reach maturity; between juvenile and mature

**Jiz:** Overall impression or 'look' of a bird, a combination of appearance and characteristic way of moving

**Juvenile:** Young bird; usually describes the first plumage after naked nestling or natal down stage; often acquired while still in the nest

**Lamellae:** Comb-like fringe to inner edge of bill; used to sieve fine food particles from water

**Lobes:** Wide flat fringes along sides of toes of water birds; an alternative to webbed toes in making the feet more effective as paddles

**Lores:** Small area of plumage or skin between base of bill and eye

**Nomadic:** Bird species that undertake wandering travels of irregular pattern in timing, direction or distance

**Nuchal:** Area of the neck

**Orbital ring:** Circle of tiny feathers or bare skin around the eye

**Order:** Grouping of families; subdivision of class

**Passerine:** Belonging to by far the largest bird group, the Order Passeriformes, song or perching birds, characterised by perching with three toes forward, one back

**Phase:** Of plumage – a difference between individuals that is not related to race or age; sometimes confined to one sex

**Primaries:** Outermost very large flight feathers of the wing; attached to the 'hand' part of the wing

**Vagrant:** Found in a location out of the usual range; usually of rare or erratic occurrence in that locality

**Washed:** A watercolour painting term, suggesting a weak tint of transparent colour, usually white, washed over to give a delicate tint

**Wattles:** Fleshy, usually colourful, growths of crown, face or neck

## BIBLIOGRAPHY

Barrett, G., et al. 2003. *The New Atlas of Australian Birds.* Birds Australia (RAOU): Hawthorn East.

Beruldsen, G. 1995. *Raptor Identification.* G. Beruldsen: Kenmore Hills.

Blakers, M., et al. 1984. *The Atlas of Australian Birds.* Melbourne University Press: Carlton.

Christidies, L. & Boles, W.E. 1994. *The Taxonomy and Species of Birds of Australia and its Territories.* RAOU Monograph 2. RAOU: Hawthorn East.

Coates, B.J. 1985–90. *The Birds of Papua New Guinea. Non-Passerines,* Vol. I. *Passerines,* Vol. II Dove Publications: Alderley.

Coates, B.J. & Bishop, K.D. 1997. *A Guide to the Birds of Wallacia.* Dove Publications: Alderley.

Davies, J., Menkhorst, P., & Rogers, D. et al. 2022, *The Compact Australian Bird Guide,* CSIRO Publishing, Melbourne.

Forshaw, J. 1981. *Australian Parrots.* 2nd ed. Lansdowne Editions: Melbourne.

Frith, H.J. 1982. *Pigeons and Doves of Australia.* Rigby: Adelaide.

Fry, C.H., Fry, K. & Harris, A. 1992. *Kingfishers, Bee-eaters and Rollers.* Christopher Helm, A & C Black: London.

Harrison, P. 1985. *Seabirds: An Identification Guide.* Revised ed. Croome Helm: Beckenham.

Hayman, P., Marshal, J. & Prater, T. 1986. *Shorebirds: An Identification Guide to the Waders of the World.* Croome Helm: Beckenham.

Hollands, D. 1984. *Eagles, Hawks and Falcons of Australia.* Nelson: Melbourne.

Hollands, D. 1991. *Birds of the Night: Owls, Frogmouths and Nightjars of Australia.* Reed: Sydney.

Hollands, D. 1999. *Kingfishers and Kookaburras: Jewels of the Australian Bush.* New Holland: Frenchs Forest.

Johnstone, R.E. 1990. 'Mangroves and mangrove birds of Western Australia'. *Rec. West. Aust. Mus.* Suppl. 32: 1–120.

Johnstone, R.E. & Smith, L.A. 1981. *Birds of Mitchell Plateau and Adjacent Coasts etc., Kimberley, W.A.* WA Museum: Perth.

Johnstone, R.E. & Storr, G.M. 1998 & 2004. *Handbook of Western Australian Birds.* Vols. I & II. Western Australian Museum: Perth.

King, B., Woodcock, M. & Dickinson, E.C. 1975. *A Field Guide to the Birds of South East Asia.* Collins: London.

MacKinnon, J. & Phillipps, K. 1993. *A Field Guide to the Birds of Borneo, Sumatra, Java and Bali.* OUP: Oxford.

Marchant, S. & Higgins, P.J. (Eds.) *Handbook of Australian, New Zealand and Antarctic Birds.* Vols. 1–6. OUP: Melbourne.

Morcombe, M. 1986. *The Great Australian Birdfinder.* Lansdowne Press: Sydney.

*National Photographic Index of Australian Wildlife: Birds.* Vols. 1–10. 1982–1994. Angus & Robertson: Sydney.

North, A.J. 1901–14. *Nests and Eggs of Birds Found Breeding in Australia and Tasmania.* Australian Museum: Sydney.

Parish, Steve 2003. *Photograph Australia with Steve Parish.* Steve Parish Publishing: Brisbane.

Pizzey, G. 1997. *Field Guide to the Birds of Australia.* Angus & Robertson: Sydney.

Reville, B.J. 1993. *A Visitor's Guide to the Birds of Christmas Island, Indian Ocean.* 2nd ed. Christmas Island Natural History Association: Christmas Island.

Schodde, R. 1982. *The Fairy-wrens: A Monograph of the Maluridae.* Lansdowne: Melbourne.

Schodde, R. & Mason, I.J. 1999. *The Directory of Australian Birds: Passerines.* CSIRO Publishing: Collingwood.

Serventy, D.L. & Warham, J. 1971. *The Handbook of Australian Seabirds.* AH & AW Reed: Sydney.

Serventy, D.L. 1977. *Distribution of Birds on the Australian Mainland.* Prepared by J.R. Busby & S.J.J.F. Davies. CSIRO.

Serventy, D.L. & Whittell, H.M. 1976. *Birds of Western Australia.* 5th ed. UWA Press: Perth.

Slater, P., Slater, P. & Slater, R. 1986. *The Slater Field Guide to Australian Birds.* Rigby: Sydney.

Storr, G.M. & Johnstone, R.E. 1985. *A Field Guide to the Birds of Western Australia.* 2nd ed. Western Australian Museum: Perth.

## BIRDLIFE AUSTRALIA

BirdLife Australia, formerly Birds Australia, was originally established in 1901 as the Royal Australasian Ornithologists Union. In 2011, Birds Australia merged with Bird Observation and Conservation Australia (BOCA) to form BirdLife Australia—a not-for-profit, member-based organisation that aims to conserve Australian birds and their habitat.

Providing information on Australian birds; access to activities, conservation reserves and observatories; discounted equipment and subscription to the quarterly magazine *Australian Birdlife* and peer-reviewed *Australian Field Ornithology*, BirdLife Australia harnesses the efforts of more than 330,000 members and volunteers Australia-wide to undertake research and citizen science field activities that help conserve and grow native bird populations. As there is great strength in numbers, membership of BirdLife Australia is one of the best ways to help save our unique avifauna. The company and assistance of more experienced observers is also one of the best ways to learn how to recognise some of Australia's more cryptic bird species.

BirdLife Australia's National office is at 60 Leicester St, Carlton, Victoria. There are branches in most States and several special interest groups to join. For more information or to join, go to the BirdLife Australia website *www.birdlife.org.au*

BIRDLIFE AUSTRALIA OBSERVATORIES & CENTRES: BirdLife Australia Discovery Centre, Broome Bird Observatory, Clarkesdale Bird Sanctuary, Eyre Bird Observatory, Gluepot Reserve.

## AUSTRALIAN WILDLIFE CONSERVANCY

THE AUSTRALIAN WILDLIFE CONSERVANCY (AWC) is an independent, not-for-profit organisation dedicated to the conservation of Australia's threatened wildlife and ecosystems. AWC acquires high conservation value properties and implements practical, on-ground programs to protect and restore wildlife populations.

Funded by donations – and partnering with Indigenous groups, governments and landholders – AWC now owns, manages or works in partnership across more than 12.9 million hectares of wildlife sanctuaries, many in remote regions, such as the Kimberley, Cape York, central Australia and the Top End. AWC operates at a 'landscape scale', protecting wildlife on the sanctuaries it runs while working with other landholders, especially of neighbouring properties, to help protect iconic wildlife and to manage the largest tracts of feral cat and fox-free land on Australia's mainland.

With staff based at sanctuaries around the country, more than 85% of AWC's operational expenditure is incurred on conservation, making a huge contribution to conserving Australia's native species in the field. AWC's programs include feral animal control, weed control, fire management and the translocation of threatened species, all based on the best available science. Visitor programs and facilities are available at several sanctuaries. For more information on AWC or how you can help, please visit the website *www.australianwildlife.org*.

## AUTHOR'S NOTE: OBSERVATION & PHOTOGRAPHY

At times, well-meaning people express concern that photography, recording or even close observation of birds might cause losses to nestlings, even when precautions have been taken to prevent harm. The impact these or similar activities have on bird populations is likely insignificant compared with the far greater, often permanent, losses caused by habitat destruction (especially of old trees with nesting hollows) and predators, both native and introduced. Millions of feral cats or foxes prey on wildlife in our bush and suburbs annually, accounting for a far greater loss of birds and small mammals than can be attributed to Australian birdwatchers and photographers combined. (An example of such a loss would be a bird deserting its nest because of disturbance around the site.) The predation of Goshawks, Sparrowhawks or Square-tailed Kites, which feed on small birds' nestlings, would similarly exceed such losses.

The many books, documentaries and articles that have built public interest, appreciation and awareness of the need to conserve bird habitat have been made possible only by careful observation, photography and videography of birds at nests, and the study of museum collections and skins. Coinciding with, and probably resulting from, such publicity, Australia has steadily increased conservation pressure, demanding that governments protect more habitat and increase funding for scientific research on endangered species. The continued growth of public awareness and resulting demands for more national parks to be set aside have slowly improved funding and budget allocations for environmental protections and sustainable ecotourism. No bird species should be put at risk; however, most populations are limited by broader constraints: lack of suitable habitat; lack of available food resources; the effects of fire; predation; and degradation or loss of habitat. Birdwatchers and nature lovers are the only army our fragile wildlife has to protect it, and all observers owe it to the birds and the environment to proceed carefully and thoughtfully when birdwatching or photographing birds, ensuring they cause no damage and help rather than hinder wildlife preservation.

RECORD OF EDITIONS

2004: Complete Compact Edition. Steve Parish Publishing. New concept, layout, including scaled identification page at head of chapter. Condensed, rearranged, revised, new maps; includes most races.

2016: 2nd Edition. Pascal Press. Genus removed from introductory pages, new maps, some genus names updated in entries and index. Cover colour changed.

2024: 3rd Edition. Woodslane Press. Updated species names and splits of subspecies to species, updated maps where required, some genus names updated in entries and index. Cover colour and font changed.

# GOULDIAN FINCH & BLUE-FACED PARROT-FINCH

**Gouldian Finch** *Chloebia gouldiae* 12–14 cm
Well-known, popular in aviculture, but endangered in the wild. Inhabits open tropical woodland with scattered trees and tall native grasses, spinifex with scattered shrubs. Never far from water, and often in vegetation along watercourses or edges of mangroves. When breeding, uses trees on low, stony ridges. Like many finches of open environs, this species is highly social; usually in flocks or small parties feeding in tall grass, rarely on the ground. Clings to strong stems of tropical grasses that may be 2 m tall, raiding seed heads. Insects, especially flying termites, are caught in flight. If flushed from feeding in grass, the birds of a flock will fly together, their bright blue rumps conspicuous, to tops of nearby trees. As contact call, a soft 'tsit' or 'ssit', sometimes a longer 'streee'. As alarm call, sharper, louder 'stret, stret-stret'. Remains moderately common in parts of the Kimberley, WA, and Arnhem Land, NT, but flocks of tens or hundreds come to waterholes, where formerly were thousands.

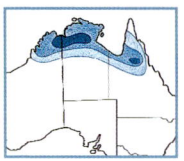

No subspecies. Facial masks of all three colours occur throughout the range of the species. About 75% are black, 24% red and 1% yellow.

**Blue-faced Parrot-Finch** *Erythrura trichroa* 11–12 cm
Habitat extends from lowlands to heights of ranges; uses grassy clearings and lush, low vegetation of rainforest margins and mangroves, as well as open grassy areas within eucalypt forest. Usually in small parties, feeding largely in rainforest trees, finding seeds in lower to mid levels. Gives chattered, slightly harsh, 'chak' notes intermixed with extremely high, thin, insect-like, 'tsiep' squeaks: 'chak-chweip, tsiep, tak-chat, tsweip'. Seasonally nomadic or migratory – on highlands through summer, descends to winter on coastal lowlands. Easily overlooked – uncommon, possibly rare.

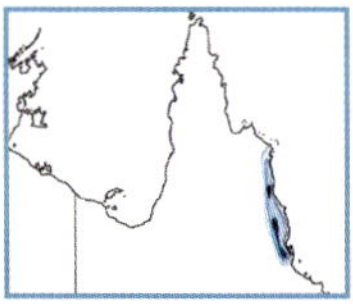

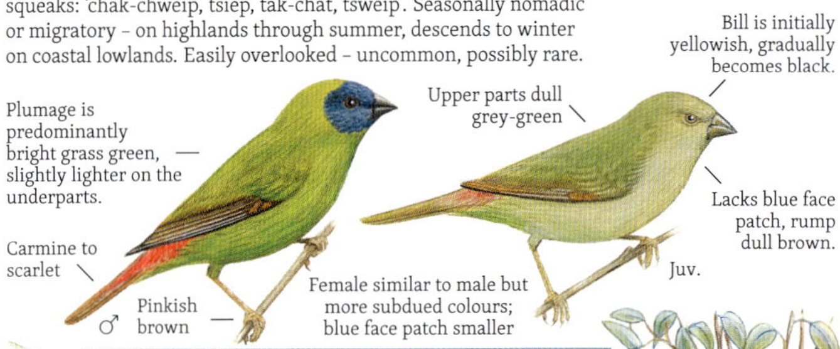

# FIRETAILS

**Diamond Firetail** *Stagonopleura guttata* 12–13 cm
Usually in small flocks, 20 to 30 birds; inhabits grassy groundcover underneath open forest; woodland, mallee, acacia scrub and timber belts along watercourses and roadsides; feeds exclusively on the ground. Tends to nest in loosely scattered colonies, after which, in autumn and through winter, larger flocks may form. Flight low with only slight undulations; large flocks travel in line lines. Usual contact call is a rather plaintive, drawn out, whistled 'tioo-whieer' or 'tioo-wheee', with the call of the female at a slightly higher pitch; male in display gives low, rasping sounds. Also soft, vibrating sounds when nest duties are exchanged. Sedentary or locally migratory; uncommon and of patchy occurrence.

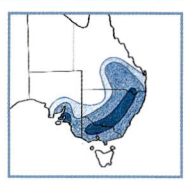

Lores slightly darker than grey of head

Named for the sharp, clear, white diamonds set in a band of black

Smoky grey

Black lores link the crimson-pink of the bill to these same colours in the eye and its encircling eye-ring.

Throat clear white

Rump and tail-coverts crimson

Flanks display neat, sharply defined 'diamond' spots of white.

♂

Adult female is similar to male; breast band is slightly narrower, lores slightly paler.

Juv.

Bill mostly black

Breast band indistinct, flanks barred dusky grey-brown

---

**Beautiful Firetail** *Stagonopleura bella* 11–13 cm
Prefers damp or swampy grassy spots in gullies, low-lying flats, woodland with dense, low undergrowth, dense thickets along creeks and rivers, coastal heaths; needs water nearby. Usually in pairs or small family parties; forages on or near the ground, often concealed beneath overhanging grass or shrubbery, finding seeds of various grasses. Also seeks seeds and insects higher in foliage of shrubs and trees. Flight is low and direct on whirring wings with only very slight undulations. The call is a single, mournful, undulating 'whee-ee-ee' in identification; soft 'chrrit' as contact; in alarm gives abrupt 'tchup-tchup' sounds. Sedentary or locally nomadic; uncommon on mainland, more common Tas.

Black of lores also encircles the eyes and links across the base of the bill.

Bill initially black, turns red early, birds in this plumage often have red on bill. Lacks black on lores and surrounds to eyes.

Upper parts dark olive brown; at close range seen to be very finely and densely barred

Rump, upper-tail-coverts and bases of darkly barred tail feathers are all glossy scarlet.

Red like adult

Imm.

Underparts closely barred olive and black; boldest across the flanks

---

**Red-eared Firetail** *Stagonopleura oculata* 12–13 cm
Habitat diverse, but always of quite heavy rainfall: heavy forest, open forest, especially thickets along creeks, winter-wet paperbark flats, dense low heath along S coast. In pairs or small family groups; secretive, low in dense undergrowth or on the ground seeking seeds of native grasses and sedges. Bulky nests may indicate presence of these birds. Although the calls are soft, they help to locate this species in its dense habitats. Usually a soft, rather mournful whistle, the shorter calls at an even, mid-level pitch, 'wheee'; longer calls slight, wavering 'wheeoo-ee'; also soft, murmured contact call. Sedentary; scarce except in extreme SW.

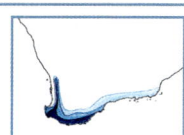

Flying up and away, displays briefly its scarlet 'fire-tail'.

Ear patch scarlet, becoming deeper on breeding males

Bill scarlet

Grey-brown with darker brown bars

Mid to deep olive-brown

Black around eyes and back to scarlet ear patch

Breast to under tail is black, heavily spotted white.

Rump, upper-tail-coverts and base of tail scarlet

Juv.: lacks scarlet ear-coverts, bill initially dark.

## STAR & PAINTED FINCHES

**Star Finch** *Bathilda ruficauda* 10–12 cm
In pairs or small flocks up to 20-odd birds; favours lush green vegetation, tall, rank grass along temporary or permanent watercourses, rushy margins of swamps, moist green crops; feeds at seed heads in low vegetation and on ground. Flight strong, swift and erratic. Most obvious and loudest call is a high, penetrating 'tseit, tseeit' given almost constantly in flight; from flock becomes a rather musical tinkling. Flocks feeding in grass keep in contact with an abrupt, softer 'tsit'. The male gives a weak twittering song. Locally common in NW in favourable seasons, but range and numbers of eastern race *ruficauda* have greatly diminished.

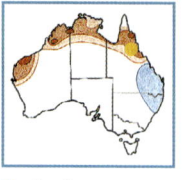

● *ruficauda*
● *subclarescens*
● *clarescens*

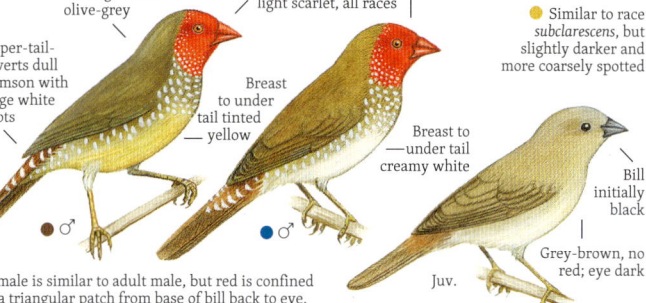

Olive-green to olive-grey

Crown to throat and bill light scarlet, all races

Upper-tail-coverts dull crimson with large white spots

Breast to under tail tinted yellow

Breast to under tail creamy white

● Similar to race *subclarescens*, but slightly darker and more coarsely spotted

Bill initially black

Grey-brown, no red; eye dark

Juv.

Female is similar to adult male, but red is confined to a triangular patch from base of bill back to eye.

**Painted Finch** *Emblema pictum* 11–12 cm
Habitat usually spinifex in rocky range country and vicinity of gorges where there are pools; also in sandplain spinifex if water is available. In flocks of typically 5–20 birds that travel with erratic, energetic, bouncing undulations. Noisy chatter, especially around waterholes. Presence is often announced by abrupt, slightly nasal 'chek' before birds are seen on the ground among spinifex. A flock, when flushed, passing overhead or arriving at water, creates a rapid, rather musical chattering: 'chek,chak-chek, chek' at random, with slightly different pitch of calls from individuals. Highly nomadic; common.

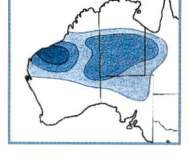

In its native haunts of rugged desert ranges and gorges, the Painted Finch, even with such bold plumage colours, becomes a part of the landscape.

Bill quite slender; crimson only at centre of lower mandible

Less red on face than male; none on throat

Crimson face with splashes of colour down over velvety black of breast

Upper parts, including wings, are deep cinnamon-brown.

Black underparts, crimson streak, heavily spotted flanks

Juv.: similar to female, but dull; red only on rump

♀

Star Finch builds in tall grass or a small shrub amid grass; rounded shape, no entry spout.

Red-eared Firetail builds in dense clump of foliage of forest tree, or low in coastal heath. Nest is large, a football size and shape with long entry spout.

Painted Finch builds in large clumps of spinifex, a flattened-oval with short entry spout; looks like the natural debris at the centre of an old clump.

# CRIMSON & RED-BROWED FINCHES

**Crimson Finch** *Neochmia phaeton* 13–14 cm
In its moist, lush habitats feeds rarely on the ground, usually higher amongst seed heads of grass a metre or more tall. Habitats include pandanus, canegrass, paperbark, lush grasses, crops. Pairs or family parties rather than large flocks; perches on strong grass stems, working at the large seed heads, fluttering between clumps. Calls are high, penetrating, squeaking 'tseit-tseit-tseit'; sharp, fine, single or double 'tsit-tsit' in contact; high 'tchieep' in alarm. When feeding birds are flushed, silvery tinkling contact calls. Sedentary; common.

- *phaeton*
- *phaeton*: E population
- *evangelinae*

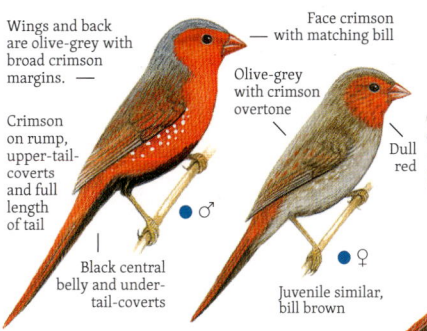

The deep crimson is best seen in direct sunlight; in shade appears deeper, dusky, rather dull red.

**'White-bellied Crimson Finch'**

**Red-browed Finch** *Neochmia temporalis* 11–12 cm
Undergrowth of forests, coastal scrubs and heaths, mangroves, canefields. Highly sociable, in close-knit flocks; forages on the ground, but occasionally perches on grass stems to reach seed heads. When flushed, the flock departs with slightly undulating flight. Voice is extremely high, almost inaudible, a drawn out squeak, 'tseee' and 'tseet'; in alarm a more abrupt 'tchip'. Sedentary or locally nomadic; common.

- *temporalis*
- *minor*
- *loftyi*

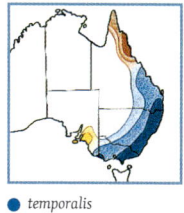

Sometimes listed as a subspecies, but differs only slightly, probably not significantly, from those of SW Vic.

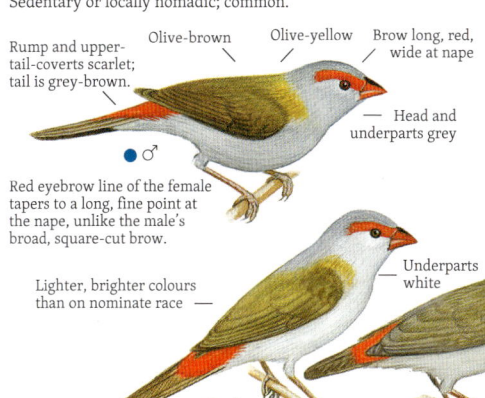

Red eyebrow line of the female tapers to a long, fine point at the nape, unlike the male's broad, square-cut brow.

Crimson Finch often builds into cavities between the stiff, serrated leaves of pandanus.

Red-browed Finch builds a large, untidy oval nest with short side spout entrance in foliage of large shrub or small tree.

## LONG-TAILED, BLACK-THROATED & MASKED FINCHES

**Long-tailed Finch** *Poephila acuticauda* 15–16 cm
Also known as 'Blackheart' for the black throat patch. Usually in open woodland, grassland with scattered trees, shrubs or pandanus; never far from water. Forages on or near the ground, pulling down grass to strip seed. Insects occasionally taken, sometimes in flight. Strongly social, in pairs or flocks. Brief bobbing appeasement rituals maintain flock harmony. Contact call, often in flight, is a soft 'tek'; also has a slow, rather mournful, descending 'whieeeuw' and loud, whistled 'whirrr'. Sedentary; common.

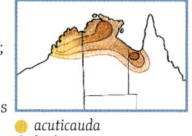

● *acuticauda*
● *hecki*

**Black-throated Finch** *Poephila cincta* 9–11 cm
Highly gregarious in small flocks of 10 to 30 birds, across open woodland, scrubby plains, pandanus flats with deep cover of grasses; never far from water; forages on ground for grass seed.
Contact call a soft 'tek', often in flight; also a whistle similar to that of the Long-tailed Finch, but lower, more hoarse. Sedentary or locally nomadic; common in N of range, but southern white-rumped race declining and shrinking in range.

● *cincta*
● *atropygialis*
● Intergrade

**Masked Finch** *Poephila personata* 12–14 cm
Grassy tropical woodland of eucalypt or paperbark, grassland with scattered shrubs; never very far from water. In pairs or small flocks, forages on ground for fallen grass seed. Flocks move to water morning and evening; at times thousands gather in trees around water where the excited birds flick tails sideways, chatter and preen. Dull grey-brown without black mask; black bill. Voice is loud, brassy, a nasal 'tziat' similar to that of the Zebra Finch. Also a nasal 'twet-twet' and soft chatterings. Sedentary; common.

● *personata*
● *leucotis*

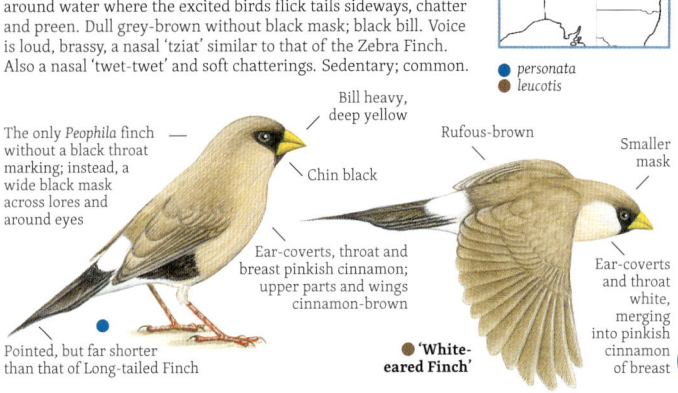

## PLUM-HEADED FINCH, & YELLOW & RUSSET MANNIKINS

**Yellow-rumped Mannikin** *Lonchura flaviprymna* 11–12 cm
Usually near water or damp spots; woodland with tall seeding grasses, also in reed beds and mangroves. Highly social; in flocks often intermixed with larger numbers of Chestnut-breasted Mannikins. Flocks are swift – birds, in unison, abruptly change direction before dropping back into tall grass. Individual birds in short, low flights have an undulating flight typical of grassfinches. Call like that of the Chestnut-breasted Mannikin: a sharp, clear, almost bell-like 'tseit' of varying length. Locally nomadic; generally uncommon, but locally abundant, especially in irrigated crops.

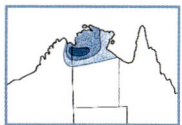

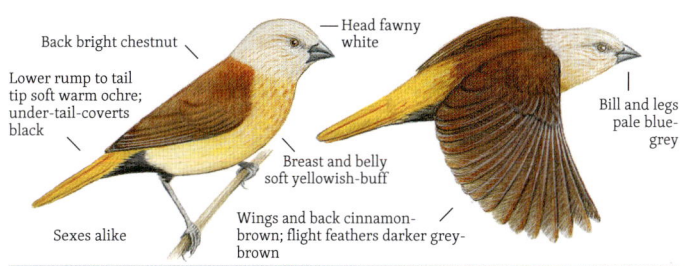

**Plum-headed Finch** *Aidemosyne modesta* 11–12 cm
Typically near swamps or rivers, in tall grass and shrubbery of nearby floodplains, lush grass of shallows, reed beds and cumbungi. In flocks, often large; feeds on the ground and climbs on taller grass to reach seed heads; goes often to water. As close contact call, gives a soft 'tlip' or 'tleip', in flight, a sharper 'tee-ip' or 'tleeip'. From flocks, calls combine to make a loudish tinkling. Similar is the Scaly-breasted Munia, but underparts scalloped rather than barred, face and throat uniformly brown, no white spots on wings. Migratory and locally nomadic; seasonally irruptive, then abundant, generally rather uncommon.

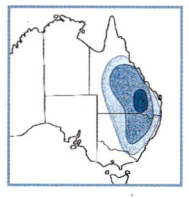

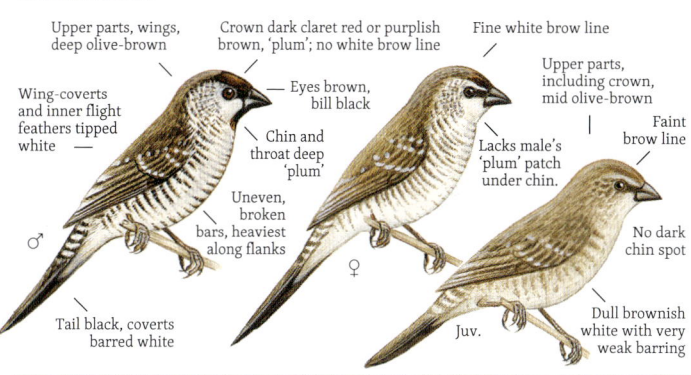

**Scaly-breasted Munia** *Lonchura punctulata* 11–12 cm
Feral species, also known as Spice Finch; native to S Asia from India to S China and SE into the Philippines and Indonesia. Probably became established in Aust. when aviary escapees colonised the E coast, but kept to areas of human activity. Usual habitat grassland – prefers wetter areas, disturbed land, roadsides, crops on moist farmland, lantana. Is seen in fast-flying flocks that make sharp turns in unison; drop suddenly into the grass. Contact call is a soft 'tchp'; identity call is a high 'kh-teee', the first part very soft, the 'teee' penetrating; alarm call is a harsh 'krek-krek'. Introduced; common and expanding its range.

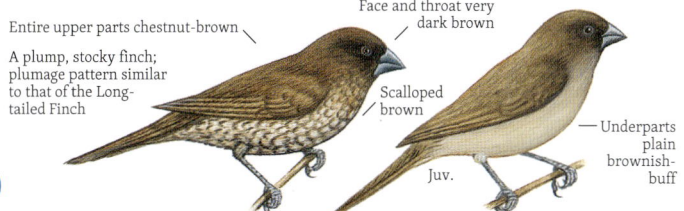

# ZEBRA FINCH, BLACK & CHESTNUT MANNIKINS

**Zebra Finch** *Taeniopygia guttata* 10 cm
A bird of drier inland regions in spinifex, mulga and other semi-arid scrub, grassland, saltmarsh, open grassy woodland and cleared land, but never very far from water. Most widespread finch; seen in pairs, small parties and huge flocks, especially when coming to water; noisy with bouncy, undulating flight. Forages on ground seeking fallen seeds rather than pulling down and stripping seeds from grass-heads; occasionally takes flying insects, especially when feeding nestlings. Has a loud nasal or brassy twanging 'tiarr' and an abrupt 'tet, tet' from flocks in flight. Sedentary or seasonally nomadic; common.

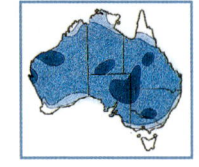

Cheeks orange, bill reddish-orange

Female and juvenile share male's vertical black and white facial streaks, but not the orange cheek patches.

Female and juvenile have white rump and barred tail.

Black bill

Flanks orange-chestnut, spotted white

White rump and heavily barred tail are conspicuous in flight up and away.

**Chestnut-breasted Mannikin** *Lonchura castaneothorax* 12 cm
Clumps of rank, tall grass in damp environs, wet grassland, swamp margins and swampy heath, mangroves, cane fields, rice fields. In small to large flocks. Even in the breeding season the flock keeps fairly close, forming a loose breeding colony. When feeding, birds climb strong grass stems to snip away the seed heads. As contact call, a clear, bell-like, quick 'tlit'; when used among a feeding flock, a slightly longer 'tleit'. The song is a long series of fine, clear notes in many variations: sounds such as 'twee, tee-oo, chie-ook, chee-ing'. Sedentary or seasonally nomadic; common.

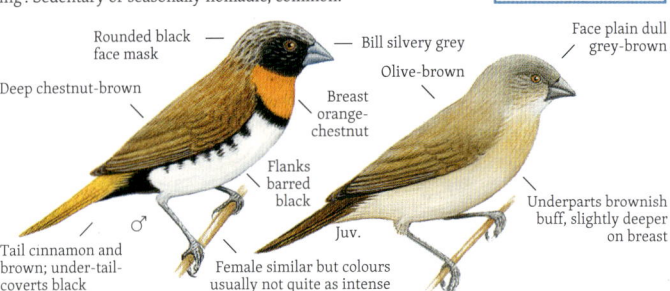

Rounded black face mask

Bill silvery grey

Face plain dull grey-brown

Deep chestnut-brown

Olive-brown

Breast orange-chestnut

Flanks barred black

Underparts brownish buff, slightly deeper on breast

Tail cinnamon and brown; under-tail-coverts black

Female similar but colours usually not quite as intense

**Pictorella Mannikin** *Heteromunia pectoralis* 11–12 cm
Better adapted to the arid environment than other mannikins; highly nomadic, moving out from the coast to country that has received rain. With the return of arid conditions, moves again towards coastal grassland. Inhabitant of open country with sparse scattering of trees, groundcover of grass or spinifex, grassy flats along creeks. Although the Pictorellas visit water at least once a day, they will travel a considerable distance to do so. In pairs to large flocks; seek seeds and insects. Contact call within flocks is a low 'tsip'; identity call is a loud, sharp 'tliep' or longer 'tlee-ip'. Nomadic; moderately common.

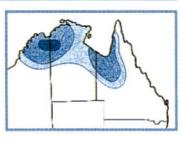

The mannikins have bottle-shaped nests, some with entry spout, usually built low in grass.

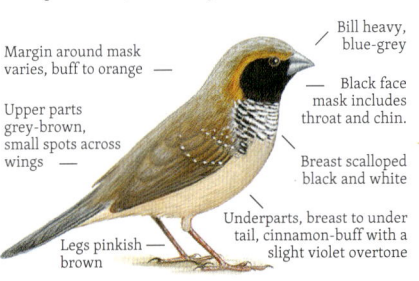

Margin around mask varies, buff to orange

Bill heavy, blue-grey

Black face mask includes throat and chin.

Upper parts grey-brown, small spots across wings

Breast scalloped black and white

Underparts, breast to under tail, cinnamon-buff with a slight violet overtone

Legs pinkish brown

# GRASSFINCHES

Current treatments of Australian grassfinches differ considerably: they are variously placed in the family Passeridae, into their own family, Estrildidae, or into subfamily Estrildinae within the Passeridae. In guides it is common to arrange families, genera and species so that 'look-alike' species are together for comparison. In this book the Australian grassfinches are together as family Estrildidae, while the sequence in the quickguide follows the Birds Australia Checklist and the main text places species of pattern and colour together.

Scarlet Robin = 12 cm

## estrildidae
page 358

Zebra Finch · Crimson Finch · Beautiful Firetail · Chestnut-breasted Mannikin
Double-barred Finch · Star Finch · Red-eared Firetail · Pictorella Mannikin
Long-tailed Finch · Plum-headed Finch · Painted Finch · Blue-faced Parrot-Finch
Black-throated Finch · Red-browed Finch · Scaly-breasted Munia · Gouldian Finch
Masked Finch · Diamond Firetail · Yellow-rumped Mannikin

## DOUBLE-BARRED FINCH

**Double-barred Finch** *Stizoptera bichenovii* 10–11 cm
Dry grassy woodland, open forest, grassy dry scrub, farmland, never far from water. Typically in close flocks; feeds on ground under seeding grasses; has a bouncing, undulating flight. As contact call, a brassy, drawn out 'tzeeaat, tzeeaat' like that of the Zebra Finch, but each call longer, more plaintive; similar sounds, weaker, may form a more continuous passage of song. In close contact, an abrupt, low 'tat, tat'. Nomadic; common, but towards extreme SE of its range only an occasional visitor.

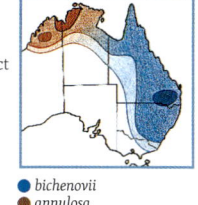
● *bichenovii*
● *annulosa*

An 'owl-faced' finch, white face encircled by a dark rim that continues across the throat as one of the two diagnostic bars —

● Upper-tail-coverts white; conspicuous when flying up and away —

Crown to lower back, cinnamon-brown

● Upper-tail-coverts black

Bill always blue-grey

Juv.: dull; bars around face and across fawn underparts are much paler, indistinct grey.

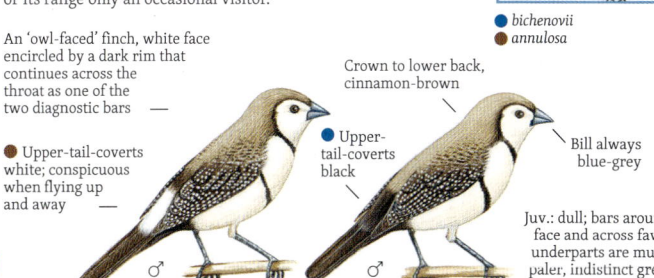

## YELLOW WAGTAILS

### Citrine Wagtail *Motacilla citreola* 16–18 cm
The few Australian records have been from around the coastline at widely separated localities. Usually on wet grassland, muddy margins of wetlands, saltmarshes, floating vegetation. Forages for insects in boggy spots or takes flying insects. Within usual range this species is gregarious, in small groups to large flocks. It occasionally wanders S to Indonesian islands, and, rarely, Aust., where it has been recorded both in summer and winter. The call is a husky, wheezy 'dzzeip' and, less frequently, a high, sharp, less rasping 'tsieeow'.

### Eastern Yellow Wagtail *Motacilla tschutschensis* 16–18 cm
Open country near swamps, salt marshes, sewage ponds, grassed surrounds to airfields, bare ground; occasionally on drier inland plains. Seeks insects, spiders, small molluscs, especially at water's edge at wet sites. Terrestrial foraging, walks briskly with nodding head, teetering tail. In flight, a wheezy 'chzeeip' or 'tsweeip', once with each bounce of the undulating flight. Rare but regular visitor around Aust. coast, especially the NW coast, Broome to Darwin.

● *similina*
● *taivana*
● *tschutschensis*

Other potential vagrants: *lutea*, yellow head; *macronyx*, white throat. Any race at any site.

### Grey Wagtail *Motacilla cinerea* 18–19 cm
European and Asian species. Migrates in winter S to Indonesia and NG. Rarely reaches Aust., then usually near fresh streams, but also on mown grass, ploughed land or near sewage ponds. Runs briskly, chasing small creatures of the stream edge or wetland margins, occasionally darting up to chase a flying insect. Its teetering movements wag the long tail up and down. Flight undulating, alternate flapping and folding of wings. Call, in flight, is a sharp 'zichiep' or 'zittick'. Rare accidental vagrant.

## BLACK & WHITE WAGTAILS

**White Wagtail** *Motacilla alba* 18–20 cm
Open country, often near water, paddyfields, open beaches, margins of effluent ponds and natural wetland, and wet grassland. Spends much of the day foraging across open ground, busily active, running and walking briskly, head jerking back and forth, with frequent pauses when tail wags up and down incessantly. Within normal range, quite gregarious: in small parties or flocks, darting about grassland or ploughed fields. Roosts at night in large gatherings in dense tree foliage or reedbeds. In frequent flights has a strongly undulating flight path. The usual southern limits of migration are the peninsulas and islands of SE Asia, about as far S as Borneo, where White Wagtails, race unknown, are reported as being quite common. Apparently very few reach Aust. Most are seen down W coast, but also sighted at Cape York and SW Vic. Call is a sharp, rather hard and harsh, disyllabic 'chizzik' or 'tichizzik' given in flight. Common within usual range; extremely rare vagrant to Aust.

● *ocularis*
● *leucopsis*
● *baicalensis*
Any race at any site

● Breeds across northern Siberia.
● Breeds in far E, mainly SE China.
● Breeds around Mongolia and Lake Baikal.

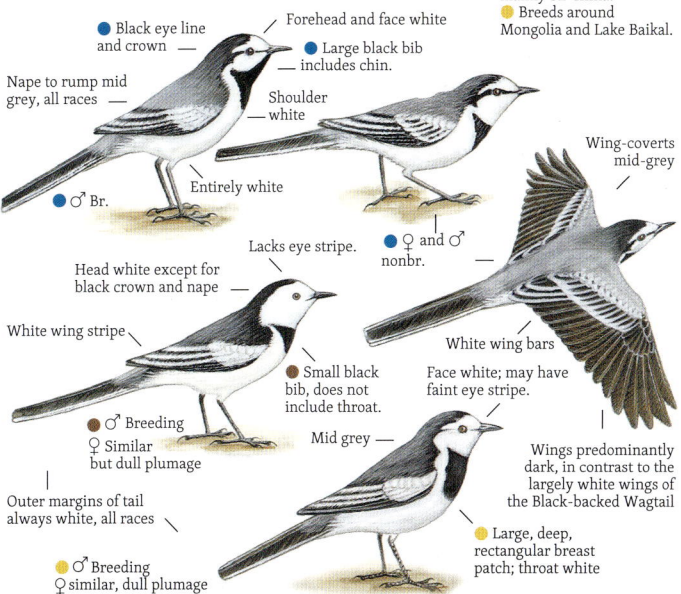

**Black-backed Wagtail** *Motacilla lugens* 16–18 cm
Breeds in NE Asia, migrates to S Japan, S China and Taiwan, with some travelling much further south. Reach Aust. coast on rare occasions. Habitat and behaviour like White Wagtail, from which not easily separated even in breeding plumages. Not only are there several races of the White Wagtail that may reach Aust., each with a slightly different pattern of black and white across head and breast, but these plumages vary with season, sex and age of birds. Voice like that of the White-backed Wagtail. Rare accidental visitor; a probable adult female recorded Fraser Is., Qld, 1987, and a record from Derby, WA.

May be a subspecies of White Wagtail (above).

# HORSFIELD'S BUSHLARK & PIPITS

**Horsfield's Bushlark** *Mirafra javanica* 12–15 cm
In flocks or solitary; open woodland, tussock grassland, saltbush. Forages on ground; runs without bobbing. If flushed, flutters low; short swoops make flight jerky; shows rufous in wings, white in tail. Many plumage colour variations, often matching soil colour. Song varied, not as rich or strong as Skylark, but interwoven with shrill trilling, rich, melodious sounds, mimicry. In breeding season performs song flights, rises high to hover while singing loudly. Migratory; locally abundant.

- *woodwardi*
- *halli*
- *forresti*
- *melvillensis*
- *soderbergi*
- *rufescens*
- *athertonensis*
- *horsfieldii*
- *secunda*

Colour varies, sandy to blackish brown, but pale edged plumage pattern is consistent.

Buff brow line — Rufous margins — Rufous wing patch
Upper parts are heavily streaked grey.
Rufous wing margins — Streaked rufous-brown
Breast mottled
Dark individual — Light — Edges white

Many races have been proposed. They differ only slightly, varying through shades of russet-brown and grey, reflecting a number of influences: sometimes similar to local soil colour; darker in more humid coastal areas; dull when feathers are heavily worn.

**Australian Pipit** *Anthus australis* 16–18 cm
A typical bird of grassland, forest clearings, grassy woodland, semi-open scrub, beaches and hind-dunes, grassy roadsides. On outback tracks, often run out or flush from the grass beside and between wheel ruts. A strongly terrestrial bird that feeds, roosts and nests on the ground. Rather scattered flocks of up to 100 birds wander locally, then break into breeding pairs in spring. Call is a brisk, cheery, rather abrupt 'chirrip' or 'ch'rip' and 'tsweip'. Courting male has a song flight from a low perch, undulating but gradually rising higher, each downward dip accompanied by a quavering, trilled 'tiz-wee-ir'. Sedentary or locally nomadic; common.

- *australis*
- *bilbali*: SW WA
- *bistriatus*: Tas.
- *rogersi*: coastal N Aust., Kimberley to Cape York; darkest race

Upper parts light tawny brown, softly mottled and streaked dark brown

● Breast sparsely streaked, flanks and belly unstreaked
● ● Breast, flanks, upper belly heavily streaked ● Intermediate

Tail moderately long, white-edged; wagged vertically with teetering body movements

Off-white brow and throat encircle brown ear-coverts; fine black line down each side of throat from base of bill to join dark streaks of breast.

Long legs, alert upright posture, runs fast, erratically.

Imm.: similar to adult or slightly more buff; finer markings

Flight low, fluttering, undulating with tail frequently spread to display white edges

**Red-throated Pipit** *Anthus cervinus* 15–16 cm
A northern hemisphere species. Habitat, within usual range, usually open, dry, bare or grassy sites, coastal to highlands, but also cultivated paddyfields and near water. Voice distinctive, a high 'pseeiew', 'tsee-itz' or 'pseeeip'. Winters south from Eurasia to Borneo and the Philippines, vagrant to Sulawesi, Christmas Is. Only rarely to N Aust.

Compared with Australian Pipit, has smaller, shorter legs and a more horizontal carriage of body.

Most likely to be seen in the northern winter nonbreeding plumage
— Face reddish buff
Flanks and breast are creamy buff heavily streaked dark brown.
Nonbr.
Imm.: similar to nonbreeding plumage
Red, face to throat
Heavy black streaking
Br.
Flanks streaked
Pale buff bars across wings
Nonbr.
Rump streaked
Tail white-edged like that of Australian Pipit, but slightly shorter

355

## FERAL SPECIES: COMMON GREENFINCH TO SKYLARK

**European Goldfinch** *Carduelis carduelis* 12–14 cm
Feral species. In Aust., around towns and settlements, farms, roadsides and wasteland. Gathers in large flocks in autumn and winter; feeds on seeds. If disturbed, flocks fly up to perch on exposed top twigs of a tree. Makes canary-like trills and chirpings, 'tswit-tsiewt-tswit-swit-zwee-tzwee'; in flight, a shrill 'tsieew, tseew'; tinkling sounds from feeding flocks. Locally nomadic; common.

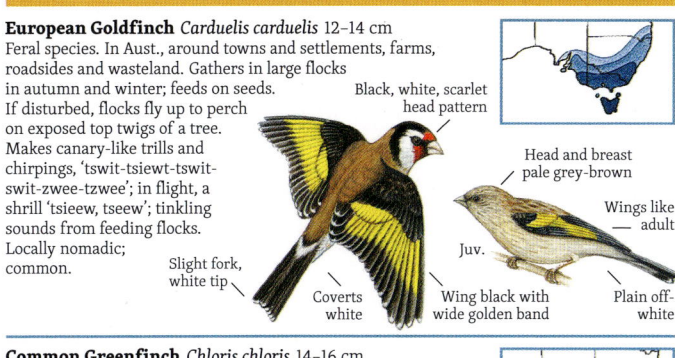

**Common Greenfinch** *Chloris chloris* 14–16 cm
Feral, mostly in areas not under native vegetation: parks, gardens, roadsides, farmland. In small flocks, foraging for seeds on the ground; flight undulating. After the breeding season, gather into much larger flocks. Makes a rather harsh, chittering sound, 'chwitit', and 'chwitchit-chit-chit-chit'. In flight gives a nasal, buzzing 'tzwee-tzwee-tzwee', and, as alarm, abrupt 'zweet!'. Sedentary; patchy distribution, most common near coast.

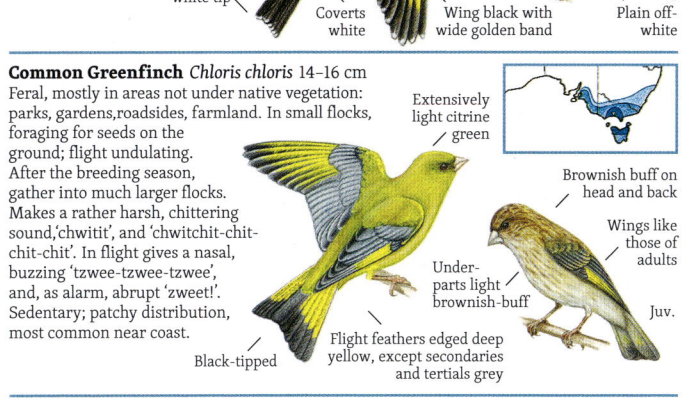

**House Sparrow** *Passer domesticus* 14–16 cm
Feral species; highly sociable, gregarious. In and around cities, country towns and farm buildings. Needs nest holes in buildings or trees. Usually in small colonies, but gathers in large flocks, at times many thousands, after the breeding season. Feeds on grains, fruit and insects, and scavenges for food scraps. Voice is almost constant, monotonous 'chirrup' and 'chissik'; harsh chatterings. Calls in alarm or excitement. Widespread, mainland and Tas.

**Eurasian Tree Sparrow** *Passer montanus* 13–15 cm
Feral; in and around cities, towns, farms, keeping mostly to vicinity of buildings and surrounding agricultural lands. Occurs in small flocks even in the breeding season – nesting may be loosely colonial; at other times in larger flocks. Call is a metallic 'tchik', 'tchit-tchup' and high chittering; in flight a softer 'tek'. Sparse to locally common.

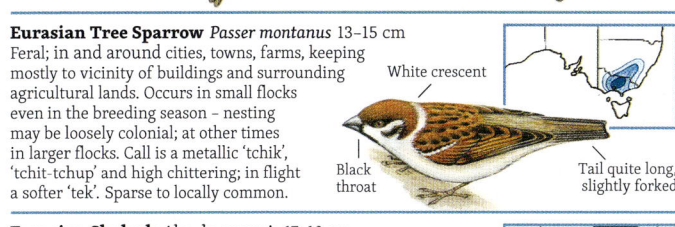

**Eurasian Skylark** *Alauda arvensis* 17–19 cm
Feral species. Lengthy outpourings of rich sound, a clear, very attractive and musical song that is often given in flights; often gives a quick, mellow 'chirrup'. Uses open country including pasture, heath, swamp margins, rarely shores and coastal dunes. Nomadic or part migratory; common in SE.

## BROWN SONGLARK & CISTICOLAS

**Brown Songlark** *Cincloramphus cruralis* 18–19 cm
Male conspicuous in breeding season, delivering his song from exposed posts, limbs, or, most vigorously, in display flight. A bird of open spaces, grassland, plain, open woodland, sparse inland scrub, saltbush plain and farm paddocks. The song is loud and clear, cheerful, pleasant, distinctive. Some sounds have been described as like the dry squeaking of a wooden wheel turning on a rusty axle – 'skzit-kotch-zzweiler, chweeip, kzzeech-kotch-aweiler' – finishes with a musical trill and whipcrack. Highly nomadic, moving to regions that have received rain; locally scarce to abundant, but generally quite common.

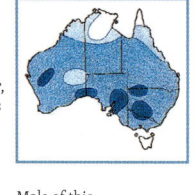

**Golden-headed Cisticola** *Cisticola exilis* 9–11 cm
Breeding males attract attention with song flights and calls from elevated perches – tall grass stems and fences. After calling, drops suddenly to cover again. Wetlands and their vicinity are the usual habitat; includes densely vegetated margins of lakes and swamps, tall wet grass and rank herbage, irrigated pastures, samphire. The call, given almost incessantly in spring and summer, is a drawn out metallic buzzing, this interspersed with quick, clear, musical 'teewip' calls. Also has a higher, faster, vibrant 'tizzzeip' and 'weezzz, whit-whit'. Sedentary; common.

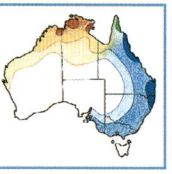

- *exilis*
- *lineocapilla*
- *diminuta*
- *alexandrae*

Slight variations:
- Back deep tawny rufous with heavy streaking
- Back mid tawny or rufous, medium-narrow black streaking
- Back pale sandy buff with light and narrow streaks

**Zitting Cisticola** *Cisticola juncidis* 9–11 cm
Behaviour is like that of the Golden-headed Cisticola. Habitat includes grassy swamps that are seasonally damp or shallowly flooded: usually coastal plains or saltpan country dominated by sand-couch; margins of mangrove swamps. When breeding, the male has a persistent zitting call. Flying high in display, he almost hovers in slow, undulating flight into the wind, giving a 'tzip-zip, tzip-zip' call. Migratory; uncommon.

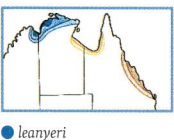

- *leanyeri*
- *normani*
- *laveryi*

## TAWNY GRASSBIRD, SPINIFEXBIRD & RUFOUS SONGLARK

**Tawny Grassbird** *Cincloramphus timoriensis* 17–19 cm
Tawny rufous and brown inhabitant of lush vegetation: bullrushes, adjoining lush, wet grass, cumbungi swamps, tall moist crops. In pairs, small parties, loose colonies. In breeding season, male is conspicuous and bold in display flights – flutters, almost hovering. Voice has some similarity to that of the Rufous Songlark. Calls mostly in early morning and towards evening. Song is a loud, varied, initially high, squeaky sequence of descending, reeling notes, finishing with harsh, deep chuckling notes. As alarm or warning, gives a single, sharp 'tjik'. Nomadic and migratory; locally common but generally uncommon.

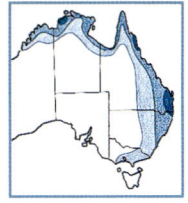

**Spinifexbird** *Poodytes carteri* 14–16 cm
Long-tailed, brownish, concealed within large clumps of needle-leaved spinifex. Favours large clumps along watercourses or lower slopes of ranges; often intermixed with shrubs and small trees. Makes brief low flights, long tail trailing behind. Often climbs a tall spinifex stem for a brief look at an intruder before scuttling down to vanish again in the spinifex. In alarm, 'tjik' and 'tjuk', like two rocks struck together. More attractive is a high, quick 'cheeryit' or 'cheery-wheit', or slower, downward 'cheer-y-a-roo'. Sedentary; locally common.

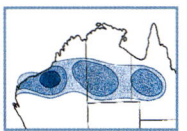

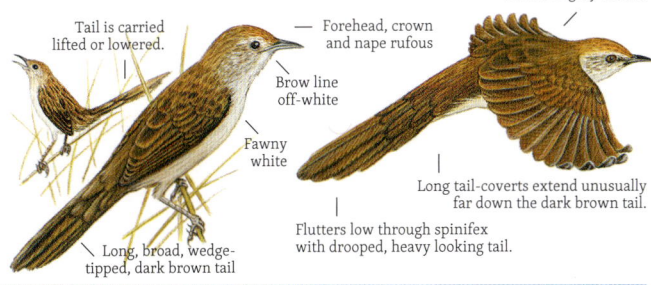

**Rufous Songlark** *Cincloramphus mathewsi* 16–19 cm
On open grassland, grassy open woodland and mulga, farmland. Attracts attention during breeding months when the male calls almost continuously from perches and in song flights. Full song is a joyous rollicking 'whitcher-whitcher, a-whitchy-wheedle-whitch', in notes rather scratchy yet musical. Has also a single, ringing, almost whipcrack, 'whitcher'. The female undertakes the entire nest-building, incubation and feeding of young. After breeding season may form small groups, but then quiet, inconspicuous. Common; migratory, moving S to breed over summer.

## REED-WARBLERS & LITTLE GRASSBIRD

**Australian Reed-Warbler** *Acrocephalus australis* 16–17 cm
Occupies almost every reedbed, large or small, but not easy to observe unless at edge of reeds, or in brief low flight across open water. Song through the breeding season is rich and powerful: 'kwitchy-kwitchy', 'kwarty-kwarty-warty'; in alarm, an abrupt 'tchuk, tchuk'. Migratory into the SE and SW of its range to breed in summer; some remain through winter, but are quiet. Common.

Unstreaked, long-tailed inhabitant of reedbeds. Sexes alike.
Juv.: slightly darker lores

Crosses open water between reedbeds in swift, low, undulating flight.

- *australis*: E Aust.
- *gouldi*: WA, slightly larger, darker and has longer bill.

Tail quite long; rounded tip

Upper body mid tawny brown
Deeper walnut-brown

Underparts tawny buff to off-white
Deeper rufous-buff to cream

Darker through the eye; pale buff brow line

Clings to reed stems, often moving down to water level seeking small aquatic prey.

When singing, lifts the feathers at the rear of the crown as a small peak.

**Oriental Reed-Warbler** *Acrocephalus orientalis* 18–20 cm
Typical habitat includes reeds, cane grass, cumbungi, sedges, mangroves. Breeds Japan and China; winters as far S as Indonesia and NG; recorded within Aust. on rare occasions. Unlike the rich, musical sounds of Australian Reed-Warbler, the Oriental makes metallic noises, commonly a loud, grating 'krrek'. Song is a harsh, churring, chattering 'kratch-kraich, krok-krok-quorrark'. Voice valuable in identification, but it is unlikely to call until in its northern hemisphere breeding territory and season. Rare; vagrant or migrant to Aust. N and E coasts.

Bill heavier, shorter

Upper parts mid to light cinnamon-brown

Oriental and Australian Reed-Warblers may be inseparable except by examination in hand.

Larger than Australian Reed-Warbler

Underparts off-white with a few faint streaks on sides of throat and upper breast. Imm.: not streaked

**Little Grassbird** *Poodytes gramineus* 13–15 cm
Secretive, skulking inhabitant of dense vegetation, in or beside wetland with rushes, lignum swamp, cumbungi, cane grass, tidal marshes, mangroves. Usually allows only brief glimpses; then flutters away into dense concealing vegetation. The call is often the only indication of the presence of grassbirds in their typically dense habitat. It is a mournful whistle of three notes, the first soft, low and brief, the second and third at high and even pitch, drawn-out, 'whp-wheeee-wheeeee' and 'whp-whiooo'. In alarm, a harsh, rattling chatter. Sedentary; locally common, dispersive.

- *gramineus*
- *goulburni*
- *thomasi*

Compared with Tawny Grassbird, upper parts are less rufous; black streaking includes head and underparts.

Olive-brown, moderately streaked black

Forehead has russet tint.

Long whitish brow line

Dark brown, pointed

Flanks light rufous-olive, plain or lightly streaked

Underparts off-white, usually with sparse fine dusky spots or short streaks

♂♀
Tail dark, edged olive or buff; lacks any barring or white tips; in flight is often spread.

- Upper parts deep brown heavily streaked black; underparts quite strongly dark spotted, but only lightly on flanks; russet on forehead
- Upper parts mid brown streaked black; crown to upper back tinted rufous; underparts, including buff flanks, heavily streaked black

# REED-WARBLERS TO PIPITS

This Group contains quite closely related families whose members prefer similar habitats with typically low vegetation from sparse grass to dense reedbeds, but otherwise open. Any trees or large shrubs are widely spaced or absent. Most forage on the ground or in the lowest vegetation, of which reeds are about the tallest. Some species do make use of tree hollows as nest sites, and high branches for song or display. In this family group are also five feral species that mostly use urban lawns and rural paddocks in preference to natural habitat.

Robin = 12 cm

## sylviidae
page 351

- Australian Reed-Warbler
- Spinifexbird
- Oriental Reed-Warbler
- Golden-headed Cisticola
- Arctic Warbler
- Zitting Cisticola
- Tawny Grassbird
- Little Grassbird
- Rufous Songlark
- Brown Songlark

## alaudidae
page 355

- Horsfield's Bushlark
- Skylark

## passeridae
page 354

- House Sparrow
- Eurasian Tree Sparrow

## motacillidae
page 356

- Citrine Wagtail ♂ Br.
- Australian Pipit
- ♂ Nonbr. / ♀
- ♂ Br.
- Red-throated Pipit
- ♂ Nonbr. / ♀
- Eastern Yellow Wagtail
- ♂ Br.
- Grey Wagtail
- ♂ Nonbr.
- White Wagtail
- ♂ Br.
- ♂ Nonbr. / ♀
- Black-backed Wagtail

## fringillidae
page 354

- European Goldfinch
- Common Greenfinch

## ARCTIC WARBLER

**Arctic Warbler** *Phylloscopus borealis* 11–12 cm
A small leaf-warbler that breeds across N Eurasia to NE Siberia and Alaska; winters S and SE Asia, usually as far S as Indonesia, but occasionally to the Kimberley coast, reaching Ashmore Reef, Scott Reef and Broome, WA. Within its usual range, it uses open wooded country, forest edges and mangroves and actively forages through foliage of trees and shrubs. The voice is rather harsh and scolding: 'tzrick, tzrick'; a husky 'tszzic'; also has a reeling, buzzing trill. Calls from about Mar., before returning to breeding grounds. A rare vagrant to W Kimberley and offshore reef islets.

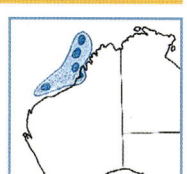

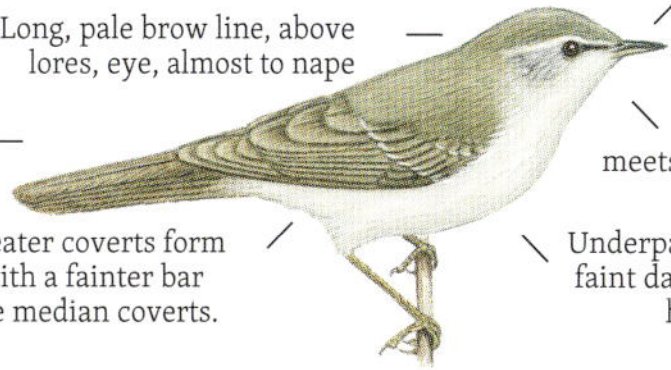

- Long, pale brow line, above lores, eye, almost to nape
- Upper parts are soft mid-olive-grey.
- Small white tips to greater coverts form a thin pale wing bar with a fainter bar further forward on the median coverts.
- Bill slender and dark with pale orange-tan along the lower mandible
- Thin dark line through eye meets streaked grey ear-coverts.
- Underparts white to off-white with faint dark streaks on breast; flanks have slight buff-brown tint.

## YELLOW WHITE-EYE, SUNBIRD & MISTLETOEBIRD

**Yellow White-eye** *Zosterops luteus* 11–12 cm
A mangrove specialist, but moves out to forage in adjoining monsoon scrub, dense river-edge vegetation, coastal scrub, paperbark swamp and coastal gardens. In pairs to large flocks, forages through the yellow-olive foliage of the mangroves and occasionally on mud exposed at low tide. Has strong, high-pitched, whistled contact or flocking calls, possibly louder than those of the Silvereye, and tinkling notes from foraging flocks. The song is strong, varied and tuneful with musical warbling and trills. Sedentary, or very localised wanderings; common.

- luteus
- balstoni

balstoni NW coast WA: variable, relatively dull to greyish colours, smaller bill

Upper parts mid yellowish-olive
Flight feathers dark grey, edged yellow
Thin black line through eye
Underparts entirely deep yellow
Flanks slightly darker olive-grey
Underside of flight feathers is plain grey-brown, but edged yellow on upper surface.

**Olive-backed Sunbird** *Cinnyris jugularis* 11–12 cm
A rainforest species, including its margins, clearings and regrowth, nearby plantations, gardens and lush watercourse vegetation, mangroves and coastal scrub. The sole Australian representative of the sunbird family, Nectariniidae, this bird is conspicuous not only for its bright colours, but also for its bold raiding of nectar from garden flowers, the pugnacious defence of its territory against other sunbirds, and its use of verandahs, from which it suspends nests. Often hovers while feeding from flowers and while taking spiders and insects. The calls are rather weak, squeaky, scratchy, metallic, rising 'tssee-ik', 'twieeik', a stronger, harsher, aggressive 'tzzzeeik', and lively chattering. Common.

Suspended from domed roof, nest is 30–50 cm from top to tip of very long tail.

Imm.: similar to adult female. Young males show the first sign of a blue breast as a thin blue line down the centre of the breast.

Face yellow; dusky band through eye
Bill is long, fine, strongly downcurved.
Lower breast to under-tail-coverts is deep golden-yellow.
Chin to breast is metallic blue-black with iridescent highlights.

**Mistletoebird** *Dicaeum hirundinaceum* 10–11 cm
Found where mistletoe grows: dense wet forest of coastal ranges, mangroves, woodland, mallee, mulga and semi-arid spinifex country. Alone or in pairs in foliage of tree or mistletoe clump; often high where it might not be noticed but for the distinctive calls. Flight fast, direct, with lively, sharp calls. Feeds on the berries of the mistletoes; digests the soft, sweet under-skin layer, but not the large seed. Nomadic in most parts of its range, following the fruiting of mistletoe species. Calls are loud and spirited, far-carrying – a sharp, squeaky 'tiech, tieech, ti-witch, tee-wietch-tieewietch-teewietch' and loud, sharp, rollicking 'kinzee-kinzee, perwita-perweeta-perweeta'. Nomadic; common.

Nest tiny, soft; in tree foliage

Upper parts glossy blue-black
Greyish white
Upper parts brownish-grey
Scarlet, chin to breast
Under-tail-coverts scarlet
Under-tail-coverts pink
Juv.: like female, but gape yellow

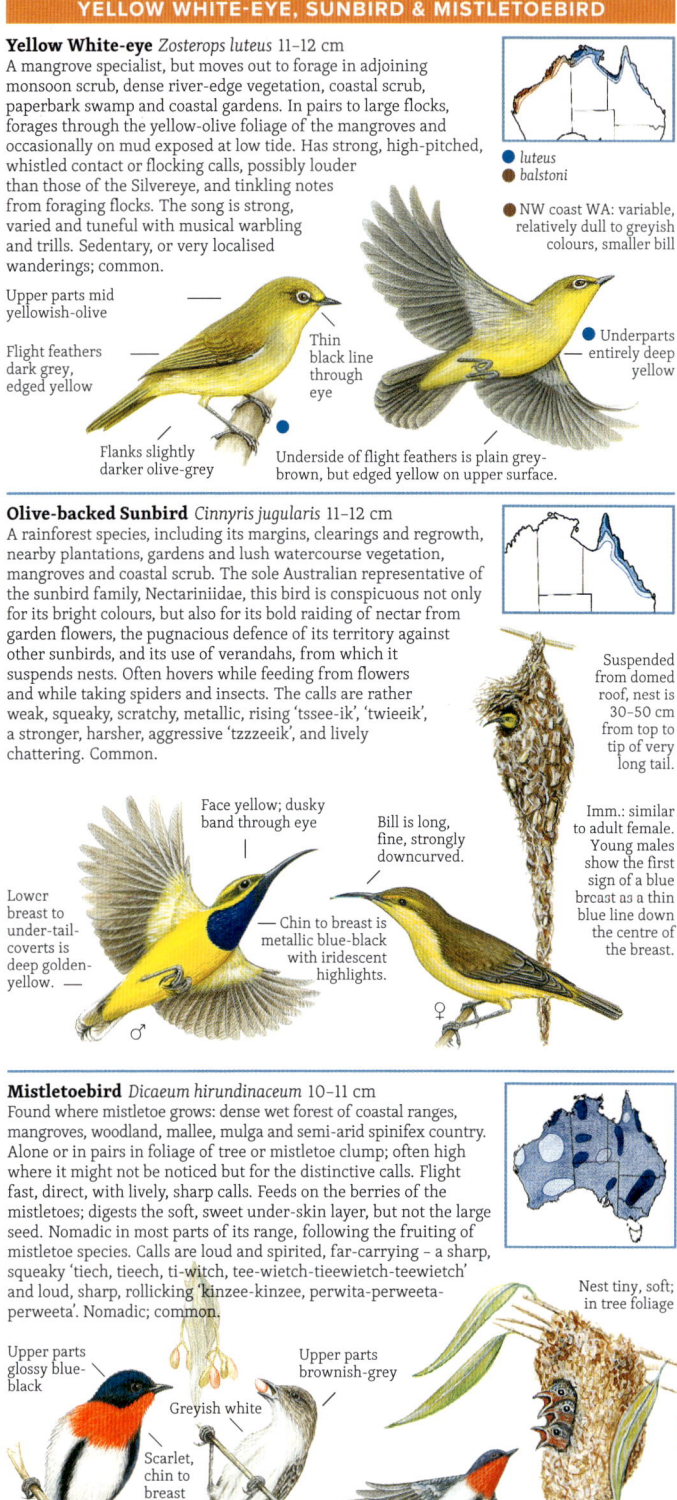

# SILVEREYE & ASHY-BELLIED WHITE-EYE

**Silvereye** *Zosterops lateralis* 11–13 cm
The many races occupy diverse habitats: eucalypt woodland, forest, coastal heath, mallee, mangroves and many other vegetation types; also can be found in gardens, orchards and vineyards. Foraging Silvereyes are lively, busily active little birds, constantly on the move; they depart to the next patch of shrubbery with brisk, bouncy flight and much calling. All races are gregarious: after breeding they gather into small parties and then into large flocks, these foraging through foliage of trees and undergrowth. Major migrations occur along Australia's E coast. The southernmost populations seem to undertake the longest migrations to escape the approaching southern winter. Common call is a clear, peevish 'tseeep'; song is a rather slow, erratic series of clear, sharp, slightly peevish notes: 'tsweeip-cheeip, peeip-a-chweip, cheeip', with many slightly different variations in rambling sequence. Calls of western race, *chloronotus*, are perhaps slightly more harsh. Migratory or nomadic; common to abundant.

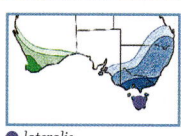

● *lateralis*
● *cornwalli*
● *chloronotus*

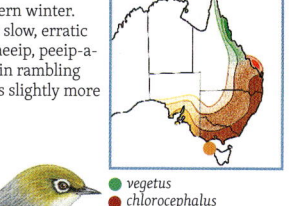

● *vegetus*
● *chlorocephalus*

**Not illustrated:**
● *westernensis*
● *ochrochrous*
● *pinarochrous*

Boundaries of races hard to define, but all have the typical conspicuous white eye-ring, emphasised by dark lores extending back as a grey or black line under and behind the white eye-ring. Sexes are alike; immatures are slightly less colourful.

Back mid blue-grey

Chestnut to cinnamon-brown flanks

White

Olive-yellow to olive-green

White

Flanks colourful, conspicuous, chestnut to dull cinnamon-brown

Under-tail-coverts white or greyish white

Citrine-yellow
Pale blue-grey
Rump pale olive
Deep, bright yellow
Flanks pale brown
Yellow tint

Head deep olive-yellow
Mantle and upper back mid grey
Rump olive
Deep yellow
Mid-grey
Deep yellow

Back olive-green rather than the grey of other Silvereye races
Chin, throat olive-yellow
Breast pale grey
Pale yellow
Flanks pale cinnamon-grey

Edged yellow or citrine, all races

Deep citrine-yellow
Mid grey
Throat yellow
Mid grey
Yellow
Brownish-buff flanks

'Capricorn White-eye' — Considerably larger than other races. Full species status has been proposed.

Yellow rump
Rounded wing-tip shape

**Ashy-bellied White-eye** *Zosterops citrinella* 11–12 cm
Occupies some islands where there is cover of scrub thickets and trees; not established on the mainland or the larger islands. Also known as Pale-bellied or Ashy-bellied White-eye. Within Aust. this species has been found only on some small, well-vegetated islands inside the Great Barrier Reef N of about Cooktown, and on small islands of Torres Str. In pairs or small flocks, it forages through foliage to seek native fruits, berries and insects. Contact call is a loud chirp. The song is a rising and falling sequence of sweetly warbled notes; less plaintive than that of the Silvereye. Sedentary; locally quite common.

● *albiventris*

Entire upper parts greenish-yellow
Forehead, chin and throat yellow
Yellow
Breast and belly off-white

This large white-eye has a relatively heavy, robust bill. Thin black lines from lores under and through eye

348

## PALE-RUMPED SWALLOWS

**Red-rumped Swallow** *Cecropis daurica* 16–18 cm
Usually open country, coastal grassland. A medium to large swallow; migrates as far S as Borneo and NG, occasionally Aust. Contact call, commonly in flight, an abrupt, slightly nasal 'tweeit'; a twittering, warbling song. Rare vagrant: records from NE Qld and Broome, WA.

● *japonica*
● *C. striolata*

- Broad chestnut collar
- Chestnut collar broken or absent
- Rump cinnamon to chestnut
- Upper parts blue-black with metallic-blue sheen
- Whitish, no band
- Under-wing-coverts pale rufous
- Under-tail-coverts black
- ● Tail long
- ● Tail shorter
- Cinnamon-rufous rump conspicuous as swallow banks and climbs
- Streaks heavier on chin, throat
- Deeply forked
- Underbody buff; lighter streaking
- Juv.: upper parts dull dark brown, rufous pale, tail short

The similar **Striated Swallow**, *C. striolata*, is a possible vagrant to Aust. The range of this SE Asian swallow extends S to islands close to NW Aust. coast.

---

**Fairy Martin** *Petrochelidon ariel* 12–13 cm
Open environs, but often near water, cliffs, caves. At most sites there are not a great many active nests, the birds inconspicuous, sjust the occasional martin darting in or out from beneath cave or rock overhang. Overhead, their flight is rather fluttering, erratic, as they pursue high flying insects. Contact call is an abrupt 'drrt, drrit'; also has a feeble, twittering song. Migratory, moving N for winter, some to NG; uncommon.

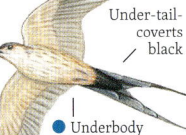

- Cinnamon-rufous
- Folded tail slightly shorter than wings, shallow fork
- Dull white
- Fine dark streaks
- Glossy blue-black
- Rump dull white, faintly streaked rufous
- Adult: sexes alike. Juv.: dull, pale crown
- Slightly forked, dark
- Grey-brown
- Underwing-coverts buffy white

Welcome Swallows also build mud nests, but of open cup shape, and usually not in colonies.

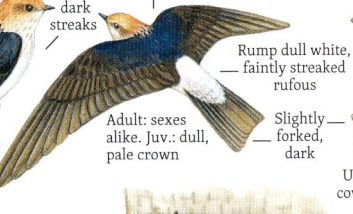
Bottle-shaped Fairy Martin nests of mud pellets cluster in colonies on ceilings of shallow caves, under ledges, boulders and bridges.

---

**Tree Martin** *Petrochelidon nigricans* 12–13 cm
Open woodland and farmland with trees, not far from water. Hunts insects in swift, twisting, turning flight around canopy of trees, low over water of lakes and rivers, and above farm pastures. Often found in small colonies where there are large trees that have many small hollows in the upper limbs. Has a slightly metallic 'tzweit' and musical, scratchy, squeaky chatter, 'chwip-chip-chzeit-chwip'. Nomadic or migratory over most of its range; common.

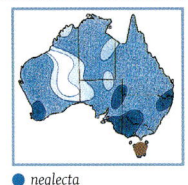
● *neglecta*
● *nigricans*

- Crown, nape and back metallic blue-black
- Forehead chestnut
- Rump dull white with dark streaks; tail-coverts slightly more heavily mottled
- Underparts off-white with fine brown streaks
- Wings and tail dull blackish brown
- Tail close to folded wing length; shallowly forked
- Tail likely to appear square-tipped when spread in flight
- Juv.: grey-brown upper parts, less chestnut

● Is found only on mainland Aust.
● Slightly larger, more rufous; confined to Tas. through spring–summer breeding; autumn and winter, it migrates N across eastern mainland, some reaching NG.

## DARK-RUMPED SWALLOWS

**White-backed Swallow** *Cheramoeca leucosterna* 14–15 cm
Dark-rumped but unique white back. Open country of drier woodland, semi-arid scrub and heath. A graceful swallow with sharply black and white plumage; white much more conspicuous than on the pale-rumped species. Usually in small flocks; tends to forage quite high; flight erratic, often swooping, fluttering with rapid wing beats. Their many calls combine as an almost continuous rattling twittering. Nomadic wanderings or migratory movements; common.

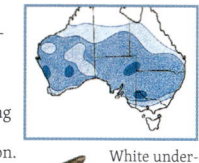

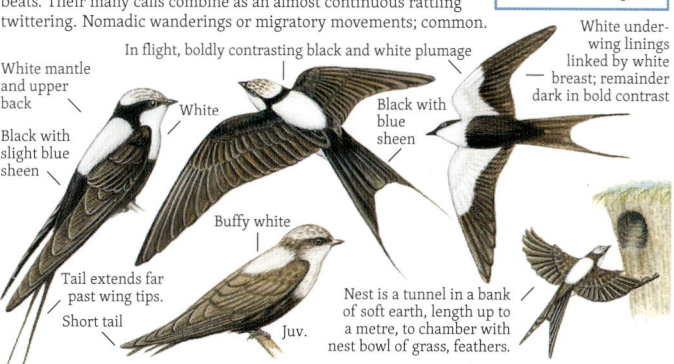

**Welcome Swallow** *Hirundo neoxena* 14–15 cm
Diverse open habitats: woodland, grassland, wetland, farms. Has adapted to artificial environment of buildings and bridges that offer sheltered sites for mud nests. Has long been a familiar nester on rafters of verandahs and sheds through almost all of inland Aust. As contact call in flight, a single 'tchek'. Song is a lively succession of squeaky chatterings. In alarm, a sharp 'tseip-tseeeip'. Sedentary, nomadic or migratory; common.

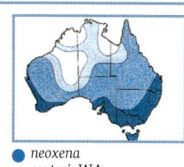

● *neoxena*
● *carteri*: WA,

**Barn Swallow** *Hirundo rustica* 14–17 cm
Summer migrant to Aust.; open sites in towns, often near water; seen on overhead wires with other swallows. Most seen in NE Qld and in far NW along coast, Darwin to Broome. Weak, twittering 'tsi-tswitt-tsee-' and soft warblings. Vagrant, but common in far N and NE.

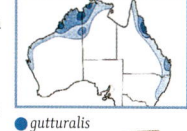

● *gutturalis*

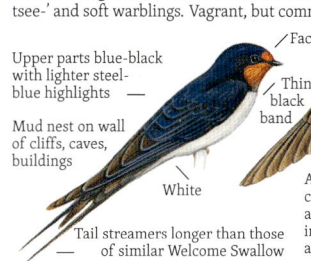

**Pacific Swallow** *Hirundo tahitica* 13–14 cm
Usually near coasts, cliffs and caves that provide nest sites. Widespread through islands of South Pacific, NG; occasional sightings Daintree, Mossman, Weipa in NE Qld. Pairs to large flocks. Usual calls a 'twzit-twzit', longer 'tzitswee'; these calls may be run together in twittering song. Extremely rare vagrant.

● *frontalis*

About 8 subspecies, W Pacific to S Asia. At E of range, dark under-parts, no white in tail. In NG and further W, races have paler underparts and, as a white bar, white spots on inner webs of tail feathers.

## STARLINGS

**Metallic Starling** *Aplonis metallica* 22–24 cm
Tropical rainforest, nearby woodland, mangroves, coastal scrub, gardens. Best known for the large and conspicuous treetop nest colonies, from which the flocks forage out across surrounding rainforest, seeking fruits and nectar. Tight flocks hurtle low above the canopy, attracting attention with their noisy passage – various nasal and wheezing calls. From individuals, 'scriaarch, scraark, scraich-scraich-scraaairch, chrak-chrak-chrak' and similar sounds; from the flock a jumbled racket of intermixed squawking and wheezing. Migrate to islands N of Aust., return in Aug. Common.

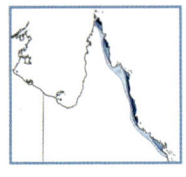

Eye of imm., brown at first, soon becomes crimson.

Eye brilliant crimson

A glossy-iridescent sheen on black plumage gives metallic blue-green highlights.

Adults, front and side views: male and female are similar.

Underparts white, streaked black; a black breast band precedes the all-black of adult.

Tail tapers to a long slender point, unlike the shorter, thicker, square-tipped tail of the Singing Starling, *Aplonis cantoroides*, of Australia's northern Torres Str. islands and NG.

The massive haystack-like mass of bulky domed nests, crowded together conspicuously in an exposed situation high in a tree can be first indication of the presence of this species; some colonies may contain well over a hundred nests, and, when young have left the nest, the flocks may build to several thousand. With the constant comings and goings of flocks, the noisy chatter of adults and calls of young, a large colony cannot be overlooked.

A second species of starling is native to Australia's island territories of Boigu and Saibai in Torres Str., close to NG. It differs most obviously in having a shorter, square-tipped tail, and in breeding as solitary pairs rather than in colonies. The nest is a hollow in a tree or cliff, or in cavities among nests of the Metallic Starlings.

**Common Starling** *Sturnus vulgaris* 20–21 cm
Feral; urban and country areas, including woodland, mallee, mulga, watercourse and roadside tree-belts, reed beds, cleared land, coast, alpine, parks and gardens. Displaces native bird species from nest hollows. Most characteristic calls are long, downwards whistled 'chwee', high 'tizz-tzz' and diverse wheezing, rattling and clicking noises; includes some mimicry. Migratory, nomadic, dispersive. Common introduced species.

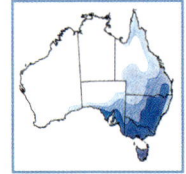

Sharply pointed; on nonbr., dull brown

Autumn's new plumage has pointed buff and white tips to black feathers.

Bill yellow, spring and summer

Fresh plumage

Summer breeding plumage: much of plumage is worn to plain glossy black.

**Red-whiskered Bulbul** *Pyctonmotus jocosus* 20–22 cm
Introduced, initially to southern Sydney suburbs about 1880, later to Melbourne. Of southern Asian origin, a trilled fluid whistle; much lively chattering; harsh scolding alarm. In Aust. has kept mostly within the alienated city and surrounding landscapes, regrowth, garden shrubbery. Locally common, Sydney, Coffs Harbour, Mackay.

345

## ROCK-THRUSH, FLYCATCHERS, BLACKBIRD & MYNA

**Blue Rock-Thrush** *Monticola solitarius* 22–23 cm
Rocky shorelines, cliffs, open stony country, coastal towns. Breeds in S Europe, N Africa, central Asia, China, Japan and N Philippines; migrates S to Sumatra, rarely to NG. Recorded in Aust. Oct. 1997 at Noosa Heads, Qld. Solitary, although occasionally in groups on rocks. Upright stance. Flushes easily; quick to disappear into concealment of rock crevices and boulders. Low clucking sounds and a pleasant, reedy song. Very rare vagrant.

**Variation**
● Nominate race, *solitarius*, is entirely soft dusky blue with dark scallopings.
● Race *philippensis* is closest to Aust., blue brighter, no scallopings.

Head, neck, throat, breast and entire upper parts mid to deep slaty blue, brightest in breeding plumage, but, in nonbreeding plumage, dull, scalloped and barred with pale grey and black

Wings blackish

♂ Perches on rocks.

Breast to under tail chestnut.

Nonbreeding: body plumage flecked and barred black and light grey; colours dulled

Upper parts grey-brown, barred

Slightly paler throat

Buffy white, scalloped darker

♀ Imm. male: like nonbreeding male, but heavier markings, darker underparts. Juv.: like female, but heavier scalloping and barring

**Blue-and-White Flycatcher** *Cyanoptila cyanomelana* 16–17 cm
Breeds China, Japan, winters S to Java. Sole Australian record an adult male dead on beach, NW coast, WA, 1995. Identified as nominate race. Usually silent out of breeding range.

Crown bright cyan blue, darker on nape, encircles black face, breast.

Wings glossy blue, except dark brown primaries ♂

Base of outer tail feathers white

Olive-brown

Wings and tail rufous-brown

Lores, face, pale

♀ Imm. male: like female, but with some blue showing on wings and tail

**Narcissus Flycatcher** *Ficedula narcissina* 13–14 cm
Breeds eastern Asian mainland, migrates S in northern winter to Java and Sumatra, where habitats include woodland, rainforest margins, gardens. Several records from NW of WA, including Barrow Is., in acacia scrub with spinifex. Usually silent on winter migration. Accidental; storm-blown.

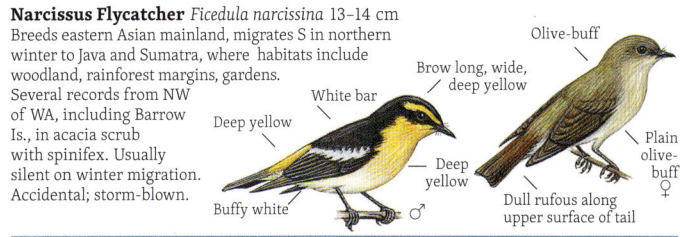

Deep yellow

White bar

Buffy white ♂

Brow long, wide, deep yellow

Deep yellow

Olive-buff

Plain olive-buff ♀

Dull rufous along upper surface of tail

**Common Blackbird** *Turdus merula* 25–26 cm
Feral; forests, roadsides, scrubs, gardens. Native Europe to S Asia. Has invaded both built and natural environs, competing with native birds. Contact call a high, fine 'tseee' and harsh chatters in alarm given in flight. Song a mellow fluting with higher trills. Abundant.

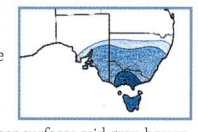

Eye-ring is deep golden-yellow, narrow.

Bill deep yellow

Plumage is entirely black with a surface sheen ♂

Dull yellow

Underparts light grey-brown mottled darker

Upper surfaces mid grey-brown, darker on wings and tail

♀

Juv.: like female, but paler underparts and streaks on crown

**Common Myna** *Acridotheres tristis* 23–25 cm
Feral scavenger; takes insects, fruits, young of other birds. Nests in crevices of buildings and in tree hollows, competing with native species. Whistles, high, vigorous 'wheeoo', harsh, rattling 'carrarrk, carrarrk' and sharp 'tseit-tseit'. Abundant.

Bill deep yellow

Distinctive bare yellow skin around the eye tapers to a point behind the eye.

Rufous-brown; darker blackish brown on head, wings and tail

White shows in folded wing.

White under tail

Outer wing is blackish brown with a large white panel, conspicuous in flight.

Tail long, dark, white-tipped

## THRUSHES

**Bassian Thrush** *Zoothera lunulata* 27–29 cm
A secretive inhabitant of dense rainforest, wet eucalypt forest and woodland, heavily vegetated gullies, gardens. Typically in localities with dense overhead canopy and thick leaf-litter layer. Solitary, or sometimes in pairs or small parties, on the ground, scratching aside the accumulated debris to expose moist soil and the small creatures on which these 'ground thrushes' feed. May sometimes be detected in the quietness of the forest by listening for rustling sounds as dry surface leaves and twigs are thrown aside. When disturbed, may run briefly, then stop, motionless, relying on the mottled brown and cinnamon plumage blending into background colours, a technique effective in the dim light at the forest floor. Usually silent, most likely to be heard at dawn or in dull weather. Song consists of three clear notes, the first level, the second rising briefly, the third again steady – 'wheeer-aoo-whooo' – and may continue as a soft, tuneful sub-song. Sedentary or dispersive; common.

Previously the ground thrush in Aust. was thought to be the Eurasian *Zoothera dauma*. Australian birds are now separated and divided giving two endemic species: the Bassian Thrush and the Russet-tailed Thrush, *Z. heinei*.

● *lunulata*
● *cuneata*
● *halmaturina*

● Underparts white with extensive rufous tint across throat, breast and flanks
● Underparts white with cinnamon-rufous tint confined to breast
● Underparts white with very faint cinnamon or off-white tint on breast only

Lores and eye-ring off-white

Has slightly longer wings and bill.

Brown to olive-brown, heavily scalloped with black

Rufous-brown, scalloped

White with cinnamon tint to breast

All races have white panel in wing.

Light to heavy black scalloping

Extensive rufous tint over throat, breast, flanks

Sexes alike.
Imm.: like adults
Juv.: buff-edged coverts

Longer tail with smaller white spot on inner web compared with Russet-tailed Thrush, below

**Russet-tailed Thrush** *Zoothera heinei* 26–29 cm
Wet eucalypt forest and rainforest, tending to be, at least in breeding season, at a lower altitude than Bassian Thrush. This slightly more rufous 'ground thrush' species inhabits mid to NE coastal forests and substantially overlaps the range of the Bassian Thrush, of which it was formerly a race: it is now recognised as a separate species. The name refers to its most obvious distinguishing feature – the overall colour, and especially that of the rump and upper-tail-coverts, is slightly more rufous. In their habits the two species appear similar, but, in mountainous areas where both occur, the Bassian occupies higher range tops above 500 m while the Russet-tailed stays below 750 m. Where the two overlap there appear to be no hybrids. The call differs noticeably from that of the Bassian – two clear whistled notes, the second lower: 'wheee-rooo'. Local seasonal movements; in some places moves from higher to lower altitude and returns; moderately common.

**Tail, Bassian Thrush**
Outermost feather on each side of tail has, near its tip, a small, dull white patch, smaller, duller than that of the Russet-tailed Thrush.

Upper parts slightly brighter coppery rufous, less heavily scalloped compared with Bassian Thrush; this is most obvious on rump and tail.

Tail slightly shorter than that of the Bassian Thrush

Underparts creamy white with light cinnamon tint on breast; lightly scalloped, but more heavily along flanks

The white in tail of Russet-tailed Thrush is on the inner web and is often not visible in the folded tail.

The Russet-tailed Thrush has more white on the inner web of the outermost pair. May not be visible in folded tail, but shows on spread tail in flight, especially at take-off.

**Russet-tailed Thrush**
Outermost feather on each side of tail has a much larger white tip.

# 24 THRUSHES TO SUNBIRD

Families Muscicapidae and Sturnidae, with species including Australian thrushes and Shining Starling, but also with ferals outnumbering native species. Of special interest are three colourful accidental visitors: a rock thrush and two flycatchers. The swallows and martins (family Hirundinidae) also include three vagrants. Australian White-eyes (Zosteropidae), Mistletoebird (Dicaeidae) and Sunbird (Nectariniidae) complete this family grouping.

Robin = 12 cm

### muscicapidae
page 343

Bassian Thrush
Russet-tailed Thrush
Narcissus Flycatcher
Blue-and-white Flycatcher
Song Thrush
Common Blackbird
Blue Rock-Thrush

### sturnidae
page 345

Metallic Starling
Common Starling
Common Myna

### hirundinidae
page 346

White-backed Swallow
Swallows: Barn
Pacific
Welcome
Red-rumped
Fairy Martin
Tree Martin

### zosteropidae
page 348

Silvereye
'Capricorn White-eye'
Ashy-bellied White-eye
Yellow White-eye

### dicaeidae
page 349

Mistletoebird

### nectariniidae
page 349

Olive-backed Sunbird

## SONG THRUSH

**Song Thrush** *Turdus philomelos* 22–23 cm
Feral species of damp woodland and forest, parks and gardens. Forages on ground, finding small prey among leaf litter and surface soil. Male sings almost continuously through breeding season, a clear, spirited sequence of notes; mimics native birds. In contact and in flight, a thin 'tsip'. Melbourne-Geelong region; uncommon, localised occurrence.

Imm.: like adult, but the feathers of wing-coverts are edged buff, and those of the mantle have fine, light buff shaft streaks.

Plain rufous brown, crown to rump, wings and tail

Underparts off-white, faintly buff on the breast; heavily streaked with dark arrow-shapes, throat almost to under-tail-coverts

In flight, underside of flight feathers brown, linings buff

## BOWERBIRDS; BOWERS

**Great Bowerbird** *Chlamydera nuchalis* 33–38 cm
Eucalypt woodland, forest; favours dense thickets, margins of monsoon and vine scrub, mangroves, tropical gardens. Twin-walled avenue-type bower (below) is decorated with bleached bones, shells and green objects, including fruits. Calls are mixed, strongly contrasting noises – wheezing, churring, raucous, harsh, grating 'scraaach, graarrk' – intermingled with clear, piping whistles, often brief snatches of a small bird's song, then switching to the demonic cackling of Blue-winged Kookaburras. Sedentary; common.

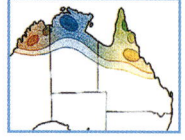

● nuchalis: at times has been split into other races in NE
● orientalis

In display at the bower, the male exhibits the pink nape crest by turning the head to face down and away.

● ♂ (Scale 2/3)

Uniformly pale fawny grey under

● Crown flecked or streaked with silvery grey
● Crown plain

● Strongly mottled, pale on dark edges

● Slightly smaller, shorter tailed; underbody colour grades from fawn at throat to creamy fawn on belly.
● The larger race; underparts uniformly pale fawn

Female: smaller, paler; usually has a greatly reduced lilac crest.

Fawn-grey at throat, fading to paler creamy fawn on belly and under tail

● ♂

Imm.: like female, except some some barring of flanks and belly

**Fawn-breasted Bowerbird** *Chlamydera cerviniventris* 29 cm
Inhabitant of coastal lowland fringes of tropical eucalypt and paperbark woodland, vine thickets, mangroves. Male noisy during spring to summer breeding season, with grating sounds, loud scratchy, rasping 'scaarrch-scarrch-scarrrch-', and longer 'sca-riarr-iarch'; almost explosive is a loud, harsh, abrupt 'tchuk,tchuk-,tchark'. In contrast are whistled notes, perhaps mimicry of part of another bird's song. Quite common within its confined range.

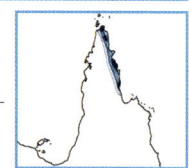

No crest

Head, neck, streaked brown and white

Fawn-breasted: unique raised base platform supporting avenue-type walls

Softly streaked with brown

Upper parts grey-brown with prominent buffy white tips and fine, pale margins

Male and female alike

Breast to under-tail-coverts is deep tawny buff.

Platform to 30 cm high

**Bowers** are often the first indication of the presence of bowerbirds, especially those of open habitat, the Western, Spotted and Great.

Regent Bowerbird: 'avenue' type, small, inconspicuous in dense rainforest

Satin: a large, thick-walled avenue with blue, violet, yellow decorations

Tooth-billed: display 'arena' of large green leaves, paler side up

Golden: very large, tall 'maypole' of twin towers, dense tropical rainforest

Great, Spotted and Western: avenue type, large, thick-walled, decorated with whitish and green objects

## TOOTH-BILLED, WESTERN & SPOTTED BOWERBIRDS

**Tooth-billed Bowerbird** *Scenopoeetes dentirostris* 26–28 cm
Dark bowerbird of upland tropical rainforest and vine scrub, altitudes 500–1500 m. Solitary or in small parties, feeding on fruits, leaves, insects. In breeding season the male attracts attention with powerful calls from perches above his 'bower', a cleared circle of ground spread with upturned leaves (p. 341). Voice extremely varied; switches from low, harsh, rasping noises to high, clear whistles and soft, mellow notes, including passages derived from songs of various other birds. Sedentary; common within its restricted range and habitat.

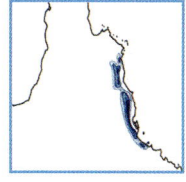

Upper parts dark olive-brown including crown, back, wings and tail

Eye brown, encircled by a narrow, pale brown eye-ring

Bill deep, heavy with a serrated cutting edge to both mandibles; grey-black

Imm.: like adults, but has brown bill and eye is lighter brownish-grey.

Underparts off-white, heavily streaked dark brown

A dull-plumaged bowerbird with robust 'toothed' bill used to snip and carry large leaves to maintain the display arena

Display stage of green leaves spread on bare ground, pale side up

**Western Bowerbird** *Chlamydera guttata* 25–29 cm
Often inhabits gorges in semi-arid regions where there are pools and native figs that grow from cliff crevices; also semi-arid scrub, especially thicker vegetation along watercourses. A slightly richer coloured version of Spotted Bowerbird, otherwise similar. Smaller, more russet-toned race *carteri* at NW Cape, WA. Calls are loud, harsh; echo far through rocky gorges. In display at bower, varied churring, grinding sounds; some like feral cat, perhaps mimicry. Sedentary in ranges near permanent water, nomadic in some localities; common.

● *guttata*
● *carteri*: centred on North West Cape, WA

The genus *Chlamydera* contains species adapted to dry open vegetation. The Spotted Bowerbird extends far into the semi-arid eastern interior, and the Western Bowerbird is found in deserts of central NT and NW of WA.

Female: crest smaller.
Imm.: no crest; black patterned throat and breast

Lilac tuft of feathers, raised in display, otherwise barely visible

Crown to rump and wings, brown with feathers tipped and finely margined cinnamon-buff

● ♂

● Smaller, richer tawny rufous, finer black feather edgings, shorter lilac ruff at nape, flanks with deeper and more extensive rufous tone

Underparts are deep buff with brown scalloping, chin to breast.

**Spotted Bowerbird** *Chlamydera maculata* 25–31 cm
Inhabitant of woodland of eucalypts, cypress pine, brigalow scrub. One of 4 similar grey or grey-brown *Chlamydera* bowerbirds that have adapted to open and mostly much drier habitats. Forages in trees, scrub thickets. Flight swift, direct, undulating, with long swoops on upcurved wings. Male maintains a display bower (p. 341). Noisy near the bower with varied churring, grinding and metallic sounds. Sedentary or locally nomadic; uncommon.

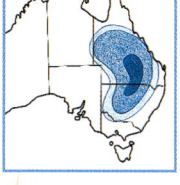

Tuft of lilac-pink, only obvious when lifted in display

Grey band: only seen on this species

An accomplished mimic, imitates diverse bird calls and other sounds.

The lilac-pink at the nape is usually only an insignificant small, flat spot of colour, but, in display, these feathers lift, giving a wide and protruding ruff or crest.

Underparts buff to off-white with dusky brown mottling and barring

♂

Female similar, but nape crest smaller, just a flat spot of colour

White stones, bleached bones and green berries decorate the bower floor.

## BOWERBIRDS

**Golden Bowerbird** *Prionodura newtoniana* 23–25 cm
Unique to a restricted area of NE Qld, confined to tropical upland rainforest, but coming lower on slopes of ranges in winter. Forages forest canopy for fruits and insects. Although a small bowerbird, it builds the largest bower of all Aust. species, a massive 'maypole' structure of sticks piled around several slender saplings (p. 341). From trees close above the bower the male gives rattling calls, a buzzing, metallic sound that entices females to his display. Sedentary; moderately common, limited habitat.

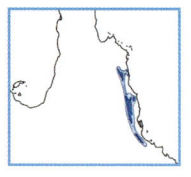

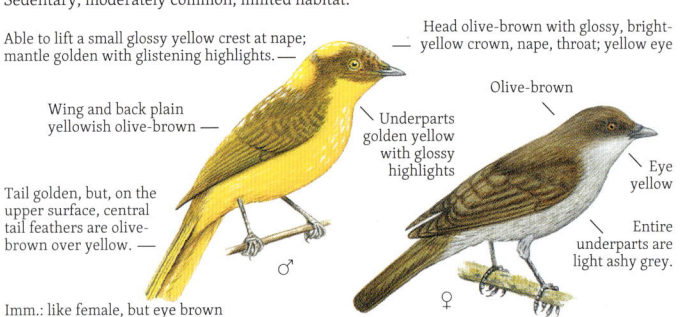

Able to lift a small glossy yellow crest at nape; mantle golden with glistening highlights.

Head olive-brown with glossy, bright-yellow crown, nape, throat; yellow eye

Wing and back plain yellowish olive-brown

Olive-brown

Underparts golden yellow with glossy highlights

Tail golden, but, on the upper surface, central tail feathers are olive-brown over yellow.

Eye yellow

Entire underparts are light ashy grey.

Imm.: like female, but eye brown

♂   ♀

**Regent Bowerbird** *Sericulus chrysocephalus* 26 cm
Spectacular in flight through rainforest with black of body all but lost against gloomy background, only the flickering gold of the wings apparent. Inhabitant of cool temperate rainforest on the highest ranges, and down into some coastal rainforests; also in rainforest regrowth and thickets. Its quiet, retiring preference for dense vegetation can make a sighting difficult. The male, in display, gives a husky, wheezy, grinding 'kzzzark-kzzaark, kzzzark-kzzaark, kzzzark', and low chattering. Sedentary, or short seasonal movements between higher and lower altitudes. Scarce, but locally common.

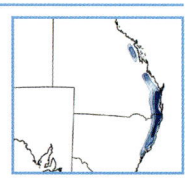

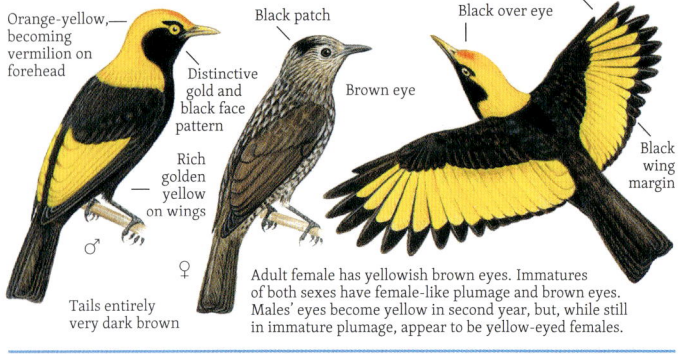

Two outer primaries black

Orange-yellow, becoming vermilion on forehead

Black patch

Black over eye

Distinctive gold and black face pattern

Brown eye

Rich golden yellow on wings

Black wing margin

♂   ♀

Tails entirely very dark brown

Adult female has yellowish brown eyes. Immatures of both sexes have female-like plumage and brown eyes. Males' eyes become yellow in second year, but, while still in immature plumage, appear to be yellow-eyed females.

**Satin Bowerbird** *Ptilonorhynchus violaceus* 28–34 cm
Occupies rainforest, wet eucalypt forest, and, in winter, open woodland. Forages widely, from treetops to ground, finding fruits, insects. Flight swift, direct, swooping undulations and strong wing beats. Noisy in breeding season, especially around bower (p. 341). Loud, harsh, grinding churring and wheezy buzzing, rapid, rollicking 'tzzarr-tzzarr-tzzarr-tzzarrtz-tzzarr-tzzarrr', whistled 'whichiew'; some mimicry. Sedentary or locally nomadic; moderately common.

● *violaceus*
● *minor*: smaller, tail and bill shorter; bill also finer

Entirely blue-black with a glossy sheen to plumage giving diffuse, soft highlights of deep blue-violet.

Eye intense lilac to bright sapphire

Bill stubby, pale, yellowish tip

Eye bright lilac

Upper parts greenish-grey ●
Or, if dull bluish-grey ●

♂ ● ●

Bright pale yellow with distinct dusky scalloping ●
Dull cream with smudgy grey scalloping ●

♀

Tail plain grey-brown

## APOSTLEBIRD & CATBIRDS

**Apostlebird** *Struthidea cinerea* 29–33 cm
Conspicuous, gregarious family groups, bustling, confident; typically 5-10, but, after the breeding season, up to 50 birds. Occupy open and rather dry country, never far from water. Calls include various harsh noises; these usually taken up in chorus by others of the group. Includes a harsh, scratchy 'scrairch-scraach', grinding, grating sounds, louder 'tz-iew, tz-iew' and rough, abrupt warning calls; garrulous participation in any flock activities. Very noisy early in breeding season, with much loud chattering. Common, but of patchy distribution.

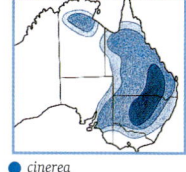

● *cinerea*
● *dalyi*: N and NW of range

Bill stubby, stout, black, superficially finch-like

Pointed pale tips make plumage look rough, shaggy.

Wings grey-brown with lighter, brighter cinnamon-rufous margins

Grey iris with pale yellow outer ring (Juv. brown eyed)

● Larger, and with relatively large bill. Underparts paler smoky grey; upper body leaden grey without brownish tint, much heavier black streaks on mantle, upper back

Dusky black tail often flicked or bounced

Imm.: as adults but brown eyed until second year

**Spotted Catbird** *Ailuroedus maculosus* 28–32 cm
Grating, wavering calls of Spotted Catbirds are a characteristic sound of Australia's rainforest, monsoon and vine scrub, riverine and paperbark forest. In pairs or small groups, forages in high to mid canopy, occasionally lower foliage. Can be difficult to sight, even when calling. Feeds on fruits, buds, insects, small reptiles, nestlings. Call is a cat-meowing, grating wailing, similar to that of Green Catbird, but probably even more nasal; the 'here-I-are' call of this species is a drawn out, wavering, grating 'hee-eir-I--aaa-ar-rr'. Sedentary or localised seasonal movements; common.

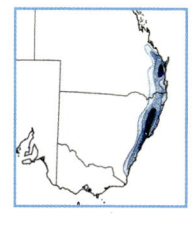

● *maculosus*
● *joanae*

Brownish black ear-coverts, crown; nape to upper back brownish, heavily spotted pale buff

Eye red

Bill deep, short, pale pinkish buff

Lower back and wings are bright olive-green. Coverts and secondaries have small off-white spots at tips.

White flecks and streaks set in olive-brown

Differences between races slight
● NE Qld, has brownish black head markings.
● Cape York, smaller, has blacker markings about the head, and underparts are deeper buff, more coarsely spotted.
Some authorities combine *melanotis* (its two races), with *A. crassirostris* as a single species, Green Catbird, of three races.

Tail bright olive-green; small white tips barely visible with tail closed

Imm.: darker dull green, iris brown

**Green Catbird** *Ailuroedus crassirostris* 24–33 cm
Subtropical and sub-temperate rainforest and paperbarks, occasionally in adjacent eucalypt forest. Forages actively in foliage of canopy, tends to be rather wary. Unlike other members of the bowerbird family, the catbirds are monogamous, with a strong pair bonding. The sexes are alike in plumage, and both contribute to raising the young. Calls are similar to those of the Spotted Catbird, and, when familiar, become a welcoming sound characteristic of many tracts of rainforest. These calls can be put into words as 'here-I-are', but typically a much more drawn out and wavering version, as in: 'heeeir-I-aaa-aarr'. Sedentary; common.

Head is greenish-brown, mottled black and finely flecked pale buff.

Iris bright red in direct sun or flash; white partial eye-ring

Slight grey tone to bill

Back, wings and rump are brilliant emerald-green with very conspicuous pure white spots at tips of feathers; on coverts these spots form two white wing bars.

Imm.

Eye dark red-brown for first 2 years

Greenish buff to dull emerald with short white streaks

Tail brownish emerald with larger white tips

Plumage is like that of adults, but dull colours.

Buff

## TORRESIAN CROW, FOREST RAVEN & CHOUGH

**Torresian Crow** *Corvus orru* 48–53 cm
Open forest, woodland, rainforest margins, coasts, farms. In arid regions, trees along rivers. In pairs, or flocks after breeding. In greeting, rapidly and repeatedly flutters wings. Glides lazily on drooped wings. Call is a nasal 'uk-uk-uk-uk-', often last notes slowest. Also harsh, aggressive 'arrk-arrk-arrk-arrk, arrrgk', evenly pitched and spaced until last deep gurgles; also a long, nasal call. Adults sedentary, juveniles nomadic. Common.

*cecilae* is only mainland race; *orru* occurs on Torres Str. islands

Glossy black with blue-violet sheen; crows have white down at base of feathers.

After landing, shuffles wings.

Bill longer than head; heavy

Throat hackle bulge not usually evident, but a smooth bulge shows during extended calls while perched, body in horizontal posture.

Tail long, square-cut, quite broad

In flight, wings broad, rounded; tail long, broad, square-tipped

**Forest Raven** *Corvus tasmanicus* 51–54 cm
Diverse habitats: high snowfields, alpine forest, islands, beaches, woodland, wet eucalypt forest, open farmland and coastal heath. Largest raven, neat, sleek plumage, heavy, sluggish wingbeat; glides on drooped, wide-fingered, rounded wings. Gives a wing-flick flight display, but calls from perch without wing flips and with tail depressed. Southern race forms loose flocks out of breeding season; may be seen in flight. Call is deep, rough, bass 'karr, karr, kar-r-r-r', last note slow and falling away. 'Returning home' call is deep, descending. Common in Tas.; uncommon and patchy on mainland.

● *tasmanicus*
● *boreus*: 'Relict Raven'

Deep, intense, glossy black; hidden grey down at base of feathers

Tail very short, broad

Massive bill, heaviest among corvids

Only small bulge at throat, most obvious while calling; hackle tips forked

Tail longer

Juv.: dark eye

● Shortest tail of all corvids relative to size of bird

— Broad, blunt tipped wings

● Longer wings and tail, more like Australian Raven

**White-winged Chough** *Corcorax melanorhamphos* 43–47 cm
Woodland, open forest, mallee, mulga, timbered watercourse margins, cypress. Forages usually on the ground, moving forward in a scattered group, probing and scratching, seeking insects and other small ground-dwelling prey. Chough flocks usually build up over several years from the offspring of one pair. Size ranges from 4 to 20 birds. Has rather mournful, piping calls, beginning as high, clear, musical whistles, descending and becoming more mellow; a pause, then rising high and clear again for the next long, downward slide: 't-i-e-e-e-ew, t-i-e-e-e-uw, t-i-e-e-uuw'. Usually given by many of a group together, calls intermingling. In alarm or aggression, harsh, rasping sound, 'ch-z-z-zark'. Locally nomadic; common.

● *melanorhamphos*
● *whiteae*: slightly smaller

Bill strongly downcurved, especially near base: gives unusual drooping look to the entire bill.

Eye orange with bright deep pink outer ring

White of wing patch usually hidden in folded wing; may show if wing is drooped.

Tail long, far beyond tips of folded wings

Very large, conspicuous white wing patches divided by black lines along leading edges of flight feathers

## RAVENS & CROWS

### Australian Raven *Corvus coronoides* 48–54 cm
Inhabits open country, natural and cleared. Behaviour and calls aid identification. In flight, it appears rakish, with backswept, tapered wings that have a slow downstroke and quick upwards flick. Glides with wings flat, tips widely fingered. Has a shallow, fluttering, 'returning to territory' flight with slow, tremulous descending calls. Voice is strong: first note of sequence rather high, loud, clear, then descending, fading to a deep, slow, muffled groan or gurgle: 'aairk, aark, aaarh, aargargh'; often wailed 'aaieerk'. Common.

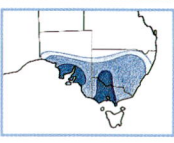

● coronoides
● perplexus

Black with purplish-greenish sheen. Ravens have grey down at base of feathers of head and neck, hidden unless ruffled by wind.

Head slightly higher towards rear of crown

Large robust bill, slightly longer than head

● Shaggy, long pointed hackles
● Shorter hackles

● Large, and heavy bill
● Smaller, finer bill

Tail length similar to folded wings

Perched territorial calling, usually from high tree: holds body horizontal, head low, and calls without any upwards flipping of wings.

Wings rather slender; shape more pointed, backswept, than those of crows

Juv.: lacks hackles; eyes are grey, brown or hazel.

### Little Raven *Corvus mellori* 48–50 cm
Flight agile; looks compact, wings tapered and only slightly fingered at tips. Only slightly smaller than the Australian Raven, with which it shares range and habitat, but has different behaviour. Inhabitant of open plains, woodland. Gathers in tight flocks of several hundred that may attract attention by synchronised aerobatics, and by massed flights to and from roosting sites. Voice a deep, guttural baritone, notes quick, clipped or abrupt, 'ok-ok-ok', then may fade away, 'ok-orhk-orrh', creakings. When perched, calls are accompanied by quick flicks of both wings together. The 'returning home' call, with wing-fluttering flight, is deep, croaking. Common.

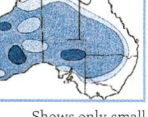

Flight agile, rapid wing action

Only slightly smaller than the Australian Raven; glossy black plumage

Tail quite long, slender

While calling, flicks closed wings upwards together.

Bill less massive, but still longer than head

While calling, shows only a small bulge of short, fork-tipped throat hackles.

Wings tapered, often backswept, but soars with wings held straight, tips broader.

Juv.: eyes brown, later hazel

### Little Crow *Corvus bennetti* 45–48 cm
Smallest Aust. corvid; plumage rather dull black, looking dusty. Open country, usually mulga, mallee, spinifex arid and semi-arid regions. Much more gregarious than the Torresian Crow; flocks vary from small to very large, often performing spectacular aerial displays. Gives only slight shuffle of wings after landing. Has a deep, nasal, 'nark-nark-nark' and slow creaking sounds. Sometimes gives a creaky call with fluttering 'returning home' or 'returning to flock' flight action. Abundant.

Shows only small hackles bulge when calling.

Plumage is black with blue-violet sheen, but many rather dull.

Tail usually longer than folded wings; variable

Smallest bill, shorter than head

When calling from a perch or ground, does not conspicuously flick the folded wings.

When giving territorial flight calls, misses a beat by pulling wings close to body, giving an undulating flight.

Short legs

### House Crow
*Corvus splendens* 42–44 cm
Native from S Iran to W China. Occasional accidental arrival; not established. Sociable, large flocks gather in noisy masses on trees at regularly used roosting sites. Voice a rasping, 'kza', downwards 'kzow', lower 'kiowk'.

Variable pale nape or collar

Black

Grey-brown may extend to sides of breast

Some almost entirely black

Eye dark brown; Aust. corvids are white-eyed (juveniles are brown-eyed)

# RIFLEBIRDS

**Paradise Riflebird** *Ptiloris paradiseus* 28–30 cm
Subtropical, temperate rainforest, nearby paperbark swamp, wet eucalypt forest; adjoining eucalypt woodland. Call is a slow, drawn out rasp: starts low, builds to loud, harsh grating and rasping; fades to soft hiss: 'scraarsh, scraaarsh', or 'yaaarss'. Also has long, upwards whistle and soft churring in displays. Sedentary; uncommon.

The three Australian riflebirds have similar plumages, differing in detail that might not be evident when, from far below, one is seen as a silhouette against the sky. But once recognised as 'a riflebird', its location, checked against the riflebird maps, may be the best guide to species.

In flight, plumage of male makes rustling sounds.

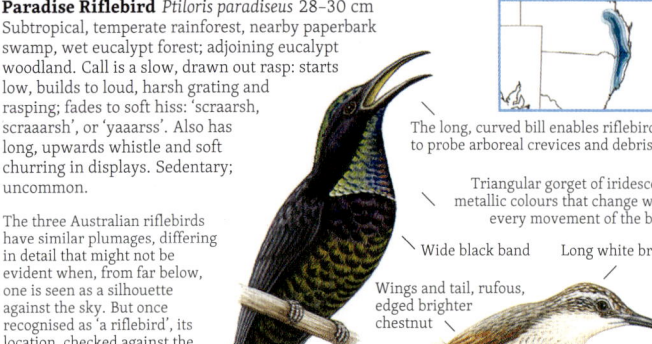

The long, curved bill enables riflebirds to probe arboreal crevices and debris.

Triangular gorget of iridescent metallic colours that change with every movement of the bird

Wide black band    Long white brow

Wings and tail, rufous, edged brighter chestnut

Very short, black

Unmarked off-white

Pale buff with crescents of dark brown

♂    ♀

---

**Victoria's Riflebird** *Ptiloris victoriae* 23–25 cm
Tropical rainforest, mostly on ranges above 500 m. Closely resembles the Paradise Riflebird, but smaller, and has a wider and more intensely black breast band. Like other riflebirds, male has a display perch on a bare branch or stump, where he calls and parades his courtship routines and poses. Calls almost identical to those of the Paradise Riflebird; a similarly harsh, rasping call, longer versions slightly undulating: 'scraark, scraar-aark'. Sedentary; moderately common.

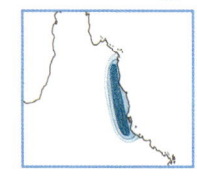

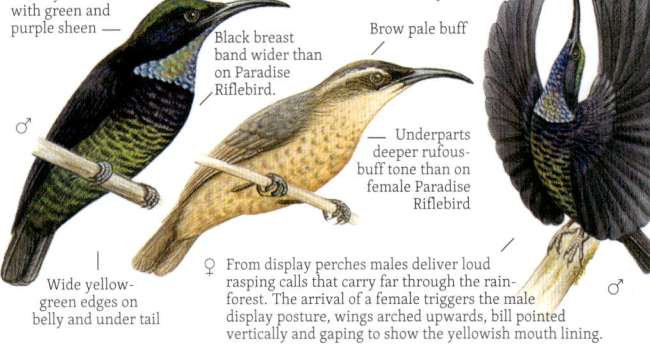

Velvety black with green and purple sheen

Bill black, long, downcurved. Inside of mouth is yellow.

Black breast band wider than on Paradise Riflebird.

Brow pale buff

Underparts deeper rufous-buff tone than on female Paradise Riflebird

Wide yellow-green edges on belly and under tail

♂    ♀ From display perches males deliver loud rasping calls that carry far through the rainforest. The arrival of a female triggers the male display posture, wings arched upwards, bill pointed vertically and gaping to show the yellowish mouth lining.    ♂

---

**Magnificent Riflebird** *Ptiloris magnificus* 28–33 cm
Rainforest; monsoon and vine scrub along rivers. In these dense, wet forests, this riflebird frequents the mid to upper strata, usually keeping to the tree trunks and thickest limbs, where, feeding among the clumps of ferns and mosses that festoon the bark, it is difficult to sight from below. Males give a deep, powerful 'awoo-arr-WHEET', the sound swooping low through the 'woo', a growling 'arr', then rising to a powerful, sharp, drawn out, whipcrack 'WHEET'. Sedentary; fairly common in restricted range.

Narrow black band, above an iridescent, changing, gold, green, bronze band

Underparts off-white with dense, fine, dark speckling and barring

Colours of iridescent plumage, especially the triangular shield across throat and upper breast, change with every move; at one moment black, then iridescent turquoise or purple, often with yellow and bright green highlights.

Tail rust-brown, short, yet longer than folded wings

♂    ♀    Imm.: like adult female; males 2 yrs to adult plumage

335

# RIFLEBIRDS TO BOWERBIRDS

This group contains several families with fascinating plumage and behaviour: Australia's 'Birds of Paradise', riflebirds (Paradisaeidae); and bowerbirds (Ptilonorhynchidae), which include *Chlamydera* genus birds, adapted to country much drier than the rainforest habitat of other bowerbirds. Two related families are the crows and ravens (Corvidae), and the mud-nest builders (Corcoracidae).

Magpie = 42 cm

## paradisaeidae
page 334

- Trumpet Manucode
- Paradise Riflebird
- Victoria's Riflebird
- Magnificent Riflebird

## corvidae
page 336

- Torresian Crow
- Little Crow
- Forest Raven
- 'Relict Raven'
- Little Raven
- Australian Raven
- House Crow

## corcoracidae
page 337

- White-winged Chough
- Apostlebird

## ptilonorhynchidae
page 338

- Satin Bowerbird
- Tooth-billed Bowerbird
- Spotted Catbird
- Western Bowerbird
- Green Catbird
- Spotted Bowerbird
- Golden Bowerbird
- Great Bowerbird
- Regent Bowerbird
- Fawn-breasted Bowerbird

## TRUMPET MANUCODE

**Trumpet Manucode** *Phonygammus keraudrenii* 28–32 cm
Inhabitant of rainforest, vine scrub, eucalypt forest and woodland. Forages for fruits in forest canopy, often high, where it can be difficult to see. Male has a very long windpipe, coiled between the skin and the muscles of the breast, enabling him to produce his loud, remarkably deep, guttural, reverberating, trumpet-like blast of sound, and an abrupt, inwards gulping 'owwgk'. Sedentary; moderately common.

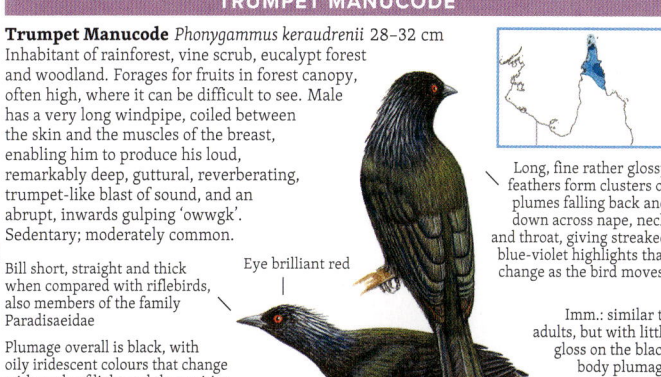

Long, fine rather glossy feathers form clusters of plumes falling back and down across nape, neck and throat, giving streaked blue-violet highlights that change as the bird moves.

Bill short, straight and thick when compared with riflebirds, also members of the family Paradisaeidae

Eye brilliant red

Imm.: similar to adults, but with little gloss on the black body plumage

Plumage overall is black, with oily iridescent colours that change with angle of light and the position of the bird, giving a shifting dull turquoise or emerald sheen across parts of the plumage.

Tail long, rounded

# GREY CURRAWONG

**Grey Currawong** *Strepera versicolor* 45–50 cm
Habitats used are many, spread across regions and climates ranging from southern coastal to semi-desert. Included are forest, woodland and mallee, larger, denser coastal thickets, heath. Across this diverse range, the Grey Currawong's plumage varies considerably; at present six subspecies are justified. All Grey Currawongs, however, have the same distinctive clinking call, and, at close range, a distinctive bill shape. Rather than the thick-pointed bills of the Black and, to a lesser extent, Pied species, the Grey's bill has straighter edges, tapering evenly to a somewhat lighter, finer pointed tip. But, like other currawongs, the Grey is omnivorous, taking small animals such as birds, their eggs and young, rodents, frogs, insects, carrion, and seeds and fruits. It hunts both on the ground and in trees, where it probes crevices of bark and timber. In flight it has a slow, languid, shallow, uneven, looping, semi-closing wing action. The best-known of many calls is a loud, metallic 'kling-kling-kling' or 'chring-chring', and various softer mewings and squeaks. In SW Aust., a clear, ringing 'tiew-tiew-tiew', while in Tas. gives a metallic 'kier-kier-killink'. Common, especially in forested SE and SW; uncommon to rare in arid regions.

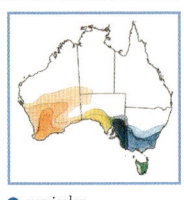

- *versicolor*
- *melanoptera*
- *intermedia*
- *arguta*

Not illustrated:
- *plumbea*: 'Clinker'
- *halmaturina*: 'Kangaroo Is.'

Darker lores and around the bright yellow eye

White across primaries

Widespread, variable species; plumage varies, grey to dusky black; six races.

Less massive; straighter taper to finer point

White tips to primaries and tail

White of wing patch may show in folded wing.

Under-tail-coverts white

White-tipped flight feathers

Mid to dark smoky-grey plumage of south-western race is most like the above race, as are its large white wing and tail markings. It differs in its slightly browner cast and heavier bill. Becomes darker, dusky brownish black, in extreme south-west corner.

**'Black-winged Currawong'**
White patch at base of primaries is absent or tiny and dull.

White

All races have white under-tail-coverts.

**'Brown Currawong'**
Brown form of Grey Currawong, Eyre and York Peninsulas, SA

Less massive than bill of Black Currawong

**'Clinking Currawong'**
Largest, darkest race. Is confined to Tas. All races of the Grey Currawong have a white tipped tail.

No white at wing tips, or greatly reduced and dull

White
White may show on folded wing.

Imm.: dull grey-brown to dusky plumage; faintly streaked on breast; pale yellow gape; brown eyes

Under-tail-coverts white, unlike Tasmania's Black Currawong

Has white in wing, unlike the other Tas. species, the Black Currawong.

## PIED & BLACK CURRAWONGS

**Pied Currawong** *Strepera graculina* 42–50 cm
Diverse habitats including woodland, forest, from coastal to alpine, rainforest, scrub, farmland, gardens. Conspicuous, often noisy; forms large flocks that move nomadically over large distances in autumn and winter. Prey includes insects, small reptiles, birds, carrion, berries; in suburbs, bold scavengers. In flight, a distinctive, slow, uneven, rowing or lapping, semi-closing wing action, distinct from the steady, purposeful beat of crows and ravens. At dusk, often congregates in communal roosts. Call is a slow, rather rollicking series of mellow, often gurgling sounds: 'kurrok, kurrowk'; 'curra-currow-currowk'; 'carrow-carrow-currawowk'. In flight, gives a distinctive, wailing, raucous, descending 'kirrair-kirrair-kirrowk'. Seasonally nomadic; abundant.

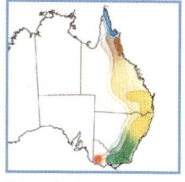

- 🔵 *magnirostris*
- 🟤 *robinsoni*
- 🟡 *graculina*
- 🟢 *nebulosa*
- 🔴 *ashbyi*

Bills of currawongs are similar to those of the butcherbirds, but proportionately larger, and with stronger convex curvature to both upper and lower margins. The appearance is heavy, even massive, relative to the head of the bird, and contributes strongly to the visual character or 'jiz' of currawongs.

Differences between races slight. From N to S there is increasing overall size of bird, but bills become relatively smaller; there is diminishing size of white markings on wings and tail, and plumage fades from black to sooty or slate grey.

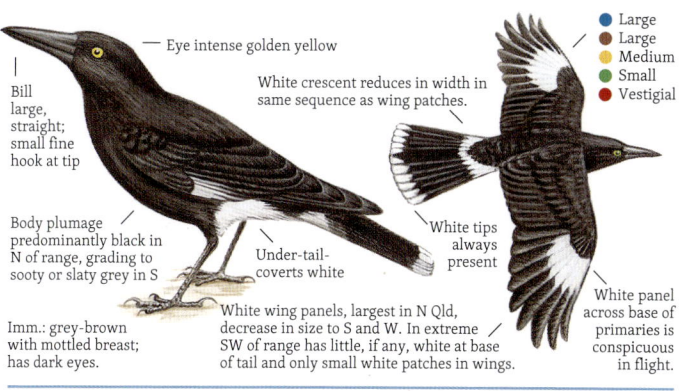

- 🔵 Large
- 🟤 Large
- 🟡 Medium
- 🟢 Small
- 🔴 Vestigial

Eye intense golden yellow

Bill large, straight; small fine hook at tip

White crescent reduces in width in same sequence as wing patches.

Body plumage predominantly black in N of range, grading to sooty or slaty grey in S

Under-tail-coverts white

White tips always present

White panel across base of primaries is conspicuous in flight.

Imm.: grey-brown with mottled breast; has dark eyes.

White wing panels, largest in N Qld, decrease in size to S and W. In extreme SW of range has little, if any, white at base of tail and only small white patches in wings.

**Black Currawong** *Strepera fuliginosa* 47–49 cm
Diverse habitats: alpine moor, dense wet forest, drier open woodland, farms, orchards, suburbs, coasts, islands. Most conspicuous autumn and winter as flocks move from heavily forested ranges to lowlands and open country. Has loosely floating flight typical of currawongs; wing beats slow, relaxed, with shallow, uneven, semi-closing action. Flocks keep high when travelling longer distances. Bird usually forages on the ground among leaf litter and low vegetation, and, on coasts and islands, fossicks on beaches among seaweed and rocks. Also, more than other currawongs, searches limbs of trees, probing crevices and ripping bark. Takes small birds, reptiles, mice, insects, carrion, berries, orchard fruit. Noisy, loud calls, often in flight; usually long, rollicking, wailing, yet rather musical 'kiarr-weeeik, weeeik-yarr'. Also short metallic croaks. Common.

- 🔵 *fuliginosa*
- 🟤 *colei*
- 🟡 *parvior*

Bill is massive, heaviest among currawongs. Has a strong convex curvature to both upper and lower profiles, and a deep base; its black silhouette merges smoothly to the black head, which appears relatively small.

Races differ in size and proportionate length of tail.

- 🔵 Large, long-tailed
- 🟡 Mid-size, mid-tail
- 🟤 Small, short-tailed

White tips to primaries

Eye intense golden yellow

Under-tail-coverts black

Only one white bar across, that being at the tail tip

Tail white only at tip. Race *melanoptera* of the Grey Currawong also has black upper-tail base and lacks white wing panels, but differs in having white under-tail-coverts.

Small white tips; never white at base of primaries

Imm.: paler, browner, dark eye, yellow gape. Like adults, has heavy billed appearance.

## BUTCHERBIRDS

**Black Butcherbird** *Melloria quoyi* 38–44 cm
Dense, gloomy tropical vegetation: rainforest, monsoon forest, vine thickets, paperbarks, mangroves. Inconspicuous in dark plumage, wary and skulking. Flight between trees swift and direct. Hunts from perches; takes prey – a small reptile, rodent, frog or invertebrate – in swooping dive to the ground. Call is a deep, clonking 'kwok' or 'kwow'; in NT, it is a currawong-like 'k-wonk' and harsh 'carr-kark'. Song has typical butcherbird character, deep, abrupt, yet musical, first note high, lilting, flute-like; two deep and mellow, then last note rising: 'kowk, koork-koork, kowk'. Also a long, mellow, bubbling yodel 'kwowk-coor-coor, kwowk-coor-coor, kwowk-coor-'. Sedentary; common.

- *spaldingi*
- *rufescens*
- *jardini*

- Has immatures that may be rufous-buff, or entirely black. Both plumages may occur within one brood; adults are black. Other Aust. races have only black immatures.

Adults entirely glossy black, all races

Bill heavy, hooked, pale, tipped black

Eyes brown

Dark scallopings, faint or fine

No white tips on tail

Imm. and Juv., rufous phase

**Grey Butcherbird** *Cracticus torquatus* 26–30 cm
Margins of rainforest, paperbark swamp, eucalypt woodland, mulga. Voice rich, varied, including musical and mellow as well as harsh sounds, mimicry. Songs include slow, deep, mellow notes, 'quorrok-a-quokoo', and deep, bubbly to loud, clear 'kworrok-a-chowk-chowk-chowk-chowk'. Loudest, often given in wing-quivering flight, are vigorous piping notes that may develop a harsh, strident quality: 'quayk-quayk-quiak-qraik-qzaaik-kzaaik-kzzaik'. Sedentary or locally nomadic; common.

- *torquatus*
- *leucopterus*
- *cinereus*: in Tas.

White spot
Narrow partial collar, white throat

Mid to light grey back, wide white rump band

Both darker grey on backs and have narrow white rump band.

Broad white tail tip
Narrow white tip

White-tipped
White spot at lores, between bill and eye

**Silver-backed Butcherbird**
*Cracticus argenteus* 28 cm
Now considered distinct from the Grey Butcherbird, the Silver-backed is distinguished by its silvery grey back and lack of white spot on the lores. It shares plumage features with the Black-backed Butcherbird.

Silvery grey
No white at lores
Silvery white
Pattern as adult, but brown and dull fawn
Imm. all races

- *argenteus*
- *colletti*

- Larger, partial black breast band, large white tail tips
- Smaller, no breast band, small white tips

**Black-backed Butcherbird** *Cracticus mentalis* 26–28 cm
Open forest, woodland. Hunts from perches, takes prey – small birds, their eggs and nestlings included – from ground, branches, foliage. Flight swift, direct, tree to tree. Calls deeper, slower than Grey Butcherbird; includes duets, mimicry. Deep, mellow notes intermingle with sharper sounds, 'whorrr-kwiiek, whorrr-kwiiek', and 'quorr-quorr-kweeik, quorr-quorr-kweeik'. Also a call similar to the Grey, a rapid, rollicking, slightly harsh, 'kwarr-kawiek, kwarr-kwiek, kwarr-kwiek'. Sedentary; common.

White throat and wide partial collar, the white broken only by a narrow black strip from nape to mantle and back

Black lores
Pale partial collar

Large area of white on rump extends well up towards lower back.
White-tipped

White-tipped
Pattern as adults, but in brown and pale buff
Imm.
Extensive white margins and scalloping

## WOODSWALLOWS

**White-breasted Woodswallow** *Artamus leucorynchus* 17 cm
Above and around forest, paperbark and mangrove swamp, rarely far from water; wanders far inland along pools of tree-lined rivers. Highly sociable, in small flocks or groups up to fifty, rarely several hundred. An aerial feeder, flutters and soars over water or forest canopy in pursuit of flying insects. Brisk calls: often, as a contact chatter within the flock, a rather brassy 'aerk, aerk-aerk,aerk'. In alarm, calls are louder, harsher, more strident. Sedentary in far N, seasonally nomadic elsewhere; common.

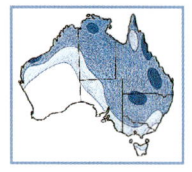

Pure white, breast to under tail

Sharply defined border

Upper parts dark brownish slate-grey, except white rump

White rump prominent on dark upper parts

Streaked, speckled rufous buff

Tail, dark, plain, both sides

Stark white rump

Barred buff on white

Juv.

Square, entirely dark; all other woodswallows have white tips

**Masked Woodswallow** *Artamus personatus* 19–20 cm
Inhabits open forest, woodland, heath, roadside and farm tree belts. Dusty tones of grey, black and white typical of woodswallows, which are among the few passerines that have powder down through the plumage. Masked Woodswallows are gregarious, soaring and travelling high for hours in large flocks, often intermixed with White-browed Woodswallows, hawking insects with flutter-and-glide wing action. The voice is rather nasal, a querulous 'chrrt' and 'chak', a soft 'chrrup', occasional mimicry. Common.

Masked Woodswallow and White-browed, below, are the only species with marked difference between sexes.

Silvery grey

Sharply defined, pale-edged black face mask

Mask indistinct

Slightly browner than male

♂  ♀

Juv.: similar to female, slightly browner, flecked and mottled buff

**White-browed Woodswallow** *Artamus superciliosus* 19 cm
Usually above forest, woodland, heath, spinifex, farmland, suburbs; often in high-flying flocks. Hundreds, sometimes thousands, of birds flutter, wheel and glide, showing chestnut underbody in colourful contrast against blue of sky and white of wings. Flocks stay aloft for hours; may descend suddenly to forage for nectar and insects among foliage. Flocks tend to travel N in autumn, wintering in central Qld and NT, then go S in spring to breed in scattered colonies. Contact call is a high, rather musical, descending 'tchip-tchep'. From large flocks this creates a constant yapping chatter. The song is a softer twittering that includes some mimicry; in alarm a harsh, scolding rattle. Highly nomadic; common.

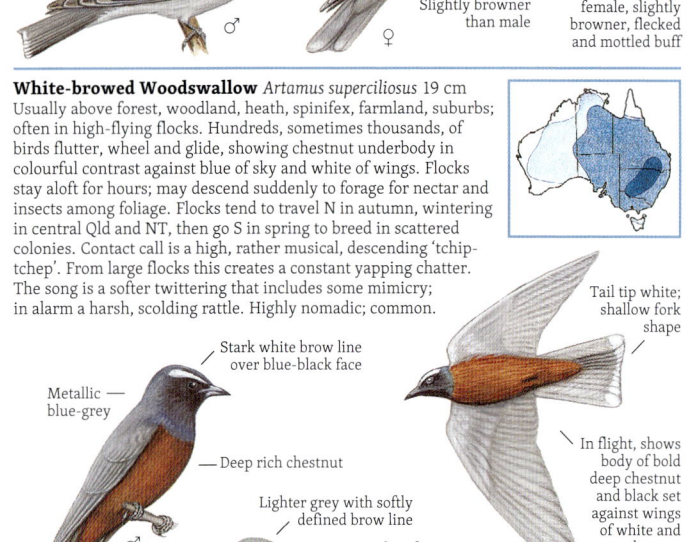

Stark white brow line over blue-black face

Metallic blue-grey

Deep rich chestnut

Lighter grey with softly defined brow line

Brow line faint

Paler rufous-buff

White tip

Tail tip white; shallow fork shape

In flight, shows body of bold deep chestnut and black set against wings of white and pale grey.

Pale grey-buff, heavily streaked brown

♂  ♀  Juv.  ♂

# WOODSWALLOWS

**ashy, dusky or dark-bodied, rather plain-faced**

Black-faced — Races with black under-tail-coverts: 🔵 🟢

Black-faced — Races with white under-tail-coverts: 🟤 🟡

Dusky

Dusky

Little

Little

**bold, contrasty head or body**

White-breasted

Masked

White-browed

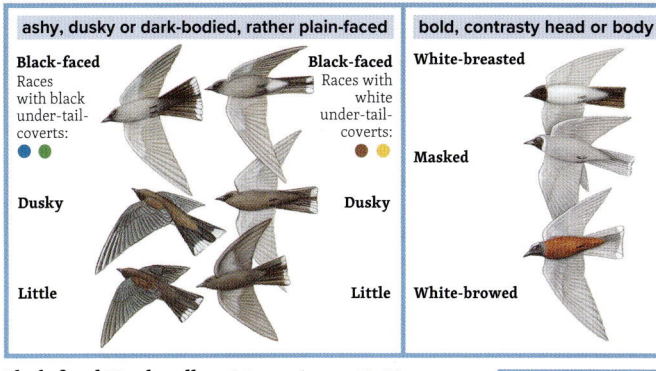

**Black-faced Woodswallow** *Artamus cinereus* 18–20 cm
Open country, the drier woodland, mulga, spinifex, gibber plains, samphire; avoid heavier forest of SE and SW Aust. In small flocks of several pairs, darts out from perch on bare limbs, fences or poles to chase insects in swift, powerful flight. Occasionally hovers; drops to ground to take small prey. Also feeds among foliage, seeking nectar and insects. Call is a subdued, slightly scratchy, 'tchif', 'tchif-tchiff' or 'tchif-tchap', the song a brief, soft twittering. Has a harsh alarm call. Nomadic; common.

🔵 *melanops*
🟤 *dealbatus*
🟡 *normani*
⚪ *cinereus*

Back ashy mid grey
Wings grey-brown
Variable size triangular patch across face
Juv.
Buff streaked and spotted, heavily on back
Large white tail tips
Black chin and face
Black under tail
Underwings dull white, but silvery in sunlight
White under-tail-coverts

**Dusky Woodswallow** *Artamus cyanopterus* 17–18 cm
In small flocks, hawking insects through clearings and above canopy. Often gather closely along a limb, waving and rotating tails, preening. At night, members of flock cluster closely in a fork or shallow hollow. Contact call, in flight and perched, is a brisk, vibrant 'tseit-tzeit', softer 'zut-zut'. A soft, quiet song often includes mimicry. In defence of nest, a harsh, scolding chatter. Migratory, common.

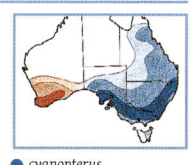

🔵 *cyanopterus*
🟤 *perthi*

Smoky or dusky browns and greys
Head dark grey-brown; blackish lores
Dusky grey
Juv.
Streaked wings like adult
White streak at edge of wing: 🔵 wide; 🟤 narrow
White tips all large
Outer tips smaller
White wing-edge: wide on 4 outer primaries 🔵 on 2 outer only 🟤

**Little Woodswallow** *Artamus minor* 12–14 cm
Rocky ranges, gorges, open arid scrub, tropical woodland. Groups often cluster on dead limbs. Darting away after insects, they show entirely dark wings that lack the white streak of the similar Dusky Woodswallow. Contact call is a brisk, high, chirping 'peit-peit', often given in flight; the song is a soft twittering. Sedentary around gorges of ranges; nomadic on open plains.

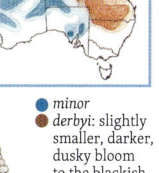

Body dusky dark brown, wings very dark blue-grey
Lores dark
Wings plain blue-grey without white edges to outer primaries
Dark, white tips

🔵 *minor*
🟤 *derbyi*: slightly smaller, darker, dusky bloom to the blackish brown body

Juv.: like that of Dusky, smaller, lacks white streak along wing.

# AUSTRALASIAN FIGBIRD

**Australasian Figbird** *Sphecotheres vieilloti* 28–29 cm
Diverse habitats, including rainforest edges, mangrove and paperbark swamp, eucalypt forest and woodland, orchards, tropical gardens. Figbirds are colonial, in small groups through breeding months when several nests may be in quite close proximity. After breeding, tend to congregate in larger flocks of 20–40, occasionally many more, these feeding in various fruiting trees, especially figs, both wild and cultivated. In the treetops, they clamber about in parrot-like manner, clinging and hanging among foliage and twigs to reach ripe fruits, insects and nectar. Calls are many and varied: commonly a rising and falling, loud, penetrating yet quite musical 'tsee-chieuw', or faster 'tsichiew'. In each case, the beginning 'tsee' or 'tsi' is very brief and high, just a lead in to the long, descending, strongly whistled 'chieuw'. A few calls from within the group or flock are usually answered by others, the many whistles intermingling in a confusion of sound with the individual calls seeming to be given faster, the calling more excited: 'tsee-chieuw, tseechieuw, tsichiew-tchiew'. Also has various erratic notes, both clear and harsh. The song is mellow but strong: 'too, too-heer, too-tooheer' and similar sequences including mimicry of other birds. Immatures nomadic, adults sedentary; common.

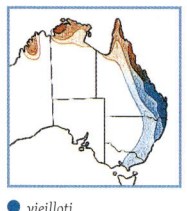

● *vieilloti*
● *flaviventris*
● *ashbyi*

Many subspecies through Indonesia and NG to Aust. where just three races occur.

- Abrupt change from black nape to olive-citrine of mantle
- Bare skin around eye; variable intensity of red or orange
- Throat to belly bright yellow
- Outer tail feathers pale grey, whitish towards tips
- Male: bare skin around the eye is buff to bright red.
- Intermediate between ● and ●, has black of head grading gradually through grey to olive of back.
- Black, males of both races
- Dull white
- **'Yellow Figbird'**
- Chin to breast and collar, grey
- Variable extent of yellow in hybrids
- Primaries black, males of all races
- Grey throat and collar
- Black grading gradually through grey collar to olive on back
- Grey collar and breast
- White or creamy tips, outer 3 or 4 pairs
- White
- Tail black with larger white tips; white along outer tail margin
- Bare skin around eye is grey.
- Imm.
- Like female but more distinct pale markings
- Olive-brown with dusky streaking; fine pale margins on wings
- ♀ Streaked underparts
- ● Heavy dense streaks and variably dusky throat
- Immature males show traces of adult male patterns and colours.
- Fine pale tips

## ORIOLES

**Green Oriole** *Oriolus flavocinctus* 26–28 cm
Tropical rainforest, monsoon and gallery forest, paperbark and mangrove swamp, lush gardens, plantations. Inconspicuous and rather elusive birds whose colour merges into surrounding foliage. Solitary or in pairs, they forage through mid and upper strata of gallery and monsoon forest, slowly and methodically seeking fruit. Most likely to be noticed when moving between trees, in direct but deeply undulating flight, or when making their characteristic bubbling sounds. Males may call for long periods from one or other of the trees within their territory, the easily recognisable calls indicating their location, although these birds are always fairly difficult to see if high in the canopy. The voice is rich and mellow with liquid notes, delivered in rapid rollicking sequence, 'chiowk-chiowk-chorrok', a deeper 'chiok-chok-chorrok', and gurgling 'chorr-chor-chorr-chorrok'. Also a high, sharp, clear 'teoo-tieeew'. Sedentary; common.

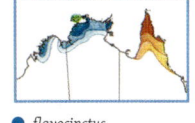

- 🔵 *flavocinctus*
- 🟤 *flavotinctus*
- 🟡 *kingii*
- 🟢 *tiwi*

🔵 Largest, slightly more olive-citrine, sexes alike

🟡 Medium size, brighter citrine, underparts more yellowish, much lighter streaks, especially males

🟤 Large, intense golden citrine; streaks light, some males unstreaked

🟢 Small, dull olive-citrine, more heavily streaked, especially female

**Olive-backed Oriole** *Oriolus sagittatus* 25–28 cm
Widespread inhabitants of eucalypt forest, woodland, mallee and taller inland scrub; occasionally rainforest. Difficult to sight while it is foraging for fruit in the foliage of the canopy unless attention is drawn by the musical, bubbling calls or by flight between trees. Usually solitary or in pairs, but is found in small, loose flocks through autumn and winter. Calls mellow and resonant, not the very deep bubbling of the Green Oriole, but mostly higher, clearer and more musical. A frequently repeated sequence is a rollicking 'orry-orry-orriole', quite clearly announcing itself as an 'oriole'. The song is a prolonged version of the call, wandering through similar sequences of notes, often with mimicry of other birds, 'quiee-kwee-kworri-kworriole'; also varied querulous squawking and rasping sounds. Common in N; migrant to SE.

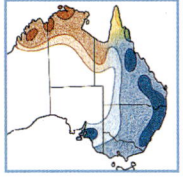

- 🔵 *sagittatus*
- 🟤 *affinis*
- 🟡 *grisescens*

🔵 Largest, rich colours, large white tail tips, moderate streaking, white underwing, short bill

🟡 Smaller, greyish olive, heavily streaked, mid-size white tail tips, cinnamon underwings, long bill

🟤 Small, pale citrine-olive, lightly streaked, small white tail tips, cinnamon underwings, long bill

## BARRED & GROUND CUCKOO-SHRIKES

**Barred Cuckoo-shrike** *Coracina lineata* 24–26 cm
Also known as 'Yellow-eyed Cuckoo-shrike'; habitat coastal rainforest and vine scrub, nearby eucalypts, paperbarks, plantations and tropical gardens. A strikingly plumaged cuckoo-shrike closely associated with rainforest where it feeds predominantly on native fruits. Gregarious, usually in flocks, often 10–20 birds, occasionally up to 50. Flocks travel between food trees with characteristic undulating flight on rather downswept wings: bursts of rather fluttering action interspersed with glides; flicks wings. Calls are sharp, brassy, two or three notes, 'caw-airk-awk' and 'cwairk-awk', often given in flight. Also a single, nasal 'quairrn' and other soft chattering. Locally nomadic; common in far N, uncommon in S.

Only the nominate race *lineata* occurs within Aust.

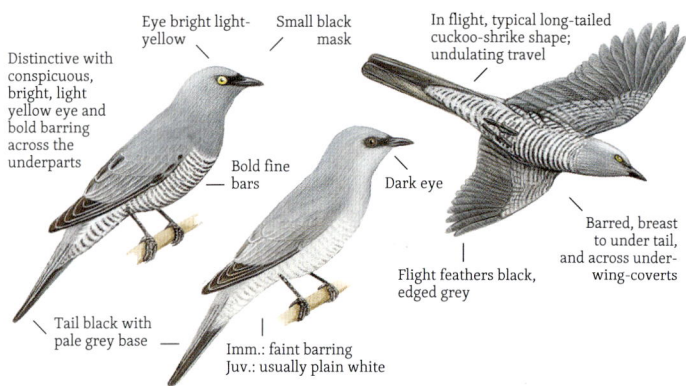

Distinctive with conspicuous, bright, light yellow eye and bold barring across the underparts

Eye bright light-yellow

Small black mask

In flight, typical long-tailed cuckoo-shrike shape; undulating travel

Bold fine bars

Dark eye

Barred, breast to under tail, and across under-wing-coverts

Flight feathers black, edged grey

Tail black with pale grey base

Imm.: faint barring
Juv.: usually plain white

**Ground Cuckoo-shrike** *Coracina maxima* 34–37 cm
Inhabitant of sparse open woodland, mulga and similar semi-arid scrub; spinifex with scattered small trees; river and roadside tree belts; dry-land farms. Foraging is much more terrestrial than other cuckoo-shrikes. Usually in small flocks, covering the ground quickly with long-legged, head-bobbing strides, pausing to peck at small prey. When flushed, flies up to trees showing white rump between black wings. Flight is quite direct. The contact call is far-carrying, usually given in flight: a high, metallic, piercing, drawn out squeak following on from a short, soft note that may not be audible at a distance, 'chr-EEEIP' or 'gr-EEEIP'. Also a rapid sequence, 'weeip-weeip, weeip-weeip' and a musical, vibrating trill, 'tr-r-rweeip, tr-r-r-reeip, tr-r-r-rrrp'. Nomadic movements triggered by rain; moderately common.

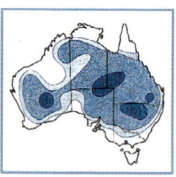

A large, pale cuckoo-shrike of the drier regions; more terrestrial than other species – legs are longer, feet are larger, heavier. There are no subspecies, but those in N Aust. are often paler.

Long, loose erectile feathers of lower back contribute to a long-tailed appearance.

Light grey; often white in sunlight

Eye bright, light yellow

Mid grey

Fine, dark, scalloped barring

Under-tail-coverts white, extend well down beneath the long tail.

Wings black

Crown, mantle, upper back silvery grey, or almost white under direct sunlight

Eye dark

More extensive brownish barring

Juv.

Tail black; tip slightly forked

Lower back, rump and upper-tail-coverts are white, finely barred brownish black.

## BLACK-FACED & WHITE-BELLIED CUCKOO-SHRIKES

**Black-faced Cuckoo-shrike** *Coracina novaehollandiae* 30–36
Found almost throughout Aust. in suitable habitat: rainforest, eucalypt forest and woodland, tree-lined rivers of interior, farms, gardens. Widespread familiarity has given rise to local names including 'Blue Jay' and 'Shufflewing'. Strongly undulating flight and shuffling of wings after landing aid recognition. Presence made more obvious by its tendency to use exposed limbs as perches. Solitary, in pairs or small family parties, or in large flocks. Call is loud, sharp churring, harsh yet rather musical. Often starts high then falls, 'churrieer', or descending 'quarieer-quarieer-quarieer'. Harsher sounds in aggression. Sedentary or nomadic; common.

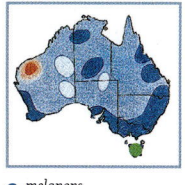

- *melanops*
- *subpallida*
- *novaehollandiae*

**White-bellied Cuckoo-shrike** *Coracina papuensis* 26–28 cm
Inhabitant of rainforest, gallery forest, eucalypt forest, woodland, mangroves, riverside tree belts. Flight undulating, long glides alternating with bursts of wingbeats. On landing, each wing briefly flicked up, then folded. In pairs or small family groups, forages for insects and fruit among foliage and branches. Voice distinctive, but reminiscent of some calls of Black-faced Cuckoo-shrike. Contact call, often while in flight, a scratchy, peevish, rather piercing, unmusical 'queeik-quisseeik', 'quirreeik, quirreeik'; also softer churrings. Locally nomadic; common in N, becoming sparse towards SE.

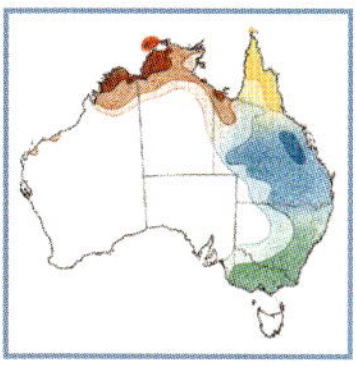

- *artamoides*
- *hypoleuca*
- *oriomo*
- *robusta*
- *apsleyi*

# TRILLERS & COMMON CICADABIRD

### White-winged Triller *Lalage tricolor* 16–19 cm
Open forest, woodland, tree lines along rivers of semi-arid regions, nearby scrub. Hunts from exposed perches, chasing flying insects and taking similar small prey on the ground. Through spring and early summer the cheery chattering trill is a prominent bush sound. Has vigorous chatter or trill, long sustained, loud, clear and far-carrying, intermingled sharp and mellow sounds, overall quality cheerful, tempo fast and rollicking: 'chwipa-wipa-wipa-wipa-wipa-'; 'chiffa-tiffa-tiffa-tiffa-'; often switches to a slightly softer but rapid, long trill,'chif-chif-chif-chif-tif-tif-tif'. In N, sedentary or locally nomadic; common. In S, summer breeding migrant; uncommon.

### Varied Triller *Lalage leucomela* 18–20 cm
An inconspicuous inhabitant of dense and closed vegetation, rainforest, monsoon forest, deciduous vine scrub, paperbark swamp, tropical eucalypt forest and woodland, river-edge thickets. Although not uncommon, may be overlooked unless calls attract attention. While foraging, contact calls are given frequently. The sound is distinctive: a rolling, swelling and fading, mellow rattling, rising and strengthening for several seconds, 'trrreee-', then falling, '-iurr', giving an undulating 'trreee-iurr, trreee-iurr, trreee-iur-'. Sedentary; common far N, uncommon to rare towards SE.

● *leucomela*
● *ruficentris*
● *yorki*
● *macrura*

### Common Cicadabird *Edolisoma tenuirostre* 24–26 cm
Uses foliage canopy of diverse forest and woodland, including mangrove and paperbark swamp. Feeding is arboreal; pairs work unobtrusively to find various insects. Can be hard to find unless calling, then elusive, easily put to flight: travels away swiftly in undulating flight through the treetops. Usually quiet, but through the breeding season males give a persistent, harsh metallic buzzing; cicada-like 'tzzzeit-tzzzeit-tzzzeit-', repeatedly. Race *melvillensis* has high pitched 'trzeee, trzeee, trzeee-'. Females give abrupt 'choyik' or 'chewik'. In far N, sedentary and moderately common; in SE a migratory breeding visitor, Aug.–Oct.

● *tenuirostre*
● *melvillensis*

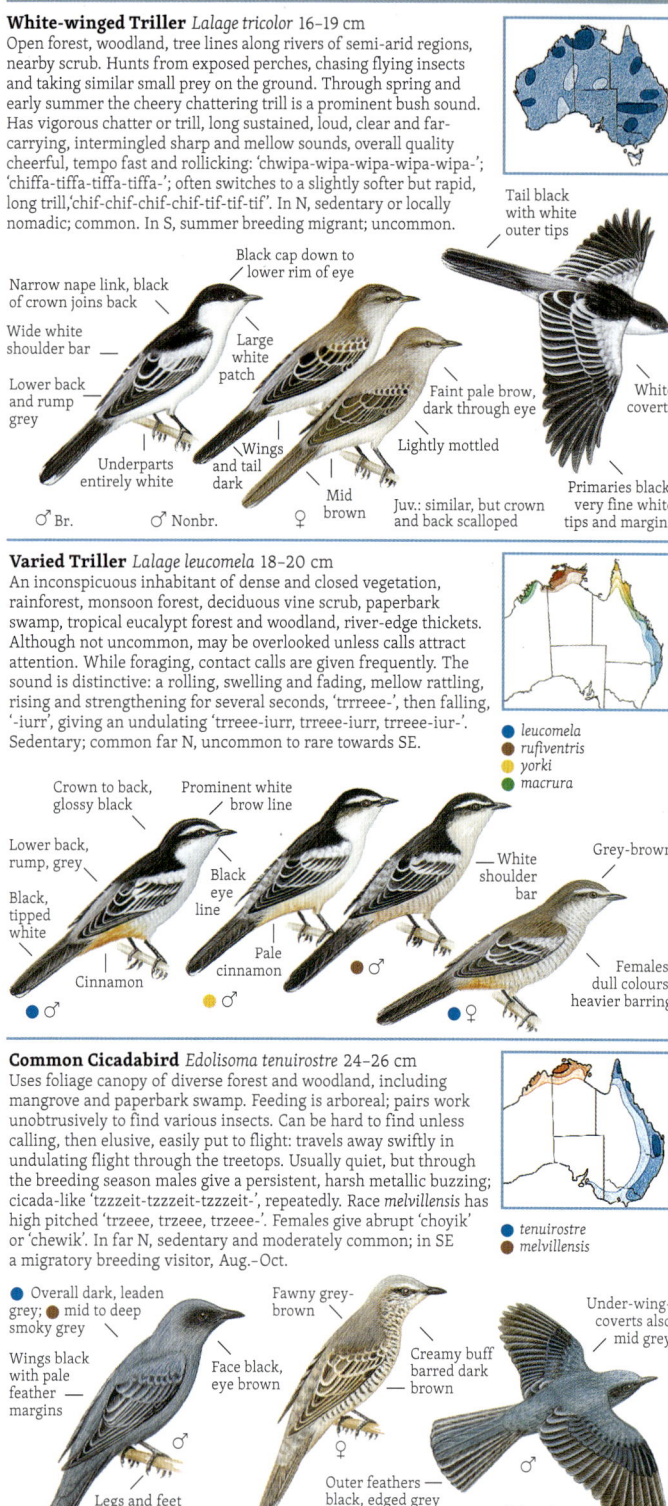

# TRILLERS TO CURRAWONGS

This group includes families Campephagidae (Trillers and Cuckoo-shrikes), Oriolidae (Orioles and Figbird) and Artamidae (Woodswallows, Butcherbirds, Australian Magpie and Currawongs). These range from quite small – the woodswallows at 14–20 cm long – to medium-large – the Magpie and Currawongs. They hunt varied small prey from on the ground to in trees and airspaces. They breed as pairs, in cup nests of fine grasses, bark or large. sticks.

WillieWagtail = 20 cm

## campephagidae
### page 323

- White-winged Triller
- Varied Triller
- Common Cicadabird
- Black-faced Cuckoo-shrike
- White-bellied Cuckoo-shrike
- Barred Cuckoo-shrike
- Ground Cuckoo-shrike

## oriolidae
### page 326

- Green Oriole
- Olive-backed Oriole
- Australasian Figbird

## artamidae
### page 328

- Imm. Black Butcherbird
- Grey Butcherbird
- Silver-backed Butcherbird
- Black-backed Butcherbird
- Pied Butcherbird
- White-breasted Woodswallow
- Masked Woodswallow
- White-browed Woodswallow
- Black-faced Woodswallow
- Dusky Woodswallow
- Little Woodswallow
- Australian Magpie
- Pied Currawong
- Black Currawong
- Grey Currawong

## RUFOUS, ARAFURA & GREY FANTAILS

**Rufous Fantail** *Rhipidura rufifrons* 15–16 cm
Rainforest, dense wet eucalypt and monsoon forest, paperbark and mangrove swamp, riverside vegetation; open country while on migration. Flits and dances about the mid level vegetation of the dark, shadowy forests, often with long tail widely spread. At times sunbeams set alight the glowing rufous-orange of its tail. The call is a very high, weak 'tsit' or 'tsi-tsit'. A song may follow as a brisk, extremely high, undulating tinkling, 'tsit, si-sit-tswit, tsit-tseit-tswit, tseit-tswit-tseit'; higher than notes of Grey Fantail. In NE Aust., common; less common in SE, a summer breeding migrant.

● *rufifrons*
● *intermedia*

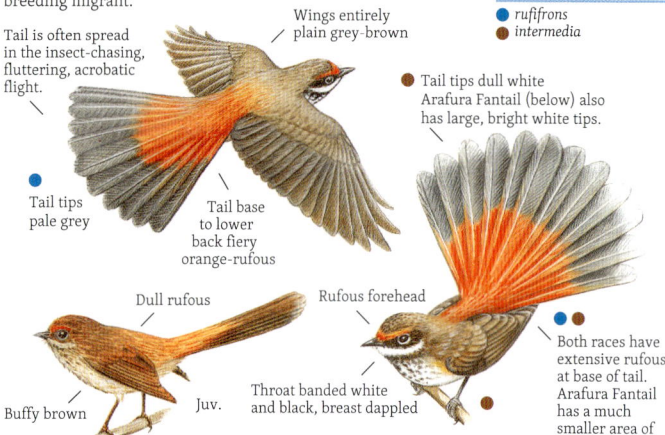

**Arafura Fantail** *Rhipidura dryas* 14–15 cm
Previously race *dryas* of Rufous Fantail, the Arafura Fantail (or Wood Fantail) is now considered a distinct species. It has a smaller rufous portion of tail and much larger, brighter white tail tips than other rufous-tailed forms.

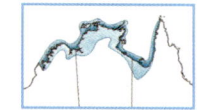

**Grey Fantail** *Rhipdura albiscapa* 14–17 cm
Confined to mangrove swamps. One of best-known small birds, its many races occupy most regions and habitats. Chases insects of forests and woodlands, fluttering about the lower levels of canopy and through undergrowth. When perched, rarely still, tail swinging, often widely fanned. Squeaky scratchy sounds, high, some almost inaudible, intermix with lower chatterings, 'twitch-twitchit, tsweeit-tseet, chit-twit, tswit-chat, tsweeit'. Most races quite common.

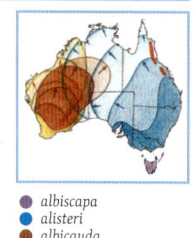

● *albiscapa*
● *alisteri*
● *albicauda*
● *keasti*
● *preissi*

Winter movements of races: ● is seasonally variable.

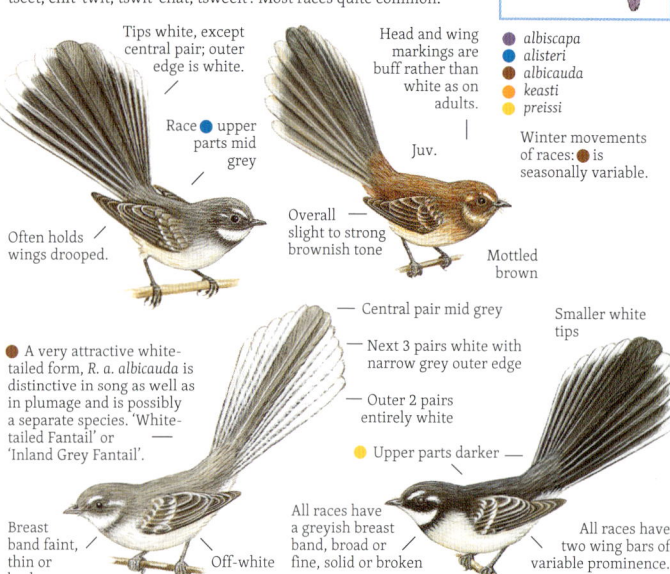

321

# WILLIE WAGTAIL, NORTHERN & MANGROVE FANTAILS

**Willie Wagtail** *Rhipidura leucophrys* 19–22 cm
The name 'wagtail' is appropriate: the tail is waved incessantly rather than fanned. Always restless, swings body side to side, gives sudden flicks of the wings to flush hidden insects. Prefers open country, farms; avoids dense forest. Aggressive near nest, attacks predators far larger than itself. The voice is brisk, strong, lively; a pleasant musical chatter with sudden switches between sharp and low notes. Sings through still moonlit nights, with repeated 'whichity-wheit', 'witch-i-wheit', whitchit'. In attack or defence gives harsh, loud, metallic, ratchetting chatter. Sedentary; widespread, common.

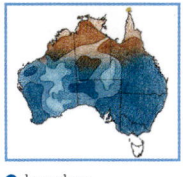

- ● *leucophrys*
- ● *picata*
- ● *melaleuca*

The long tail is entirely black, often fanned, and restlessly swung back and forth.

Only slight differences between races:
● is smaller than the southern nominate race and has slightly longer bristles radiating from base of bill;
● of Torres Str. and NG is largest, with longest bristles, at least to tip of bill.

Brow white, size variable

Wing entirely brownish-black, lacking white or buff feather tips and margins shown by the smaller fantails

Black throat

Widespread, generally familiar. Although a fantail, and so alike that it is placed with them in the genus *Rhipidura*, it is larger and of heavier build.

Juveniles have rufous edges to wing feathers and brow line; tail initially short. Immatures are similar to adults but have pale tips to wing-coverts.

Legs proportionately much longer than those of other fantails, reflecting its frequent pursuit of small prey on the ground

---

**Northern Fantail** *Rhipidura rufiventris* 19–22 cm
Usually around margins of rainforest, monsoon and vine scrub, paperbarks and mangroves, also in adjoining eucalypt forest and woodland. Rather sedate, even sluggish, in its movements compared with the hyperactive Grey Fantail. Has a more upright posture, usually keeps tail down, rarely fanned. Typically waits quietly on a perch, darts directly out to take insects that fly past. Call is an abrupt 'kek', or 'dek', or rather musical 'chunk', repeatedly over long periods. An ascending buzzing call is often used: 'zziop--i-deet'. The song is similar to that of the Grey Fantail, but at lower pitch, lacking the highest squeaks of that species, yet quite sharp, pleasant. Sedentary or locally nomadic; moderately common.

Race *isura* in mainland Aust.; *gularis* in Torres Str. and to N

Dark face, white chin and throat, streaky dark breast band

Upper parts dark grey

Tail usually held down, not often fanned or waved; has white tips and outer margins.

Pale buff

---

**Mangrove Fantail** *Rhipidura phasiana* 15–16 cm
Inhabitant of mangrove forests, both tall and stunted, and forages over adjoining samphire flats, occasionally acacia or melaleuca scrub nearby. Agile and active in aerial pursuit of insects; flutters about the mangrove foliage, and low across mud and pools. Contact call is an abrupt 'tek, tek'. As song, a series of short, twittering squeaks, 'tsit-tsit, tsit-chit-chit, chitty-chit,' distinctly different from the Grey's lively, cascading, silvery, squeaky song. Common.

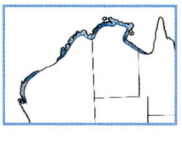

The Mangrove Fantail was regarded as a race of the Grey Fantail until studies showed that it differed not only in details of appearance, but also had a distinctive song and differed in aspects of ecology and breeding.

Upper parts, crown to tail, paler than on Grey Fantail

Larger pale buff or white brow line

Broad white tips to outer four pairs

Tail feathers have white shafts; outer vane narrowly edged white.

Thick white wing bars

Bill slightly longer than Grey Fantail

Throat bar very thin

Imm.: like adult, but most white replaced by buff

## MONARCH FLYCATCHERS

### Shining Flycatcher *Myiagra alecto* 16–18 cm
Inhabitant of mangroves, rainforest, pandanus, creeks, paperbark swamps. Typically near or over water, among stilt roots of mangroves or around pools in rainforest; drops to take prey from mud or ground more than from foliage or in flight. When perched, swings and flicks tail, lifts crest. Diverse calls – clear, musical whistles to frog-like buzzing and croaking sounds. Common call is a rapid series of clear whistles – starts softly, increases in volume, 'whit, whit, whit' or more rapid 'whit-whit-whi-'. A buzzing call begins low, increases suddenly, 'zzzreEOW, zzzreEOW', both high, clear and low buzzing, 'kwit-zzzur, kwit-zzzur'. Sedentary; common.

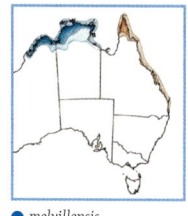

● *melvillensis*
● *wardelli*

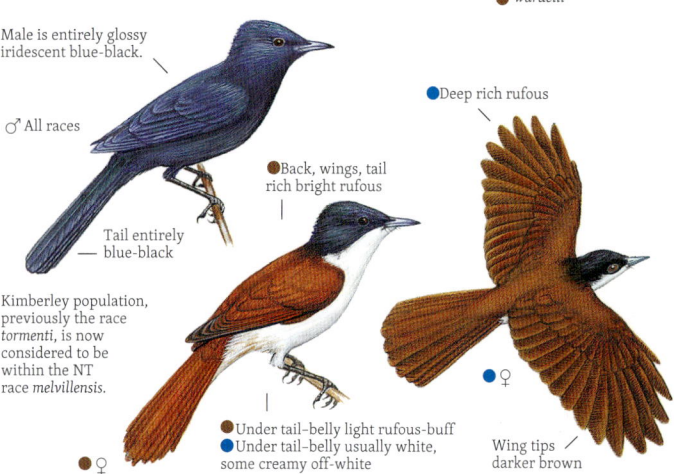

Male is entirely glossy iridescent blue-black.

♂ All races

Tail entirely blue-black

Kimberley population, previously the race *tormenti*, is now considered to be within the NT race *melvillensis*.

● Back, wings, tail rich bright rufous

● Deep rich rufous

● Under tail-belly light rufous-buff
● Under tail-belly usually white, some creamy off-white

Wing tips darker brown

● ♀

### Restless Flycatcher *Myiagra inquieta* 16–21 cm
Open forest, woodland, farmland, inland scrub. Hovers over grass, shrubbery, tree foliage, branches, logs, often with grinding calls; flight appears erratic, light and buoyant. Restless when perched, waves tail, soon off after a flying insect or to a new perch. Makes some unique and distinctive sounds. Common call a high, clear, musical 'toowhee-toowhee-toowhee', 'twheee-twheee', or, with slight buzzing, 'tzweet-tzweet'. In contrast, a grating, harsh 'grrzziek'. Most remarkable is a call given while the bird is hovering: it is a rising series of metallic grinding noises that reverberates far through forest and woodland, 'kzzzzrrk', or a long sequence, 'kzowk! kzowk-kzowk, kziok-kziok, kzeeek-kzeeek, kzeeik-kzeik-'. Sedentary or nomadic; common.

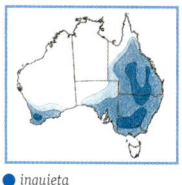

● *inquieta*

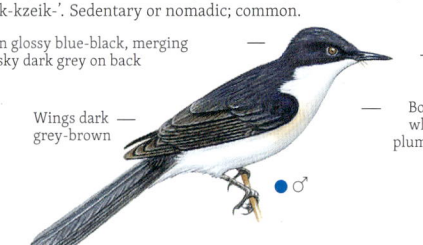

Crown glossy blue-black, merging to dusky dark grey on back

Wings dark grey-brown

● ♂

Bill longer, more slender than Paperback Flycatcher, with only sparse bristles at its base

Both Restless and Paperback have white underparts, which, in fresh plumage, may have a slight buff tint on breast, lost as fine buff feather tips wear.

Slightly larger than the Paperbark Flycatcher.

### Paperbark Flycatcher *Myiagra nana* 16–21 cm
Formerly a subspecies of Restless Flycatcher, the Paperback Flycatcher of northern Aust., is now a species distinguished by a smaller, slightly shorter, broader bill. Calls also differ, the Paperbark having the musical whistled 'toowhee' call but rarely the grating, grinding, scissors-sharpening sound of the Restless. Breeding ranges abut in NE Qld without evidence of interbreeding.

● Black extends further down back.

● *nana*

● Shorter bill is also proportionately broader, longer bristles at its base.

Females have grey rather than black lores. Juveniles are duller than adults, and have creamy buff on throat and breast.

● ♂

# MONARCH FLYCATCHERS

**Satin Flycatcher** *Myiagra cyanoleuca* 15–17 cm
Inhabitant of forest and woodland, mangroves, coastal heath scrub, but avoids rainforest. In breeding season favours dense, wet gullies of heavy eucalypt forest. Energetically active in mid to upper levels of forest; darts out from perch to snap up flying insects; flutters about the foliage. On perches, restless with swaying, quivering tails; crests often erected. Calls are sharp, metallic: 'chwee-wip, chwee-wip, chwee-wip' and faster, high, sharp 'cheeip-cheeip-cheeip-cheeip'. Harsh, grating grinding: 'grzzz-urk, grzz-urk'; 'gzzirk, gzzirk'. Summer breeding range SE Qld to Tas.; winter migration range extends to NE Qld and NG. Uncommon.

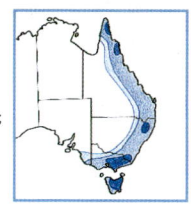

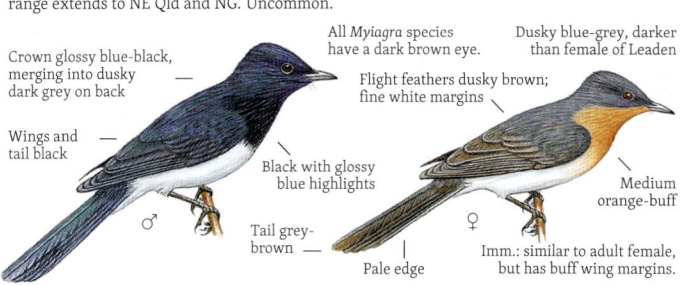

**Broad-billed Flycatcher** *Myiagra ruficollis* 15–16 cm
Usually in mangroves, but also monsoon forest, paperbark swamp and riverine forest. In pairs or solitary; actively forages about the crowns of mangroves or other dense vegetation. Flutters about the foliage; comes down to lower levels around forest margins; often darts out to take flying insects. Rather quiet, but not shy; when perched, rapidly quivers its tail. Call is a clear, far-carrying, ringing, musical 'chiewip-chiwip-chwip-chwip'. In contrast are some of its other noises: harsh, grinding, buzzing, 'tzzzeep, tzzeep, tzzeep'; abrupt, metallic, buzzing 'tzwip-tzwip-tzwip'; also gives soft churring sounds. Sedentary; common.

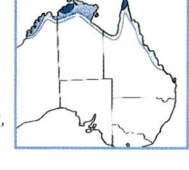

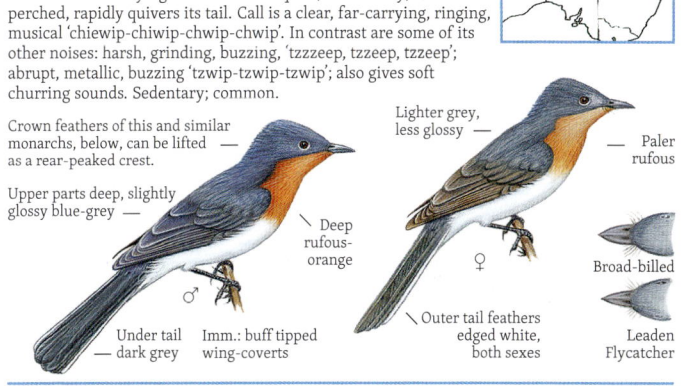

**Leaden Flycatcher** *Myiagra rubecula* 15–16 cm
In eucalypt forest, woodland, scrub, rainforest margins, mangroves, inland rivers. Busily active in mid to upper forest strata; hovers about the foliage; gleans small insects from leaves, snatching many in flight. Usually solitary, occasionally in pairs; quivers tail while perched; frequent calls attract attention. Calls are clear, carrying 'whit-ee-eight, whit-ee-eight'; similar 'whee-ity, whee-ity','too-wheit, too-wheit'; 'wheeit-wheeit, wheeit-wheeit'. Also, harsh, nasal buzzing: 'tzzzeep'; 'scrzzarch'. Sedentary, common in N; uncommon summer migrant to SE.

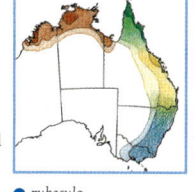

# DRONGO & MAGPIE-LARK

**Spangled Drongo** *Dicrurus bracteatus* 28–32 cm
Diverse habitats: open woodland, margins of rainforest or vine scrub, mangrove and paperbark swamp, riverside thickets, gardens. The Drongo's habit of perching in the open on a bare limb or wire, its acrobatic flights and its loud, metallic, tearing calls combine to make its presence obvious. Rips into bark for spiders, beetles and other small prey, or harasses other birds, including raptors. Insects are taken in flight after twisting, turning aerial pursuit, and brought back to the perch. Noisy, with harsh, brassy, tearing sounds: loud, rasping, nasal 'k-zark, kzairk-kzairk-', 'korrk-korrk--'; metallic 'wheit-', 'kairk-kiairk'; high 'kierk'; whistles, rackety chatterings. Migratory or sedentary; common.

- 🔵 *bracteatus*
- 🟤 *baileyi*
- 🟢 *atrabectus*
- 🟠 *carbonarius*

Eye brilliant red
Fine crest at nape, inconspicuous
Heavy bill with bristles around the base
Plumage entirely black with a glossy greenish blue metallic sheen
Highly reflective, iridescent spots give a glinting or 'spangled' appearance.
Eye brown
🔵 Has underside of wings with many large white spots. These small and sparse on 🟠 🟢; medium-size and dense on 🟤.
Juv.
Scattered white tips, mostly on breast feathers
Unique among Australian birds is the distinctive, out-curved 'fishtail' shape of these long feathers.
Under-tail-coverts retain pale tips on young adults.
Under-tail-coverts white tipped or white scalloped
Legs and feet small, thin
🟠 Has tail more forked, and with outer feathers much more strongly up-curved than seen on the Aust. mainland races.
Immatures are dusky black rather than the brownish black of juveniles, and begin to show adult's iridescent greenish sheen. White tips are lost from breast, but are retained on under-tail-coverts.

**Magpie-lark** *Grallina cyanoleuca* 26–30 cm
The 'Peewee'. Habitats diverse, coastal to semi-desert, almost anywhere with the water and trees needed for mud nests and for foraging along margins of lakes, rivers or swamps. Has distinctive, slow, buoyant flight with deep, uneven, lapping beats of broad, rounded wings. Calls are mellow, liquid yet clear; often a ringing, carrying 'tiu-weet, tiu-weet' and liquid 'cluip-cluip, cluip-cluip'; as alarm, gives a strident 'treee-treee-!'. Pairs, together, give a closely synchronised duet, one a musical, mellow 'qwoo-zik', the other immediately following on with a sharp, yet rather harsh, 'wheeik'; in duet, 'qwoo-zik, wheeik...qwoo-zik-wheeik-', wings lifting with each call. In flight, together, a softer, liquid, piping, 'qwoo-whik'. Nomadic or sedentary; very common. Has adapted well to environments, especially where there is now permanent water.

- 🔵 *cyanoleuca*
- 🟤 *neglecta*

♀ Vertical black line down through eye; white forehead and throat
Black subterminal band, much wider on central feathers
Adult: pale eye
Imm: dark
♂ Horizontal black line through eye; black throat and forehead
♂
Uneven white bar from shoulder back across the folded wing
Rump white
Imm.
Tail white with wide black band
Imm. combines the female's vertical band and white throat with male's horizontal band and black forehead.
Aust. races similar except:
🔵 larger, with proportionately short bill and long tail;
🟤 smaller, with proportionately longer bill and shorter tail

## MONARCH FLYCATCHERS

**Black-faced Monarch** *Monarcha melanopsis* 16–20 cm
Rainforest, mangroves, eucalypt forest and woodland. Monarch flycatchers are rather sedate and slow moving. Take insects from foliage; forage in denser parts of mid-level forest; occasionally hawk for flying insects. Gives a series of deliberate whistles, a distinctive mellowness in lower notes; slightly scratchy high whistles, rise then fall – 'wheech-iew, whieeeuw', 'whee-awhit, whieeeuw' as single call, or widely spaced, 'wheit-chiew'. Also scratchy, husky, rising 'shreeeit, shreeeit, shreeeit' and harsh, scolding 'scraach'. Sedentary in NE, summer migrant further S.

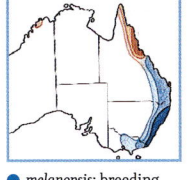

- 🔵 *melanopsis*: breeding
- 🟤 *melanopsis*: migratory nonbreeding range

Head, back, most of folded wing and breast are mid blue-grey.

Pale bill

Black 'face', forehead, through lores down to throat

Rufous-buff

Tail blue-grey similar to rump and back; no white at tips

Imm.

Dark bill

Face grey, lores pale

Flight feathers are darker dusky grey.

Rufous-buff wing linings

Shows some variation within the breeding range, gradually becoming paler, smaller northwards, but differences slight, and now considered insufficient to justify a subspecies in N part of range. In addition to this, there is a migration path of southern birds, extending N of breeding range and reaching NG.

**Black-winged Monarch** *Monarcha frater* 17–19 cm
Rainforest, mangroves, adjacent eucalypt woodland, where active in mid to upper levels of foliage. Gleans insects from foliage, and by aerial flycatching. Often in feeding parties of mixed species; attracts attention with noisy calls. Voice similar to Black-faced, but slightly more harsh, a rising and falling 'whewheit-whiew, whewheit-whiew' and scratchy, husky, nasal 'wheeit-chow, wheeit-chow'. Also a single, clear, rising whistle at regular intervals: 'wheeeit, wheeeit, wheeeit'. Summer migrant from NG; breeds Cape York where it is moderately common; vagrants as far S as Cairns.

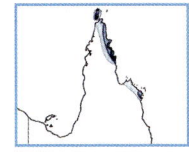

Much less common than the Black-faced; usually confined to Cape York, occasionally seen near Cairns

The Black-winged Monarch differs from the much more common Black-faced and Spectacled Monarchs most obviously in its much darker wings and entirely very dark tail.

Grey paler, more 'pearly'

Dark bill

Wings mostly brownish black

Rufous-buff

Tail entirely plain grey-black

Imm.

Black close around eye; joins small patch on throat.

Head entirely very light pearly grey

**Spectacled Monarch** *Symposiarchus trivirgatus* 15–16.5 cm
Usually in rainforest, mangroves, moist gloomy gullies of dense eucalypt forest. Slightly smaller, more active than *Monarcha* species; darts and hovers about foliage and limbs, tail often spread and waved to display the broad white tips. Has lively calls, the most distinctive a series of whistles, beginning with a somewhat buzzing quality, each note drawn out and rising strongly to a clear, high finish: 'zreee-e-e-e, zreee-e-e-e, zreee-e-e-'. Also scratchy chatterings, harsh scoldings, and a brief, squeaky warble. NE Qld resident; summer breeding migrant further S.

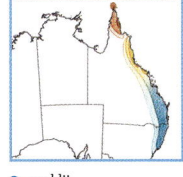

- 🔵 *gouldii*
- 🟤 *albiventris*
- 🟡 *melanorrhoa*

Distinctive are white belly, white tail tips and black ear-coverts.

Black 'spectacles' extend to ear-coverts and throat.

Grey ear-coverts and throat

🟡 Similar, but darker upper parts

🟤

Tail dark grey, outer 3 pairs with large white tips

🔵 Imm.

## BLACK & WHITE MONARCH FLYCATCHERS

**Frill-necked Monarch** *Arses lorealis* 15–16 cm
This species and the Pied are quite spectacular as they forage in rainforest or adjoining eucalypt forest, hopping and fluttering up tree trunks and limbs. Wings and tail often partly spread, they take insects from fissures of bark or flutter about foliage and pursue flying insects. In display, white frill and crescent are widely spread. Usual call is a harsh, upward, rasping 'tzeeeeit, tzeeeeit'; can be a single call, or several slow, widely spaced calls, or given very rapidly, as long series with buzzing, cicada-like effect, 'tzeet-zeet-zeet-zeetzeetzeet'. Also has harsh, squabbling scoldings or chatterings. Sedentary; common within its restricted range.

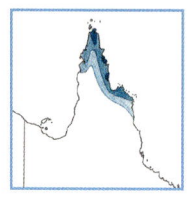

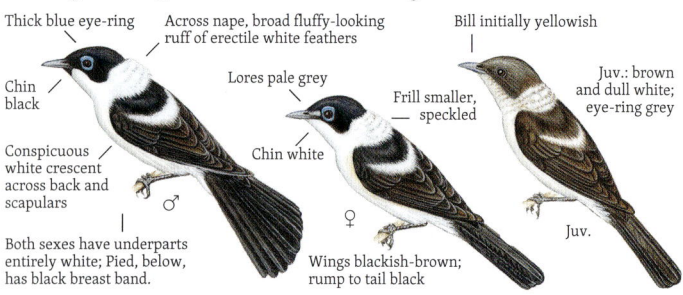

**Pied Monarch** *Arses kaupi* 14–15 cm
Confined to tropical rainforests and vine scrubs. Forages with lively activity, fluttering and running up and down trunks of trees, treecreeper fashion, but more lively; often half-spreads wings, fans tail to dart out and take flying insects. Call is a harsh 'tzeeeeeit', beginning as a low, rasping, buzzing sound, rising to a high, sharp finish; calls widely and unevenly spaced. Also a long series of clearer, whistled notes, rising steadily higher throughout, 'wheit-wheit-wheit-wheit-' and 'whit-whit-whitwhitwhitwhit'. Sedentary; moderately common.

● kaupi
● terraereginae

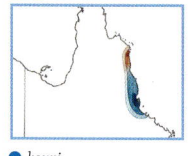

**White-eared Monarch** *Carterornis leucotis* 13–14 cm
Inhabitant of rainforest, mainly coastal, also mangroves, swamps, watercourse thickets. Very active about canopy; takes insects in flight, and from leaves. Call is a loud, clear sequence of whistles, each rising, then falling away, 'twei-tsieeew, twei-tsieee...'; also has a series of long, piercing, descending whistles, 'tsieeeeuw, tsieeeeuw, tsieeeeuw'. Much softer, a musical, chattered song, brief, but rapidly repeated, 'chzit-chit-ch-weeeit, chzit-chit-ch-weeeit', with first notes soft, slightly buzzing; the last much louder, higher, sharper. Sedentary, some locally nomadic or migratory movements, probably altitudinal; uncommon.

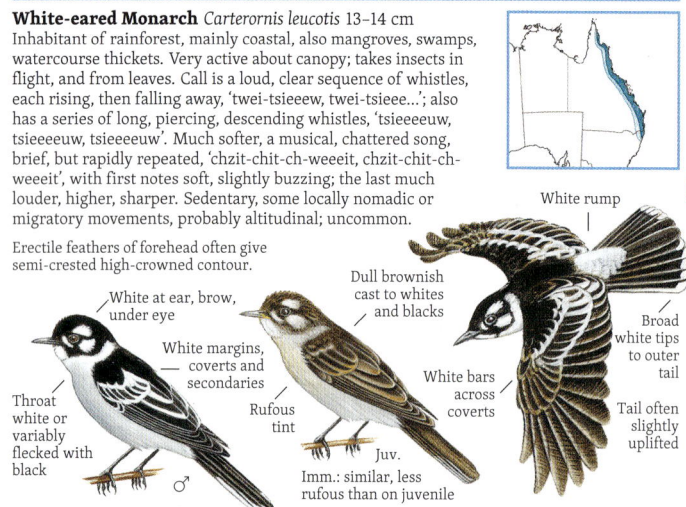

# 21 MONARCH FLYCATCHERS & DRONGO TO FANTAILS

The monarch flycatchers, the drongo and the fantails together make up the family Dicruridae. Most are insectivores that take their prey on the wing, twisting and turning, their long tails like a rudder. Long bristles around the bill help in the final capture. There are several exceptions. The frilled monarchs (genus *Arses*) mostly seek insects in bark crevices, and the Magpie-lark forages on the ground. Its relationship to flycatchers shows in its slow, buoyant flight.

Scarlet Robin = 12 cm

## dicruridae
page 314

- Yellow-breasted Boatbill
- White-eared Monarch
- Black-faced Monarch
- Black-winged Monarch
- Spectacled Monarch
- Frill-necked Monarch
- Pied Monarch
- Leaden Flycatcher
- Satin Flycatcher
- Broad-billed Flycatcher
- Shining Flycatcher (♂, ♀)
- Restless Flycatcher
- 'Paperbark Flycatcher'
- Spangled Drongo
- Magpie-lark
- Willie Wagtail
- Northern Fantail
- Rufous Fantail (May include new species 'Arafura Fantail'.)
- Mangrove Fantail
- Grey Fantail

## YELLOW-BREASTED BOATBILL

**Yellow-breasted Boatbill** *Machaerirhynchus flaviventer* 12 cm
Rainforest, vine and swamp thickets, nearby woodland. Alone or in pairs, forages actively around foliage of mid to upper levels; darts out after flying insects. Calls clear, sharp, rising; drawn out whistles followed by much softer, rapid chuckling: 'wheee-wheeet, chuk-chuk-chuk'; similar 'wheit-wheit-wheit, tsi-tsi-tsi-tst–'. Also a softly trilled 'whit-wh-wh-wh, wheeeee-whit'. Sedentary; uncommon.

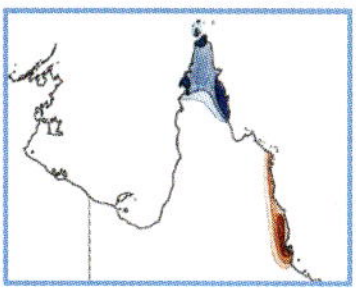

- ● *flaviventer*
- ● *secundus*

- Back blackish, like crown; flight feathers edged white
- Long yellow brow
- White
- Olive
- ♂
- Outer three tipped white
- ♀
- Underparts greyish yellow
- ● Olive back contrasts with black crown
- Wide 'boat' bill shape
- Brownish olive
- Juv.
- Buff wing margins and underparts
- White wing bars and margins
- ● Flight feathers edged yellow
- Broad flat bill
- Bright yellow, both races

# GREY SHRIKE-THRUSH & CRESTED BELLBIRD

**Grey Shrike-thrush** *Colluricincla harmonica* 22–25 cm
Diverse habitats: coastal open forest, woodland to arid mallee and mulga of interior. Takes small insects, spiders, lizards, frogs, mice, nestlings and eggs of small birds, seeds and fruit. Forages on tree trunks, branches, finding prey in bark crevices; also among foliage, undergrowth, fallen logs and on ground. Flight swift, direct, undulating. Has an extremely rich and varied repertoire of calls and songs – high, clear and often loudly ringing whistles intermingled with mellow, musical notes and deep, rich bubbly sounds. A common sequence is 'quorra-quorra-quorra, WHIEET-CHIEW', beginning with mellow throaty sounds and finishing with a high, clear, ringing whistle; similarly, 'wheit-wheit-quor-quor-quor, WHIEET-CHIEW'; 'wheit-wheit, WHIEET-chiew'. Also high, whistled 'whee-it' and 'wheit-whiet-wheeeit'. Calls sometimes confused with those of Rufous Whistler, but more varied and mellow. Widespread; common.

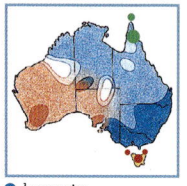

- *harmonica*
- *rufiventris*
- *superciliosus*
- *strigata*

- Has many regional forms or races, upper parts light grey, mid grey and mid brownish-grey

Grey-brown to dull rufous

Juv.: rufous lores, brow

Juv: dark streaks on underparts

Grey, light streaks on throat

♀

Rufous-buff under tail coverts

- **'Western Shrike-thrush'**
Many regional forms, previously races, including *kolichisi* (NW Cape), *whitei* (Eyre Peninsula) and *murchisonii* (NE to Centre)

Male has blacker bill, white rather than pale grey lores.

- Local forms of nominate race haveat times been recognised as races: *brunnea* (Kimberley, NT), *pallescens* (Qld), *halmaturina* (NSW).

---

**Crested Bellbird** *Oreoica gutturalis* 20–22 cm
The deep, mellow, liquid and slightly melancholy voice of the Crested Bellbird is one of the characteristic sounds of mulga and mallee scrub and semi-arid woodland. The song, usually begun softly by the male with female contributing, builds louder. There is a ventriloquial quality. Has a regular, strong rhythm with unique soft, liquid 'ook' inserted among whistled notes. Variations include a low, husky sequence – rather slow rhythm, mellow whistles and deep waterdrop effect: 'dee, dee, dee-ook, dee, dee, dee-ook'; higher, clearer, whistled 'whit, whit-whit-quiook'; low 'plonk-plon-plonka'. Forages on ground, stumps, logs, in shrubs and lower branches of trees. In pairs during breeding season, then briefly in family groups, but solitary most of the year. Through interior, nomadic, common; nearer coasts, sedentary and uncommon.

- *gutturalis*
- *pallescens*

The southern and northern subspecies merge together in a broad zone from south-western Aust. to north-eastern Qld.

Males of northern race *pallescens* are slightly smaller, considerably paler on upper parts, slightly paler beneath, and with a larger white face patch extending higher on forehead. Females are slightly paler than those of the nominate race.

Black-topped crest may be raised slightly higher than shown or tightly down as a black line.

Combination of rounded black bib and overall sandy brown tone identify the male.

Male has unique facial pattern and bright orange-red eye.

Female has only a black line as crest.

Eye brown, throat white

♀ Underparts pale buff, slightly darker on breast

Imm.: like female, but with buffy margins to wing feathers; male has mottled dark throat.

Tail slightly darker brown

In flight, male's crest folds flat; shows as black crown streak.

Upper parts sandy brown to rufous-brown, primaries darker

Paler buffy margins

## SHRIKE-THRUSHES

**Arafura Shrike-thrush** *Colluricincla megarhyncha* 17–19 cm
A widespread inhabitant of monsoon and vine thickets, coastal woodland, lowland rainforest, mangroves. Usually quiet, unobtrusive, wary; alone or in small groups. Uses strong bill to probe bark, debris-entangled vines, epiphytes or ground litter. Populations in N are so different from those of the E that they have at times been listed as separate species. Song often begins with high, drawn-out, weak notes, then stronger, rollicking, deep, liquid-sounding notes: 'tseee-tsee-tsee, chok-ok, chok-a-wheet'; 'twheit-twheit-twhit-wheeee, chok-korr'; 'whicha-whieew'; other varied, clear, calls; harsh wheezy sounds. Sedentary; common.

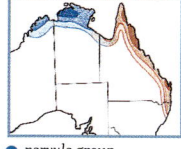

- *parvula* group
- *rufogaster* group

Parvula group: single race, *parvula*

Rufogaster group: comprised of race *rufogaster* and five other similar, overlapping, intergrading races – *gouldii*, *synaptica*, *griseata*, *normani*, *aelpetes*.

Each group was, in the past, considered to be a full species and may be again.

Head greyish; only slightly paler brow and lores
Olive-grey to russet-brown
Face buffy white, faint streaks
Cinnamon; deeper rufous in NE Qld
Conspicuous white brow and lores
Upper parts deep rufous-brown
White
Pale sandy or fawny buff

**Rufous Shrike-thrush**   **Arafura Shrike-thrush**
Juv.: secondaries and their coverts: broadly edged rufous

**Bower's Shrike-thrush** *Colluricincla boweri* 19–20 cm
Confined to mountain rainforest above about 400 m altitude. Where rainforest continues unbroken to coastal lowlands, some may move to lower altitudes in winter. Rather slow, deliberate foraging; works middle and lower strata of forest, searching bark, foliage, entangled debris. Solitary or in pairs, rich song, but through winter quiet and unobtrusive. Voice is extremely varied; similar to that of well-known Grey Shrike-thrush. High, clear first notes are followed by deep, mellow, bubbling notes: 'trip-trip-shrieee, quorr-quorr-quorrot'. Also loud 'chuk!', chirps, harsh scoldings. Common.

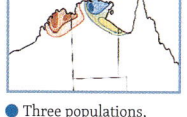

Stoutly built, short-tailed, appears to have large head, heavy bill.
Slaty dark grey
Short-tailed, heavy-bodied appearance
Streaked
Wings and tail grey to grey-brown, edged pale rufous
Olive-buff to deep rufous
Female has rufous lores, brow and eye-ring; bill is grey.
More heavily streaked than the Little Shrike-thrush
Imm.: like female but breast more heavily streaked, back browner

♂ Male has pale grey lores, brow and eye-ring; the bill is black.   ♀

**Sandstone Shrike-thrush** *Colluricincla woodwardi* 24–26 cm
Escarpments, ravines of sandstone ranges, plateaux. Usually visible only briefly, plumage blends with stone. Alone or in pairs, probing crevices, finding small lizards, grasshoppers and spiders. Gives high, clear, drawn-out, rising whistles, quickly followed by abrupt, rich, mellow, at times bubbling notes. Among these, 'pieeer-pieeer, chokka-chok-chok'; 'wheeeit-wheeeit, quorr-quorr-quorr, queet-queet'. Long, descending trill and single, quick 'cheewhip! Sedentary; common within specific habitat of sandstone ranges and gorges.

- Three populations,
- previously
- subspecies

Differences now considered to be slight, clinal. In W, it is slightly paler, more streaked and with shorter bill than those further E.

Large, slender, long-tailed; typically on or near sandstone cliffs; briefly perches in adjacent trees.
Lores pale, throat buffy white
Rufous margins to wing-coverts; more apparent on immatures
Back sepia-brown to olive-brown
Breast grey-buff, lightly streaked; lower underparts pale rufous-buff
Grey head merges into brown back.
Long slender body and long tail
Slightly deeper rufous-buff on Kimberley population

# WESTERN & GOLDEN WHISTLER & CRESTED SHRIKE-TIT

## Western Whistler
*Pachycephala fuliginosa* 16–18 cm
Formerly a subspecies of the Golden Whistler and with similar lifestyle but found only in WA. It is darker grey on the back than Golden with a dark-grey, black-tipped tail.; females are more rufous. Juveniles are similar.

- 🟢 *fuliginosa*
- 🔵 *pectoralis*
- 🔵 *queenslandica*
- 🟡 *youngi*
- 🟣 *glaucura*

## Golden Whistler
*Pachycephala pectoralis* 16–18 cm
Widespread with four subsp. in Aust. Rainforest, eucalypt forest, mallee, brigalow scrub. Males call in spring with bold posturing, undulating flights about territories. Often solitary except when breeding; forage in trees and taller shrubs. Ringing whistles in a long sequence: 'whit-whit-whit-whiet-whiet-wheet-quWHIT'; also repeated 'whit-whit-whew-WHIT', 'chwit-chwit-CHEW-WIT'. Contact a rising 'tseeip'. Sedentary, local migration in SE; common.

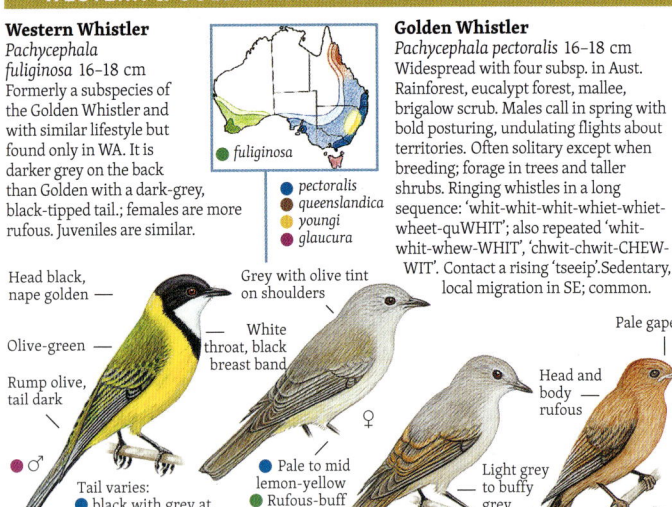

Head black, nape golden
Grey with olive tint on shoulders
Pale gape
Olive-green
White throat, black breast band
Head and body rufous
Rump olive, tail dark
♂
♀
Light grey to buffy grey
Tail varies:
- 🔵 black with grey at base; 🟢🟡 dark grey, tipped black; 🟣 entirely deep olive-grey
- 🔵 Pale to mid lemon-yellow
- 🟢 Rufous-buff
Imm.
Wings edged rufous
Juv.

## Crested Shrike-tit *Falcunculus frontatus* 15–19 cm

Open forest, woodland, mallee, riverside and watercourse trees, cypress pines, banksia woodland. Unobtrusive; alone, in pairs or family groups, usually quite high. Easy to overlook; most likely to reveal its presence with sound of tearing bark as it seeks small prey in crevices. Voice soft, rather mournful, undulating, seeming always from afar. Starts softly, strengthens through several sequences of calls, with the last note of each set lowest, strongest and, finally, slightly up-slurred: 'whiert, whi-whit, wheeeir-whiert,whi-whit, wheeeir'. Also has a sharper, downwards call, 'peeeir, peeeir, peeeir'; scolds with a hard, harsh, rattling chatter. Sedentary or locally nomadic. Generally rare; in optimum habitat, locally common.

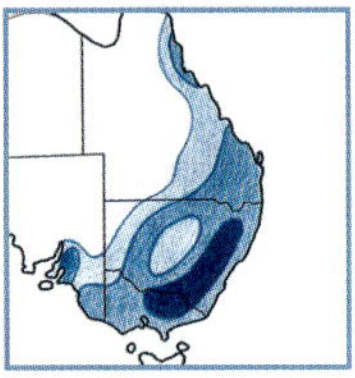

Massive bill, shaped for stripping bark

**'Eastern Shrike-tit'**: mantle, back and rump olive-green to olive-brown.
Low crest
♂
Males have black throat.
♂
Light bright yellow
Cinnamon to olive-brown
♀ All species, throat olive to olive-brown
Pale yellow
Creamy off-white
Juveniles of all species are apparently similar.

## Western Shrike-tit
*Falcunculus leucogaster*
Habitat, ecology, calls like Eastern, of which, at times, it is listed as a race. Uncommon.

Unique head pattern constant between species; males always have black throat.

♂
Only species with white belly

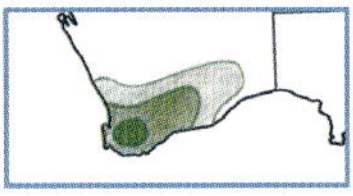

Wing-covert margins dull yellow

## Northern Shrike-tit' *Falcunculus whitei*
Smaller and entirely yellow from breast to under tail. Similar in habits to Eastern. Differences between the three shrike-tits include size, wing and tail shape, as well as plumage colours.

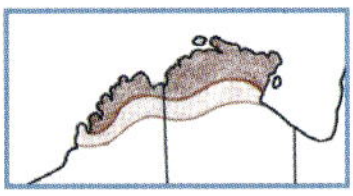

Calls differ, alternating between lowering and rising sequences.

♂
Belly bright, light yellow

## MASKED WHISTLERS

**Rufous Whistler** *Pachycephala rufiventris* 16–17.5 cm
Open eucalypt forest, woodland, mallee, mulga, watercourse vegetation, gardens. Forages in pairs or alone; moves methodically through tree and shrub canopy; occasionally hovers in its search for insects. In spring and early summer, there are loud outpourings of song as rival males bob, bow and chase. Gives a long succession of ringing notes, often 20 to 35 without pause: 'cheWIT-chWIT-chWIT-chWIT' or 'joey-joey-joey-'. Also a call with drawn out beginning, powerful, ringing whip-crack finish: 'eeee-CHIEW!' and 'eee-CHONG!' Migratory or sedentary; common.

- 🔵 *rufiventris*
- 🔴 *falcata*
- 🟡 *pallida*
- 🟢 *minor*

Upper parts mid-grey
Wings grey-brown, pale margins
♂
Deepest cinnamon-rufous in S of range, paler in N and, as feathers wear; often has whitish area on belly.

Black breast band continues up across ear-coverts to lores, but rarely forward of ear-coverts on other races.
♂
Becomes paler inland and northwards.

Females (all races) variably streaked, mid cinnamon to almost plain cinnamon-buff
♀ Streaked buff to mid cinnamon

Faint streaks
Almost plain, pale cinnamon

Immature males are similar to adult females of their respective races.

---

**White-breasted Whistler** *Pachycephala lanioides* 18–20 cm
Inhabitant of mangroves, usually thickets with low, dense canopy; also ventures into other similar dense waterside vegetation. A large species with bill more typical of shrike-thrush than whistler. Usually solitary, quiet and unobtrusive; forages among foliage, stems and on mud at low tide when the large bill is used to take small crabs and other marine creatures. Voice distinctive, but not unlike that of the Rufous Whistler. Most notes deep, mellow, somewhat husky nasal, with frequent higher, clear, piercing whistles. Sedentary; moderately common.

- 🔵 *lanioides*
- 🔴 *carnarvoni*
- 🟡 *fretorum*

Subspecies uncertain; differences slight, possibly not sufficient to justify recognition as subspecies

Bill heavy, slightly hooked
Black crown edged rufous at nape
Mid grey-brown
Blackish tail-coverts
♂
Juv. and Imm.: like female, but underparts more heavily streaked

White throat patch bordered by a black, chestnut-edged breast band. 🟡 Has much narrower chestnut band than black band.
🔴 Back paler sandy brown; darker grey-brown on other races

Eye deep red
Wings dark brown, pale margins
♀

---

**Mangrove Golden Whistler** *Pachycephala melanura* 16 cm
Inhabitant of mangroves on coasts, estuaries, tidal rivers, but also in rainforest, monsoon forest and adjoining dense vegetation. Voice is deeper, richer than that of the Golden Whistler, ranging from low sounds to clear, sharp, whip-crack whistles, 'chwieop-cheiop-chweip-wheit-WHIT'. A frequent shorter call is 'whit-wheeta-WHIT' and single 'wheeeeit'. Calls in breeding season; otherwise usually quiet and unobtrusive. Sedentary; moderately common.

- 🔵 *melanura*
- 🔴 *robusta*
- 🟢 *spinicauda*

Golden collar broader at nape than on Golden Whistler
Bill longer, heavier
🔵 Tail slightly shorter than Golden Whistler
♂
Wings blackish; margins of coverts grey or yellow
Juv.: like that of Golden Whistler

🔴 Back olive-brown
Yellow slightly deeper tone
♀
Yellow tint under tail

🟢 Back olive
♀
Yellow tint, throat to belly and under-tail-coverts, where brightest

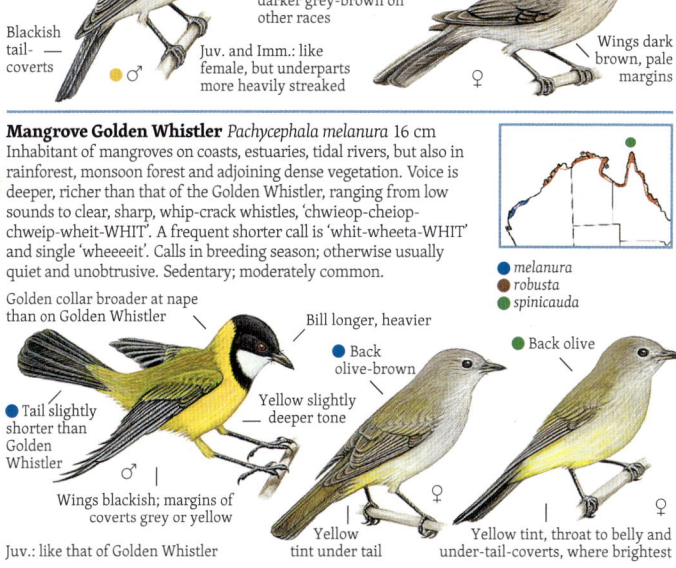

## GREY-BROWN WHISTLERS

**Olive Whistler** *Pachycephala olivacea* 20–22 cm
Northern race above 400 m in rainforest; S race in eucalypt forests, tea-tree, or in heath. Keeps low in undergrowth or on ground; often solitary or with robins, scrubwrens. Easily overlooked but for loud whistles: sharp, or deep and mellow. In S, whip-crack 'wheeeow, WHIT, whit, whit' and 'chee-o-whit, wheeeow'. In N, calls high, then abruptly low, 'whiee-whuooo', and low-high 'cheeowhit, wheee'. Sedentary or local movements; uncommon.

- olivacea
- macphersoniana
- apatetes
- hesperus
- bathychroa

Nominate race darker
Eye brown
Colour slightly brighter, yellower; throat less heavily mottled
Scalloped dark grey
Breast darker
Mid tawny buff (deeper on nominate race)
Only race with male unlike female (has richer colours)
♀ Has plain throat, overall dulled colours

**Grey Whistler** *Pachycephala simplex* 14–15 cm
Lives in rainforest, monsoon and riverine forest, mangroves, paperbark swamps and adjoining eucalypt woodland. Often in mixed foraging parties where this very small whistler looks and behaves more like a flycatcher. At other times, may sit hidden for long periods among foliage. Gives clear, penetrating whistles, slow and deliberate, or vigorous, first notes rising, last descending, 'whit, whit, wiet, wheat, whieew'. Sedentary; common.

- simplex
- peninsulae

Forehead grey-brown
Crown and nape grey
Eye dark brown
Pale brow line extends forward to bill; fine white eye-ring.
Chin and throat white, very faintly streaked
Olive-brown
Breast and flanks tinted grey-buff
Pale yellow, darker grey-brown wash over breast

**Red-lored Whistler** *Pachycephala rufogularis* 19–21 cm
Mallee, broombush, native pine and banksia heaths with dense low shrubs and spinifex. Secretive most of year, in breeding season has distinctive calls. Gives series of deliberate, swelling yet abrupt whistles of even pitch and strength, 'wheit-wheit-wheit' or 'whieot-whieot-whieot-'; in distance, 'tchiopt-tchiop-tchiop'. Song a wistful, husky, scratchy sound, rising, falling, rising: 'wheit-chi-u, wheit-chi-u, wheit-chi-u'; 'wheit-chu, wheit-chu'. Sedentary; rare.

Imm.: similar to female, except lores and throat are whitish, flight feathers edged buff; eyes brown

Forehead, lores, face to throat orange-buff
Head and upper parts mid grey-brown
Facial colours paler than male
Olive-brown
Wings dark grey-brown with fine olive margins to flight feathers
Underparts paler cinnamon-buff

**Gilbert's Whistler** *Pachycephala inornata* 19–20 cm
Dry woodland, mallee, mulga with shrubby understorey, abundant litter. Usually inconspicuous in dense thickets; feeds in shrubs and on the ground. Has distinctive, carrying sequence of husky, mellow, deep whistled calls, starting low, building louder: 'cheop-cheop-cheiop-CHEIOP-CHEEIOP-CHEEIOP'. Also scratchy 'eechowk, eechOWK, eeCHOWK' and clear, strong 'eew-WHIT, ew-WHIT, ew-WHIT'. Sedentary or locally nomadic; uncommon.

- inornata
- gilberti (May not differ sufficiently to be retained as a subspecies.)

Lores dark
Deep grey
Bill deep and stubby
Eye deep red, both sexes
Small, rich rufous-buff throat patch
Throat pale
Pale grey with light cinnamon tint
Imm.: similar, eye brown, flight feathers buff edged

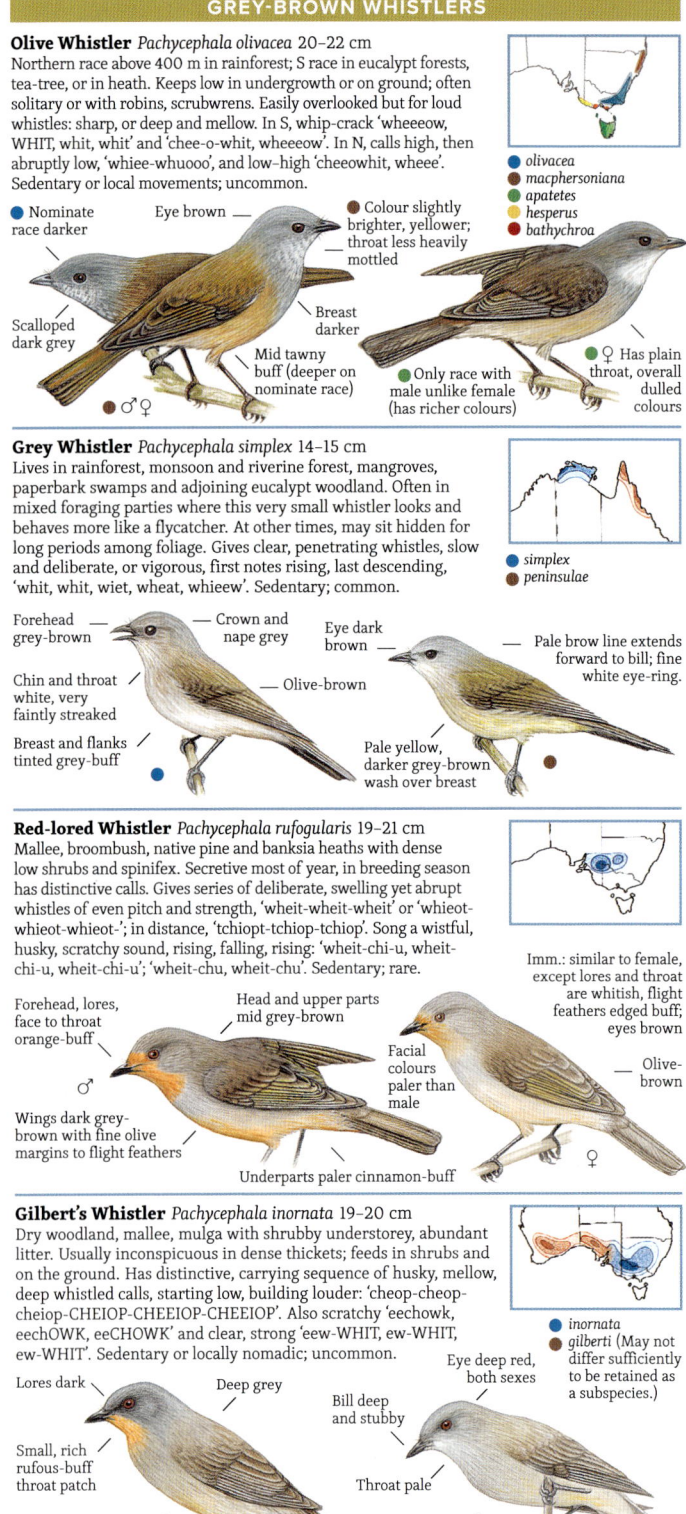

# VARIED SITTELLA

**Varied Sittella** *Daphoenositta chrysoptera* 11–13 cm
Small birds of upper branches: often unnoticed but for almost incessant sharp twittering. Flight strongly undulating, rather erratic, with outspread wings showing bands of white, cinnamon or orange. Forage downwards on branches and tree trunks, probing for insects in crevices of bark. Through spring and summer live in smaller breeding flocks, typically a breeding pair and several immatures of the previous season. Contact call is a rapid, sharp, 'chwit' and a fast, chattering 'tchwit-tchweit-tchwit-tchweit'; call strongly in flight from tree to tree. Five races with long-established common names, at one time all full species:

● **Varied Sittella**, nominate race. Orange wing stripe, dark cap, streaked. In eucalypt forest, woodland, mallee, shelter belts, roadside trees. Sittellas, all races, not usually seen in rainforest. Sedentary nearer coast, nomadic in arid regions. Common.

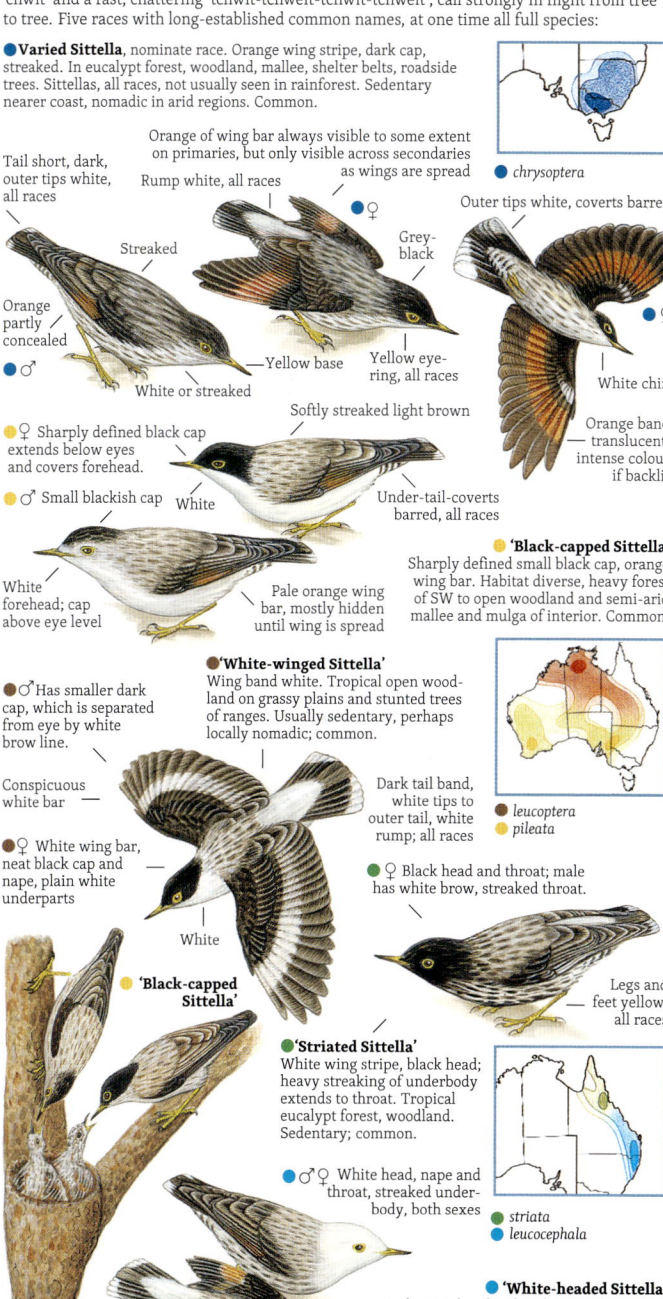

● *chrysoptera*

● 'Black-capped Sittella'
Sharply defined small black cap, orange wing bar. Habitat diverse, heavy forest of SW to open woodland and semi-arid mallee and mulga of interior. Common.

● 'White-winged Sittella'
Wing band white. Tropical open woodland on grassy plains and stunted trees of ranges. Usually sedentary, perhaps locally nomadic; common.

● *leucoptera*
● *pileata*

● 'Striated Sittella'
White wing stripe, black head; heavy streaking of underbody extends to throat. Tropical eucalypt forest, woodland. Sedentary; common.

● *striata*
● *leucocephala*

● 'White-headed Sittella'
Combines white head with orange wing bar; both sexes. As with other sittellas, the wing bars are barely visible until wing is spread. Forests and woodlands along the Great Divide, E coastal Qld, and inland on more open grassy woodlands. Common.

## CINNAMON-PLUMAGED QUAIL-THRUSHES

### Nullarbor Quail-thrush
*Cinclosoma alisteri* 18–21 cm
Former race of Cinnamon, but slightly smaller and upper parts, especially the crown, brighter cinnamon. Uncommon.

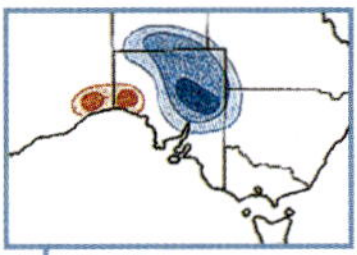

● *alisteri*   ● *cinnamomeum*

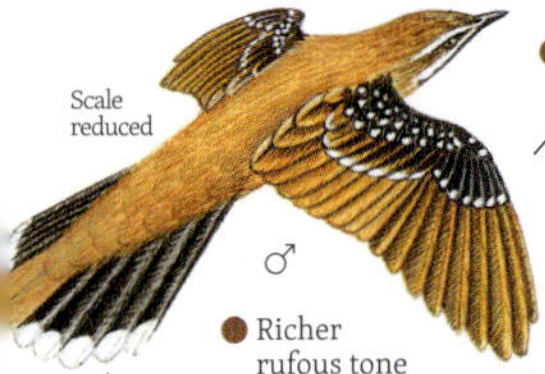

### Cinnamon Quail-thrush
*Cinclosoma cinnamomeum* 18–21 cm
Arid stony plains with sparse low shrub cover. Feeds only on the ground. Foraging birds wander, hunched low, pecking at insects, seeds; run swiftly, squat low in cover of bush/rock if disturbed. Put to flight, burst upwards with quail-like take-off. Contasct call a weak, high, piping whistle, uniform pitch, 'tsit, sit-sit'. Song, mostly at dawn: 'seit, sit, sieee', 'tseit, see-seeeit'; 'tssi-sieee, seei-eit'. Locally nomadic; common.

*Scale reduced*

♂ Richer rufous tone

White wing bands across black coverts

Central tail feathers are plain dusky cinnamon, outers blackish, tipped white

Finely flecked brown

Wing-coverts buff-brown

Buff, mottled brown

Juv., both species

Crown rufous

Black unbroken from chin to breast

● ♂

Wing-coverts white, tipped black

Back rufous-cinnamon, darker cinnamon-brown on crown

Central tail hides darker, white-tipped outers.

Long down-curving white brow line

Black broken by broad white band

● ♂

● ♀

Breast mid-grey, flanks cinnamon

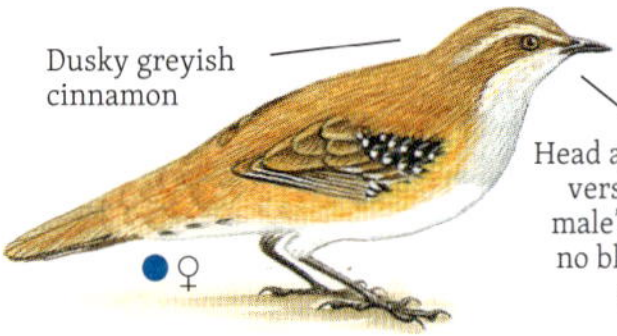

Dusky greyish cinnamon

● ♀

Head a paler version of male's, and no black at throat

---

### Western Quail-thrush
*Cinclosoma marginatum* 20–24 cm
Formerly considered a subspecies of *C. castaneothorax*, its habitat, diet, call and lifestyle are similar, although it is now considered a distinct species.

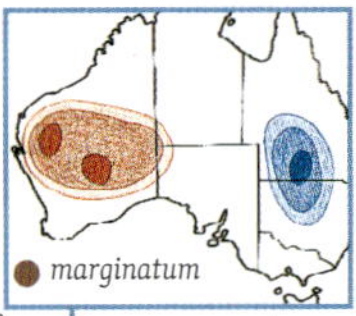

● *marginatum*   ● *castaneothorax*

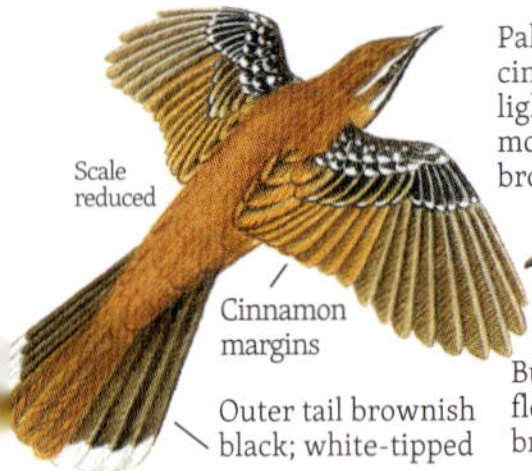

### Chestnut-breasted Quail-thrush
*Cinclosoma castaneothorax* 20–24 cm
Low scrub on hard, stony plateaus and ranges. Pairs or families; takes insects, fallen seed. Runs if disturbed, using vegetation as cover; may squat to conceal, or quail-like, burst into flight showing white tail tips. Contact/alarm calls high-pitched, weak, piping with a slight rise in pitch, 'tseit-tseit, tseit-tseit'. Song is soft, tremulous 'whit, whit-a-whittee'; husky, soft, 'wheit, wht, a whteee'. Locally nomadic; uncommon.

*Scale reduced*

Pale cinnamon, lightly mottled brown

Cinnamon margins

Outer tail brownish black; white-tipped

Buff, finely flecked brown

Juv., both species

● ♂ Speckled shoulder at times concealed

Rump rufous-chestnut, darker on tail;

Lighter cinnamon-buff

♂

white tail tips usually hidden

White brow down-curved; wide white throat-edge streak

Brownish, merging to chestnut on rump

Crown olive-brown

Throat black

Cinnamon-chestnut

♂ ●

● ♀

Cinnamon-grey merging to cinnamon flanks

Brow and throat light buff

● ♀

Grey-brown

307

## GREY-BROWN QUAIL-THRUSHES

**Spotted Quail-thrush** *Cinclosoma punctatum* 26–28 cm
Inhabitant of eucalypt forest, woodland, favouring rocky ridges with sparse shrubs, tussocky grass and abundant bark, leaf, twig litter, fallen logs. Largest quail-thrush and sole forest species. Forages on ground; crouches low; scuttling movements as it pecks at seeds, insects and other small prey. Very shy; it relies on camouflage, until almost underfoot, bursts up on whirring wings. Flight undulating; fanned tail displays white tips. Drops to cover 50 to 100 m away; then may be very timid, flushing at a far greater distance. Song a penetrating double whistle, 'whee-it, whee-it, whee-it', the initial 'whee' with a mellow ringing quality, the final 'it' sharply higher, giving undulations throughout the long sequence. At times a faster, higher, 'sweeit-sweeit-sweeit'. Also a long, whistler-like sequence, 'wheet-wheet-wheet' and high twittering. Males often call from low perches in trees. Sedentary; uncommon.

● *punctatum*
● *dovei*
● *anachoreta*

Differences between races slight, some perhaps indefinite or variable

**Chestnut Quail-thrush** *Cinclosoma castanotum* 22–26 cm
Wide range across Aust., diverse habitat. Uses open semi-arid woodland of eucalypt or cypress pine, mallee or mulga, with sparse shrub layer and litter debris; heath, coastal tea-tree thickets. In pairs or small family parties of 3–5, fossicking in leaf and twig ground litter. Usual call a rapid, very high rush of squeaked whistles at uniform pitch, 'swit-swit-swit-swit-swit-swit'. Song by male, from a high perch, is an even series of rich, mellow whistles, 'wheit-wheit-wheit-wheit'; may be repeated at varied tempo and pitch, including a higher, quicker 'whit-whit-whit-whit-whit'. Nomadic in arid regions, sedentary towards coasts; uncommon, patchy. E of Spencer Gulf–Lake Torrens, SA, race *castanotus* mostly uncommon and of patchy distribution; western race *clarum* widespread, rather scattered, generally uncommon but in places quite common.

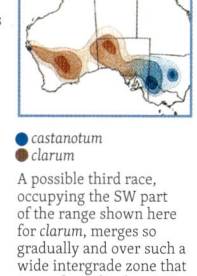

● *castanotum*
● *clarum*

A possible third race, occupying the SW part of the range shown here for *clarum*, merges so gradually and over such a wide intergrade zone that it may be a clinal variation rather than a distinct subspecies.

## WHIPBIRDS & WEDGEBILLS

### Eastern Whipbird *Psophodes olivaceus* 25–30 cm
Inhabitant of dense thickets, margins of rainforest, wet eucalypt forest, heath. Keeps to thickets, fossicking in ground litter; climbs through shrubbery; flutters weakly across open spaces. Loud ringing whip-crack call of this species is one of the most common, widely recognised eastern bird calls. From male, a long whistle building up to an explosive whip-crack; instantly answered with sharp 'tchew-tchew' from female. Sedentary; common.

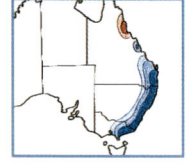

- *olivaceus*
- *lateralis*: smaller, olive-brown with brownish tips to tail

No white cheek streak until about second year

Bold white streak down cheek and side of neck

Olive-green

Black with variable amount of patchy, streaky white

Imm.

On folded tails, white outer tips may be concealed by all-dark central feathers; obvious on fanned tail or from beneath.

### Black-throated whipbird
*Psophodes nigrogularis* 22–24 cm
Rare, elusive, hard to see in dense heath, mallee, broombush. Pairs forage in leaf and twig debris; if alarmed, vanishes under cover. Call: squeaky with repetitive rhythm, a musical, lilting, 'WHIT-chee-a-WHEER-chwit'. Female often quickly answers 'chwik-it-up'. Also sharp, downward, rattling finish, 'whit-chi-a-tr-r-r-r-r-t!'. As contact, 'chrrk!' Locally common in protected sites, generally rare.

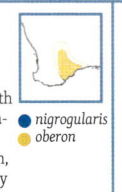

- *nigrogularis*
- *oberon*

Small crest, often not erected

Chin and throat black

Underparts olive-grey

Black along upper edge of white chin streak.

Dark subterminal band across outer tail

### White-bellied Whipbird
*Psophodes leucogaster* 22–24 cm.
Formerly a race, 'Mallee Whipbird', of *P. nigrogularis* and very similar, the White-bellied is now a separate species with its own race, *P. l. lashmiri*.

- No black strip along top margin of white on chin.
- *leucogaster*
- *lashmiri*

Pale centre

Outer tips white

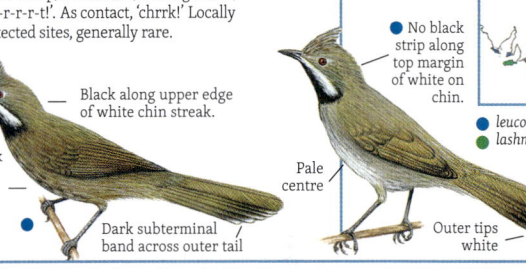

### Chiming Wedgebill *Psophodes occidentalis* 20–22 cm
Arid mulga and similar scrub, broombush, mallee. Wary; if singing from a shrub-top perch, goes silent at sight of intruder, dropping to cover, or gliding out of sight. Song: far-carrying, musical, wistful, rather lonely; the first 3 notes high, ringing, metallic, the last much deeper, metallic, 'tchip-chipity-chiep, tchonk', repeatedly, vigorously. Also, high-low, ringing, metallic 'WHI-whiet-WHI-wheit-'. Nomadic or dispersive; possibly most common near W coast.

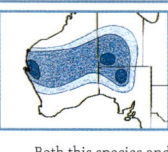

Stubby bill

Tall dark crest curves well forward.

Breast plain, very pale grey-buff

Imm.: like adult, but bill and legs brownish

Both this species and Chirruping have tail that may be uplifted, held straight, or lowered.

White tipped

### Chirruping Wedgebill *Psophodes cristatus* 19–21 cm
Inhabitant of arid, open country with sparse acacias, eremophilas, saltbush, bluebush, lignum. The two wedgebill species appear almost identical, but differ in song and behaviour. This species is less shy; allows closer approach to see a calling bird or watch a flock. Small groups; runs rather than hop. Call is a high squeak and strong, vibrating, descending trill, 'tsiep-TSIEEER', repeatedly, usually as male-female call-answer. Adults sedentary; more common in N, rare in S of range.

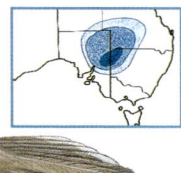

When perched, looks like Chiming Wedgebill, but has a faintly streaked breast.

Often departs shrub-top perch in long, low glide, giving a glimpse of conspicuous white tips of widespread tail.

305

# 20 WHIPBIRDS & QUAIL-THRUSHES TO BELLBIRD

Families Cinclosomatidae (whipbirds, wedgebills, quail-thrushes), Neosittidae (sittellas) and Pachycephalidae (whistlers, shrike-thrushes, shrike-tits and Crested Bellbird). Small to medium birds, insectivorous, foraging in foliage, on bark and ground. Breed as pairs, except sittellas, which breed cooperatively, assisted presumably by young of previous year. The Crested Shrike-tit has three distinctive subspecies, at times regarded or proposed as full species.

Scarlet Robin = 12 cm

## cinclosomatidae
page 305

- Eastern Whipbird
- Black-throated Whipbird
- Chirruping Wedgebill
- Chiming Wedgebill
- Chestnut Quail-thrush
- Spotted Quail-thrush
- Cinnamon Quail-thrush
- Nullarbor Quail-thrush
- Chestnut-breasted Quail-thrush
- Western Quail-thrush

## neosittidae
page 308

- Varied Sittella
- 'White-winged Sittella'
- Olive Whistler
- Red-lored Whistler ♂ ♀
- Gilbert's Whistler ♂ ♀
- Grey Whistler
- 'Brown Whistler'

## pachycephalidae
page 309

- Golden Whistler & Western Whistler
- Arafura Shrike-thrush
- Mangrove Golden Whistler
- Bower's Shrike-thrush
- Rufous Whistler
- White-breasted Whistler
- Sandstone Shrike-thrush
- Crested Shrike-tit 'Eastern Shrike-tit'
- Grey Shrike-thrush
- 'Western Shrike-tit'
- 'Northern Shrike-tit'
- Crested Bellbird

## BABBLERS

**Hall's Babbler** *Pomatostomus halli* 22–24 cm
Habitat of dry rocky slopes, ridges with mulga and other large shrubs, or open plains of sparse, stunted trees. Superficially like White-browed Babbler. In flocks of up to 20; forages mainly on ground. Voice like White-browed, but slightly deeper, more nasal; similar grating 'chweip-chweip-chur-r-r-r-'. Locally nomadic, moderately common.

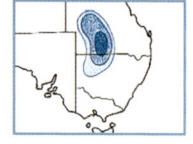

Long bold white brow line, thickest towards the nape

White bib has a more sharply defined edge than on the White-browed Babbler.

Narrow white tips to central pair

Broad white brows curve in at nape.

White band continues completely around spread tail, though only narrowly on central pair.

Similar to nest of widespread White-browed Babbler, but generally smaller and of sticks, giving a neater exterior.

**Grey-crowned Babbler** *Pomatostomus temporalis* 26–29 cm
Open forests, woodland. Highly gregarious, in noisy family flocks of about 15. Calls softer, more nasal than other babblers. Breeding adults give low–high duet: female a nasal 'awark'; male adds a high, clear 'tiew'; other birds of group join in – cacophonous calls, yet mellow, like distant yapping of many small dogs. As contact, a soft, mellow 'tchuk'. Sedentary; common in NW, uncommon in SE.

● *temporalis*
● *rubeculus*

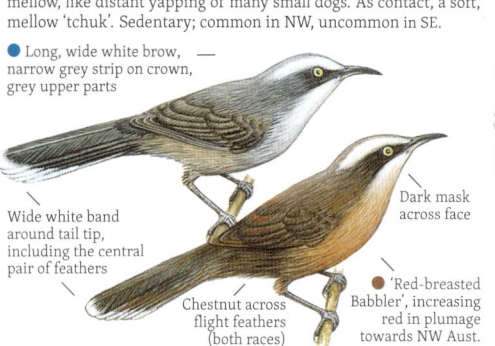

Long, wide white brow, narrow grey strip on crown, grey upper parts

Wide white band around tail tip, including the central pair of feathers

Dark mask across face

Chestnut across flight feathers (both races)

'Red-breasted Babbler', increasing red in plumage towards NW Aust.

Nest 40–50 cm diam.

**Chestnut-crowned Babbler** *Pomatostomus ruficeps* 21–23 cm
Inhabits sparse mulga, mallee, lignum, saltbush on plains and stony ranges. Probes ground litter and bark crevices seeking insects. Contact call a whistling 'tsee-tsee, tsee-tsee'; much chattering among members of family group; whistling calls intermixed with 'tchak-tchak-tchak', which becomes louder, more rapid in excited play, quarrels, alarm. Territorial song a rather strident piping. Sedentary; generally uncommon.

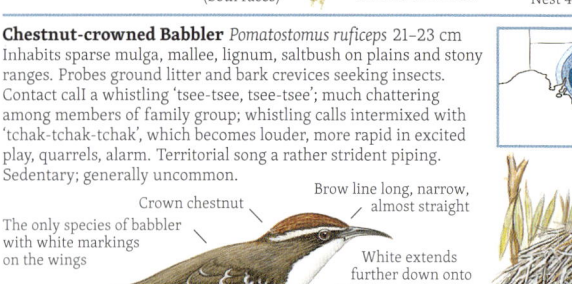

Crown chestnut

Brow line long, narrow, almost straight

The only species of babbler with white markings on the wings

White extends further down onto belly.

Large white tips to outer feathers; tapers to very narrow white on central pair.

Juv.

Juvenile similar, but with rufous-buff where adult is white; it is only briefly in this plumage.

Nest up to 100 x 50 cm

# LOGRUNNER, CHOWCHILLA, WHITE-BROWED BABBLER

**Australian Logrunner** *Orthonyx temminckii* 18–21 cm
Subtropical rainforest, mainly on higher ranges. Forages on forest floor; scratches with large, strong feet: props against broad, stiff tipped tail while throwing debris aside with feet, leaving a trail of small bare circles. Pairs, small family parties; unobtrusive yet not shy. Noisy at dawn with rapid, bubbly, sharp 'qwikit-qwikit-qwikit'; and sharp, metallic 'queeik-queeik-queeik'; deeper, bubbly, mellow 'quokkit-quokkit-'. Sedentary; sparse in S, more common in N.

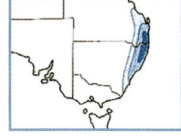

The male, below, is less colourful than the female, at right.

Tail broad, dark, tipped with stiff feather shaft 'spines'

Male white-throated

Colourful, but merges into rainforest leaf litter: gold and rufous of fallen leaves; dark browns of moist earth and decaying leaf litter.

Bold pattern of grey, black, orange

Female has orange throat and richer colours overall, including brighter rump.

Imm.: darker than adult, underparts dull white, scalloped and blotched brown

**Chowchilla** *Orthonyx spaldingii* 27–29 cm
Tropical rainforest. Forages through leaf litter; runs swiftly when alarmed; flies briefly on short, rounded, whirring wings. Presence may be revealed by rustling sounds or patches of soil bare of litter – inconspicuous. Usually in family groups, whose chorus gives a barrage of sound, often at dawn and dusk: a loud 'chowk-chowk, chiowk-chiowk, chuck-chiow, chweik, chwieeik-chowk'; also softer musical warblings and low growlings. Sedentary; common.

● *spaldingii*
● *melasmenus*: smaller, but with relatively larger bill

● Head black, mantle and back deep olive-brown, both sexes

Bluish-white eye-ring

Tail has stiff, bare, feather shaft 'spines'.

Male white, chin to belly

Short, rounded wings

Flanks olive-brown

● Female orange, chin to upper breast (● chestnut)

While scratching with one foot in leaf litter, carries weight on other leg and on the stiff tail.

Legs and feet large, black

Imm.: dull rufous-brown finely patterned with cinnamon

**White-browed Babbler** *Pomatostomus superciliosus* 19–22 cm
Open forests and dry woodlands, mallee, tree-lined inland watercourses. Noisy, gregarious behaviour has resulted in diverse descriptive names, for example, the 'Twelve Apostles'. Almost always found in small family parties with restless, follow-the-leader activity in shrubs, on ground, spread out or clustered in their busy foraging. The passage of a family flock is typically marked by incessant churring and wheezing chatter. Have varied nasal, squeaky and wheezy chatterings, cluckings, miaows: 'squarrk-squarraik, wheeit-wheeit, chur-r-r-r-r'; clucking 'tchuk' contact calls; alarm whistles. Sedentary; common.

● *superciliosus*
● *gilgandra*
● *centralis*
● *ashbyi*

White brow is long, quite narrow, dark edged, and curves back from base of upper bill to a fine point at nape.

Dark band or mask across dark eye

Subspecies differences slight, possibly clinal; probably not able to be differentiated in field.

Upper parts, including crown, dusky mid to dark brown

Bill large, strongly downcurved

Throat and breast white, merging to buffy brown flanks and belly

White tips to all except central pair, entirely dark

Nests are large, conspicuous, built of quite large sticks, and lasting several years, a useful pointer to the presence of babblers.

30–40 cm diameter

# FLYCATCHERS

## Jacky Winter *Microeca fascinans* 12–14 cm

Extremely widespread; well known in rural districts. Prefers open woodland, farm paddocks with scattered trees, mallee, mulga. Uses perches on branches, stumps or posts from which to scan the ground; drops on small prey more like a robin than a flycatcher. Its acrobatic pursuits of flying insects are typical of flycatchers; but its habit of hovering just above the grass, perhaps to flush out flying or hopping insects, is unusual. On landing, wags tail side to side revealing the white outer feathers, these are also obvious in flight. Has a clear, far-carrying, ringing, whistling call, a rapid 'chwit-chwit-chwit-queeter-queeter-queeter', slower 'cheweet, cheweet'. Sedentary or nomadic; common.

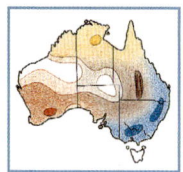

Subspecies: three, or perhaps four; darkest in S, paler in far N and interior

- *fascinans*
- *assimilis*
- *pallida*

## Lemon-bellied Flycatcher *Microeca flavigaster* 12–14 cm

Inhabits tropical savannah woodland, rainforest margins, monsoon and vine scrub, riverside vegetation, paperbark swamp, pandanus, mangroves. Actively pursues insects in flight; flutters about the foliage, low to upper canopy, at times close to or on the ground; often uses a perch in the open as a vantage point. Usually quiet and unobtrusive; alone, in pairs or small family parties. Performs song flights when breeding; rises above treetops. Call is a cheery, vigorous, musical whistling. Notes vary abruptly from squeaky, sharp to mellow, varied 'chiwi-chiwi-chweeip-chip-chrup' and 'quieee-chirrup-chi'. Also metallic, vibrant scolding sounds. Sedentary; common.

- *flavigaster*
- *flavissima*
- *laetissima*
- *tormenti* – often now considered a species, Kimberley Flycatcher, *Microeca tormenti*.

**Kimberley Flycatcher** or 'Brown-tailed Flycatcher'

## SMALL YELLOW ROBINS & YELLOW-LEGGED FLYCATCHER

**Pale-yellow Robin** *Tregellasia capito* 12–13.5 cm
A small robin of the rainforest that often hunts like a typical robin – clinging to tree trunks, dropping to ground to take prey – but also like small flycatchers, actively capturing insects in flight and from lower to mid-level foliage. It favours dense undergrowth, often lawyer vines, but ventures into nearby dense eucalypt forest. Gives bursts of three or four piercing, metallic squeaks: 'chweeik-chweeik-chweeik'; also harsh, scolding 'scraich-'. Sedentary; common.

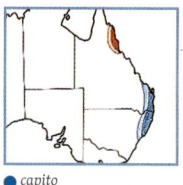

● *capito*
● *nana*

Upper parts olive-grey; wings and tail grey-brown

Large white patch at lores joins white throat.

● Has pale rufous-buff on lores and eye-ring.

Pale yellow, darker on sides of breast

White throat patch larger on this race

Bright yellow; flanks olive

Legs and feet dull pale pink

Juv.: streaked rufous-brown, and some rufous on wing-coverts.

---

**White-faced Robin** *Tregellasia leucops* 12–13 cm
Inhabits tropical rainforest, drier lowland vine scrub; forages in undergrowth of lower to mid-levels. Appears not to use open woodland adjoining rainforest. Often perches on vertical stems, drops to take small prey on ground, pursues flying insects. Usually quiet, unobtrusive, and seems to call only in breeding season. The song is a cheery, rising whistle, 'whitia-whittik, whitia-whittik'. As contact, a soft 'tsip'; grating 'scraich' in alarm or aggression. Sedentary; moderately common.

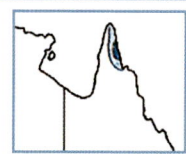

One subspecies, *albigularis*, occurs in Aust., many others in NG.

Upper parts greyish olive

Head black, except white on forehead, chin, and around the eyes

Juv.

Rump olive-yellow

The rather comic expression given by the white face mask is most evident in front view.

Underparts yellow, olive tinge to breast

Juv.: rufous-chestnut, streaked white
Imm.: like adult, but dull colours

---

**Yellow-legged Flycatcher** *Kempiella griseoceps* 11.5–12.5 cm
A restlessly active, very small flycatcher of tropical rainforest, favouring outer canopy, margins and edge of adjoining eucalypt woodland, paperbark swamp. Often hovers to glean from foliage, darts after flying insects. Has strong, resonant, whistled calls, but hesitant pauses between most notes: 'wheeit, wheeit, wheewit, whee-whi-wi-wi-whit, whit'; also high, clear, rapid, piping trill; buzzing 'scraich'. Sedentary; common within restricted range.

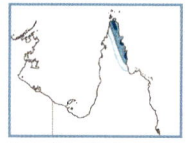

One subspecies, *kempi*, is found in Aust., others in NG.

Head appears high and rounded compared with the sleeker, lower head profile of the Lemon-bellied Flycatcher.

Bill small but broad and flat with pale yellow lower base

When perched, often raises tail.

White throat merges into light grey breast and pale lemon remainder of underparts.

Juv.: brown, spotted buffy white on upper parts

Unique among small flycatchers and robins for its bright orange-yellow legs and feet

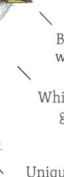

## YELLOW & WHITE-BREASTED ROBINS

**Eastern Yellow Robin** *Eopsaltria australis* 15–16 cm
Inhabits eucalypt and rainforest, woodland, banksia heath, drier mallee, acacia scrub. Usually pairs to small family groups; scans ground from a branch or tree trunk; watches with an occasional flick of wings; drops to ground to take prey. Flights generally short and undulating, yellow rump conspicuous in final glide to landing. Voice a clear, even, piping whistle, 'tchiep-tchiep-tchiep'; and loud double noted 'tchweip-tchweip'. Sedentary with some local seasonal movements; widespread, common.

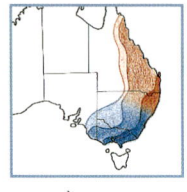

- *australis*
- *chrysorrhoa*

- In SE Aust. has olive-green rump.
- Upper parts, face and chin, grey
- Unlike the Western Yellow Robin (below), the Eastern is yellow from throat to under-tail-coverts.
- Juv.: streaked brown and white
- Yellow rump
- Pale wing bar around base of flight feathers

**Western Yellow Robin** *Eopsaltria griseogularis* 15–16 cm
The western equivalent of the well-known Eastern Yellow Robin differs most obviously in having a grey breast. Behaviour is similar; unobtrusive but not shy. Habitats include forest, woodland, mallee, mulga. Calls include a clear piping, 'whit-whit-whit-whit-'; a slower, high 'chwip-chwip-chwip'; a slow, deliberate, 'chiOWP-chiOWP, chiOWP-chiOWP'. Some differences between calls of races. Some local seasonal movements; widespread, common.

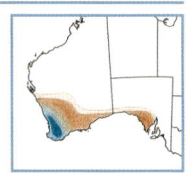

- *griseogularis*
- *rosinae*

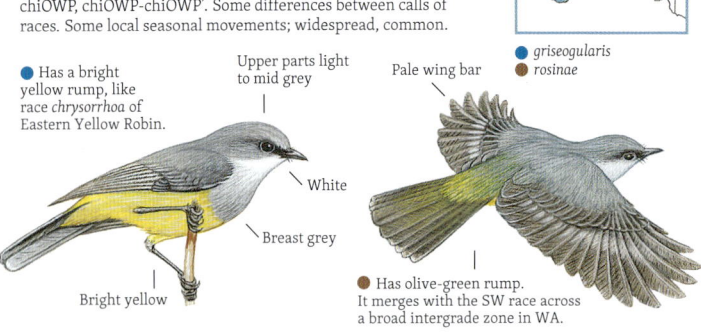

- Has a bright yellow rump, like race *chrysorrhoa* of Eastern Yellow Robin.
- Upper parts light to mid grey
- Pale wing bar
- White
- Breast grey
- Bright yellow
- Has olive-green rump. It merges with the SW race across a broad intergrade zone in WA.

**White-breasted Robin** *Quoyornis georgiana* 14.5–16 cm
In S of range, inhabits tall, dense, damp undergrowth of heavy karri and jarrah forest, especially creekside thickets; in N of range, occurs in drier but dense coastal sandplain scrub. Clings to tree trunks, low branches; occasionally drops to ground to take prey. Has mellow, liquid, yet abrupt 'wi-CHWEK' at intervals of several seconds. Sound carries well, and, in distance, just 'chwek' is heard. The song is a higher, more lively 'tchiew-tchiew, whiet-siew, whiet-siew'; in alarm, a hard 'tchek, tchek'. Endemic to SW; common in S of range, uncommon elsewhere.

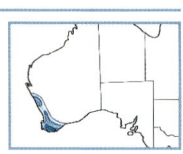

- Black lores
- Upper wing has pale bar, hidden when wing is folded.
- Upper parts smoky mid-grey blending softly into pale-grey breast
- Pale wing bars around base of flight feathers; white underwing primary coverts
- Underparts white with pale grey across the breast
- Wings and tail dark grey

Birds of an isolated population in coastal thickets at the extreme NW of the range of the species are smaller, and have slightly warmer brownish grey cast to plumage.

## HOODED, DUSKY & MANGROVE ROBINS

**Hooded Robin** *Melanodryas cucullata* 15–17 cms
Woodland of eucalypt, mallee, mulga; heath, inland and drier parts of the coast; semi-cleared farmland. Perches motionless, scanning ground; occasionally darts down to take prey. Moderately common. In flight, displays conspicuous white markings. Mellow notes, the first highest, strongest, then descending, fading: 'CHIERP, chwep-chep-chep'. Adults sedentary, immatures dispersive.

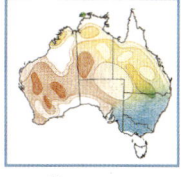

- cucullata
- westralensis
- picata
- melvillensis

Black hood, unbroken by markings

Bold white shoulder bar

White wing bar

Bold, neat black and white pattern; hood extends down as rounded black bib. Female similar, but black of hood and back are dark grey, white markings are less conspicuous.

White wing bar and margins

No hood; uniform grey

White towards base of tail

White mantle margins

Differences between races are slight. On mainland, broad intergrading between races suggests merely a gradual, clinal variation. From time to time the number of recognised races has changed: currently four, previously, from time to time, two, one and none.

Conspicuous white bar along centre of wing; follows base of flight feathers, tapering towards outer wing; does not reach two outermost primaries.

**Dusky Robin** *Melanodryas vittata* 16–17 cm
Temperate rainforest to open woodland, farmland. Sits motionless for long periods on stump or low branch, dropping occasionally to ground for small prey. Call is a plaintive, monotonous, rather carrying, soft whistle, 'twei-twoo'; first part louder and clearer, the second lower, muffled. Song a rush of mostly squeaky notes, 'chwot-chwert-chwee-chweeeit-chweeeit-chwert'. Locally nomadic; common.

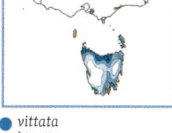

- vittata
- kingii

Upper parts dull olive-brown

Throat pale

White at bend of wing

Indistinct wing bar across base of secondaries

Tail dark olive-brown with pale margins towards base

Subspecies on King Is., Bass Str., has warmer brownish plumage, the flanks having a russet tone.

Juv.: heavily mottled, brown on white, darker on upper parts

**Mangrove Robin** *Peneothello pulverulenta* 15–16.5 cm
Inhabitant of mangroves; prefers the shrubbier landward and seaward margins. Clings to trunks and branches, occasionally flicking wings; watches ground, dropping to probe the mud for tiny crabs and other creatures; also forages among foliage. Most typical call is a husky, mournful, descending 'whieer-wherrr...whurr' and much lower, level 'whurrr-whurrr'. Song is a higher, clearer 'whitch-a-whitchu' and lively chattering. Sedentary; moderately common.

- cinereiceps
- alligator
- leucura

Variation around the northern coast is slight; opinions differ whether subspecies should be recognised.

Lores, through eye to ear-coverts, black. Fades away to grey towards nape.

Rather large head and long straight bill give a whistler-like appearance.

Upper parts mid to dark slaty grey

Large white panel each side of spread tail

White throat

Underparts white, lightly tinted blue-grey

Outer tail feathers white towards base

Faint pale wing bar around base of flight feathers is hidden when wing is folded.

## FLAME, RED-BREASTED & RED-CAPPED ROBINS

**Flame Robin** *Petroica phoenicea* 11–13 cm
Inhabits rainforest, wet eucalypt forest, woodland. Winters in open woodland and farmland, when, unique among Aust. robins, it gathers in flocks. Perches conspicuously on stumps and boulders; usually takes prey on ground and flying insects in forest habitat. Cheery, clear, sharp, piping trill, 'chrip-a-chip, chrip-a-chip, chirripa-tirrrrip'. Slower, 'whit, whit, whit-'; harsh scolding. Migratory; common.

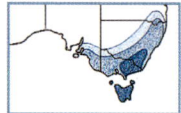

● Breeding range
○ Winter dispersal range

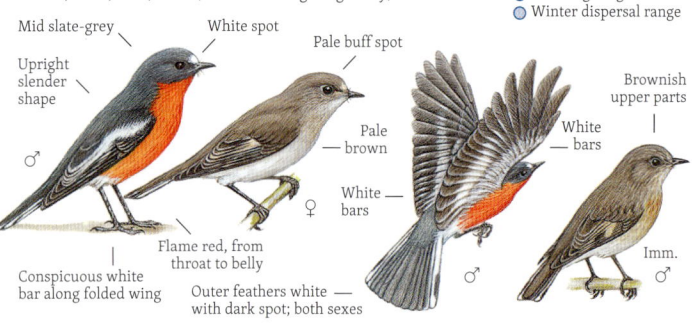

**Scarlet Robin** *Petroica boodang* 12–14 cm
Forest, woodland, suburbs; heavier forest in spring and summer breeding months. Waits, clinging to tree trunk or branch; drops to ground to take prey. In flying up and away, bold white markings on black make male distinctive. Call: a shrill yet pleasant, cheery, rippling, musical trill, 'tirrrit-tirrrit-tirrrit, tirrrit-tirrrit-tirrrit'; contact a quiet, tapping 'tik'. Locally migratory or dispersive; common.

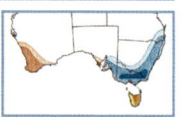

● *boodang*
● *campbelli*
● *leggii*

**Red-capped Robin** *Petroica goodenovii* 11–12 cm
Dry open woodland, mallee, semi-arid mulga. Frequently drops from low perches to ground, then off to another perch, all the while giving quick flicks of tail or part opening wings to flush out insects. Usually in pairs or, after breeding, in small family parties. Has a cheery, rapid, metallic, ticking trill, one of the most characteristic sounds across vast areas of the inland and drier regions. Although not loud, the song seems to carry far: 'did-dit-d-wier, did-dit-d-wier, did-dit-d-wier'. Contact call, abrupt 'tchek, tchek'. Common.

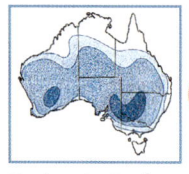

No subspecies. Females from inland and drier regions are slightly paler.

# WHITE-BROWED, ROSE & PINK ROBINS

**White-browed Robin** *Poecilodryas superciliosa* 15–17 cm
Inhabitant of monsoon forest, vine thickets, mangroves. Often holds tail uplifted like a scrub-wren, but its habits of drooping the wings and flicking the wings and tail are typical of robins, as is the clinging to sides of trees while scanning the ground below. Gives clear, piping whistles: 'whiet-wheit-whiet'; 'wheet-t-wit, whiew'; 'whieeeet-whiet'. Also a quick 'chokok, queitchiew' and 'chokoc-chiew'. As alarm, a harshly chattered 'chrok-chrorrok'. Generally scarce, but locally common in optimum habitat.

- *superciliosa*
- *cerviniventris* – sometimes now considered a full species, Buff-sided Robin, *P. cerviniventris*.

**White-browed Robin**, NE Qld rainforests
Two white wing bars
Both have a conspicuous long white brow and an upcurved white band from throat to ear-coverts.
**Buff-sided Robin**, W Kimberley to NW Qld
Upper parts mostly deep olive-brown
Bold white bar along wing
Buff flanks
Boldly patterned; tail is often cocked upwards, but the drooped and restless flicking of wings and tail are behaviours typical of robins.
White tips
Second faint bar midway down leading edge of each flight feather

**Rose Robin** *Petroica rosea* 11–13 cm
Rainforest, wet eucalypt forest, especially gullies of ranges and in understorey of acacia trees. Forages actively about the upper canopy, flitting about the foliage more like a flycatcher. In winter, disperses over lower, more open country. Has a cheery chattering trill, 'chwit, whit-wit-wit-'; slower, 'whit, whit, whit-'; as contact, a repeated 'tic'. Usually takes prey on ground. Uncommon.

- Breeding range
- Winter dispersal range

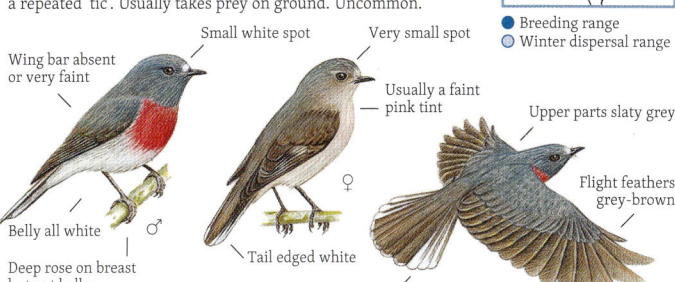

Wing bar absent or very faint
Small white spot
Very small spot
Usually a faint pink tint
Upper parts slaty grey
Flight feathers grey-brown
Belly all white ♂
Tail edged white
Deep rose on breast but not belly
Outer three mainly white
♀

**Pink Robin** *Petroica rodinogaster* 11–13 cm
Rainforest, wet eucalypt forest, in tree-fern gullies of ranges; winters in more open country. Less active than Rose Robin, waiting patiently to drop to ground for prey. Voice a cheery chattering trill, may seem more like call of fairy-wren than robin, a rather metallic rattling or tinkling, yet pleasant. After first note, a slight pause, then descending, 'chwit, whit-wit-wit-'; slower, 'whit, whit, whit-'. Usually takes prey on ground. Dispersive; uncommon.

- *rodinogaster*: breeding
- Nonbreeding
- *inexpectata*
- Striped area = overlap out of breeding season

Small white spot above bill; buff on female
Plain sooty brown
Sooty brownish-black
Warm olive-brown
Faint tan-buff wing bars
♂
Pink down onto belly
Cinnamon-buff, with pinkish tint
Brownish underwing has pale coverts and may have faint wing bar.
♂
Pink extends far back under belly.
♀
Tail plain, dark brown

## SCRUB-ROBINS & GREY-HEADED ROBIN

**Southern Scrub-robin** *Drymodes brunneopygia* 21–23 cm
Dense mallee, acacia scrub with layer of leaf litter. Although appearing shy, often quietly sneaks close, behind low scrub or debris. Gives only fleeting glimpses, scurrying behind vegetation and litter. Has persistent, brisk 'chwip-chWIPpee' and deeper, mellow 'chep-whep-wheeip'. Song is a briskly whistled, cheery 'wheet-d-wheeeit, wheet-d-whii-it', rising sharply. Scolds with harsh, chattered 'scraach-chak-chak-chak'. Uncommon.

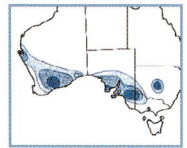

No subspecies. Slightly paler in N of WA and SA

- Distinctive facial pattern with dark line down through eye
- Face pale grey, with white eye-ring
- Tail-coverts and rump rufous
- Tail is often uplifted when bird is on ground, but drooped down when on perch.
- Tail long, dull grey-brown, feathers edged rufous
- Often fluffs out the feathers of breast and belly, appearing quite rotund.
- Narrow white tips to outer feathers may be hidden when tail is closed.
- Legs quite long, suited to foraging in ground litter

**Northern Scrub-robin** *Drymodes superciliaris* 21–23 cm
Tropical rainforest and vine scrub. Locally quite common. Gives a strong, drawn out, rising whistle, 'wheeeeeiit', but may be steady, 'wheeeeet', undulating, 'wheeeur-eeeit', or interrupted, 'wheee-wheet'. Also a descending 'screeearch' or undulating 'screee-air-arsch'. Forages on rainforest floor; quickly bounds away on long, strong legs into concealment of vegetation.

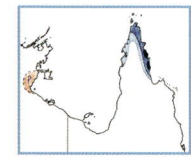

- • *superciliaris*
- • *colcloughi* (Probably never existed; two specimens, supposed to be from the Roper River, are more likely from Cape York.)

- Conspicuous bold black band down through the eye
- White tips to dark wing-coverts form twin pale bars
- Nape to back cinnamon-rufous, becoming brighter rufous on the rump and upper-tail-coverts
- Juv.: throat and breast flecked darker brown with plumage of crown and mantle edged dark brown
- Flanks buff
- Legs very long; feet large, pale, fleshy pink

**Grey-headed Robin** *Heteromyias cinereifrons* 16–18 cm
Softly colourful robin of lower levels of forest. Inhabits tropical rainforest, mostly at higher altitude, exception being where forest extends unbroken from ranges to coastal lowlands. Clings to trunks; drops to ground to take prey or forage in litter. Has a soft, piping whistle, rather mournful, notes at even pitch, 'whiet-whiet-whiet', or high-low 'whiet-whit, whiet-whit'. Territorial song a long, high whistle followed by several lower, shorter notes. Sedentary; common within restricted habitat.

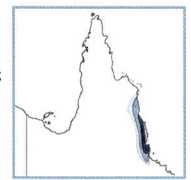

- Crown and nape light grey, finely flecked darker
- Heavy, pale-tipped bill
- Mantle olive-brown
- Eye underlined white
- Rufous-chestnut
- Short, uneven, dull, white bar
- Flanks buff
- Tail tipped white
- Conspicuous white wing bar around base of flight feathers; faint trailing-edge wing bar

# 19 SCRUB-ROBINS & ROBINS TO BABBLERS

Families Petroicidae (Scrub-robins, Robins, Flycatchers), Orthonychidae (Logrunner, Chowchilla), and Pomatostomidae (Babblers). Mostly insectivorous, some species taking flying insects (flycatchers and some smaller robins), but many dropping from perches to take small terrestrial prey (robins). Others spend time on the ground, foraging in leaf litter and on bark of logs and tree trunks (scrub-robins, Logrunner and babblers). Most breed as pairs, but the babblers are gregarious and have many helpers at bulky nests.

Splendid Fairy-wren = 12 cm

## petroicidae
### page 295

Southern Scrub-robin

Rose Robin

Western Yellow Robin

Northern Scrub-robin

Pink Robin

Eastern Yellow Robin

White-breasted Robin

Flame Robin

Hooded Robin

Scarlet Robin

Mangrove Robin

Dusky Robin

Red-capped Robin

Lemon-bellied Flycatcher

Grey-headed Robin

Pale-yellow Robin

Yellow-legged Flycatcher

White-browed Robin

White-faced Robin

Jacky Winter

## orthonychidae
### page 302

Australian Logrunner

Chowchilla

## pomatostomidae
### page 302

White-browed Babbler

Hall's Babbler

Grey-crowned Babbler

Chestnut-crowned Babbler

## ORANGE & YELLOW CHATS; GIBBERBIRD

**Orange Chat** *Epthianura aurifrons* 10–12 cm
Very low but dense shrubbery of saltbush, samphire around lakes, sparsely vegetated plains. This chat's habit of perching atop low vegetation on flat expanses of mudflats makes the gold male visible from afar. If approached, tends to be elusive, keeping low; especially wary near nest. After breeding may be encountered in small groups. Insects and other tiny prey are taken from ground and vegetation. Vibrant, metallic 'tang'; softer 'tchek, tchek'. Nomadic; moderately common.

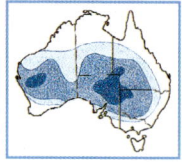

No subspecies: nomadic wanderings maintain uniformity across range.

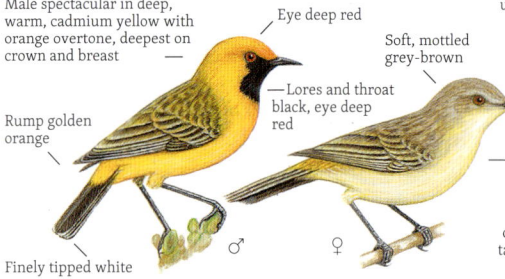

Male spectacular in deep, warm, cadmium yellow with orange overtone, deepest on crown and breast
Eye deep red
Soft, mottled grey-brown
Rump golden orange
Lores and throat black, eye deep red
Pale yellow throat, whitish brow line
Underparts softer fawny yellow
In flight, shows golden orange rump and blackish tail narrowly tipped white. Imm.: like adult female
Finely tipped white

**Yellow Chat** *Epthianura crocea* 10–12 cm
A rare chat, usually recorded from coastal grassy swamps and lagoon margins in vegetation of reeds, saltbush. Inland, recorded around bore overflows, where it uses cane grass, cumbungi, lignum or bluebush. May be in pairs or small parties, occasionally 20–50 birds in a locality of optimum habitat. Call is high, piping 'tee-tsue-tee' with the middle 'tsue' usually slightly lower. Males give displays: low undulating or dipping flights with metallic, piping 'tee-tee-tee-' of 2–15 notes. In alarm or threat near nest, low, harsh churring, scolding sounds. Nomadic; rare.

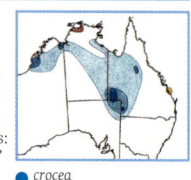

● *crocea*
● *tunneyi*
● *macgregori*

Male almost as brightly plumaged as the Orange Chat, but without such a warm orange overtone, rather an intense, lemon yellow

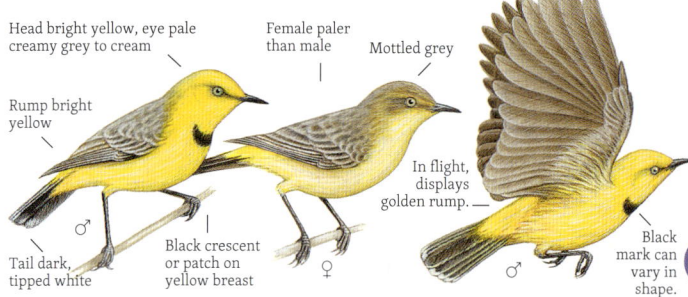

Head bright yellow, eye pale creamy grey to cream
Female paler than male
Mottled grey
Rump bright yellow
In flight, displays golden rump.
♂
Tail dark, tipped white
Black crescent or patch on yellow breast
♀
Black mark can vary in shape.

**Gibberbird** *Ashbyia lovensis* 12–13 cm
Inhabitant of arid gibberstone plains; sparse low vegetation. Like a small yellowish pipit, with similar upright stance, teetering and tail-wagging movements, abrupt short dashes across ground. In pairs or small parties, occasionally larger flocks, at times in company with pipits; searches ground for small prey, occasionally flies up to snatch a flying insect. Gives clear 'tswee-tswee-tswee' calls and a rather musical chatter. In alarm, several piercingly sharp notes. Nomadic; uncommon.

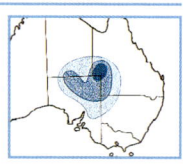

Although similar to the chats, the Gibberbird has sufficient differences to justify placement in a separate genus, *Ashbyia*.

Light buffy grey-brown rump and back; feathers pale-edged
Eye pale yellow
Slightly longer, more slender than chats; often in tall, upright posture
Female has a smudgy grey-brown band across paler yellow breast.
♂
Blackish tail with white tips to inner webs of all except central pair

## CRIMSON & WHITE-FRONTED CHATS

**Crimson Chat** *Epthianura tricolor* 11–12 cm
An inhabitant of semi-arid mulga and other acacia scrub, dry open woodland, typically with low sparse groundcover of bluebush, samphire, other very low shrubs and tussocks, and often the very open surrounds of inland salt lakes. Usually in small flocks, working across herbage of open ground seeking small prey, perhaps occasionally nectar at low shrubs; flight bouncy and undulating. When breeding, the male often attracts attention with a high, silvery, even 'tseee-tseee-tseee' given in high song flight display over nest site, or from nearby high perch. As contact call in flight, 'tchek, tchek', also a softer 'chikit, chikit'. Nomadic and irruptive; generally common but vary locally by season, scarce to abundant.

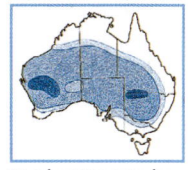

No subspecies: nomadic wanderings maintain uniformity across its range.

Forehead and crown crimson
Yellow eyes conspicuous in black face mask
Throat white
Coverts, secondaries, edged buffy white
Breast to belly brilliant crimson
Crimson forehead
♂
White tips conspicuous across widespread tail
♀
Crimson rump; tints and patchy smudges of crimson across most of underparts
Imm.: like females, but lack pink and have buff edged wing feathers.

**White-fronted Chat** *Epthianura albifrons* 11–13 cm
Open country – inland salt lakes, estuaries, salt marshes where there is low, often sparse samphire, swamp margins, open low heath, remnant low vegetation on farms. A familiar bird in S Aust. Many local names reflect this chat's call, 'Tin-tack', 'Tang'; or its looks, as in 'Moonface', 'Baldy-head'. Usually conspicuous, but secretive near nest; gives distraction displays. Gregarious; usually in small parties to large flocks. Seeks insects on ground and in low bushes. Flies with bouncy undulations, often rising high, calling with metallic, nasal 'tang'. Nomadic in arid parts, sedentary in wetter S of range.

Adult female is dull grey-brown with indistinct markings. Imm.: similar to female, but breast band very faint or missing

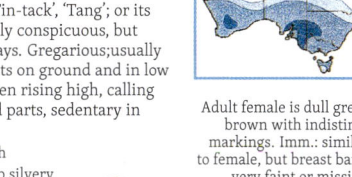

Eye yellowish
Back, rump silvery mid grey
White is encircled by black band from crown down to hind crown, nape and breast band.
Wings, coverts, dark brown
Wings, coverts, dark brown
Underparts white with broad black breast band
♂
Reduced breast band
♀
White tips on dark brown; conspicuous in flight
White-tipped

**Mistletoebird** *Dicaeum hirundinaceum*
Superficially resembles Crimson Chat, Red-headed and Scarlet Honeyeaters. Several comparative illustrations here, but see also main text and illustrations on p. 349.

Red from chin to upper breast
Upper parts iridescent blue-black
♂
Black central streak
♀
Bill dark grey
Imm.: like females, but yellow at base of bill and gape
♂

**Olive-backed Sunbird** *Cinnyris jugularis*
Like honeyeaters in habits and appearance, but Sunbirds are in family Nectariniidae. Main text and illustrations p. 349.

Bill finer, longer, than all but the spinebills
Iridescent blue-black
♀
♂
Underparts entirely yellow
Often hover to probe flowers or take small prey.

## PIED, SCARLET, RED-HEADED & DUSKY HONEYEATERS

**Pied Honeyeater** *Certhionyx variegatus* 15–18 cm
Arid woodland, mallee, acacia scrub, spinifex, drier heath. In display flights, rises almost vertically, showing white of wings and spread tail; gives high, far-carrying calls then drops vertically with wings closed. Silent except in breeding season, when males seem to call incessantly – clear, carrying whistles at almost uniform pitch, 'tieeee, ti-tiee, tieeeee', and tieee, ti-tieee, tieeee-tieeee-'. Nomadic; common, but of erratic occurrence.

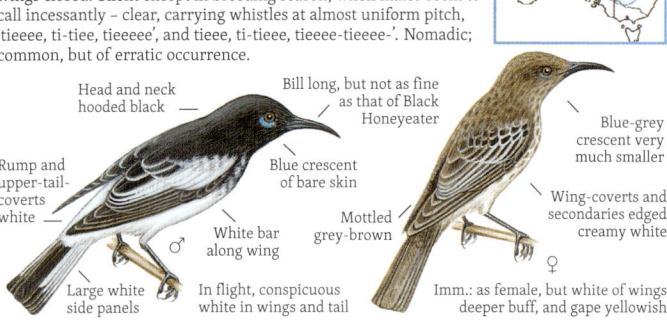

Head and neck hooded black
Bill long, but not as fine as that of Black Honeyeater
Rump and upper-tail-coverts white
Blue crescent of bare skin
Blue-grey crescent very much smaller
White bar along wing
Mottled grey-brown
Wing-coverts and secondaries edged creamy white
Large white side panels
In flight, conspicuous white in wings and tail
Imm.: as female, but white of wings deeper buff, and gape yellowish

**Scarlet Honeyeater** *Myzomela sanguinolenta* 10–11 cm
Rainforest margins, eucalypt forest, woodland, paperbark swamp, heath, acacia scrub. Gives piercing 'tseeip, tsip, tseeip'; higher, weaker, squeaky 'seep-seep'. Male has brisk, cheery song, a silvery, tinkling whistle, erratically rising and falling, 'chwip, swit-sweet-switty-swit-switty', often with chattering 'chawak-chwakity' at end. Sedentary in N of range, erratic migratory visitor to S, elsewhere nomadic; common.

Head, upper body and breast are entirely glossy scarlet.
Grey-brown
Pink tint
Brownish-buff
Lores black
Juv.: like adult female
Black wings with scarlet body make males not only spectacular, but unique and easily recognised.
Flight feathers black, edged white
Males develop red first on head and back, last on breast.
Imm. ♂

**Red-headed Honeyeater** *Myzomela erythrocephala* 12 cm
Mangroves, paperbark swamp, nearby tropical monsoon and riverine forest, woodland. Small, agile; hovers around flowers and foliage, snatches insects in acrobatic aerial pursuits. Abrupt, sharp, scratchy, metallic 'tchwip-tchwip-tchwip' or softer 'swip-swip-swip'. Also buzzing 'tzzip, tzzip-tzzip, tzzip' and scolding 'charrk-charrk-'. Locally nomadic; common.

● *erythrocephala*
● *infuscata*: red further up onto lower back; black breast band wider, belly to under tail darker grey

Scarlet extends to lower back.
Red tint, forehead and chin
Lores black
Scarlet plumage glossy, reflecting fine highlight streaks under bright light
Light grey
♀
Juv.: like female, but paler base to bill and olive-grey iris
● ● Races merge on Cape York.

**Dusky Honeyeater** *Myzomela obscura* 12–15 cm
Rainforest, paperbarks, mangroves, watercourse thickets, nearby woodland. Sedentary or locally nomadic. Although of sombre plumage, can attract attention by its brisk foraging; flits about the upper storey, often hovering to reach nectar or insects. Call is a quite musical 'whik-it, whikit-whikachuk'. Common in N.

Dark grey-brown
● *obscura*
● *harterti*: darker brown, and bill of male slightly shorter

Dark brown, often with a slight scarlet undertone

Sexes are alike. Imm. and juv.: similar, but paler brown, yellowish at gape and base of lower bill

## SPINEBILLS; BANDED & BLACK HONEYEATERS

**Eastern Spinebill** *Acanthorhynchus tenuirostris* 14–16 cm
Forest, woodland, heath. Colourful honeyeater of understorey; attracts attention with 'flop-flop' sound of wings, flash of white tail. Often at flowers of banksia and callistemon. Call a sharp 'chip-chip-chip-' at even, high pitch. Song a clear, cheery 'cheer-whit cheerwhit'. Common.

- *tenuirostris*
- *cairnsensis*
- *halmaturinus*
- *dubius*: rich, deep colours

Small, long-billed; female almost identical, but slightly paler grey on crown

Upper surface of wings dark grey-brown

Plain; paler below eye

Bill very long, fine, strongly downcurved

Tail dark blue-grey

Eye red

Juv.

White outer tips

Unique bold pattern of black, white, chestnut

Cinnamon-buff

Underparts pale olive-buff

**Western Spinebill** *Acanthorhynchus superciliosus* 12–15 cm
Forest, woodland; sandplain heath. Lively, colourful, unique to SW. Probes deep flowers; takes insects in acrobatic flight. Call sharp, clear 'chwip, chwip, chwip'; faster 'chip-chip-cherip'. Song a musical 'chri-chri-chri-chri-', often in flight. Endemic to SW WA; locally nomadic; common.

Small, very long, fine bill; males colourful; females relatively plain

Red eye conspicuous in black facial mask, set between bold white lines.

Lines and colours pale, indistinct, but eye red

Nape rufous

Bill long, fine, strongly downcurved

Broad white band across outer tips

Bright rufous

Underparts pale grey-buff, darker from chin to breast

White and black breast bands

Juv.: like female, but plainer, lacks rufous nape; eyes brown.

**Banded Honeyeater** *Cissomela pectoralis* 12–14 cm
Eucalypt forest, woodland, mangroves, dry tropical scrub. Feeds in larger trees – eucalypts and paperbarks. Darts about foliage of flowering trees with bouncy, undulating flight. Calls loud, metallic, vibrant: 'tzziep, tzziep'; higher, buzzing 'chewip-chiewip-chrzieep'; song a softer 'queeit-quitchie-quitchie'. Nomadic; common.

White extends up to reach chin and eye.

Band pale or absent

Upper parts gingery brown, mottled or roughly streaked darker brown

Black breast band joins black of back and wings.

Juv.

Bill rather short, heavy compared with most other very small honeyeaters

Rump white

Rump white

Immatures become more like adults, but with some juvenile plumage; finally similar to adults but duller, back grey-brown.

**Black Honeyeater** *Sugomel niger* 10–12 cm
Semi-arid regions. Solitary, pairs or small groups. Moves nomadically in large numbers to where rains have brought flowering and prolific insect life. Call is soft, metallic 'chwit, chwit' or louder 'tieee' of high, even pitch, at intervals of several seconds. In song flights, double noted 'tieee-teee'. Locally common.

Entirely black

Pale brow

White 'hook' between black of breast and shoulder

Bill fine, strongly downcurved

Mottled brown

Rump black

Black down centre of breast as long tapering point

Sexes unlike: male boldly pied; female dull grey-brown

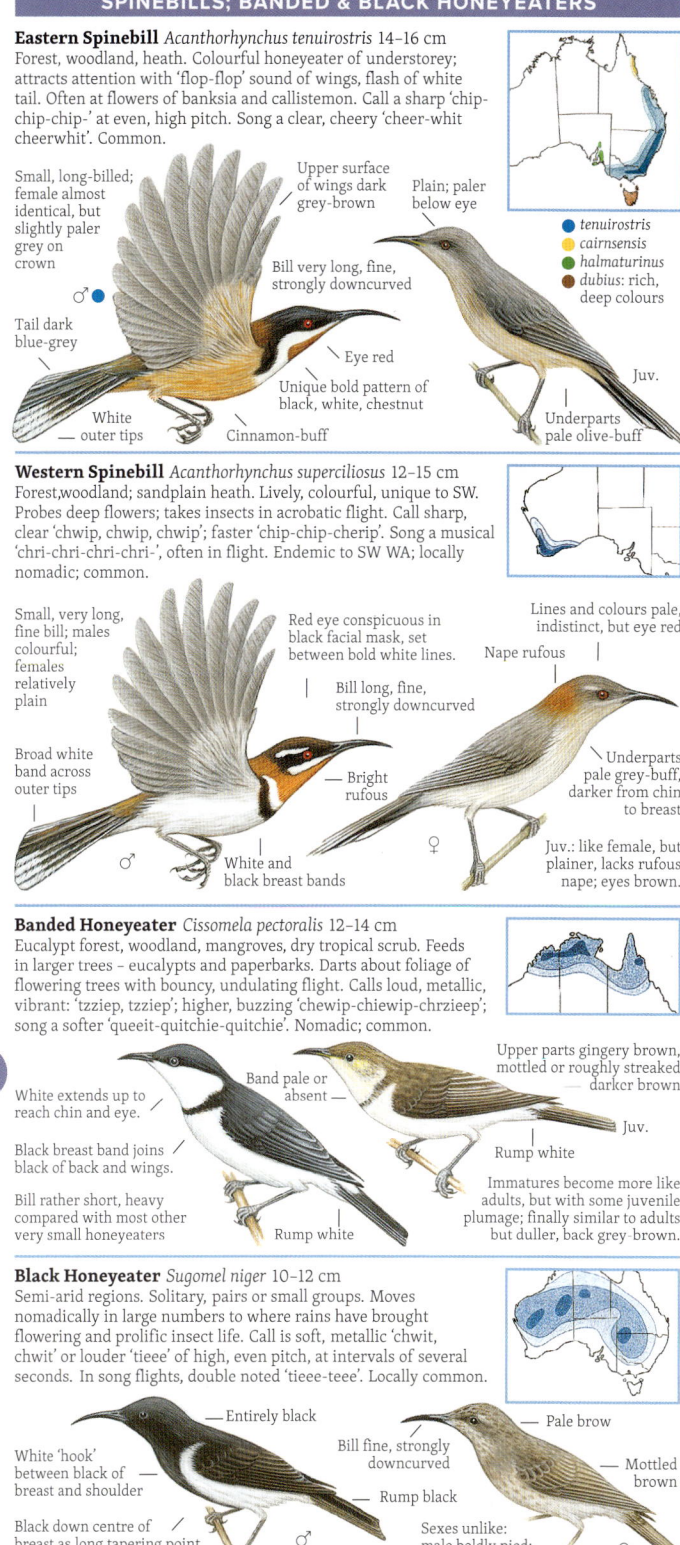

## WHITE-NAPED & BLACK-HEADED HONEYEATERS

**Black-chinned Honeyeater** *Melithreptus gularis* 15.5–17 cm
Forest, woodland of eucalypts, paperbarks, inland tree-lined watercourses. Active, noisy, forages in canopy, usually in small parties, calling to maintain contact; works quickly through foliage, moving on with swift, direct flight. Eastern *gularis* has vibrant, ringing 'chi-chrrrip-chrrrip'; scolding 'quorrip-quorrip'. 'Golden-backed' has a clear, ringing, musical 'trrirrip-trrirrip-trrirrip-', rapid 'trip-trip-trip-trip'. Nomadic; common in NW, uncommon SE.

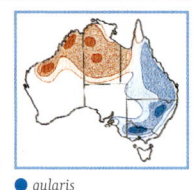

- *gularis*
- *laetior* – now sometimes considered a full species.

**White-throated Honeyeater** *Melithreptus albogularis* 14 cm
Paperbarks in forest and woodland, rainforest margins, riverside vegetation, mangroves, watercourse trees in arid regions. A small honeyeater of outer foliage. Attracts attention by calls, sudden aerial pursuits. Contact call is a sharp 'pit' or 'tsip'. Has also, in contrast, a peevish, rasping 'querrk, querrk, querrk' and quick, higher 'quierk-quierk--quierk'. Song is a rapid, piping whistle, accelerating: 'pit, pit, pit pit-pit-pitpit'. Sedentary, migratory or locally nomadic.

- *albogularis*
- *inopinatus*: (E coast below Cape York) back dull green rather than golden green

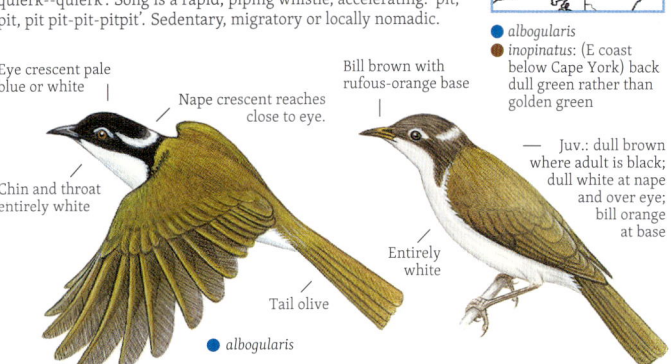

**Black-headed Honeyeater** *Melithreptus affinis* 13–15 cm
Eucalypt forest, woodland, heath; avoids rainforest and similar wet vegetation of SE Tas. Busy, active, noisy; forages in upper foliage; small flocks move tree to tree with swift undulating flight. Calls harsh, rasping 'kherrk', like White-naped; harsher, 'scraak-scraak', faster in aggression or alarm. More distinctive high, sharp, clear whistle: 'tseip, tseip, tseip'; similar, quick, double-noted 'tsip-tsip, tsip-tsip, tsip-tsip' in flight. Nomadic flocks after spring–summer breeding; common.

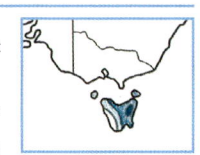

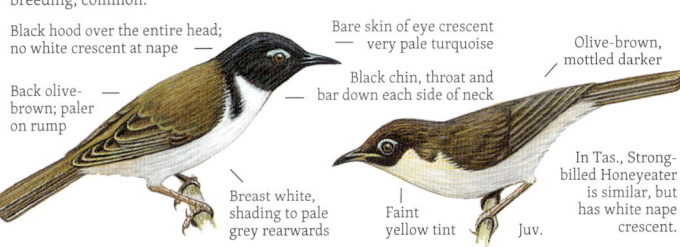

## WHITE-NAPED HONEYEATERS

**White-naped Honeyeater**
*Melithreptus lunatus* 13–15 cm
Inhabits eucalypt forest, woodland. Distinctive husky contact calls attract attention as individuals or flocks forage in the canopy for nectar, insects, manna gum and honey-dew secretions of leaf insects. Contact call a husky 'shierk' or 'szerrk', abrupt in flight; also a clear whistle, 'shierrk-WHIET', and loud, clear, musical 'wherrt-wherrt-wherrt'. In alarm, a high, penetrating, rapid 'chwip-chwip'. Most likely to be confused with other species of this genus.

**Gilbert's Honeyeater**
*Melithreptus chloropsis* 13–15 cm
Former race of White-naped. Eucalypt forest, woodland. Flocks forage in canopy for nectar, insects, manna, honey-dew. A husky 'shierk' or 'szerrk' in flight; a clear whistle, 'shierrk-WHIET', and loud, musical 'wherrt-wherrt-wherrt'.

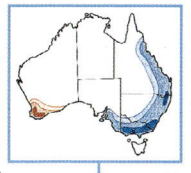

Labels (White-naped/Gilbert's):
- White
- Imm.
- Eye crescent, bill base, both orange
- Juv.: browner-backed than imm.; eye crescent weak tint of adult colour; bill base more orange.
- Slightly larger, bolder nape crescent
- Longer bill
- White of nape does not reach eye.
- Eye crescent red
- Olive-green
- Eye crescent greenish
- Olive-green
- Rump golden-olive
- Black extends down slightly beneath bill, onto chin.
- Sides of neck black, throat to under-tail-coverts white
- ● lunatus
- ● chloropsis

**Strong-billed Honeyeater** *Melithreptus validirostris* 15–17
Rainforest, eucalypt forest, woodland, coastal heath. As it searches for insects and other small prey it energetically probes, prises, rips bark from tree trunks and limbs; insects are rarely pursued in flight, and nectar appears to be of relatively little interest. Gives a regular, sharp, treecreeper-like 'tip, tip, tip, tip' and rapid, sharp, scratchy 'whitch-whitch-whitch'. The song is a rather rollicking 'chwip-ckukachip-kukachip-kukachip'. Also has a rather harsh 'kraach-kraach-'. Sedentary; common.

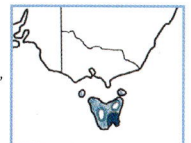

Labels:
- Eye crescent pale blue
- Eye-ring and bill orange-buff
- Crown brownish-black
- Nape crescent tinted yellow
- Sides of neck black
- Yellow tint, throat to breast
- Bill large, almost straight, rather heavy, triangular profile
- Rump rufous-olive
- Imm.
- In Tas., only one similar honeyeater, the Black-headed, but that species has a black hood without white at nape.

**Brown-headed Honeyeater** *Melithreptus brevirostris* 13–14
Forest, woodland, mallee and heath. Forages in canopy around flowers, leaves, bark, finding insects, manna gum, pollen, nectar. In flocks or breeding communities, up to 20 birds. Scratchy 'chwik-chwik-chwik' call; individuals and small flocks flying tree to tree give a lively 'chak-chak-chak'. Locally nomadic; common.

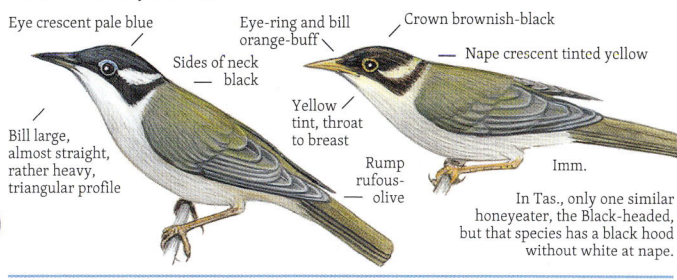

Labels:
- Crown mid to dark brown. Dull white nape crescent meets a completely encircling, buffy white eye-ring.
- Has a dull appearance: brown and grey where similar species are black and white.
- ● brevirostris
- ● leucogenys
- ● pallidiceps
- ● magnirostris
- ● wombeyi
- Chin grey
- Upper parts dull olive-grey with ochre tone to rump
- Dull grey-buff
- Tail grey-brown with yellow ochre tint along margins
- Race ● overall slightly paler; others are darker, darkest ●.

## YELLOW-WINGED HONEYEATERS

**Painted Honeyeater** *Grantiella picta* 15.5–16 cm
Forest, woodland, dry scrub, often with abundant mistletoe. Bright pink bill and yellow feather margins justify the name 'Painted'. Differs from most honeyeaters in looks and in its dependence on mistletoe berries, although nectar and insects are also taken. Alone or in groups; breeds in loose colonies. Often makes erratic, towering display flights from high perches. Best-known call a carrying, rising, falling 'wheeit-whiu, wheeit-whiu, …' and, alternatively, 'whiu-wheeit'. Also a brisk, high 'chiewip-chiewip-' and fast 'whit-whew'. Nomadic or migratory; uncommon.

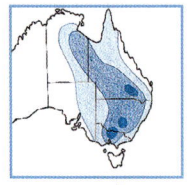

Upper parts blackish; females and immatures paler, dusky brownish black

Wings edged yellow and white

Bill deep pink, sharp, triangular

Short droplet shaped streaks or spots along flanks; females and juveniles, plain white

♂

White tips usually concealed with tail folded; obvious when spread in flight

♂

**New Holland Honeyeater** *Phylidonyris novaehollandiae* 18 cm
Forest and woodland with undergrowth, heath, mallee, coastal thickets. Aggressively competes for nectar and insects, often taken in flight. Gives abrupt, metallic 'tjik!', or 'chwik!'; long whistled 'tseee', often in flight. As alarm, loud, piercingly sharp 'chwiep-chwiep-chwiep'; harsh, scolding 'tjuk, tjuk', slowly or in rapid bursts with sharp alarm calls. In flight, shows yellow in wings and tail, and white tail tips. Sedentary or locally nomadic.

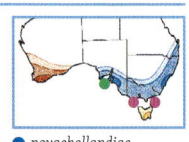

- *novaehollandiae*
- *longirostris*
- *canescens*
- *caudata*
- *campbelli*

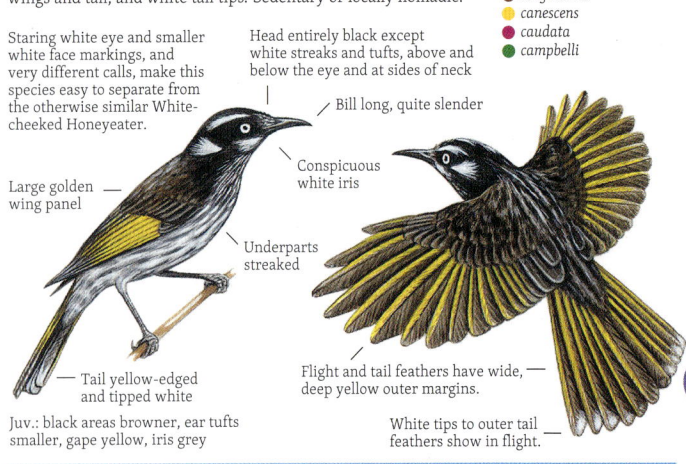

Staring white eye and smaller white face markings, and very different calls, make this species easy to separate from the otherwise similar White-cheeked Honeyeater.

Head entirely black except white streaks and tufts, above and below the eye and at sides of neck

Bill long, quite slender

Large golden wing panel

Conspicuous white iris

Underparts streaked

Tail yellow-edged and tipped white

Flight and tail feathers have wide, deep yellow outer margins.

White tips to outer tail feathers show in flight.

Juv.: black areas browner, ear tufts smaller, gape yellow, iris grey

**White-cheeked Honeyeater** *Phylidonyris nigra* 16–19 cm
Forest, rainforest margins, heath undergrowth areas of open forest, woodland, sandplains. Travels nomadically in search of nectar. Distinctive yapping call: 'chwikup, chwikup'; 'chwipit-chwipit'; 'chakup-chakup-chakup'. Song is a more musical, less sharp 'chwippy-choo, chwippy-choo' and lilting 'twee-tee, twee-tee, twee-tee', often given in display song flight when breeding. Locally migratory or nomadic; common.

- *nigra*
- *gouldii*

Long, tapering brow line

Eye brown

Large golden wing panel

Bill long, obviously heavier than that of New Holland

Very large white cheek patch

Yellow-edged, no white on tips

Underparts heavily streaked

● Illustrated; eastern race has broader, fan-shaped white cheek patch, bill shorter, and appears thicker.

Juv.: gape and brow yellow, plumage dusky

# YELLOW-WINGED HONEYEATERS

**White-fronted Honeyeater** *Purnella albifrons* 16–18 cm
Semi-arid woodland and scrub, typically with scattered shrubs of acacia, eremophila, grevillea, hakea. Superficially resembles the New Holland Honeyeater, and has a similarly wide range, but mostly through the dry interior. Wanders in search of the wealth of flowers that follows rain in the arid regions; often gathers in large numbers, busily seeking nectar and insects, darting about the bushes with almost constant calling. When nesting, tends to be quiet, wary, skulking. Common call a quite high, nasal, metallic 'kzeeip' or rapid 'kzeip-kzeip-'. In song adds deep, mellow, musical notes, 'kzeip, chrrok-chrrok-chrrok' and higher 'pzeip-pzip, chrreik-chrreik'. Nomadic; moderately common.

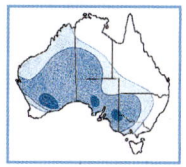

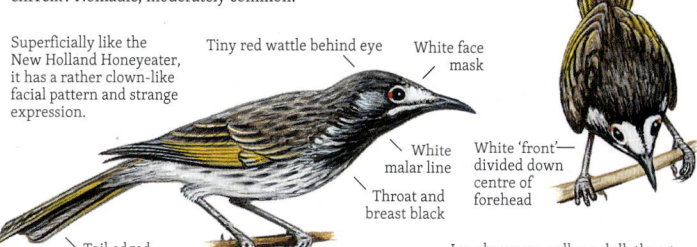

Superficially like the New Holland Honeyeater, it has a rather clown-like facial pattern and strange expression.

Tiny red wattle behind eye
White face mask
White malar line
Throat and breast black
White 'front'—divided down centre of forehead
Tail edged yellow

Juv.: browner, yellows dull, throat mottled, lacks red behind eye.

**Crescent Honeyeater** *Phylidonyris pyrrhopterus* 15–16 cm
Eucalypt forest, alpine woodland; prefers dense undergrowth, including tall heath. Uses wet forest of ranges; forages actively for nectar, sweet sap, insects; moves about forest in swift, undulating flight. Travels to lower altitudes in winter. Has a loud, far-carrying 'tee-chiep'; variations include 'tee-cheop', 'ti-chi-cheop, ti-chi-cheop', 'tchop-tchop-tchip'; rollicking 'tchwop-it, tchwop-it, tchwop-it'. Higher, but not sharp, a musical 'tcheip-tcheip-'; sharp, rapid 'chip-chip-chip-' alarm. Locally migratory; common.

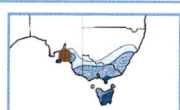

● *pyrrhopterus*: darker
● *halmaturina*: smaller; yellow of wings and tail very dull

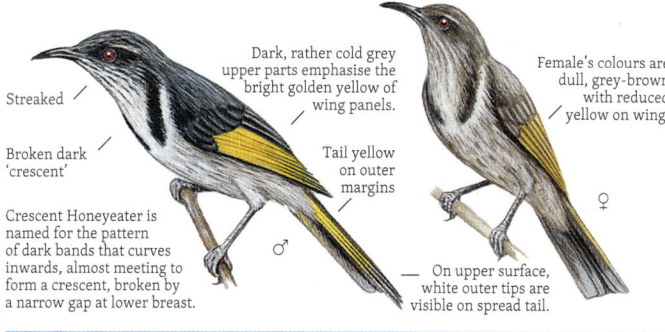

Streaked
Broken dark 'crescent'
Crescent Honeyeater is named for the pattern of dark bands that curves inwards, almost meeting to form a crescent, broken by a narrow gap at lower breast.

Dark, rather cold grey upper parts emphasise the bright golden yellow of wing panels.
Tail yellow on outer margins

Female's colours are dull, grey-brown with reduced yellow on wing.

♀

On upper surface, white outer tips are visible on spread tail.

**White-streaked Honeyeater** *Trichodere cockerelli* 16–19 cm
Monsoon and riverine rainforest, vine scrub margins, melaleuca swamps and heath, and adjoining eucalypt woodland. Forages actively in crown and lower foliage of flowering paperbarks and other trees; solitary, pairs or small flocks; competes noisily with other small birds for nectar and insects. Flight swift, undulating, often in agile aerial pursuits of insects flying around foliage. The call is a harsh, grating 'scrairrk, scrairrk' and faster 'scraik-scraik-scraik'. Song is a musical piping: 'chwip-cheweep-chwip-cheweep'; rapid 'wip-wipwipwip'. Locally nomadic; moderately common.

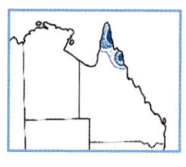

Imm.: similar to adult, but lacks streaky lanceolate plumes, throat to breast being plain with a faint yellow tint; gape is yellow, eye red.

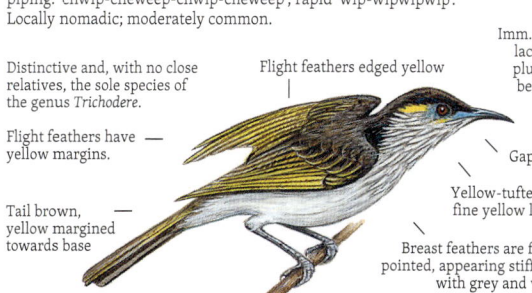

Distinctive and, with no close relatives, the sole species of the genus *Trichodere*.

Flight feathers edged yellow

Flight feathers have yellow margins.

Tail brown, yellow margined towards base

Gape line pale blue-grey
Yellow-tufted ear-coverts, fine yellow line under eye
Breast feathers are finely, sharply pointed, appearing stiffly bristle-like with grey and white streaks.

## BAR-BREASTED TO TAWNY-CROWNED HONEYEATERS

**Bar-breasted Honeyeater** *Ramsayornis fasciatus* 13–15 cm
Paperbark, eucalypt woodland; tree-lined watercourses, floodplain, paperbark swamp. Rather quiet, alone or in small groups. Large, domed nest suspended low over water. Sharp, metallic, 'chwit-chwit' call; faster 'chwee-chwee-chwee'; brisk, cheerful: 'chawak-chawak-chawakety-'. Locally nomadic; common.

- ● *fasciatus*
- ● Has at times been listed as race *broomei*, but differences slight.

Black, scalloped white, appearing finely speckled or barred

Light to mid rufous-brown, softly mottled darker

Fine black line

Prominent black scalloping or barring on breast and flanks

Pinkish brown

Juv.: streaked rather than barred

**Rufous-banded Honeyeater** *Conopophila albogularis* 12–13
Melaleuca and riverine swamp, monsoon forest, mangroves. In dry season, inland waterways; coastal swamps in wet season. Call sharp, piercing, up-slurred 'shrieee' and 'szieep'; song: 'shrip-shrip-SHRIEEE-shriee-shrip-SZRIEEE-shrip-SHRIEE-szriee'. In disputes, a scolding 'squaak' or 'scraach'. Nomadic; common.

Head grey

Back chestnut-brown, flight feathers edged deep yellow

Dull rufous-brown

Dull grey-brown

Clearly defined white chin and throat

Broad cinnamon-rufous breast band, paler tint along flanks

Juv.

Underparts dull grey-buff with rufous-brown tone to breast

**Rufous-throated Honeyeater** *Conopophila rufogularis* 14 cm
Preference for riverine paperbark forest and woodland; also drier, open eucalypt woodland, mangroves. Often forages in flocks, or mixed-species gatherings in flowering trees or low shrubs. Insects are taken in addition to nectar, the agile birds darting out to take them on the wing. As contact call, a sharp, slightly harsh 'chziep-chziep-' or 'chzip-chzip-'. Nomadic; common.

Slight clinal variation: paler and browner towards W and inland

Light grey-brown

Flight feathers, and tail towards base, conspicuously bright yellow along outer edges.

Clearly defined rufous-orange patch on chin and throat

Breast to under tail white, faintly shaded grey on breast, brownish on flanks

Thin pale eye-ring

Juv.

Throat same dull off-white as rest of underparts

**Tawny-crowned Honeyeater** *Glyciphila melanops* 17 cm
Heathlands, open eucalypt heath and woodlands. Wildflowers provide nectar, but also hunts insects. Secretive; keeps low, occasionally calls from higher perch. Noticeable in spring, the lilting song drifts across the heathlands, flute-like notes, wistful, mellow 'quip-peeer, pieer-pieer-piier', faster 'quip-pip-pip-pip'; with many variations, from softly muted to loudly ringing. Nomadic; patchy, common on extensive heathlands.

- ● *melanops*
- ● *chelidonia*: russet cast to plumage

Crown tawny cinnamon-buff

Pale grey to grey-brown; blends with slightly darker upper parts.

Bold black mask through eye and down sides of white throat and breast

Wing linings tinted pale cinnamon, adult and juv.

● Imm.: face mask blurred, brownish

Underparts white, flanks softly streaked grey-brown

## SMALL, PLAIN-PLUMAGED HONEYEATERS

**Green-backed Honeyeater** *Glycichaera fallax* 11–12 cm
Rainforest, monsoon forest, vine scrub, adjoining eucalypt woodland. Active about upper canopy, but small, inconspicuous. Solitary, pairs, small groups. Call, as contact, a high, weak 'piep'; a whistled 'whiet' is slightly husky, often in brisk, even bursts of 4–10 notes: 'whiet-whiet-whiet-wheit'. Sedentary; quite common.

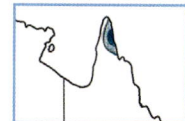

- Crown grey; olive-green—nape and back
- Eye pale grey
- Wings grey-brown, flight-feathers edged yellow
- Bill, for a honeyeater, rather short and almost straight
- Looks and jiz more like flycatcher or white-eye than honeyeater
- Pale yellow, with grey tint at sides of breast mostly hidden under folded wing
- Juv.: similar, but dark eyed

**Brown Honeyeater** *Lichmera indistincta* 12–16 cm
Forest, woodland, heath, mulga, other arid scrub, watercourse trees, mangroves, gardens. Song clear, ringing, musical, delivered in erratic bursts: 'whit, whit, whitchit'; hesitant 'whit, whit, whit, quorrit, quit'; vigorous 'quorrit-quorrit-quorrit'; rapidly trilled 'whitchit, whit, whit-whit-witiwit...'; hesitant, disjointed pre-dawn version. Sedentary, locally nomadic; common.

- Principally a nectar-feeder; long bill well adapted to feeding at deep flowers
- Small yellow triangle behind the eye
- Upper parts mid olive-grey to mid olive-brown
- ● *indistincta*
- ● *ocularis*: slightly larger; darker, less olive
- ● *melvillensis*: larger and darker
- Bill is long, fine, and gently downcurved.
- Flight feathers and tail edged yellow
- Gape black on breeding males, otherwise a fine yellow line; imm. yellowish on and around gape, paler bill

**Grey Honeyeater** *Conopophila whitei* 11–12 cm
Mulga and other similar arid scrub, arid woodland. Busily gleans surface of foliage, often hovering; captures flying insects, when white or pale tail tips are most evident. Has a piercing, metallic, quick double squeak: 'chirraWIEK-chirraWIEK' or 'tsi-WIEK-chirraWIEK'; also a weak, twittering, tinkling song. Not well known. Probably nomadic; uncommon to rare.

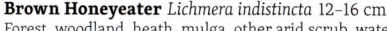

- Small plain honeyeater; often associates with thornbills, gerygones, to which it is similar in looks and activity.
- Quite short, yet longer than bills of thornbills or gerygones
- Obvious pale eye-ring
- Grey, becoming grey-brown in worn plumage
- Chin to breast grey to grey-brown; colour merges into similar tone of upper parts.
- Weak buff tint on throat and cheeks
- Pale tips
- Yellowish tint to feather edges
- Juv.

**Brown-backed Honeyeater** *Ramsayornis modestus* 12–13 cm
Swampy paperbark woodlands, river and rainforest edges, mangroves, nearby eucalypt woodlands. Captures insects around flowers, frequently on the wing; finds insects and spiders in bark crevices. Contact call a sharp 'chwiet'; softer 'chwit' as flight call. Similar scratchy, sharp notes, slightly descending, slowing and fading: 'chwit-chwit-chwich-chwich-chwech-chwech'. Nomadic or migratory; moderately common.

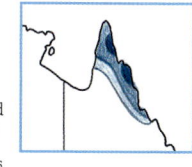

- Light brown with darker flecks
- Rufous-brown back, appearing dark brown in shade
- Bill pinkish-brown
- Buff edges to wing feathers
- Face darker, bill to ear-coverts; white line under eye
- Underparts dull white, faint brownish bars and streaks
- Tail plain brown
- Legs pinkish-brown
- Imm.: gape area buff, breast streaked

## YELLOW-PLUMED & WHITE-PLUMED HONEYEATERS

### Fuscous Honeyeater *Ptilotula fuscus* 15–17 cm
Forests, woodlands, rainforest margins, drier inland mallee and acacia scrubs, coastal heaths. Calls include sharp 'twietch' followed by a rapid, hard rattle, 'twietch-chuka-chuka-chuka-', and sharp, abrupt 'tchwip, tchwip-tchwip, tchwip'. Songs include a ringing, clear, rhythmic 'teechu, teechu, tichu'; in song flight a faster, lilting, clear 'tichu-tichu-tichu'. Sedentary, locally nomadic, seasonal movement in SE; common.

- Small yellow plume
- Grey-brown to olive-brown
- Wings, tail, yellow margins
- Bill black
- Fine yellow eye-ring
- Dark eye-ring
- Imm. even smaller yellow plume
- Pale buffy grey
- Bill has yellow base and gape.
- Juv.

● *fuscus*
● *subgermanus*: yellow plume at side of neck longer, more pointed; lores, throat tinted citrine; slightly smaller

Nonbreeding adult similar, but less yellow on eye-ring, gape and base of bill

### Yellow-tinted Honeyeater *Ptilotula flavescens* 15–17
Open forest and woodland, favouring vicinity of rivers, watercourse tree belts, occasionally mangroves. Active, pugnacious; often in small flocks. Swift undulating flight; acrobatic pursuit of flying insects. Calls vary greatly, from clear, high, even 'tiew-tiew, tiew-tiew', to harsh, rather raucous, complaining 'chaerp, chaerp'. A cheerful call is a clear, high, whistled 'whi-whit, whi-whit' and quick, abrupt 'quorr-quorr, quorr-quorr-a-whit'. Locally nomadic, follows flowering of trees; moderately common.

● *flavescens*
● *melvillensis*

- Head yellow, becoming grey towards nape
- Upswept black line underlined by matching yellow plume
- Light olive to brownish grey
- Chin to throat, pale yellow
- Heavier streaking
- Slightly darker than mainland race
- Wing feathers yellow along outer margins
- Tail edged yellow
- Underparts off-white to yellow tinted, lightly streaked grey
- Melville Is., NT
- Juv.: paler with fainter streaks

### White-plumed Honeyeater *Ptilotula penicillata* 15–17
Woodland, mallee, inland rivers. Widespread, conspicuous, noisy, gregarious; small loose flocks forage for nectar and insects. Rather pugnacious, always alert, quick to give alarm calls that ripple bird to bird at first hint of danger. Call is a clear, quite sharp 'whitch-a-whee, whitch-a-whee-whit', 'whit-a-wheeit, whita-wheeit'. In flight, song is a musical 'whee-whit-a-wheioo'. In alarm, raucous, nasal 'cak-ak-ak-ak-ak-ark'. Usually sedentary; common.

● *penicillata*
● *leilavalensis*
● *carteri*
● *calconi*

- Head yellow
- Black line and white plume
- Upper parts, western and central Aust. races, paler grey-brown
- Head yellow, grey tinted towards nape
- Darker olive-brown
- Wings and tail edged yellow, all races
- Western race has brighter, more extensive yellow extending as pale tint to underparts.
- Becomes darker in SE part of range, paler with more extensive yellow in NW and interior.

## YELLOW-PLUMED HONEYEATERS

**Purple-gaped Honeyeater** *Lichenostomus cratitius* 16–19 cm
Mallee, open woodland, heath. Some calls have distinctive, metallic, vibrant quality, evident in many notes, but especially a 'chairk, chairk, chairk'. Also a clear, pleasant 'quitty' followed by contrasting louder, harsh, peevish calls: 'quitty-quitty-KAIRRK-KAIRRK'. In alarm or aggression, abrupt 'chek-chek-chekchekek'. Locally nomadic; moderately common.

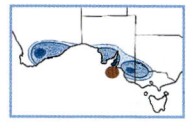

- occidentalis
- cratitius: larger, but shorter bill; crown darker grey, breast deeper olive-grey

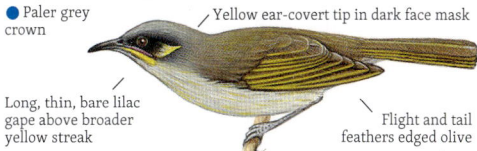

- Paler grey crown
- Yellow ear-covert tip in dark face mask
- Long, thin, bare lilac gape above broader yellow streak
- Flight and tail feathers edged olive

Imm.: plumage is dull; bare gape yellowish.

**Grey-headed Honeyeater** *Ptilotula keartlandi* 15–16 cm
Rocky gorges and escarpments of arid ranges with scattered trees, shrubs, spinifex; less often in tree belts of larger rivers. Busy, active, noisy species among flowering trees and shrubs. Voice a mellow, musical 'ka-towt'; clear, whistled 'whei-wheit-wheit-'; also a harsh, nasal 'kraak'; rapid rattling alarm trill. Nomadic; common.

No subspecies; no apparent variation in size or plumage across range

- Crown grey merging to pale fawny grey-brown on back
- Wings grey-brown, edged yellow
- Ear-coverts oval-shaped, dark grey or almost black, underlined deep yellow
- Imm.
- Facial mask is lighter grey, underparts more lightly streaked.
- Underparts pale yellow, softly streaked brown
- Tail edged yellow

Often gathers in large numbers at sites of profuse flowering; breeds at almost any time of the year after heavy rains.

**Yellow-plumed Honeyeater** *Ptilotula ornata* 15–17 cm
Woodland, open forest, mallee, heath. Forages in foliage of trees and shrubs. Has loud, clear, cheery, rollicking calls: 'WHIT-chier, WHIT-chier, whit-whit-CHIER' and 'CHWIEP-ier'. Alarm call a loud, high, penetrating, musical trill; carries far through bushland, quickly taken up by others. Also a harsh, hard 'chzak-chzak-chzak'. Nomadic; common.

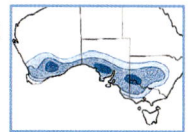

- Upswept, narrow, yellow ear plume
- Imm.: dull; base of bill and eye-ring yellowish
- Juv.
- Underparts dull white, heavily streaked dark grey-brown
- Wing and tail feathers olive-brown, edged yellow
- Tail edged yellow; lacks white tips.

A noisy, pugnacious species at flowering trees, often driving away other smaller birds

**Grey-fronted Honeyeater** *Ptilotula plumula* 15–17 cm
Mallee, dry woodland, mulga and other scrub; watercourse tree belts. A busy, active honeyeater, seeking nectar, snatching insects in flight. Call a quick, mellow, musical 'kwit, kwit-kweeit, kwit-kwit' and rather peevish 'queeeit-queeit-t'. Also a much sharper, more penetrating, loud, whistled 'peet-peet-peet' and lower, persistent 'chwok, chwok'. Uncommon.

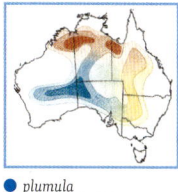

- Crown greenish grey
- Wings dark greenish-grey, edged yellow
- Pointed, black-edged plume across ear-coverts ahead of a broad yellow plume
- Tail grey-green, edged olive yellow
- Underparts creamy white, softly but extensively streaked grey
- Under-tail-coverts off-white

- plumula
- planasi: slightly smaller, paler
- graingeri: larger and darker

Sexes alike. Juv.: dull colours, pale eye-ring, yellowish base to bill

## WHITE-EARED, YELLOW-TUFTED & YELLOW-THROATED

**White-eared Honeyeater** *Nesoptilotis leucotis* 19–22 cm
Forest, woodland, heath, mallee and dry inland scrub. Actively searches for insects, spiders; often works over bark in treecreeper fashion. Prefers trees where peeling, flaking bark harbours insects. Nectar and fruits taken when available. Usually solitary, but at times in small family groups. Voice deep, mellow but slightly metallic 'chwok, chwok', 'choku-whit, choku-whit' and 'kwitchu, kwitchu'; very sharply scratchy, metallic 'chwik!' Sedentary, or local migratory travels; moderately common.

- *leucotis*
- *novaenorciae*
- *thomasi*: Kangaroo Is.

Upper parts, including wings and tail, are olive-green.

Conspicuous white ear patch

Black of face extends to chin, face, upper breast.

Eye brown

● Is slightly smaller than nominate, *leucotis*; the latter race also differs in having brighter olive-yellow on lower breast and belly.

Imm.: similar to adults but lacks bold contrast of adult head pattern; dull white and dusky grey-brown.

**Yellow-tufted Honeyeater** *Lichenostomus melanops* 17–23
Eucalypt forest and woodland with shrub undergrowth; also mallee, brigalow, cypress. Race *cassidix*, 'Helmeted Honeyeater', is confined to swamp-gum woodland with melaleuca and tea-tree undergrowth. A noisy, active species; often in colonies of a few to a hundred or more; aggressive around flowering trees. Feeds on nectar, sap flows, fruit, and insects taken in flight or from flowers or foliage. Often forages rather like treecreepers, over bark of trunks and limbs. Usually active in mid to upper storeys, occasionally in shrubs of understorey. Calls are whistled 'wheit, wheit' and 'tseit, tseit'; harsher, scolding 'chzeet'. Infrequent whistled song.

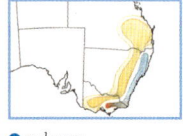

- *melanops*
- *cassidix*
- *meltoni*

Yellowish, without crest

Black mask

Throat yellow with central dark streaking

Variable dusky olive-brown or grey-buff

Tail longer, slightly tipped white

Deep olive-brown

Tail olive-brown, paler tips

Golden-tufted 'helmet'

Broad black band across red-brown eye; yellow ear tuft

● Yellow-tufted Honeyeater nominate subspecies, coastal side of Great Divide; race ● common on inland side of Divide.

● 'Helmeted Honeyeater', a most distinctive but rare, endangered subspecies; now only near Yellingbo, Vic

**Yellow-throated Honeyeater** *Nesoptilotis flavicollis* 18–21
Confined to Tas. and Bass Str. islands. Wet forest to open woodland, mallee and heath. Solitary, in pairs, or, briefly after nesting, in small family parties. Voice is rich, mellow, rhythmic, musical: 'whit-chorr, whit-chorr', 'chorr, chorr-chorr' and slightly higher 'whit-cheow'. Aggressive, rattling, sharper 'chop-chorr-chor-r-r-rr'. Sedentary; common.

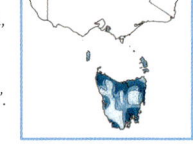

Head mid grey, darker at forehead, lores and border to yellow throat, pale on yellow-tipped ear-coverts

Throat bright golden yellow

Upper breast dark grey, almost black at border with yellow throat

Upper parts olive-green; outer edges of wings and tail olive-yellow

Under-tail-coverts olive-grey

Juv.: similar, but overall dull colours, plumage of softer, fluffier texture, darker eyes

## STRIPE-FACED HONEYEATERS & YELLOW HONEYEATER

**Singing Honeyeater** *Gavicalis virescens* 18–22 cm
Dry scrubby woodlands, mallee, mulga and other inland scrubs, sandplain and dune thickets, mangroves. In pairs or small parties, often solitary. Seeks berries, nectar, insects in shrubs, trees, on ground. Call typically an abrupt, musical 'prrip, prrip' at intervals of several seconds; faster if agitated, very fast at higher pitch when alarmed, 'prrrrrrrrt'. Song is best in dawn chorus; includes high, clear, and deep, mellow notes, varied, 'cheewip-chip-quorri-cheep-quorit-chiwip'. Sedentary or nomadic; abundant.

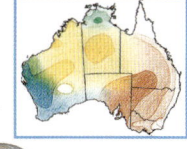

Grey-brown to olive-brown

Bold black streak through eye, underlined by yellow and white

Differences between races are slight; generally palest and smallest in central arid regions, darkest and largest near coast:
- 🔵 *virescens*
- 🟤 *sonorus*
- 🟢 *cooperi*
- 🟡 *forresti*

Underparts dull off-white streaked grey-brown, often with a slight yellowish tint

Wings, tail, brown, edged yellow

**Mangrove Honeyeater** *Gavicalis fasciogularis* 18–21 cm
Coast and offshore islands, usually in mangroves, but visits nearby eucalypt and paperbark woodland. Pairs or small parties forage for insects, upper foliage down to mangrove mud. Song has first note loud, clear; those following mellow, lower, 'CHWIP, chwiew, chwiwy, chwuu'. Repeated rapidly by many birds, becomes distinctive rollicking chorus. Also softer 'chwiek, chwouk, chwouk'. Common.

Black through eye, underlined yellow or white

Eye blue-grey

Wings and tail brown with narrow yellow margins

Throat pale yellow roughly scaled dark grey

Blackish or grey-brown zone across breast

Juv.: similar to adult, but slightly paler, dull with less streaking, barring or mottling

**Varied Honeyeater** *Gavicalis versicolor* 18–21 cm
Mangroves, coastal woodland and gardens. Roosts, nests, feeds in mangroves, descends to mangrove mud to take tiny crustaceans. Travels with swift, undulating flight, sometimes to offshore islands. Active, pugnacious at gatherings at flowering trees. Calls are sharp, brisk, cheerful 'whitchu, whitchu' and piercing 'tchiew', often combined with deep, mellow, chuckling 'quorrit-iew'. Locally nomadic; common.

Black and white streaks down sides of yellow neck

Black and yellow bands across face

Wing feathers edged olive-yellow

Yellow, dark streaked across almost the entire underparts

Imm.: like adult, but dull with paler streaks, brown eye

**Yellow Honeyeater** *Stomiopera flava* 17–18 cm
Rainforest, riverine and swamp vegetation, eucalypt and paperbark forest, mangroves, adjoining woodland. Almost constantly active in foliage; pairs or small groups, occasionally larger flocks at flowering trees. Calls include high, slightly husky, whistled, 'whee-whiew' and a scolding 'chzuk-chzuk, chuk-chuk'. Locally common.

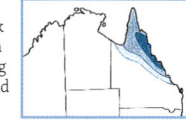

Crown brownish; rear of brow brighter yellow

Upper parts olive-green, flight feathers brown, edged bright yellow

Darker through eye and gape

Overall has rather plain yellowish, small-headed, plump-bellied jiz.

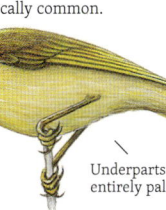

Underparts entirely pale yellow

## BRIDLED & SIMILAR HONEYEATERS

**Bridled Honeyeater** *Bolemoreus frenatus* 19–22 cm
Mountain rainforest, lowlands in winter; visits flowering trees of nearby forest, woodland. Often in noisy, busy flocks in canopy of flowering trees, seeking nectar, insects. Call loud, penetrating, rapid, slightly harsh 'chwip-a-whip, chwip-a-whip', and, in sharp, clear, descending notes, 'chiep-cheep-chierwiep-chier-chiew', also scolding 'scratchy-scratchy'. Locally nomadic; common.

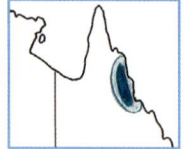

Named for the distinctive yellow and white 'bridle' from bill to rear of eye

Yellow and black bill

Fringed 'bridle' encircles eye, beneath and behind.

Flight feathers dull brown, edged olive

Dark olive-brown

Mottled

Yellow at base of bill extends to whitish gape.

Ear-coverts black, finely encircled and tipped with yellow

Juv.: slightly more rufous

---

**Eungella Honeyeater** *Bolemoreus hindwoodi* 17–19 cm
Confined to small area of plateau rainforest in Clarke Ra., Qld. Active in canopy, rainforest margins, low shrubs. Forages into nearby riverine and open forest. At times noisy around flowering trees, but often wary, hard to locate. Calls of high, clear, sharp, or almost scratchy notes, often preceded by low, abrupt sounds: 'tchup, tchup, chwiep-chiep-cheep-chip-a-wiep'. Locally migratory; common.

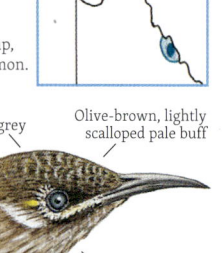

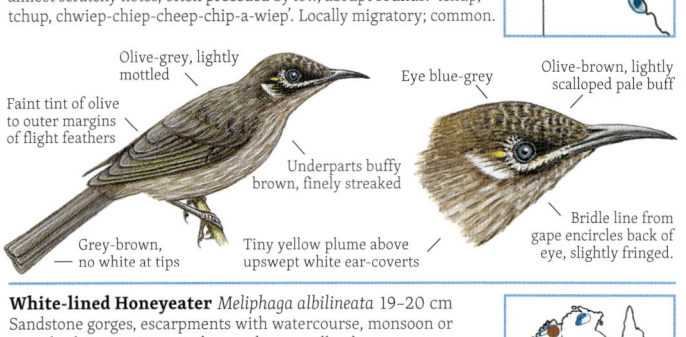

Olive-grey, lightly mottled

Eye blue-grey

Olive-brown, lightly scalloped pale buff

Faint tint of olive to outer margins of flight feathers

Underparts buffy brown, finely streaked

Grey-brown, no white at tips

Tiny yellow plume above upswept white ear-coverts

Bridle line from gape encircles back of eye, slightly fringed.

---

**White-lined Honeyeater** *Meliphaga albilineata* 19–20 cm
Sandstone gorges, escarpments with watercourse, monsoon or paperbark vegetation; nearby eucalypt woodland. Usually rather quiet, inconspicuous. Whistled calls, slightly melancholic or brisk, cheery; often amplified by cliffs: 'tiew, tiew'; 'twieoo, tioo'; 'whee-a-whit, whee-a-whit'; chwip, chwip'; 'whiew-wirrit-whirroo-wheeit-tiew'. Sedentary; common.

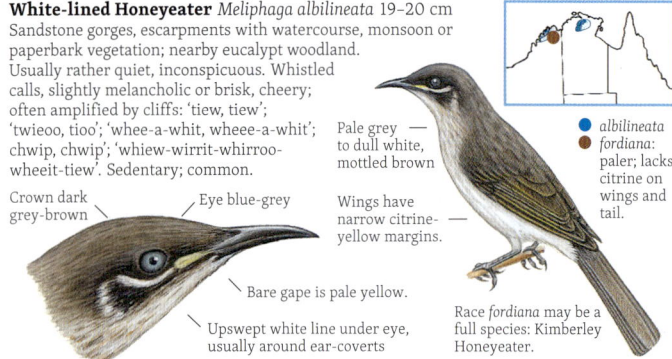

Pale grey to dull white, mottled brown

● *albilineata*
● *fordiana*: paler; lacks citrine on wings and tail.

Crown dark grey-brown

Eye blue-grey

Wings have narrow citrine-yellow margins.

Bare gape is pale yellow.

Upswept white line under eye, usually around ear-coverts

Race *fordiana* may be a full species: Kimberley Honeyeater.

---

**White-gaped Honeyeater** *Stomiopera unicolor* 19–22 cm
Mangroves, paperbark swamps, pandanus. Seeks insects, spiders on foliage as well as nectar. Alone or small groups, occasionally larger gatherings; may be quite noisy, aggressive. Call a clear, sharp whistled 'tchiep, tchiep'; slightly husky 'wheit'; scratchy, scolding 'tzzeip'. Fast rollicking song, 'tchicka-tchicka-WHEIT-T'. Common NT–Kimberley, sparse Qld.

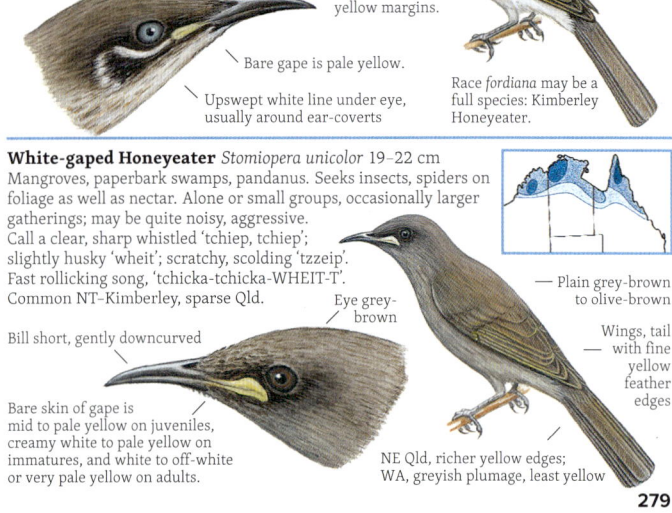

Plain grey-brown to olive-brown

Eye grey-brown

Wings, tail with fine yellow feather edges

Bill short, gently downcurved

Bare skin of gape is mid to pale yellow on juveniles, creamy white to pale yellow on immatures, and white to off-white or very pale yellow on adults.

NE Qld, richer yellow edges; WA, greyish plumage, least yellow

## YELLOW-SPOTTED HONEYEATERS

### Lewin's Honeyeater *Meliphaga lewinii* 19–22 cm
Rainforests, but in far N, usually above 200 m; elsewhere, coast to 1000 m and in wet eucalypt forests, woodlands, heaths. Forages in pairs or alone about the higher canopy; at times hovers to take insects from foliage, spiralling around trunks and branches, probing bark; also takes fruit and nectar. Undulating flight with audible wing beat. Usual call sharp, harshly chattered 'chak-ak-ak-ak-ak'; varies from slow, deliberate to rapid, machine-gun rattle; harsh, raucous scolding: 'kairrk'. Abundant.

● *lewinii*
● *mab*: smaller
● *amphochlora*: smaller, yellower

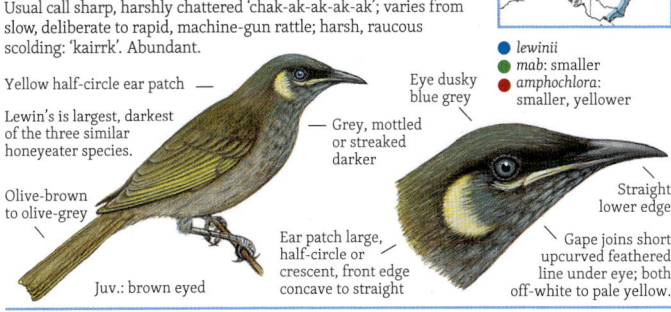

### Yellow-spotted Honeyeater *Meliphaga notata* 17–19 cm
Rainforests, coast, ranges, usually below 500 m; also monsoon forests, mangroves. Tends to perch with body semi-horizontal. Noisy, aggressive; solitary, pairs or small parties; forages low to high among foliage. Flight direct, swift, undulating. Calls a loud, staccato 'plizk-plizk-plik-plik-', a plaintive, descending 'kiaak-kiaak-', harsher, descending 'kriaak-kriaak-'. Localised wanderings; common.

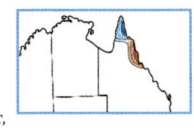

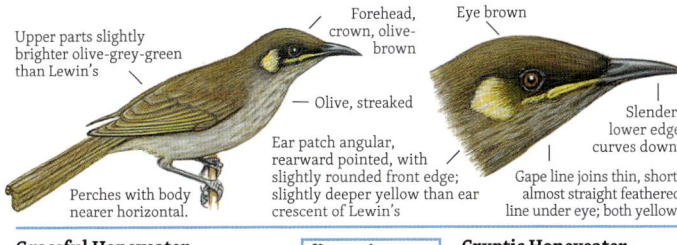

### Graceful Honeyeater
*Meliphaga gracilis* 14–17 cm
Rainforests, lowlands, ranges to ~500 m. Slender, graceful; perches with upright posture; restless, flicking wings. Forages at all levels. Call abrupt, musical 'tchip', also clear, ringing 'whit-whit-whit-whit-', weaker 'tsip-tsip-tsip'. Sedentary; common.

● *gracilis*  ● *imatrix*

### Cryptic Honeyeater
*Meliphaga imatrix* 14–17 cm
Smallest of the genus. Formerly a race of Graceful. Back is darker, bill more curved and underparts olive-green. Call is a sharp, repeated 'tchip'.

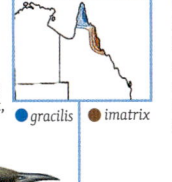

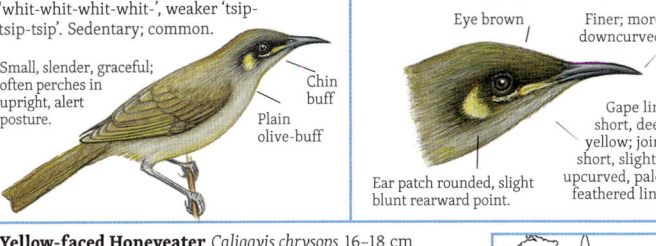

### Yellow-faced Honeyeater *Caligavis chrysops* 16–18 cm
Forest, woodland, heath, mangroves. Busy, active forager in foliage, taking nectar and insects. Has cheery ringing calls, 'whit, whit, whit' and 'chwikup, chwikup'. These often combine in brisk, rollicking song: 'WHIT-chiwit-chiwickup', with many variations. In flight, flock contact with 'clip, clip' and peevish 'kree'. Migratory; moves N and S with seasons. Common.

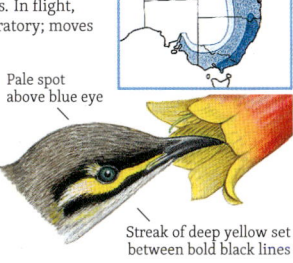

# MINERS

**Yellow-throated Miner** *Manorina flavigula* 26–28 cm
Noisy, gregarious, conspicuous; often in large pugnacious flocks; feeds more on insects taken from foliage, bark or ground than on nectar. Calls are loudest in alarm or aggression, a sharp nasal 'kiek-kiek-kierk', 'kweek-kewk' or 'kieerk-kieerk'. Also softer, plaintive contact calls: 'chwip' and 'chwiep'. Woodlands, heaths, arid scrubs, grasslands. Locally nomadic; common.

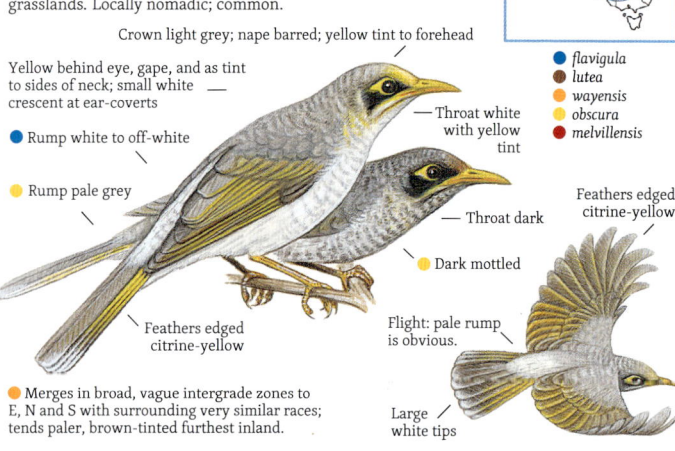

- *flavigula*
- *lutea*
- *wayensis*
- *obscura*
- *melvillensis*

**Black-eared Miner** *Manorina melanotis* 26–28 cm
Dense mallee, remnant farmland mallee. Currently a full species, but could be reduced to subspecies; may be no more than darkest variation of Yellow-throated Miner population. Endangered by land clearing, which has opened access by Yellow-throated Miners, these now competing closer or within the Black-eared's habitat. Possible interbreeding may have reduced the population of the Black-eared form. Positive field identification is difficult; reputedly very wary. Voice more harsh, complaining than other miners. Endangered; sedentary.

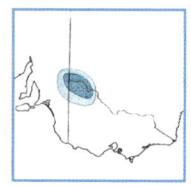

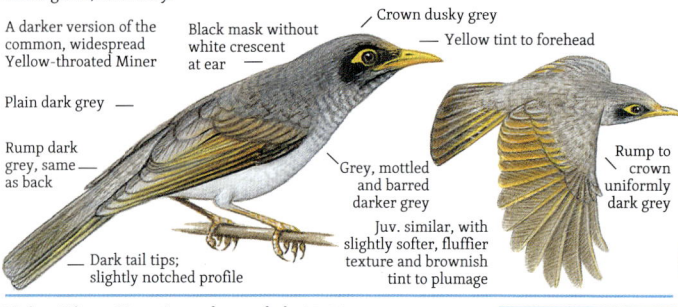

**Noisy Miner** *Manorina melanocephala* 25–28 cm
Open grassy forests, woodlands; in colonies, often large. Forages treetops to ground; aggressive towards other birds. Some calls clear, mellow, and quite musical, 'chwip-chwip, chwip-chwip', but others loud, raucous, complaining, noisy when taken up by others of flock: 'kairk-kairk-kairk'; harsher 'karrk-karrk-karrrk'. Territorial call, often given in flight, a musical 'tiew-tiew-tiew'. In breeding season has a clear musical song in dawn chorus. Locally nomadic; common.

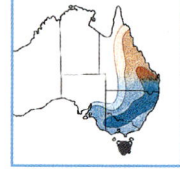

- *melanocephala*
- *lepidota*: paler, smaller, blacker crown, darker mottled beneath
- *leachi*: larger, darker, back dusky olive-brown, breast plainer

## HONEYEATERS & BELL MINER

**Macleay's Honeyeater** *Xanthotis macleayanus* 19–21 cm
Rainforests, monsoon and riverine forests, adjoining woodlands, mangroves, where inconspicuous in mid to upper canopy. Usually solitary or in pairs; probes for insects or debris suspended in high vine tangles and epiphytes, rotting wood, bark crevices, foliage; nectar and fruits a smaller part of diet. Call is a clear, brisk, rather strident 'tsuweit-weet-weet, tsuweit-weet', undulating, with 'weit' each time louder, sharper, the following 'weet' lower, softer. A similar species is the Tawny-breasted, paler, further N. Sedentary; common.

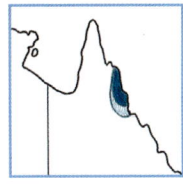

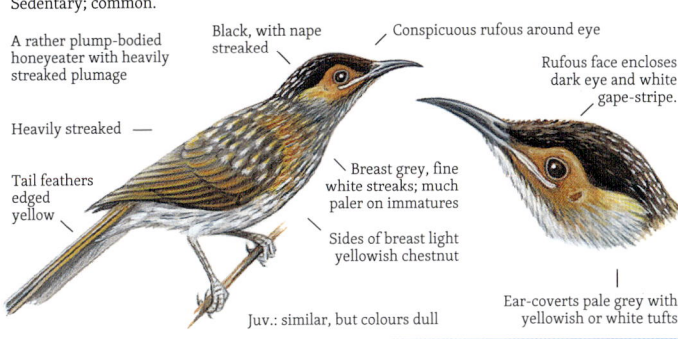

**Tawny-breasted Honeyeater** *Xanthotis flaviventer* 18–21 cm
Rainforests, gallery forests; rarely river-edge vegetation, mangrove and paperbark swamps, woodlands. Uses high and mid levels of rainforests; finds insects among clumped epiphytes, tangled vines, bark crevices, but also takes nectar, some native fruits. Hard to see in high canopy, but undulating flight is often lower through forest. Gives series of strong whistles, the first loudest with longer, sharper ending followed by several lower, more abrupt, descending notes: 'WHEEEIT-whit-whut, WHEEEIT-whit-whut'; 'wht-wheeit-whit-wht'. Sedentary; uncommon.

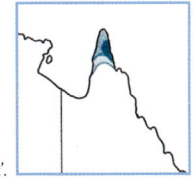

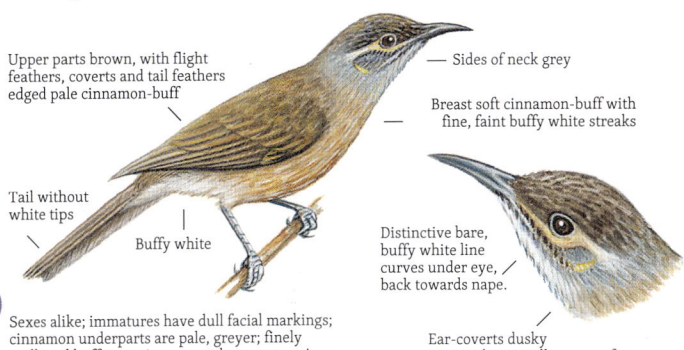

**Bell Miner** *Manorina melanophrys* 18–20 cm
Typically in dense, tall, wet eucalypt forests, rainforest margins, creek thickets. Famed for clear bell-like calls; difficult to sight in forest canopy. Feeds high, seeking lerps and other insects on leaves, flowers and bark. Call is an abrupt, clear, penetrating 'peep', which in the distance becomes a 'tink'; the calls of the hundreds of birds in a large colony make a continual tinkling. Also has a harsh 'chak, chak'. Usually sedentary; abundant.

## LITTLE FRIARBIRD & LARGE HONEYEATERS

**Little Friarbird** *Philemon citreogularis* 25–30 cm
Open forests, woodlands, river edges and swampy woodlands, mangroves. Pugnacious, noisy; nomadic flocks follow flowering of trees. Calls vary from clear, musical 'chip-wheeoo', 'chip-weik-weeoo' and 'chowk, chowk-ik' to raucous and guttural croaking noises. Often these are combined: 'chwip-weik-skrarrrch-weeoo, chwip-weik-skraarrch-weeoo'; a common sequence sounds like 'rackety crook-shank'. Nomadic in N; summer migrant to SE Aust.

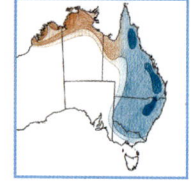

Smallest, most widespread of the friarbirds

Plain grey-brown

Lacks knob on bill; bare facial skin is deep greyish blue.

Grey-brown

Bare facial skin of immatures is dull grey brown.

Imm.: plainer breast, pale buffy scalloping on upper parts, throat lightly tinted yellow

● *citreogularis*
● *sordidus*: longer, heavier bills; plumage appears slightly more mottled than on eastern nominate race.

Tail mid-brown with narrow white tips

**Blue-faced Honeyeater** *Entomyzon cyanotis* 25–31 cm
Open forests and woodlands of eucalypts, paperbarks; also in river-edge vegetation, pandanus, margins of monsoon scrub, mangroves. Large, gregarious, aggressive; feeds on nectar, fruits, insects; has undulating flight. Noisy, often with a husky 'kieeerk, kieeerk', repetitive, sharp contact calls, 'whik, whik, whik', and softer 'quorriek, quorriek'. Also clear, whistled 'quorrieek'; begins low, finishes with loud, high shriek. Abundant, locally nomadic in N; migratory, uncommon in S.

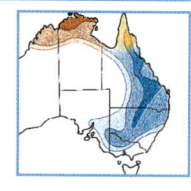

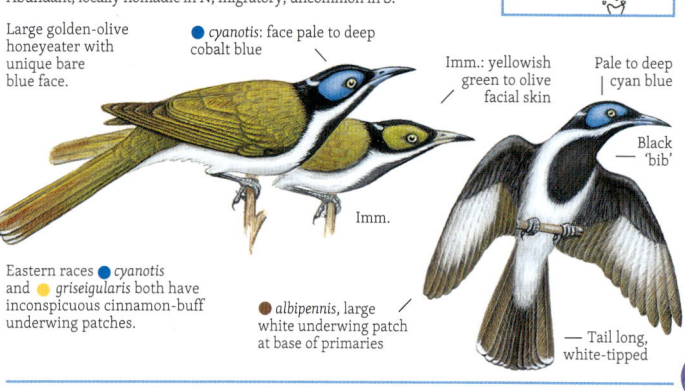

Large golden-olive honeyeater with unique bare blue face.

● *cyanotis*: face pale to deep cobalt blue

Imm.: yellowish green to olive facial skin

Pale to deep cyan blue

Black 'bib'

Imm.

Eastern races ● *cyanotis* and ● *griseigularis* both have inconspicuous cinnamon-buff underwing patches.

● *albipennis*, large white underwing patch at base of primaries

Tail long, white-tipped

**Regent Honeyeater** *Anthochaera phrygia* 20–23 cm
Preference for ironbark, but also in forests and woodlands of box, yellow gum, swamp mahogany, river oak. Extremely active bird of upper foliage, busily moving among trees with audible flutter of wings. Calls with clear bell-like notes, some quite sharp, others deep, mellow, musical: 'quip-quorrip, quip-kip, quorrop-quip'; and sharper 'chlink, chlink'. Some mimicry of other birds; aggressive harsh scoldings. Reduced in range and numbers; endangered.

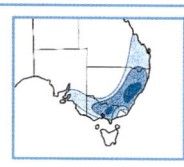

Wide, deep yellow bar across outer primaries; broad yellow outer margins to most flight feathers

Boldly patterned honeyeater with bare, rough, salmony pink facial skin

Scalloped white and pale yellow

White under tail

Diamond pattern with black margins

Sexes are alike, except female is slightly smaller.

Tail broadly tipped and edged pale to deep yellow

## LARGE FRIARBIRDS

### Helmeted Friarbird
*Philemon buceroides* 32–37 cm
Margins of rainforests, monsoon forests, vine thickets, mangroves, eucalypt woodlands, sandstone escarpment vegetation in NT. Noisy and aggressive to other birds on flowering trees, where it seeks nectar, fruit, and insects. Calls are varied: harsh cackles, squawks and clanking 'chlank'. Some musical notes intermixed with harsh 'warruk, kuk-kaowww, warruk-kuk-kaowww'; 'kaowww-kuk, kaowww-kuk'; a higher 'kewik, kewik' and variations. Nomadic, follows seasonal flowering; common.

- *gordoni*: casque small, rounded; gorget silvery, 'scaly'
- *ammitophila*: casque tiny, neck gorget silvery, 'scaly', rough

Nominate race and others on islands of Indonesia and New Guinea

### Horned Friarbird
*Philemon yorki* 32–37 cm
Formerly a subspecies of Helmeted. Rainforest margins, monsoon forests, vine thickets, mangroves, eucalypt woodlands. Noisy, aggressive. Eats nectar, fruit, insects. Call similar to Helmeted. Nomadic; common.

Silvery crown; nape has upturned white tufts.
Eye red
Largest, most elongated casque.
Dark bare skin has rounded rear edge.
Imm.
Knob usually barely visible; eye brown

- *P. yorki*: casque large, neck gorget more prominent silvery greyish-fawn

### Silver-crowned Friarbird *Philemon argenticeps* 28–32 cm
Tropical forests and woodlands, usually of eucalypts and paperbarks; river-edge tree lines, mangroves, parks and gardens. Follows the flowering of trees in noisy gatherings; aggressively dominates feeding sites, chasing other birds. Voice loud, raucous, rather metallic and a loud, rackety, raucous 'karrowk-kak-kuk...'. Other calls include a clear, ringing, musical 'cherrowik...', and sharp 'aik-aik'. Locally migratory; common.

Bill knob rises gently from front, drops steeply to rear.
Blackish facial skin has rearwards point behind eye.
Ad.: iris bright red
Imm.: iris brown

Forehead and crown silvery white, tufted on nape and hind neck. Throat feathers are grey-buff, silvery pointed.

Knob on bill small or absent

Back, wings and tail mid fawn-brown

Underparts buffy white; faint yellow tint on throat

Tail plain mid brown, little if any white edging to tips

Imm.

- *argenticeps*
- *kempi*, slightly larger, but with much smaller casque on bill; plumage similar to nominate

### Noisy Friarbird *Philemon corniculatus* 30–35 cm
Open forests, woodlands, swampy woodlands, watercourse trees; heathlands with banksia. Loud squabbling at flowering eucalypts makes them conspicuous; wandering flocks follow the flowering of forests. Flight between trees may be direct or undulating; their arrival announced with loud rollicking calls of deep 'owk-orrok, owk-orrok, arrowk-kok-carrowk, kor-r-r-r-owk', parts of which can sound like 'tobacco' or 'four-o'-clock'. Also sharp, metallic 'owk!, owk!'. Sedentary in N; migratory within S.

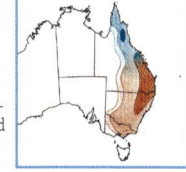

Head entirely bare, black
Knob with leading edge steeper

- *monachus* (southern race) is the larger, and has darker brown back.

Silvery white neck plumes

Mid fawn-brown
Knob very small

Mottled or scalloped

Tail brown with broad white tips

Underparts fawny white
Imm.

- *corniculatus*, northern (nominate) race: smaller, upper parts paler greyish fawn (not illustrated)

## HONEYEATERS & WESTERN WATTLEBIRD

**Striped Honeyeater** *Plectorhyncha lanceolata* 22–25 cm
Drier open forest, woodland, mallee, mulga; also heath and mangroves. Although a honeyeater, the Striped uses many food sources including insects, seeds and fruits as well as nectar. Bill is short, sharply pointed and almost straight, more suited to prising insects from crevices than probing deep flower tubes; it has the honeyeater brush tongue and takes nectar at more shallow flowers, such as eucalypts. Usually alone or in pairs, infrequently in family groups or small flocks. Calls clear, rapid, rather more mellow than sharp, with slight variations giving a rollicking rise and fall: 'quirrip-quarreep-quirrip-quarreep-quirrip-quarreep'. Sedentary, or nomadic in drier regions.

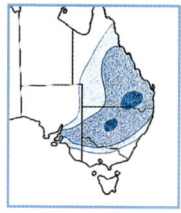

No races; gradually become very slightly smaller northwards.

Flight, fast, erratic, gently undulating, showing plain grey-brown upper surface of wings; the moderately long tail has a shallow-forked tip, becoming square-cut when partly spread.

Crown, nape, face boldly streaked

Imm.: similar to adults, but with rather dull markings

Upper parts softly striped brown and grey-buff

Bill almost straight, sharp, relatively short

Tail plain, dark brown, slightly forked

Feathers of throat and upper breast are long, lanceolate, of stiffly pointed, spiked texture; this is most noticeable at edge of neck where they overlap against dark upper plumage.

**Spiny-cheeked Honeyeater** *Acanthagenys rufogularis* 25 cm
Arid woodlands, mallee, mulga, often with groundcover of spinifex; also scrubby vegetation of drier parts of coast. Feeds on insects and small native fruits as well as nectar; in pairs, alone or in small flocks. Clear, piping, musical yet melancholic notes often reveal this bird's presence, and seem in keeping with the loneliness of its desert haunts. Often calls from high bare perches and while in high display flights. Common sequences: 'quip-kpeeer-kpeeer-kpeeer-quipip-quipip-quipip'; 'chrriee-chrriee-'; 'whiteeer, whiteeer, whiteeer'. Also a gurgling, bubbling 'quorrok-quorrok-quorrok-quok'. Nomadic; generally common.

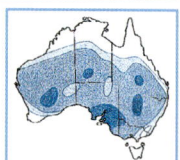

No races. Juv. and imm. are similar to adult.

'Pink lipstick' makes this honeyeater unmistakeable; its calls are one of the characteristic sounds of the dry interior.

Pink of bill extends back under the eye on bare skin of gape.

Dark coverts and scapulars, edged buffy white

Bristle-like fine 'spiny' feathers extend back from cheeks onto sides of neck.

Throat and upper breast soft cinnamon-buff

Iris blue

Rump patchy off-white and pale grey

Undulating flight shows pale rump and white tips to tail.

**Western Wattlebird** *Anthochaera lunulata* 27–33 cm
Long considered sufficiently distinctive to be a full species, but until recently remained a race of Little Wattlebird (opposite). The voice most unlike that of the Little Wattlebird: combines croaking, crackling, gargling, bill-snapping sounds typical of wattlebirds with pleasantly musical song, an undulating, 'cook-cook-cup-hook, cook-cook-cup-hook', often as duet. Unlike Little Wattlebird, which usually lays two eggs, the Western Wattlebird invariably has a single egg. Common.

Was subspecies of Little Wattlebird, i.e. *A. chrysoptera lunulata*

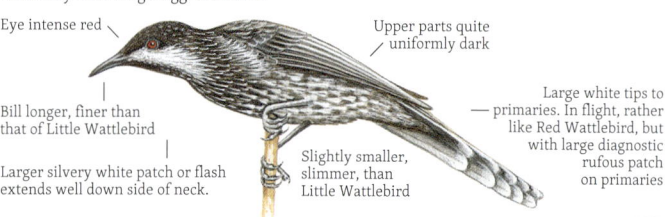

Eye intense red

Upper parts quite uniformly dark

Bill longer, finer than that of Little Wattlebird

Larger silvery white patch or flash extends well down side of neck

Slightly smaller, slimmer, than Little Wattlebird

Large white tips to primaries. In flight, rather like Red Wattlebird, but with large diagnostic rufous patch on primaries

## WATTLEBIRDS

**Yellow Wattlebird** *Anthochaera paradoxa* 38–48 cm
Forests; eucalypt and banksia woodlands, heathlands. Large, gregarious honeyeater; flocks gather in flowering eucalypts, seeking both the nectar and the insects around the flowers. Feeds mainly in forest canopy; moves among the trees with strong, undulating flight. Flocks wander from highlands to lowlands at end of summer; return as breeding pairs in spring. Loud, harsh, throaty, grating sounds: gurgling 'growgk', 'chrrowk' or short, hollow 'klok' repeated irregularly. Also a duet; female gives a gurgling sound, male quickly responds with a harsh croak. Confined to Tas. and King Is., migratory or nomadic; common except W Tas.

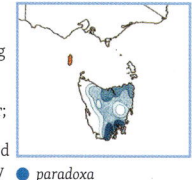

● *paradoxa*
● *kingi*

● King Island race has a smaller yellow patch on belly, slightly browner back and breast, and finer wattles.

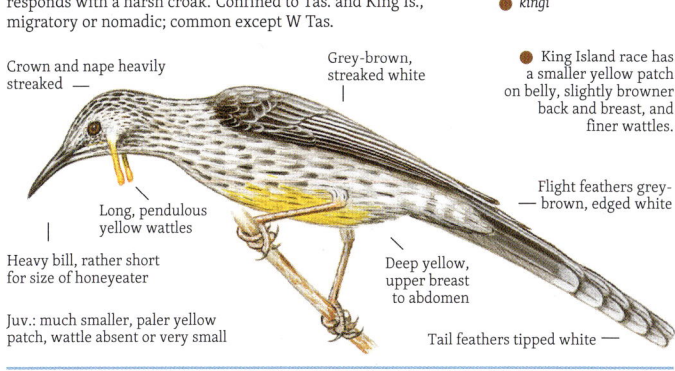

Crown and nape heavily streaked
Grey-brown, streaked white
Long, pendulous yellow wattles
Heavy bill, rather short for size of honeyeater
Deep yellow, upper breast to abdomen
Flight feathers grey-brown, edged white
Juv.: much smaller, paler yellow patch, wattle absent or very small
Tail feathers tipped white

**Red Wattlebird** *Anthochaera carunculata* 32–36 cm
Eucalypt forests and woodlands, mallee, gardens. Large, noisy, aggressive honeyeater; widely distributed; feeds on flowers of eucalypts, other trees and shrubs; often in flocks, at times in large numbers. In flight, white tips of wings and tail conspicuous. Has harsh, grating cough or bark: 'hrarrark-hrak', 'hrrak-a-yak', 'hraak, hraka-yak'; slow, grating, growled 'graarrrrk'; female a rather more pleasant, mellow 'kieuw-kieuw-kieuw', at times combined with harsher coughing noises of male. Nomadic or sedentary; common.

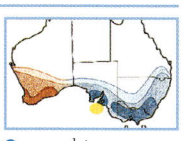

● *carunculata*
● *woodwardi*
● *clelandi*

● Slightly smaller, deeper yellow on belly

● Upper parts deep brownish grey, extensively streaked and margined with white
Crown black, unstreaked
Cheek patch silvery white; wattles red
Pale yellow under belly
Tail white-tipped except central pair
— All races: primaries have large white tips, conspicuous in flight.
● Upper parts, on average, dark grey with reduced brownish tone

**Little Wattlebird** *Anthochaera chrysoptera* 28–35 cm
Forest, woodland, banksia heath, gardens. Large, rather plain grey-brown at distance, intricately white-streaked close up. In flight, white tips to primaries are conspicuous. Noisy, often aggressive, energetically seeks nectar, berries, insects from tree tops to low shrubbery; often in flocks. Calls varied; notes range from harsh to musical. Often a mellow 'kook-' alternating with a grating 'kraagk' as 'kook-kook-kraagk-kraagk,' and 'kraagk-kook-kraagk-kook-'. Also a grating, single 'graarrrk!' Migratory or nomadic; common.

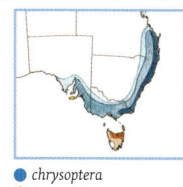

● *chrysoptera*
● *tasmanica*
● *halmaturina*

Eye varies, usually grey-brown, occasionally brighter chestnut.
Bill relatively short
Face and side of neck grey-brown with very little of Western Wattlebird's large silvery flash
Rufous on inner vanes of primaries, not usually visible on tightly folded wing
Conspicuous white tips

## HONEYEATERS & CHATS

# HONEYEATERS & CHATS

Family Meliphagidae, honeyeaters and chats. The former utilise many types of nectar-producing trees and shrubs; their bills are mostly medium to long, slender, downcurved. Tongues are partly tubular, brush-tipped to access this nectar, often held in deeply tubular flowers. For most species, insects form an important part of the diet; a few are predominantly insectivorous. In contrast, chats, while having brush tongues, seek insects on or near the ground.

Scarlet Robin = 12 cm

## meliphagidae
### page 272

- Yellow Wattlebird
- Helmeted Friarbird
- Tawny-breasted Honeyeater
- Bell Miner
- Red Wattlebird
- Silver-crowned Friarbird
- Yellow-throated Miner
- Little Wattlebird
- Noisy Friarbird
- Black-eared Miner
- Noisy Miner
- Western Wattlebird
- Little Friarbird
- Lewin's Honeyeater
- Yellow-spotted Honeyeater
- Striped Honeyeater
- Blue-faced Honeyeater
- Cryptic Honeyeater
- White-lined Honeyeater
- Regent Honeyeater
- Bridled Honeyeater
- Spiny-cheeked Honeyeater
- Macleay's Honeyeater
- Eungella Honeyeater

## CHESTNUT-BREASTED & BANDED WHITEFACES

**Chestnut-breasted Whiteface** *Aphelecephala pectoralis*
Length 10 cm. Rare whiteface inhabiting slightly elevated plains and
low flat-topped mesa formations W of Lake Eyre, largely gibber-stone
surfaced with widely scattered mulga and very sparse low cover of
shrubs, typically eremophilas, bluebush. Often on ground in single-
species groups or mixed flocks, seeking seeds and insects. Rather
timid; when disturbed these birds travel some distance in typical
whiteface undulating flight before dropping to ground again.
Call is a rapid, weak trill often given as contact call while in flight:
'trit-trit-trt-trt-trt-trt-trt-trtt-rtt'. Louder, plaintive, whistled song:
'wheeit, wheeit-weeeo'. Has a very restricted range, difficult to
locate; probably quite rare within that range: possibly nomadic.

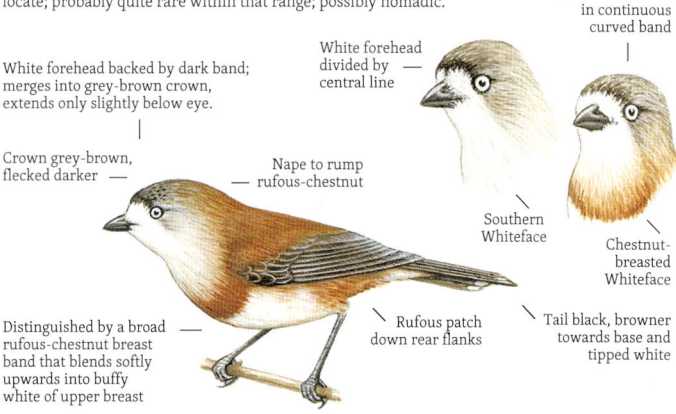

**Banded Whiteface** *Aphelocephala nigricincta* 10–11 cm
Inhabits some or most arid central regions, with sparse mulga,
eremophila, other stunted trees and shrubs; often scattered across
spinifex dunes or saltbush flats. Reportedly need not visit waterholes
and wanders nomadically in search of seasonally favoured country.
Gregarious, in small flocks, sometimes in mixed foraging parties
with other small birds, except when nesting. Fossicks busily over
ground; flies into bushes if disturbed. Gives a mellow, musical
succession of notes, rather bell-like, almost fast enough to be a trill,
usually preceded by several longer, clearer notes: 'whh-wheeee-
wheeee-wheeet-whit-whee-wheewhiwhitwhit-whit'. Also slower, clear
'whit-whit-whit-whit-whit-whit-'. In alarm, a harsher, buzzing scolding.
Breeding males give song-flights: rise steeply; flutter slowly, briefly,
singing; then drop to bushes. Seasonally nomadic; uncommon to
locally quite common.

269

## TASMANIAN & MOUNTAIN THORNBILLS, WHITEFACE

**Tasmanian Thornbill** *Acanthiza ewingii* 10 cm
A very plain thornbill, usually in cool, wet rainforest; moves to drier forest, woodland through winter. An arboreal species; forages through dense forest canopy and down into mid level vegetation, rarely very near or on ground. Contact calls are sharp, scratchy, twittering: 'tszit, tszit'; in alarm, harsh churring and buzzing sounds. Other calls richer, more musical, although disjointed rather than continuous. Has free-flowing song: 'Chip-cheerwheep. Chirr-chirrowp. Chir-r-r-r-owp!' Local seasonal movements; common.

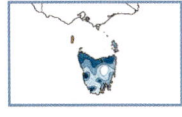

- ● *ewingii*
- ● *rufifrons*: King Is.; forehead slightly brighter rufous

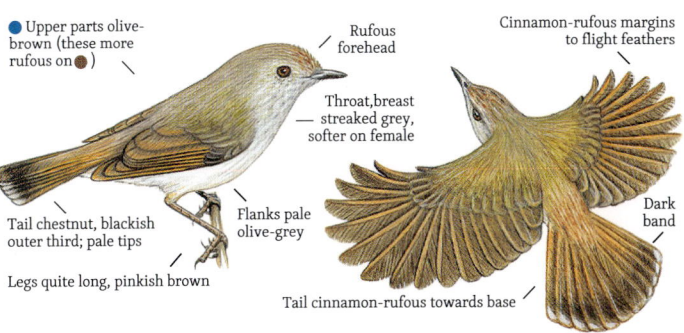

**Mountain Thornbill** *Acanthiza katherina* 10 cm
Inhabits mountain and high plateau rainforest above about 400 m altitude. Lives in dense mid and upper strata of forest, where it can be difficult to sight as it forages in outer foliage. Gregarious, gathering into small flocks after spring–summer breeding season. Intermingled quick, sharp, scratchy sounds and contrasting louder, mellow, musical, whistles: 'tsit-tsew-tseweit, tzit-tseet, toowheet, tsit-toowheet, tsit-tchip-tsew'. Sedentary; moderately common in restricted but secure range.

**Southern Whiteface** *Aphelocephala leucopsis* 10–12 cm
Semi-arid woodlands, mallee, mulga, similar dry-country scrublands. All species of whiteface have white band across forehead, dark rear border through eye. A stubby, finch-like bill for both seeds and insects. Lively, fossicks about on ground and low in shrubbery; often in small flocks that may include other species. Voice is very rapid, scratchy, twittering: 'tzip-tzip-tziptzip-' or 'tchip-tchip-chiptchipt-chipt-chip' and harsher 'tzzit, tzzit-tzzit, tzzit'. Alarm a scolding 'kzzurrrk, kzzurrk-'. Sedentary; common.

- ● *leucopsis*
- ● *whitei*: paler, may be a race, but at times included within *leucopsis*.
- ● *castaneiventris*

# INLAND & SLATY-BACKED THORNBILLS

**Inland Thornbill** *Acanthiza apicalis* 9.5–10.5 cm
Dry woodland, mostly of eucalypt or acacia. In SW of WA also in wetter coastal habitats, including sandplain heath and forest of karri and jarrah. Forages briskly through shrubby lower growth rather than on ground or high in trees; at times in small flocks with other species of small birds. Unlike Brown Thornbill, often holds tail angled upwards. Calls are lively, sharp 'tsip-chip-'; loud, clear, musically whistled 'chweeeip!' and deep, mellow, tremulously warbled 'quor-r-r-r-r-eip'. Overall effect a varied 'tseeeip-tsip-tseep-chweeeeip, chip-tzzeep-tseep-quorr-r-r-reip! tsip seep-'. Also buzzing scoldings and passages of mimicry of other birds. Sedentary, seasonally locally nomadic; common.

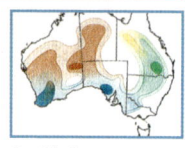

Possibly four races; differences slight
🔵 *apicalis*
🟤 *whitlocki*
🟡 *cinerascens*
🟢 *albiventris*

**Slaty-backed Thornbill** *Acanthiza robustirostris* 9–10 cm
Arid mulga scrubland with shrub understorey; low shrubbery around claypans and salt lakes. Forages from foliage of low shrubs, and occasionally on ground. Has a frequent 'tseeip' similar to Chestnut-rumped and Inland Thornbills; some sounds may be mimicry of those species. Song varies – whistled, rich, musical notes interspersed with sharp, higher twittering: 'chwip-wip, chwip-wip, cheeowheeep, chwip, chwip-chwip, cheeowheeep, chwip, whip-e-chiew'. Sedentary; generally uncommon.

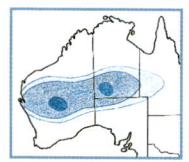

Only very slight individual variations across the wide range of this species; no races recognised

# CHESTNUT-RUMPED & BROWN THORNBILLS

**Chestnut-rumped Thornbill** *Acanthiza uropygialis* 9–10 cm
Inland and semi-arid eucalypt woodland, mulga and similar acacia scrubland, mallee, casuarina, and, occasionally, low shrubland of saltbush, bluebush, lignum. A small, busily active thornbill; seeks insects on foliage, branches, low shrubs and the ground; in pairs, singles or small parties; at times mixes with other small birds, often Weebills, pardalotes and other thornbills. Usual call is a high, peevish, squeaking: 'tsee, tsee-tseep, tseep, tseeip'. Less often a whistled, clear, penetrating 'cheweep-cheweep-cheweep', or faster, lower, a mellow, musical warbling. Sedentary or seasonally locally nomadic; moderately common.

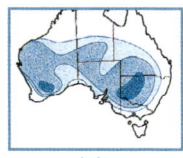

No races: slight variation in colour across range, apparently gradual and clinal. Slightly darker backed, fawn flanked in SE; gradually paler further into arid and north-central regions

Pale grey to brownish-grey

Forehead has finely scalloped pattern.

Rump and base of tail are light, bright chestnut.

Underparts white with light grey on breast.

Eye creamy white

Tail blackish-brown with pale central, white outer tips

Chestnut rump conspicuous when birds fly up from ground

The Chestnut-rumped Thornbill is distinguished from other reddish-rumped species by the adult's off-white eye and by its foraging on the ground as well as in foliage.

Imm.: like adult, but dark eyes

**Brown Thornbill** *Acanthiza pusilla* 9.5–10.5 cm
Forest and woodland with dense undergrowth, vegetation of creeks, rainforest, coastal dune thickets, mangroves. Actively seeks insects and other small prey in undergrowth and lower foliage; rarely high in canopy or on ground. Alone or in pairs, rarely small family parties. Unlike Inland Thornbill, rarely holds tail upwards. Voice is rich, musical, with wide range of notes. Very high, clear whistles abruptly cascade in strong, rippling trills to notes unusually deep and mellow for so small a bird. Among these calls, sharper scratchy sounds or mimicry of other birds may be inserted. In alarm, gives harsh, churring scoldings. Sedentary; common.

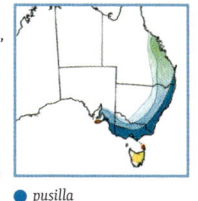

- 🔵 *pusilla*
- 🟢 *dawsonensis*
- 🟡 *diemenensis*
- 🟣 *archibaldi*
- 🟤 *zietzi*

Upper parts are warm grey-brown, but become more olive in northern population.

Eye deep red

Forehead rufous, pale-speckled

Streaked, all races

Grey to white on outer tail tips

Rump and base of tail cinnamon

🔵 White to pale buff flanks

Brown with dark band

Wings grey-brown to olive-grey

Brownish olive

🟢 Has deeper buffy yellow underparts.

🟢 Cream breast and belly
🔵 Dull creamy white to olive-grey breast and belly; also 🟡, 🟣 and 🟤

## BUFF-RUMPED TO SLENDER-BILLED THORNBILL

**Buff-rumped Thornbill** *Acanthiza reguloides* 10.5–11.5 cm
Open dry eucalypt forests, woodlands, heaths; forages in low shrubs and across debris-littered ground; northern race lives in woodlands with grassy groundcover, foraging higher in foliage of large shrubs and trees. Usually seen on ground, sometimes in foliage. Often in small parties; forages busily, restlessly; departs with bouncy flight displaying yellow rump, rather like Yellow-rumped Thornbill. Calls lively, sharp, twittering: 'tsip-twit-titch-tseip, twit-ti-twit-twit-tseip, tseip-tseip'. Song much more rapid, an unbroken, tinkling 'chipitit-chipitit-chipitit-chipitit'. Sedentary; moderately common.

- reguloides
- squamata
- australis
- nesa

- Olive-brown to olive-grey back, brownish wings
- rufous-tinted; finely scalloped pattern
- Throat, breast, flecked grey
- Pale creamy white iris
- Rump pale pinkish buff
- Pale yellow
- Throat pale lemon
- Very pale buff
- Pale tips
- Brighter yellow underparts, greener back

**Yellow-rumped Thornbill** *Acanthiza chrysorrhoa* 11–12 cm
Grassy woodlands, scrublands, farms, gardens. One of the most familiar small birds. Usually forages on ground, often in small flocks; they work across grass or leaf litter with a jerky, hop-and-peck action. Song cheery, undulating, tinkling; lasts just 3 or 4 seconds, usually ending with two clear whistled notes, the second descending: 'chip-chip-chippity-cheepity-chippity-cheepity-wheit-wheehoo'. Whole song is then repeated almost identically, a few or many times. Sedentary; abundant.

Variation slight, some gradual or clinal. Species tends paler inland and northwards. Possibly four races

- Olive-brown to olive-grey
- Forehead black, spotted white
- Eye creamy ochre
- Rump light, bright yellow
- Off-white; olive-buff along flanks
- Bouncing flight shows yellow as bird departs.
- Black with very narrow white tips
- Imm.: like adult, but with darker, brownish eyes
- Bright yellow, conspicuous as bird flies up from ground.

**Slender-billed Thornbill** *Acanthiza iredalei* 9–10 cm
Western races inhabit saltbush and samphire flats; the SE race lives in dense, low heath. Forages in low shrubs, on ground, in pairs or small parties, seeking insects, spiders and other small prey. If flushed, departs with low, bouncy undulations, with yellowish rump conspicuous against dark tail. Has a rapid, high, twittering, almost tinkling, 'tsip-tsip-sip-sip-chip-chweeet, tsip-chip-' and occasional louder 'chweeeit'. Sedentary or nomadic; nominate race secure, common, others uncommon, or only locally common.

- iredalei
- hedleyi
- rosinae

- Upper parts olive-grey
- Eye pale
- Back and wings darker olive-grey
- Rump buffy white to pale yellow
- Throat faintly mottled
- Mottling heavier, flanks deeper olive
- Palest forms: under tail pale lemon-yellow, but some almost white; flanks buffy white to buffy pale grey
- Darkest form

# WESTERN, STRIATED & YELLOW THORNBILLS

**Western Thornbill** *Acanthiza inornata* 10 cm
Tall, wet, dense eucalypt forest, but also quite arid woodlands and sandplain heaths. Busy, restless small bird of tree and undergrowth foliage, but also often on the ground. In pairs or small flocks, at times with other species. Voice is weak, erratic with broken sequences of scratchy, chittering, 'tchip-tsip' sounds, with an occasional louder 'choweep'. Mellow and rather musical 'tchip-tseeip, cheewip, tseep-tsip, choweep, tsip-cheowip, tsip, seep, choweep, tsip-'. Includes mimicry of other birds. Similar is the Slender-billed Thornbill, but usually further inland, in drier, different habitat. Sedentary; common.

No races. Slight variation N to S. Near S coast, darker with olive-brown cast to upper parts, more buff beneath. Further N, paler, greyer. Juv.: like adults, but with fainter facial markings that are even less obvious.

Western is plainest thornbill, plumaged in greys with only faintest tints of olive and buff. Has only very slight speckling about face.

Eye off-white to greyish white

Forehead browner, lightly scalloped feathers are paler.

Back olive-grey to brownish olive

Face and ear-coverts are lightly flecked buffy white.

Pale brownish-buff rump

Underparts are greyish-white with a faint buff tint, strongest towards the rump.

Broad dark tail band, pale fawn tip

---

**Striated Thornbill** *Acanthiza lineata* 10 cm
Typically uses high-rainfall, heavy eucalypt forests, but also drier open forests, woodlands and coastal mangroves. Like Yellow Thornbill, the Striated is strictly arboreal, but concentrates its foraging on eucalypts. After the breeding season, Striated Thornbills often gather in larger flocks; hunt energetically through foliage; briefly hover to take insects from leaves. Call is a soft, insect-like 'tzzt, tzzit, tzzt-tizzit', at times run together rapidly to form a short tune, almost trilled, 'tzt-tzt-tizzit-tizzit-tzt-tzt-tizzit-tizzit-'. In same habitat, Yellow Thornbills, Weebill. Sedentary; common.

● *lineata*
● *alberti*
● *clelandi*
● *whitei*

Crown rufous-brown, finely streaked white

Eye grey-brown

Shows only slight variation across range: brighter, yellower, yet quite heavily streaked towards NE (●), darkest in SE (●); bright and obviously streaked around Mt Lofty Ra., SA (●). Race ● occurs on Kangaroo Is.

Cheeks and ear-coverts heavily streaked

Back olive-green to olive-brown

---

**Yellow Thornbill** *Acanthiza nana* 9–10 cm
Inhabitant of open forests, woodlands of acacia, casuarina, melaleuca, brigalow, in preference to eucalypts. Strictly arboreal; small parties, pairs or solitary in mid to upper foliage; when high, easily overlooked but for calls. Always busily active and on the move, fluttering and hovering about foliage or flowers, often displaying dark, pale tipped tail. Voice a harsh, scratchy rather than high, insect-like sound, erratic 'chzip-chzzzip, chzip-chzip, chzeeep, chzip-chzzip-chzzzip'. Also gives a strong, harsher, scolding 'kirrzz, kirrrzz-kirrrzz'. Sedentary; common.

● *nana*
● *modesta*
● *flava*

Olive

Brow short, pale buff

In SE, ● deeper olive-green to brownish-olive

Cheeks to ear-coverts grey, streaked white

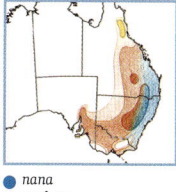

Bill quite long, fine

Some forms have slight rufous tint to chin and throat

Broad dark tail band

Birds of isolated Atherton population have brightest overall yellowish coloration, ● 'Yellow Thornbill'.

Inland, ● smaller, paler; underparts soft creamy buff

Faint pale tips

264

## YELLOW-BELLIED GERYGONES

**Fairy Gerygone** *Gerygone palpebrosa* 10–11 cm
Lowland rainforests, monsoon, riverine forests, mangroves, nearby eucalypt woodlands. Forage in foliage canopy where individuals, pairs or small parties busily seek insects and other tiny creatures; often hover, attracting attention with scratchy chatter and song. Voice typically lively and cheerful, not very high-pitched, but quite strong and penetrating, rather scratchy, metallic, repetitive: 'whit-WHITch-e-tew, whit-WHITch-e-tew, whit-WHITch-e-tew'; 'whit-a-whit-a-WHITchu, whit-a-whit-a-WHITchu'. In these and similar tunes it lacks the variety and creativity of the best of other gerygone songsters, the White-throated, Large-billed and Mangrove. Sedentary; moderately common.

● *personata*
● *flavida*
Nominate *palpebrosa* is a NG form.

● 'Black-throated Gerygone' has a dark head and is boldly patterned from face to upper breast.

Flight feathers and underside of tail grey to grey-brown

White streak each side

Coverts pale

Olive-grey

Blackish brown

♂

Under: yellow

♂

Small black streak

♂

● 'Southern Fairy Gerygone' overall pale, colours of head pattern merge softly.

White throat merges into pale yellow underparts; some males have small black streak under chin.

Females have faint white forehead spots.

♀

Females of both races lack the black facial markings of the respective males.

Little difference between females of the two races, *personata* and *flavida*

---

**White-throated Gerygone** *Gerygone olivacea* 10–11 cm
Wide range, conspicuous bright yellow underparts and exquisite song combine to make this one of the best-known gerygones. Found in woodlands, open forests, watercourse vegetation in small family parties, pairs or alone; forages busily in foliage, darting, hovering about leaves and flowers to take insects. In spring and summer, loud, clear, unmistakeable song attracts attention from afar. At other times of year is often silent, or gives only a few brief notes. Song loud, clear, carrying. Usually begins with several loud, piercing, high notes immediately followed by pure, high, clearly whistled, violin-like notes that descend in an undulating, silvery sweet cascade, at times lifting briefly, only to resume the downward, tumbling momentum. Abruptly returns to the initial louder, sharper notes to repeat the whole sequence, often with slight variations. Partly migratory, south-eastern birds moving N to winter with the sedentary northern population; common.

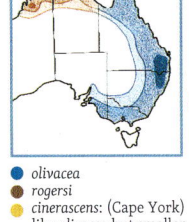

● *olivacea*
● *rogersi*
● *cinerascens*: (Cape York) like *olivacea* but smaller with paler, greyer upper parts

Upper parts grey-brown with olive tone to back

Eye red

White spot each side of forehead

Eye brown; no white on forehead

Overall dull coloration

White throat patch

Entirely pale yellow

Yellow

Imm.

Imm.: similar to both Weebill and Yellow Thornbill; different shape, behaviour and song.

● Has white patch at base of tail, but absent or grey on ●.

Tail grey-brown with large white spot at tip of each tail feather. These spots are not readily visible on upper surface of fully folded tail.

# BROWN, WESTERN & GREEN-BACKED GERYGONES

**Brown Gerygone** *Gerygone mouki* 9–11 cm
Rainforests, nearby dense wet eucalypt forests, mangroves. Attracts attention with almost constant twitterings and brisk activity; small parties or larger flocks. Seeks insects and other small creatures on foliage and bark; moves rather methodically among leaves, inspecting mossy limbs; often hovers over and under leaves of canopy. Calls are brisk, sharp, more vibrant and insect-like than songs of other gerygones: 'wit-chippitee-wit-chippitee-chippitee-chipitee'; 'whitchip-whitchip-whitchip' and other variations; scolding 'tzip-tzip-tzip'. Song differs slightly in NE Qld, where notes are given in an ascending sequence. Sedentary; common.

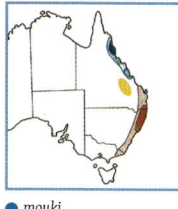

● *mouki*
● *richmondi*
● *amalia*: back olive-brown

- Face plain pale olive-grey
- Pale brow extends slightly behind eye.
- Upper parts greyish olive-brown
- Back russet-brown
- Greyish white, buff tint along flanks
- Face, sides of neck pale grey; lores black
- White spots on inner webs at tips, except inner pair
- Juv.: similar, but eye brown
- Flanks have buff-grey tint.

**Western Gerygone** *Gerygone fusca* 10–11 cm
Open forest, woodland, mulga, mallee; in SW of WA it extends closer to coast, and is also found in heavier jarrah and wet karri forests. A small, plain species; forages in the tree canopy; often hovers about outer foliage; solitary or in pairs, rarely in small parties. Clear, faint, whistled tune; a tenuous thread of high, clear notes, descending and rising, rather wistful, always ceasing abruptly before seeming complete. Sedentary; common.

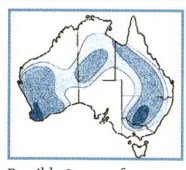

Possibly 3 races: *fusca* in SW of WA; *exsul* in E; *mungi* linking across centre and N

- Plain plumage except patterned tail when spread in flight
- Eye intense red
- Short, dull pale brow merges into eye-ring.
- Grey, some with olive-brown cast
- Fine white eye-ring
- Underparts pearly white to mid grey, slightly darker on throat and breast
- Imm.: face and underparts have a faint buffy tint; eye brown.
- Wide black subterminal band, white tips and base; central pair dark

**Green-backed Gerygone** *Gerygone chloronota* 9–10 cm
Usually in mangroves, monsoon scrubs, paperbark swamps, inland along rivers. Inconspicuous, blends into foliage; solitary or pairs, occasionally in small groups; vigorously forages in foliage for insects. Calls in sustained fast bursts: 'CHIRReep-CHIRReep-CHIRReep-' with emphasis on first sharp 'CHIRR'; or faster 'CHIRRiepCHIRRiepCHIRRiep-' in long bursts. Sedentary; common.

● *chloronotus*
● *darwini*: has yellow-olive back; faint buffy yellow tint to breast and flanks.

- Distinctive green overtone to back, wings, rump; otherwise very plain with insignificant white markings to face and tail
- Face grey with faint pale brow and eye-ring
- Brown with overlay of metallic olive-green
- Eye red
- Grey-brown, slight greenish overtone
- White with touch of dull buff along the flanks

Nest is useful in identifying the rather plain little birds that may be in attendance. Distinctive globular shape, long slender tail, and suspended among foliage, often over or near water of stream or swamp.

# MANGROVE, DUSKY & LARGE-BILLED GERYGONES

**Mangrove Gerygone** *Gerygone levigaster* 10–11 cm
Narrow strip of mangroves around most of N and E Aust. Mostly within tidal mangroves, but also visits adjacent rainforests, swamps, woodlands. Forages through canopy, occasionally hovering. Solitary, in pairs, small family groups. Identification is complicated by occasional use of mangroves by Brown and Large-billed Gerygones; there is also some overlap with Dusky Gerygone on W Kimberley coast. Song is similar to White-throated and Western Gerygones, but probably superior. Clear notes, some strong and piercing, others tenuous, clear, like miniature violin. Sedentary or locally nomadic; common within core habitat.

- *levigaster*
- *cantator*

**Dusky Gerygone** *Gerygone tenebrosa* 11–12 cm
Mangroves, along the seaward side of wider mangrove belts, but also visiting adjoining dryland scrub. (The Large-billed Gerygone more typically found on landward side of mangrove belt.) Forages for insects on foliage. Calls include plaintive, mellow notes, like those of a whistler, although softer. Quite unlike lively, high-pitched, thin, at times sharp, songs of most gerygones; rather slow, hesitant: 'queeit, toowheet, queeit, quit, queeit, whit'. Sedentary; common.

- *tenebrosa*
- *christophori*

**Large-billed Gerygone** *Gerygone magnirostris* 10–11 cm
Mangroves, paperbark swamps, rainforest streams. Forages in canopy; often hovering. Presence revealed by large slender nest suspended over water. Superb voice, powerful, rich notes in diverse sequences, high, clear, or low and mellow. Diverse tunes, creative variations: 'wheecheeapip-wheecheeapip-; cheeapip-cheeapip-cheeapip; wichee-wichee-wichee-wichee'. Sedentary; common.

- *magnirostris*
- *cairnsensis*

## FERNWREN, REDTHROAT, WEEBILL

**Fernwren** *Oreoscopus gutturalis* 13–15 cm
Confined to undergrowth of highland rainforests above 600 m altitude. Deep olive-brown plumage merges into dark tones of moist leaf litter as it sneaks among ferns and accumulated sforest-floor debris. A brief frontal glimpse may reveal the conspicuous white throat. Call is a high, fine, 'pee-pee-PEEEE', the last note strongly whistled, musical; another call is softer, brief, more even: 'pip-pip-pip-peeeee', rather like display calls of Pied Honeyeater. Also a scolding, metallic, buzzing 'pzeep-pzeep-pzzzeep'. Sedentary; moderately common.

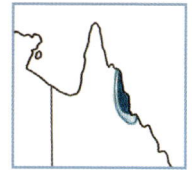

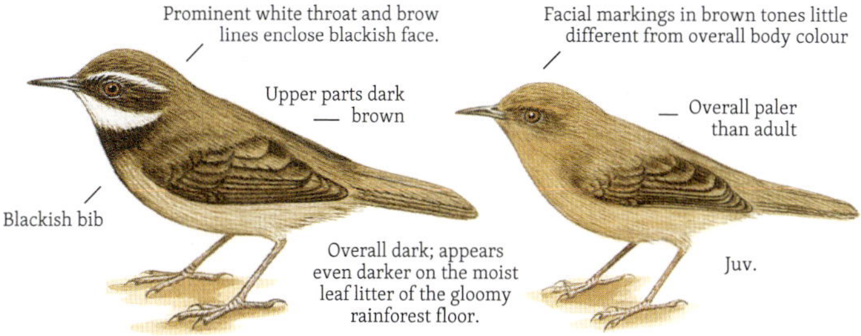

Prominent white throat and brow lines enclose blackish face.
Upper parts dark brown
Blackish bib
Overall dark; appears even darker on the moist leaf litter of the gloomy rainforest floor.

Facial markings in brown tones little different from overall body colour
Overall paler than adult
Juv.

**Redthroat** *Pyrrholaemus brunneus* 11–12 cm
Semi-arid open eucalypt woodland, mulga and mallee with scattered low shrubs, open areas with saltbush and bluebush. Small, grey-brown; sole claim to distinction a patch of bright colour, more orange than red, at throat of male; the female is more difficult to identify unless in company with the male. Forages briskly on the ground in the usually rather sparse low shrubbery of its semi-arid habitat. Undulating low flight with frequently fanned, white tipped tail. Song is a distinctive intermix of sharp, clear, whistled notes alternating with harsher, chattered sounds: 'whit-whit-chee-chee-quorr-quorr, whit-whit-chee-chee-chak-chak'; 'chi-chi-chi-chi, quirrrr'. Sedentary; uncommon to moderately common.

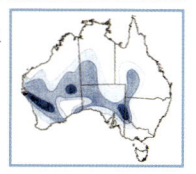

White forehead, lores
Cinnamon-orange
♂
Dull white merging rearwards into grey-buff flanks; buff under tail

Mid grey-brown
Dull greyish-white throat
♀
Tail mid brown; all but central pair white tipped

In flight, upper parts uniformly mid grey-brown; the broad white tips to outer tail are conspicuous.
Imm.: similar to female, but slightly paler

**Weebill** *Smicrornis brevirostris* 8.5–9.5 cm
Dry open eucalypt forests, woodlands, mallee. Tiny, greenish; loud, clear, distinctive 'wee-bill' calls. Inconspicuous in mid to upper canopy where pairs or small parties flutter about foliage seeking small insects. Call a strong, far-carrying whistle, slightly husky compared with clear, pure notes of gerygones. Often a single, loud 'wheetiew'; and longer 'whit, whit, WHEETiew' and 'whit, whit, WHEET-a-whit'. Sedentary, locally nomadic; common.

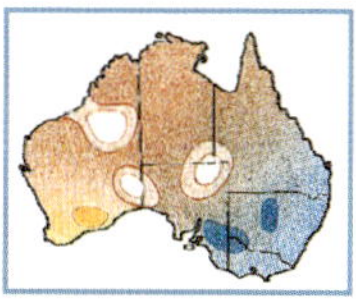

- *brevirostris*
- *flavescens*
- *occidentalis*: includes possible NW to central race *ochrogaster*.

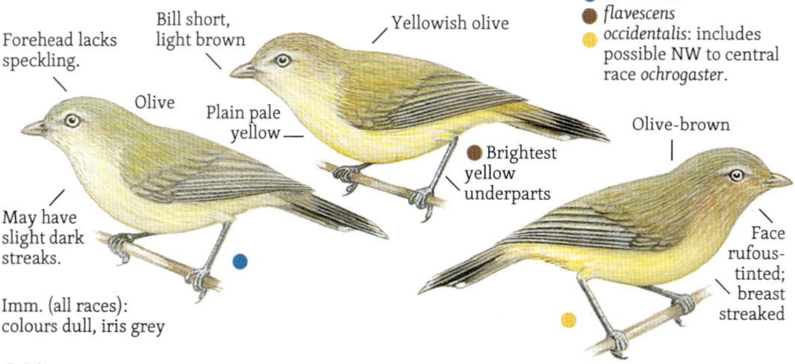

Forehead lacks speckling.
Olive
May have slight dark streaks.
Imm. (all races): colours dull, iris grey

Bill short, light brown
Yellowish olive
Plain pale yellow
Brightest yellow underparts

Olive-brown
Face rufous-tinted; breast streaked

260

## MASKED SCRUBWRENS

### Yellow-throated Scrubwren *Neosericornis citreogularis* 12–14 cm
Undergrowth, debris beneath rainforest, 600–1000 m altitude. Southern races in forests of ranges and lowlands. Forages through moist, decaying debris. Varied, cheery, whistled calls, rather like Mistletoebird – sharp notes interspersed with lower, more mellow notes: 'sieeip-chzweep-chip-sieep-chzeep'. Mimics the calls of other birds. Contact calls are sharp, abrupt: 'chiep, chiep, chiep'. Scolds harshly: 'chzzak-chzzak-chzzak'. Sedentary; common.

- Long brow, white and yellow
- Primaries edged yellow
- Underparts patchy olive-buff to creamy white
- 🟡 *cairnsi*, back darker, breast brownish, smaller
- ♂ 🔵 *citreogularis*: may merge with race *intermedius* in N of its range.
- Red eye set in black mask, bright yellow throat
- ♀ Face markings paler
- Legs pink
- Builds big greenish black nest, usually hanging above creek beds in rainforest.

### White-browed Scrubwren *Sericornis frontalis* 11–13 cm
Uses any dense, scrubby cover in damp, low undergrowth of forest, creekside thickets, dense sandplain heath. Always alert, quick to give harsh scolding alarm calls. Inquisitive – may be called up with squeaky sounds. Song is a high, clear, 'ch-weip, ch-weip, ch-weip'; 'chi-wipip, chi-wipip'; 'chp-wiep, chp-wiep'; 'cheweep-chip'; and a very high 'ts-eeer, ts-eeer', the second syllable usually higher. In alarm, scolds with hard, metallic, rattling 'chk-chkchk' or buzzing 'tzzzt-tzzt'. Often mimics other species. Sedentary; generally common; many races, all very similar.

- 🔴 *rosinae*    🔵 *frontalis*
- 🟠 *flindersi*  🟤 *laevigaster*
- ⚪ *ashbyi*      *tweedi*
- 🟢 *tregellasi*  🟡 *maculatus*
- 🔵 *balstoni*

- ♂ Black face mask, all races
- Eye yellow to pale grey, all races
- 🔵 Pale grey to black streaks on throat; rarely to breast
- Upper parts dark brown
- Flanks and belly, buff to mid olive-brown
- White tips to coverts vary with race, bright to dull.
- Paler, less contrast in facial markings
- Legs dull pink
- ♀ 🔵
- Long, thick, white brow line
- Tails dark-banded; varies with race.
- Unstreaked
- 🟤 Brightest colours
- grey-buff to creamy yellow
- Lores black, ear-coverts grey
- ♂
- 🔴 Between *laevigaster* and *frontalis*; some faintly streaked on throat
- Streaked or spotted
- 🟡 'Spotted Scrubwren'. Similar races with streaked breast: *ashbyi*, *balstoni*, *rosinae*
- All races have a brow line low across top of yellow eye, giving an intense, staring, almost scowling expression.
- ♂

### Tasmanian Scrubwren *Sericornis humilis* 12–14 cm
Dense undergrowth of wet forests. Keeps near the ground. Has scolding alarm calls, similar to those of White-breasted Scrubwren. Gives a scratchy, sharp 'chzit-chzit'; song a scratchy, squeaky 'chzeit-chzeit-chzitty-chzitty-chzitty'. Confined to Tas. and nearby islands; common.

- White brow line extends only back to eye then fades away.
- Dark face mask
- Grey-brown
- Olive-brown
- Primary coverts are lightly tipped white.
- Flanks dull olive-brown
- Long legs, large strong feet

This form is sometimes included within the White-browed Scrubwren as race *humilis*.

Imm.: like adults, but dulled colours, indistinct facial pattern

# PLAIN-FACED SCRUBWRENS

**Atherton Scrubwren** *Sericornis keri* 13–14 cm
Confined to rainforests above about 600 m altitude. In behaviour and voice similar to the White-browed Scrubwren, but differs in its preference for gloomy depths of rainforest that White-browed does not usually penetrate. Tends to forage on ground or a few metres above, very rarely high; often alone or in pairs, or, after breeding, in small family groups of up to four birds. Shy, unobtrusive, mouse-like movements. Call a sharp, scratchy, deliberate 'chi-wiep, chi-wiep-', rapid 'chiep-chiep-chiep', musical 'chippity-wiep, chippity-wiep'. Sedentary; restricted range, uncommon.

Upper parts plain mid to dark brown

Indistinct pale brow line

Bill slender, straight

Base of tail edged dull red, tail-coverts rufous-brown

Rather plain, lacking facial or other markings

Underparts white with weak buff tint, slightly darker on flanks

This and Large-billed build very different nests: Atherton's domed, hidden in ground litter; Large-billed obvious, suspended quite high.

Sexes alike; imm. has more rufous upper parts, slightly stronger yellow tint beneath.

**Large-billed Scrubwren** *Sericornis magnirostra* 12–13 cm
Tropical rainforests up to 1500 m altitude. More gregarious than the Atherton Scrubwren, usually in small flocks, actively foraging in mid levels of rainforest – on trunks and branches, over bark and among debris suspended in vines and epiphytes; often, briefly hovers and flutters around high foliage. Song is a cheerful, sharp, loud 'chiWIP-chiWIP-chiWIP', faster, scratchy 'chizip-chzipchzip', piercing 'chweeip-chweeip', scolding 'kzzip-kzzip-kzzip'; often mimics other small birds. Sedentary; quite common.

Body slightly more slender than Atherton Scrubwren

Long, slender, straight bill is inclined slightly upwards.

Races differ only slightly.
● *magnirostra*
● *viridior*
● *howei*

Back mid brownish olive rather than deep warm brown of Atherton Scrubwren

Dark eye prominent, set in plain pale face.

Rump and base of tail rufous

Pale creamy grey

Throat very pale buffy white, belly greyish white

Juv.: buffy breast, yellowish gape

**Tropical Scrubwren** *Sericornis beccarii* 11–12 cm
Tropical rainforests, monsoon forests, river-edge vine scrubs. Forages on trunks, vines, foliage of lower to mid level, ground litter; in pairs or small family groups. Calls similar to White-browed Scrubwren, a rapid, undulating 'chippity-wippit, chippity-wippit-'; also scolding 'squarr-squarr-squarr'. Tropical Scrubwren may be a subspecies of Large-billed; its plainer southern race *dubius* appears almost identical to the Large-billed Scrubwren. Sedentary; quite common.

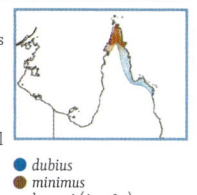

● *dubius*
● *minimus*
*beccari* (Aru Is.)

Black patch on lores and under eye; forehead edged with pale tufts

Eye bright red

Short white brow; with white lower rim forms broken eye-ring.

Female's markings less distinct

Dull rufous

♂

● Has more distinctive face pattern, deeper rufous-brown upper parts.

Underparts softly buffy white

♀

Coverts tipped white, forming two diagonal lines

Bill slender point, slightly downcurved

● More uniform cinnamon-buff; lacks bold facial markings of northern race

Imm.: plainer version of adult, slight rufous tint to flanks

♂

Wing-coverts tipped dull buffy white, making two short, faint, diagonal lines

## HEATHWRENS, SPECKLED WARBLER

**Chestnut-rumped Heathwren** *Hylacola pyrrhopygia* 13.5 cm
Heathlands, scrubby thickets in woodlands, forests. Usually keeps beneath dense vegetation; forages on or near ground. Shy, secretive, but in breeding season sings from higher perches; even then keeps partly concealed. Renowned for spirited song including mimicry. High, clear, silvery notes: 'chweep-tweep-toowheet, tweet-wheeit-wheeit, chirreeep, twitchy-twitchy-quorrop-quorreep', with many variations. In alarm, harsh chattering, scolding, 'chzzt-chzzzt!'. Sedentary; uncommon.

Possibly three races; differences slight
- pyrrhopygia
- parkeri
- pedleri

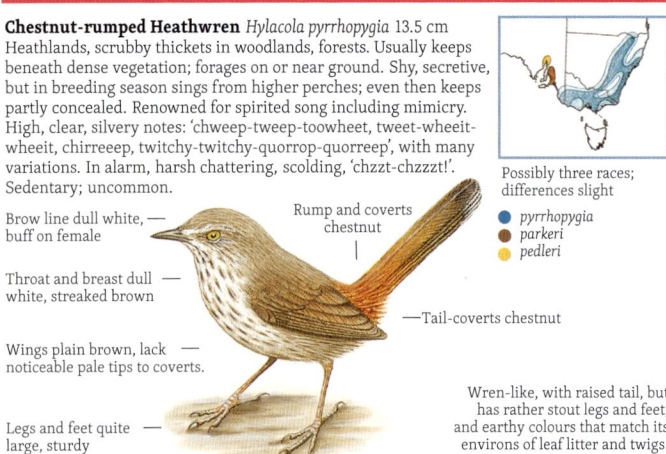

Brow line dull white, buff on female

Rump and coverts chestnut

Throat and breast dull white, streaked brown

Tail-coverts chestnut

Wings plain brown, lack noticeable pale tips to coverts.

Legs and feet quite large, sturdy

Wren-like, with raised tail, but has rather stout legs and feet, and earthy colours that match its environs of leaf litter and twigs.

**Shy Heathwren** *Hylacola cauta* 12–14 cm
Mallee and coastal thickets with dense low cover; grass tussocks on sandplains. Pairs or small parties forage on ground among low vegetation and debris. Flight reluctant, low and brief; gives but a fleeting glimpse of the bright rufous rump and white wing patches. Song a clear, sharp, musical sequence, not as sustained or varied as that of the Chestnut-rumped. Begins with several high, thin squeaks, then stronger, a touch of buzzing rattle that fades away: 'see-sree, chweip-chzeip-quirrip-quip-chip-chiep'. In alarm, harsh scolding. Sedentary; uncommon, often overlooked.

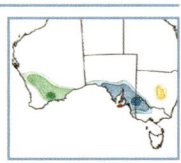

Possibly four races; differences slight
- cauta
- halmaturina
- whitlocki
- macrorhyncha

Long white brow

Rump and tail-coverts bright rufous-chestnut

Dark coverts

♂

Underparts streaked

White spot on folded wing

White bases to primaries

In flight, conspicuous bold pattern of chestnut rump and dark-banded, white-tipped tail

♂

**Speckled Warbler** *Pyrrholaemus sagittata* 12–13 cm
Open eucalypt woodlands with rocky gullies, ridges, tussocky grass, sparse shrubbery. Quiet, unobtrusive, yet not hard to see in its open habitat. Flies readily; perches in trees, easily visible. Forages on ground, often with other small birds. Song an undulating, cheery mix of clear whistles and more mellow notes, 'chwiep-cheerip-chip-weip-weip-weip-cheerip-weip-chip', with many variations; includes mimicry. Patchy distribution; scarce to moderately common.

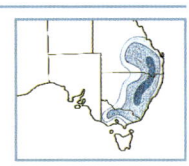

Male has black upper margin to brow.

Female, rufous brow edge

Back, wings, tail grey-brown, softly streaked darker

White brow line extends down to ear-coverts.

Plumage overall heavily streaked rather than speckled; these streaks are more conspicuous on white undersurface.

In flight, spread tail shows dark and white bands across near tail tip

Face off-white, streaked; large white patch at sides of neck

♂

♀

Holds tail horizontally; white tips of outer tail may be hidden under longer, dark tipped inner feathers.

# FIELDWRENS

**Striated Fieldwren** *Calamanthus fuliginosus* 13–14 cm
Swampy alpine and coastal heathlands, tussocky grasslands, low shrubby vegetation, margins of swamps. Usually in pairs, solitary or in small family parties; easily overlooked, flying briefly only if suddenly startled or forced. In spring male climbs to top of a shrub to deliver song, tail waving sideways as he pours forth a lively, although subdued, succession of trills and chatterings: 'whit-whit wirree, whit-whit-whitCHIER, chrrit-chrrit-chrrit'. First part is in notes clear, spirited, musical, leading to a louder 'chier', then quickly followed by a slightly harsh, chattered 'chrrit', repeated several times. Sedentary, uncommon.

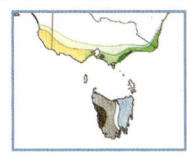

Four races, but differences slight. Perhaps clinal variations, insufficient to justify any subspecies

- ● *fuliginosus*
- ● *albiloris*
- ● *bourneorum*
- ● *diemenensis*

Black streaks across upper parts
Brow and lores white
Black centres to tertials
White tips
Dark band
Brow, lores, throat tawny buff
♂
Flanks buff, heavily streaked
♀
Buffy white with broad black striations

● and ● have heavy black streaking. Others, not shown, ● and ●, have finer and fewer dark streaks.

Grey-brown, edged buff
White tips to outer tail; dark subterminal band
Imm.: like females, but duller markings

## Western Fieldwren
*Calamanthus montanellus* 12–13 cm
A former race of the Rufous now considered a species and WA endemic, with no range overlap with the Rufous. Browner overall than the Rufous and more speckled than the Shy.

- ● *campestris*
- ● *isabellinus*
- ● *winiam*
- ● *rubiginosus*, *hartogi*
- ● *wayensis*

● *montanellus*

## Rufous Fieldwren
*Calamanthus campestris* 12–13 cm
Now separate from Western Fieldwren, one of its former races. Inhabits inland or dry country with saltbush, blue-bush, scattered low shrubs on sandplain, gibber, saltmarsh. Secretive, but in breeding season, males sing from high bare perches; drop into cover. Song like that of Striated: 'chit-whit-whit, whirrrrrreet', with first short notes whistled clear and high, the last long passage a rattled chattering. Striated Fieldwren is similar, but heavier streakings, and range into SE in wetter habitats. Sedentary; has lost much habitat.

Tail olive-grey with little or no rufous
Olive-grey, streaked black
White tips and dark subterminal band
Rufous confined to forehead, unlike on Rufous Fieldwren.
Brow and throat white, males. Buff, females.
Creamy white with narrow black striations
Legs quite large, sturdy, in keeping with terrestrial habits
♂
● **Western Fieldwren**

● Nominate race becomes greyer and more heavily streaked towards S part of its range.
● Further N is also cinnamon-rufous toned, but with reduced streaking.

● **Rufous Fieldwren**
Cinnamon-rufous
Brow and lower rim of eye white
Buffy white, finely striated
Often climbs to an exposed perch above shrubbery to sing, principally in breeding season.
♂

**Variation:** Many small populations are scattered across mostly drier parts of southern and western Aust. and are thought to represent six races, with the Western Fieldwren, *C. montanellus*, now a separate species.

# PILOTBIRD, ROCKWARBLER, SCRUBTIT

**Pilotbird** *Pycnoptilus floccosus* 16.5–17 cm
Wet eucalypt and temperate rainforest, alpine and coastal woodland in dense undergrowth with abundant debris. Often forages with Superb Lyrebird, taking small creatures exposed when litter and soil are raked about; its far-carrying calls 'piloted' bushmen who were searching for lyrebirds. Usually keeps on or near the ground; if alarmed runs swiftly on long, strong legs. Male's loud call is described as having 'piercing sweetness': 'whit-wheet-WHEE-a-wer' or fast 'whit-whit-WHEER, whit-whit-WHEER'; these and other variations have the clear musical quality of bristlebird calls. Females also may give these calls and softer 'whitta-whittee' and 'too-whit-too-whittee' responses. Sedentary; common.

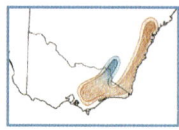

- *floccosus*
- *sandlandi*: smaller, upper parts paler, rufous; below 600 m altitude

Long, broad, wedge-tipped tail may be held angled upwards, flicked vertically, or trailed low.

Slender bill projects abruptly from a bluntly rounded face.

Occurs above 600 m altitude; upper parts entirely deep rufous-brown.

Underparts cinnamon with each feather edged darker brown, giving a softly scalloped or lightly mottled appearance

Central belly is dull white; often hidden with bird on ground.

Long legs, large strong feet; rakes among leaf litter; runs swiftly.

---

**Rockwarbler** *Origma solitaria* 13–14 cm
Forested sandstone and limestone gullies, caves, cliffs, often with streams, waterfalls. Best known for using caves, often behind waterfalls, for nest sites. Mainly terrestrial, it hops about on sandstone ledges, clings to the rock face, and takes insects from crevices and the margins of streams, only occasionally foraging on bark of tree trunks or limbs. Call a clear 'chweep-chweep' with a slightly mournful quality to the sound. Also a sharper 'chwik-chwik', and harsh, scolding 'tzzt, tzzt'. Sedentary; locally common.

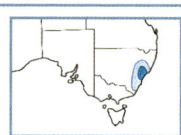

Female is similar to male; immatures have slightly dulled colours.

Olive-grey

Bill slender, finely pointed

Face plain, dark eye, faint, pale eye-ring

Tail darker, smaller, narrower, than that of Pilotbird, often flicked sideways

Throat dull white, lightly speckled

The Rockwarbler is unusual in building a nest that is suspended from the roof of a cave or the underside of a rock ledge, often near a waterfall and frequently in semidarkness.

Breast to under tail rufous-cinnamon; appears deeper cinnamon-brown seen in shadows of gullies, caves.

---

**Scrubtit** *Acanthornis magnus* 11–12 cm
Shy, inconspicuous in dim light of rainforest; scuttles mouse-like beneath undergrowth, searching with treecreeper-like hopping action on logs and lower trunks. Gives piercing whistle, extremely rapidly repeated, sharp 'chi-chi-' followed by louder, clearly whistled 'chiew' or 'chiewit', sequence varied: 'chi-chi-chi-chi-CHIEW, chi-chi-ch-chi-CHEWIT'; 'chiewit-chiewit, chi,chi,chi'; 'chiewit, chiewit, chiewit'. Also low, scolding 'chak-chak-chak'. Dense undergrowth with much debris in wet temperate rainforests, fern gullies. Common SW Tas., uncommon elsewhere.

- *magnus*
- *greenianus*: smaller, back paler brownish russet, narrow tail band

Bill slender and very slightly downcurved

Olive-brown

Face light grey with darker grey band beneath eye and expanding across ear-coverts; pale eye-ring

Narrow white tips to innermost flight feathers, visible along top edge of folded wing

White tips to outer-wing-coverts; on folded wings these make a short white bar of spots.

The Scrubtit is sufficiently distinctive to be placed in a genus of its own, *Acanthornis*. Resembles both Tasmanian Scrubwren and Brown Thornbill, which share same environs and have similar calls.

# BRISTLEBIRDS

**Eastern Bristlebird** *Dasyornis brachypterus* 20–22 cm
Wet eucalypt forests; dense coastal heaths, gaps and edges of rainforests, sedge or tussock gullies. Secretive, moving quickly through dense heath and tussocky vegetation with tail horizontal or occasionally uplifted, sometimes fanned; flights low and brief. In spring, loud clear calls. Song of varied cheerful, musical notes, some extremely high and weak, others more mellow, or strong and clear. Repeats a sequence several times, then adds variations: 'wit-wit wheerie-whEIT', with slight whip-crack finish. Also 'chi-chi-cheery-whEIT'; and 'whit-wheeeir-weeit-weeit-whit'; 'queeity-queeeir'. In alarm, a scolding 'chzzzt-chzzzt'. Now scattered remnant populations, vulnerable; sedentary.

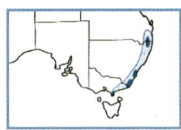

Race *monoides* at N end of range; only slight differences

- Tail long; faint fine bars barely visible
- Rump, tail-coverts rufous-brown
- Wings short, rounded, chestnut
- Pale around eye, lores
- Strong bristles visible only at close range
- Light scaly or scalloped pattern; in parts may form indistinct rough bars.
- Flanks and belly dull grey-brown

---

**Western Bristlebird** *Dasyornis longirostris* 18–21 cm
In dense, low, closed heaths, open heaths near refuge clumps of taller dense shrubbery, or dense watercourse thickets. Shy, difficult to sight; song often the only sign of its presence. Calls similar to Eastern Bristlebird; has slightly sad quality, reminiscent of the lonely moonlight singing of Willie Wagtail. Song in S of range differs in detail from that of populations further W. Intermixed sharp and mellow notes form descending sequences: 'tchip-eeee, quit-a-weer'; 'tchip-it-er-pieerit'; and 'quiee-ity-pieer, quieeity-pieer'; variations by individual birds. Rare, reduced to scattered remnant populations; vulnerable.

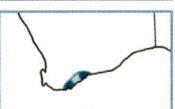

- Wings short, rounded, rufous-brown
- Rump, tail-coverts rufous-brown
- Tail long, yet slightly shorter than other bristlebirds; trails through and under low scrub, at other times may be partly uplifted.
- Crown and mantle grey-brown, mottled light grey
- Indistinct pale brow, red eye
- Undersurface feathers dark-tipped, giving mottled or unevenly scalloped rather than barred appearance
- Sexes alike. Imm.: plain olive-brown without paler spottings or scallopings

---

**Rufous Bristlebird** *Dasyornis broadbenti* 24–27 cm
Large bristlebird with rufous crown and ear coverts. Dense coastal heaths, thickets, wiregrass; in Otway Range, in undergrowth of wet forests. Elusive in typically dense habitat, often inquisitive; scurries through low vegetation, seldom flying, occasionally raising and spreading long tail. Clear, ringing whistle; if recognised, it helps to locate these shy birds. Clear, sharp notes precede a much stronger, quick, double whip-crack, the call often repeated by the female: 'chip, chip, chip, chup, CHEE-O-WHIP'. Sedentary; eastern race moderately common.

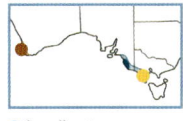

- ● *broadbenti*
- ● *littoralis*, wings dull rufous, extinct
- ● *caryochrous*, darker

- Bright chestnut, or dull brown in SE of range
- Wings short, rounded, chestnut
- Long tail trails, or is occasionally raised and spread widely, as when alarmed.
- Chin and throat white, finely scalloped and flecked with grey
- Underparts grey-brown; pale-edged feathers give strongly scalloped pattern.
- Sexes alike, except female is slightly smaller. Imm.: probably similar to adults

# RED-BROWED & STRIATED PARDALOTES

**Red-browed Pardalote** *Pardalotus rubricatus* 10–12 cm
Unusual pale-eyed, large-billed pardalote of the drier woodlands, watercourse trees, mulga scrubs; feeds in foliage of trees and shrubs; solitary or in pairs. The calls are distinctive, a song quite unlike that of other pardalotes. It is a strong, mellow whistle of five or six notes, several slow and rising, three to five higher, quicker, the result like a softly muted Crested Bellbird. Nomadic; common in NW, N, uncommon in E of range.

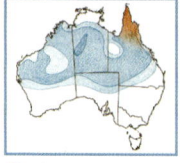

- *rubricatus*
- *yorki* with upper parts olive-brown and deep yellow breast spot

The Red-browed Pardalote combines the crown pattern of the Spotted Pardalote with the striped wing of the Striated.

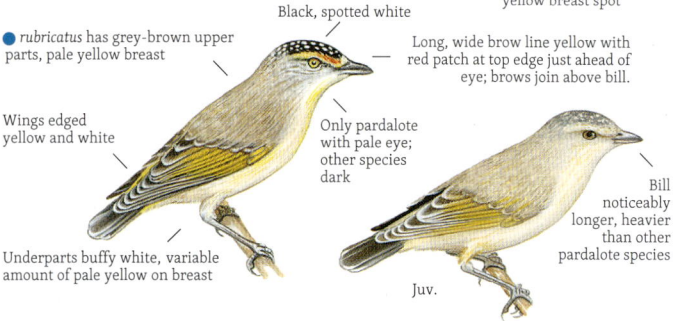

- *rubricatus* has grey-brown upper parts, pale yellow breast
- Black, spotted white
- Long, wide brow line yellow with red patch at top edge just ahead of eye; brows join above bill.
- Wings edged yellow and white
- Only pardalote with pale eye; other species dark
- Bill noticeably longer, heavier than other pardalote species
- Underparts buffy white, variable amount of pale yellow on breast
- Juv.

**Striated Pardalote** *Pardalotus striatus* 9.5–11.5 cm
This widespread pardalote lives in diverse habitats covering almost the entire continent, from wet coastal eucalypt forest, rainforest and mangroves to watercourse trees and the mallee or mulga scrubland of arid and semi-arid interior. Six races: all have black caps, conspicuous yellow and white brow line, either a yellow or red wing spot, and a white wing stripe. May be seen in pairs, alone or, except when breeding, small parties to large flocks. Forages in foliage, taking insects, especially lerps, and other tiny creatures in the leaves, bark and flowers. Call is a clear, sharp, musical 'witta-witta', the second part slightly lower, repeated regularly at several second intervals for long periods; also gives soft, low trills. Common; some races migratory, others locally nomadic or sedentary.

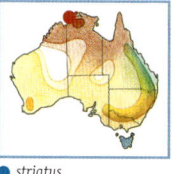

- *striatus*
- *substriatus*
- *ornatus*
- *melanocephalus*
- *uropygialis*
- *melvillensis*: most like *uropygialis*, back darker; confined to Melville Is., NT

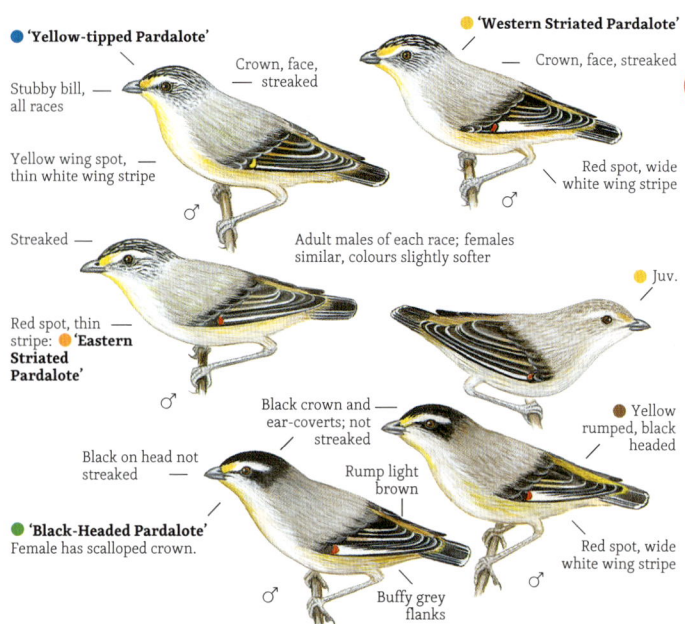

- 'Yellow-tipped Pardalote'
  - Stubby bill, all races
  - Crown, face, streaked
  - Yellow wing spot, thin white wing stripe
- 'Western Striated Pardalote'
  - Crown, face, streaked
  - Red spot, wide white wing stripe
- Streaked
- Red spot, thin stripe: **'Eastern Striated Pardalote'**
- Adult males of each race; females similar, colours slightly softer
- Juv.
- Black crown and ear-coverts; not streaked
- Yellow rumped, black headed
- Black on head not streaked
- Rump light brown
- **'Black-Headed Pardalote'** Female has scalloped crown.
- Buffy grey flanks
- Red spot, wide white wing stripe

## SPOTTED PARDALOTES

**Spotted Pardalote** *Pardalotus punctatus* 8–10 cm
Eucalypt forests, woodlands, watercourse vegetation; race *xanthopygus* in sandplain mallee and drier open woodland. These races may be kept apart by habitat preferences. One of smallest of Australian birds, leaf-sized. Often draws attention by persistent soft calls, as it takes insects from leaves and flowers of eucalypts. Their beauty is rarely apparent when high in trees, but occasionally tolerate quite close approach when feeding in low foliage; the neat colourful pattern is then visible. In pairs or small family groups. Contact call is repeated regularly for long periods; an even, musical, soft, whistled 'weep-weeip'. Song is a sequence of clear whistled notes, 'whee,whee-bee' or 'sleep, may-bee'. Yellow-rumped race has a quiet, rather mournful quality: 'chnk-whee-a-bee', soft, high, then falling away. Seasonally nomadic or migratory; common.

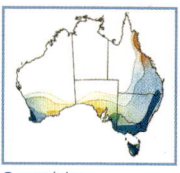

- ● *punctatus*
- ● *xanthopygus*
- ● *militaris*: rump deep olive-rufous; cinnamon tint to breast and flanks

Most colourful of the tiny birds whose clear white spots give this genus the alternative name 'Diamondbird'.

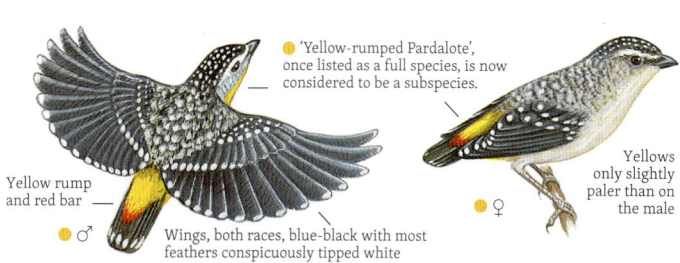

**Forty-spotted Pardalote** *Pardalotus quadragintus* 9–10 cm
One of Australia's rarest birds, evidently a declining population now fragmented into several colonies. Confined to SE Tas., where Maria and South Bruny Is. are strongholds; occasionally wanders to Hobart suburbs. Pairs or small parties forage in foliage of forest trees, slowly and methodically searching the leaves for lerps, spiders and other small creatures. Calls similar to those of Spotted Pardalote, perhaps harsher: double-noted 'wheet-whoo'; a louder, nasal 'twnnt' in breeding season. Eucalypt forests, usually relatively dry forest dominated by manna gum, *Eucalyptus viminalis*; foraging is almost exclusively confined to manna gum. Sedentary or locally nomadic; restricted range, rare, endangered.

Retains wing pattern similar to widespread Spotted Pardalote; may be remnant of a previous invasion into Tas. of Spotted Pardalotes, which, in long isolation on the island, became olive-brown on head and rump.

## WEEBILL & GERYGONES TO WHITEFACES

The family Pardalotidae is divided into three subfamilies. Pardalotinae has just one genus, *Pardalotus*, with four species of pardalotes, very small birds, some very brightly coloured. Subfamily Dasyornithinae has three large species, the bristlebirds, in one of its two genera. The third subfamily, Acanthizinae, has thirty-six species in ten genera. This includes many 'small brown birds', the rather plain species of thornbill and gerygone. Also in this family are the whitefaces, a single genus of three species.

Scarlet Robin = 12 cm

### acanthizinae (cont.)

page 260

- Weebill
- Green-backed Gerygone
- Inland Thornbill
- Yellow-rumped Thornbill
- Brown Gerygone
- Fairy Gerygone
- Chestnut-rumped Thornbill
- Yellow Thornbill
- Mangrove Gerygone
- White-throated Gerygone
- Slaty-backed Thornbill
- Striated Thornbill
- Western Gerygone
- Buff-rumped Thornbill
- Mountain Thornbill
- Southern Whiteface
- Dusky Gerygone
- Brown Thornbill
- Western Thornbill
- Chestnut-breasted Whiteface
- Large-billed Gerygone
- Tasmanian Thornbill
- Slender-billed Thornbill
- Banded Whiteface

# PARDALOTES & BRISTLEBIRDS TO SCRUBWRENS

## RUFOUS-BROWN GRASSWRENS

**Thick-billed Grasswren** *Amytornis modestus* 15–20 cm
The one grasswren needing shrubby habitat: sandplain with sparse scrub, treeless dense saltbush, bluebush, cottonbush. Also inhabits some sandy river channels with cane grass, saltbush and spinifex. Extremely shy, elusive; hides under shrubs; darts across open ground. Flights are low, fluttering; often in pairs or family groups. Eastern race has extremely high, soft 'see-see-see' and brief, squeaky trill, 'see-see, tsewit-tsewit'. Range now greatly reduced; locally common in a few sites.

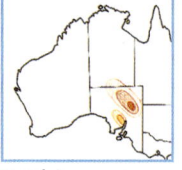

● *modestus*:
● *myall*
'Eyre Peninsula Grasswren'
(Not illustrated)

Bills of both races rather blunt and finch-like, although shape not as deep as that of the Eyrean Grasswren.

Races differ significantly. Both sexes of ● have shorter tail, plumage paler, greyer and more lightly streaked than race *myall*.

Pale fawn with rufous flanks

**Western Grasswren** *Amytornis textilis* 15–20 cm
Now considered a separate species to the Thick-billed Grasswren, this shy bird prefers the same shrubby habitat. Its calls are strong: a treecreeper-like, whistled 'cheep-cheep-cheep' that may extend to a brief, musical, rattling trill, 'chip-chip, chew-wekaweka'. Range now greatly reduced.

Drab grey-brown upper parts, heavily streaked

Female has underparts sandy fawn with deeper chestnut patch on flanks.

In the far west of its range, it is larger with longer tail and the darker plumage is more heavily streaked.

Tail of male longer than that of female

Bill is blunt and finch-like, although not as deep as that of the Eyrean Grasswren.

Scale slightly reduced.

**Eyrean Grasswren** *Amytornis goyderi* 15–16 cm
Dunes with cane grass; spinifex at base of cane-grass dunes. Bounces rapidly across sand when disturbed, or flutters low, soon dropping into concealing grass. In pairs or small parties. Contact calls are extremely high pitched, weak, insect-like squeaks, probably inaudible unless close: 'tseee' and a tinkling, squeaked 'tsip-tsip-tseee-tsip'. Song begins with these same high squeaks, changing abruptly to a fast, louder, rattling, musical chatter, 'tsee-tseee-chakachakachaka'; alarm a sharp, strong 'tzeeet'. Remote habitat, where in places it may be quite common.

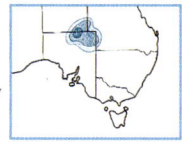

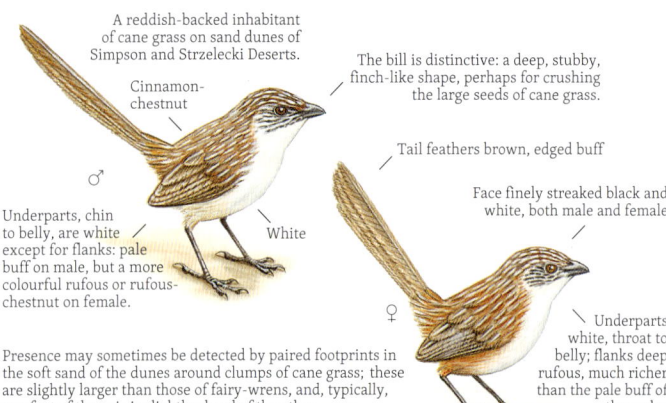

A reddish-backed inhabitant of cane grass on sand dunes of Simpson and Strzelecki Deserts.

The bill is distinctive: a deep, stubby, finch-like shape, perhaps for crushing the large seeds of cane grass.

Cinnamon-chestnut

Tail feathers brown, edged buff

Face finely streaked black and white, both male and female

Underparts, chin to belly, are white except for flanks: pale buff on male, but a more colourful rufous or rufous-chestnut on female.

White

Underparts white, throat to belly; flanks deep rufous, much richer than the pale buff of the males

Presence may sometimes be detected by paired footprints in the soft sand of the dunes around clumps of cane grass; these are slightly larger than those of fairy-wrens, and, typically, one foot of the pair is slightly ahead of the other.

## RUFOUS-BROWN GRASSWRENS

**Dusky Grasswren** *Amytornis purnelli* 15–18 cm
An inhabitant of rough, boulder covered ranges, gorges; favours large clumps of spinifex. Less timid, more inquisitive and easily seen than most other grasswrens. In family parties of 10 or 12 birds; pairs when breeding. Usually noticed bouncing among boulders and spinifex, tail held steeply upwards, but often scuttling, tail trailing low, into spaces under and between boulders; very rarely flies. More likely to be seen around dawn or dusk; shelters through hottest hours in shadowed crevices. Call is a high rippling trill, the first several notes faint squeaks, the rest an undulating low-high sequence, 'see-see, tchoo-tchoo-chee'; contact calls, weak twitters, 'tsee-tsee-tsee-'; harsher, staccato 'tchik-tchik!' in alarm. One of the more common grasswrens, easily seen in accessible ranges and gorges of central Australia.

Former subspecies *ballarae* is now considered a full species, *Amytornis ballarae*, the Kalkadoon Grasswren, described below.

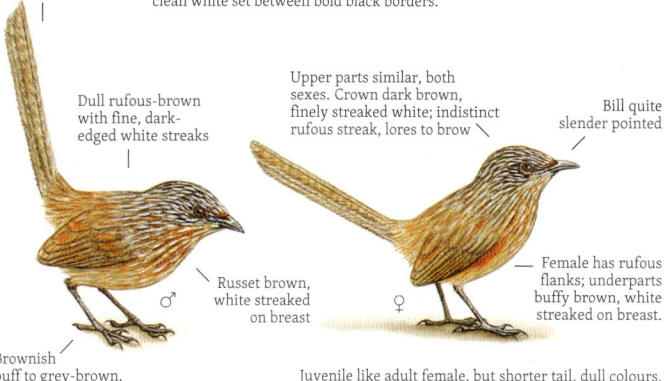

Tail quite long, dark brown with buff edges

Overall, looks rather dark and lacking distinctive markings. The one overlapping species, the Striated, has, from chin to breast, clean white set between bold black borders.

Upper parts similar, both sexes. Crown dark brown, finely streaked white; indistinct rufous streak, lores to brow

Bill quite slender pointed

Dull rufous-brown with fine, dark-edged white streaks

Russet brown, white streaked on breast

♂

Female has rufous flanks; underparts buffy brown, white streaked on breast.

♀

Brownish buff to grey-brown, belly to under tail

Juvenile like adult female, but shorter tail, dull colours, indistinct streaking, bill pale with pinkish lower base

**Kalkadoon Grasswren** *Amytornis ballarae* 16 cm
Previously a subspecies of the Dusky Grasswren confined to rough, spinifex-clad country around Mt Isa, including Selwyn Ra. and E Barkly Tableland. Widely separated from similar Dusky Grasswren. Is now considered sufficiently distinctive to be a full species. ('Kalkadoon' is the name of Aboriginal group of same region.) Pairs or small family groups; keeps low in spinifex, occasionally running or fluttering across open ground between clumps. Calls differ from those of Dusky Grasswren. Has a single loud alarm squeak, otherwise calls are weak, high peeps, mainly while foraging. Song is a series of short high trills. Probably uncommon through most of limited range, but quite common at some localities. Vulnerable to fires that eliminate preferred old, very large spinifex clumps.

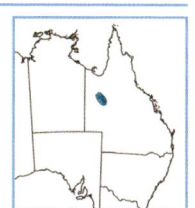

In the NW of its range the Kalkadoon Grasswren slightly overlaps the SE part of the range of the similar sized Carpentarian Grasswren, from which it differs most obviously in being grey-brown and streaked rather than stark white from chin to breast.

Rufous-brown

Throat to upper breast brownish grey, streaked

Crown dark brown, streaked; narrow rufous streak, lower forehead to brow

Head and face darker brown with white streaks

Male has plain pale grey flanks and belly.

Rufous brown with dark-edged streaks

Juv.: similar to adult female, but with shorter tail, dull streaking; male-female flank difference less distinct. Bill has pale gape and pinkish base to lower mandible.

♂

♀

Female has rufous flanks and pale buffy grey belly.

Unusual stiff, barb-tipped feathering

## STRIATED & SHORT-TAILED GRASSWRENS

**Striated Grasswren** *Amytornis striatus* 15–18.5 cm
Most widespread of all grasswrens with a broken range through much of the interior; probably three subspecies. Occurs in spinifex country with sparsely scattered trees, on sand dunes or rocky ranges. Like other grasswrens, skilled at hiding in low dense cover, contriving always to keep spinifex between themselves and observer. Most active early, bouncing onto exposed rocks and up into more open shrubs; later, through heat of day, likely to be silent and hidden. Probably more vocal, active and conspicuous in breeding season; at other times of the year locating these birds is far more difficult. Contact calls are high, scratchy squeaks and trills, 'tsee, tsee, tzee, chweep-chweep-chweep'; in alarm a harsher, louder 'tzirrr!'. Song begins with similar high squeaks, quickly followed by rapid, loud, clear, rippling, musical notes, 'tsee-tsee, piew-piew-piew'. Widespread, patchy, in places quite common, but difficult to see.

- ● striatus
- ● whitei
- ● rowleyi

Former race, now a full species, *Amytornis merrotsyi*, Short-tailed Grasswren

Tail quite long, rounded, dusky grey-brown edged buff

● Grey-brown coloration of SE Aust.; sexes similar except flanks

Streaks on upper parts mostly white, each streak finely edged with black or dark brown

Dull cinnamon-brown, extensively streaked on all races

Wings grey-brown, buff-edged

Female has rufous flanks.

Flanks pale grey-buff

Nominate race has widest colour variation, from greyer birds of the mallee of NW Vic (above) to reddish birds of sandy deserts of interior of WA.

● Most colourful race, bright rufous and golden buff

Tail rufous, with central feathers darker

Brow bright rufous

Cinnamon-rufous

Bold black moustache line separates streaked face from white throat on all races.

Flanks pale grey-buff

Flanks, lower breast and belly golden buff; Female has deeper rufous patch on flank.

Central Australian reddish toned form of race *striatus*

● Occurs in the Pilbara region of NW WA; its bright rusty rufous and deep buff tones match the red rock and spinifex environs.

**Short-tailed Grasswren** *Amytornis merrotsyi* 15–16 cm
Previously included within Striated Grasswren, this form occurs in Flinders and Gawler Ranges, SA; now has full species status. Habitat in spinifex with scattered small trees on rocky ranges. Contact call a high 'tsee-tsee', and a sharp squeak in alarm when flushed – both similar to calls of Striated. Other calls unknown. Habits and behaviour probably similar to Striated. Status uncertain; apparently several small and perhaps insecure populations.

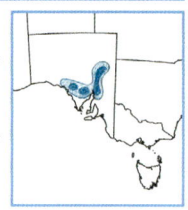

Tail much shorter than those of Striated and Thick-billed Grasswrens

Back rusty red with fine, white, dark edged streaking

Bill slightly heavier than that of Striated

Black facial line broken with white streaks and flecks, much less prominent than on the Striated Grasswren

Female has rufous flanks where males are pale buff.

Legs slightly longer than those of Striated, possibly to cope with the extremely stony habitat

Juveniles, all species, are rather short-tailed, and so are more difficult to separate: these are brighter rufous than juv. Striated or Thick-billed, finer-billed than those of the latter.

# CARPENTARIAN & GREY GRASSWRENS

**Carpentarian Grasswren** *Amytornis dorotheae* 16–17 cm
On sandstone ranges and ridges, or rough gullied plains, often where broken into gorges, ledges and boulders. Vegetation of spinifex with sparse, stunted trees. These grasswrens can be very elusive, bouncing away to disappear behind rocky bluffs or dense spinifex clumps. May take flight on first encounter; flutter away in slow, emu-wren-like flight, long tail trailing, to drop back to cover, or vanish beneath densely concealing clumps of spinifex. Contact within groups is maintained with a high, cricket-like 'tseeit'. A harsh, buzzing 'tzzeit, tzzeit' in warning. Song is a rapid, cheerful, canary-like warbling, a musical intermingling of squeaky, high trills and deeper, stronger, prolonged churring sounds: 'chrrip-chewip-chur-r-r-r-r-p'. Uncommon; probably in sparsely scattered populations.

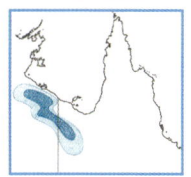

No subspecies; variation is confined to possibly very slightly finer dorsal streaking in part of range.

Upper parts, forehead to tail, similar for both sexes, with female almost as intensely patterned as male

Smaller and paler than other far northern grasswrens, the Black and the White-throated

Rufous-chestnut with fine white, dark-edged streaks

Crown black, streaked; brow rufous

Sexes differ in colour of flanks: male is tawny buff on flanks and belly.

♂

White extends to belly; no dark breast band.

Bold black streak back from bill, gives a strongly defined edge to white of chin, throat, breast

Female has chestnut flanks and belly.

White, no breast band

♀

---

**Grey Grasswren** *Amytornis barbatus* 17.5–20 cm
Inhabits river overflow swamps, with dense clumps of lignum and cane grass, sedges, saltbush, where striped plumage makes these birds hard to see among tangled stems of dense vegetation. Best sighting times are early morning and towards dusk, when birds may venture out on open mudflats or dunes to feed. In parties of up to 15 birds, or pairs in breeding season. Contact between birds is maintained by high pitched, quick 'tsip' calls, given by one, answered by others in the group – an almost unbroken, twittering rippling of metallic, squeaky 'tsit-tsit-tsit, tsit-tzit, tsit' calls that continue for several seconds. The alarm is a high, squeaked 'tseeep'. Generally uncommon and of restricted range, but may be locally common.

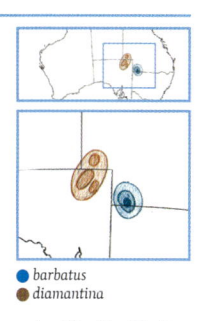

● *barbatus*
● *diamantina*

The unique facial pattern of the Grey Grasswren simplifies identification. Differences between the two races, and between male and female, are slight.

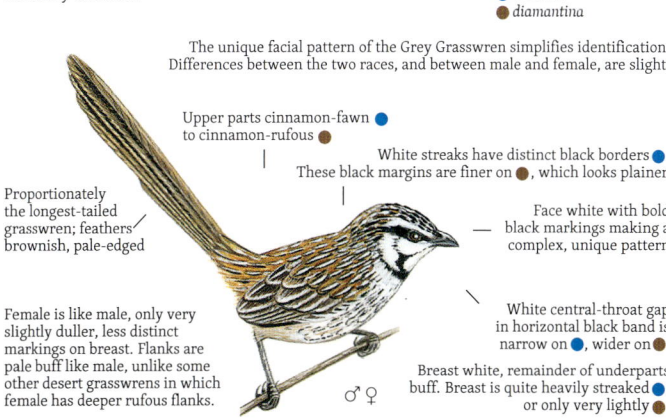

Upper parts cinnamon-fawn ● to cinnamon-rufous ●

White streaks have distinct black borders ●. These black margins are finer on ●, which looks plainer.

Proportionately the longest-tailed grasswren; feathers brownish, pale-edged

Face white with bold black markings making a complex, unique pattern

Female is like male, only very slightly duller, less distinct markings on breast. Flanks are pale buff like male, unlike some other desert grasswrens in which female has deeper rufous flanks.

White central-throat gap in horizontal black band is narrow on ●, wider on ●.

♂♀

Breast white, remainder of underparts buff. Breast is quite heavily streaked ●; or only very lightly ●.

## LARGE, DARK GRASSWRENS

**Black Grasswren** *Amytornis housei* 19–21 cm
Inhabitant of rugged ranges, gorges with spinifex, scattered trees and shrubs. In pairs or parties of three or four in breeding season; at other times, larger groups of six to nine birds forage in and around large clumps of spinifex among tumbled boulders. Calls distinctive, quite loud; song is harsher, stronger than that of the much smaller fairy-wrens of region. Gives a rather metallic, slow, rattling sequence, into which it inserts low, husky and high, scratchy, squeaky sounds and contrasting brief, clear, mellow, musical, rising whistles: 'chaka-chuk-seeip-chuk-zwiEEEP, chak-chukchuk-chewiEEEP! chuk-chuk-seeip'. Restricted range; sedentary.

**White-throated Grasswren** *Amytornis woodwardi* 20–22 cm
Inhabitant of rough sandstone ranges and escarpment valleys with spinifex, sparsely scattered shrubs, trees. Scurries among boulders, through spinifex; occasionally bounds to a higher rock to call. Flocks keep together with sharp, scratchy 'chwiep' contact calls; occasional harsher 'tzzrrrt!' alarm calls. The song, perhaps only in breeding season, is loud, rich, varied; in this respect rather like that of the Reed Warbler; stronger, not high and squeaky like fairy-wrens. It is a musical intermix of both sharp and deep rich notes, 'chziep-chiep-quorrop, chirrip-chok, chwiep-quarriep-'. Sedentary or locally nomadic; uncommon; patchy occurrence within restricted range.

# EMU-WRENS

**Southern Emu-wren** Stipiturus malachurus 15–19 cm
In dense, low cover, damp heaths, sedges, sand-dune and sandplain heaths, spinifex shrublands, buttongrass plains. Usually stays under cover; occasional flights with long tail streaming behind. Calls similar to fairy-wrens, but higher, weak, scratchy, yet carry well over open heath. The squeaky calls have an undulating, rollicking variation of volume and pitch, 'siee-seit-tzeet-tzeet-siee-seeit-tzeeit-tzeet-seee-seeit'. Contact call is a faint squeak, 'tsiee'; alarm a sharp 'trrrit!'. Locally common, but some populations endangered.

- malachurus
- halmaturinus
- westernensis
- intermedius
- polionotum
- parimeda
- hartogi
- littleri

**Mallee Emu-wren** Stipiturus mallee 13–15 cm
Difficult to sight; keeps low in spinifex. Habitat low sand dunes where old spinifex forms large clumps under low mallee and native cypress; also tall heaths with tea-tree and broombush. Calls like Southern Emu-wren, but weaker. Song is a feeble trill, squeaky, insect-like. Contact call thin, drawn out 'sreeep'; although weak, carries some distance, gives chance of locating birds. Sharper, harsh 'tirrit' in alarm. Rare; in localised colonies vulnerable to fire.

**Rufous-crowned Emu-wren** Stipiturus ruficeps 12–13 cm
Spinifex with scattered shrubs on dunes or plains. Calls similar to other emu-wrens, but weaker. Contact call a cricket-like, squeaked 'tseeet-tseeet-tzeeet-'. Song a rapid succession of scratchy, squeaky notes, rather like those of Southern Emu-wren, but extremely high, often beyond observers' hearing unless very close. Probably quite common, but not easily found.

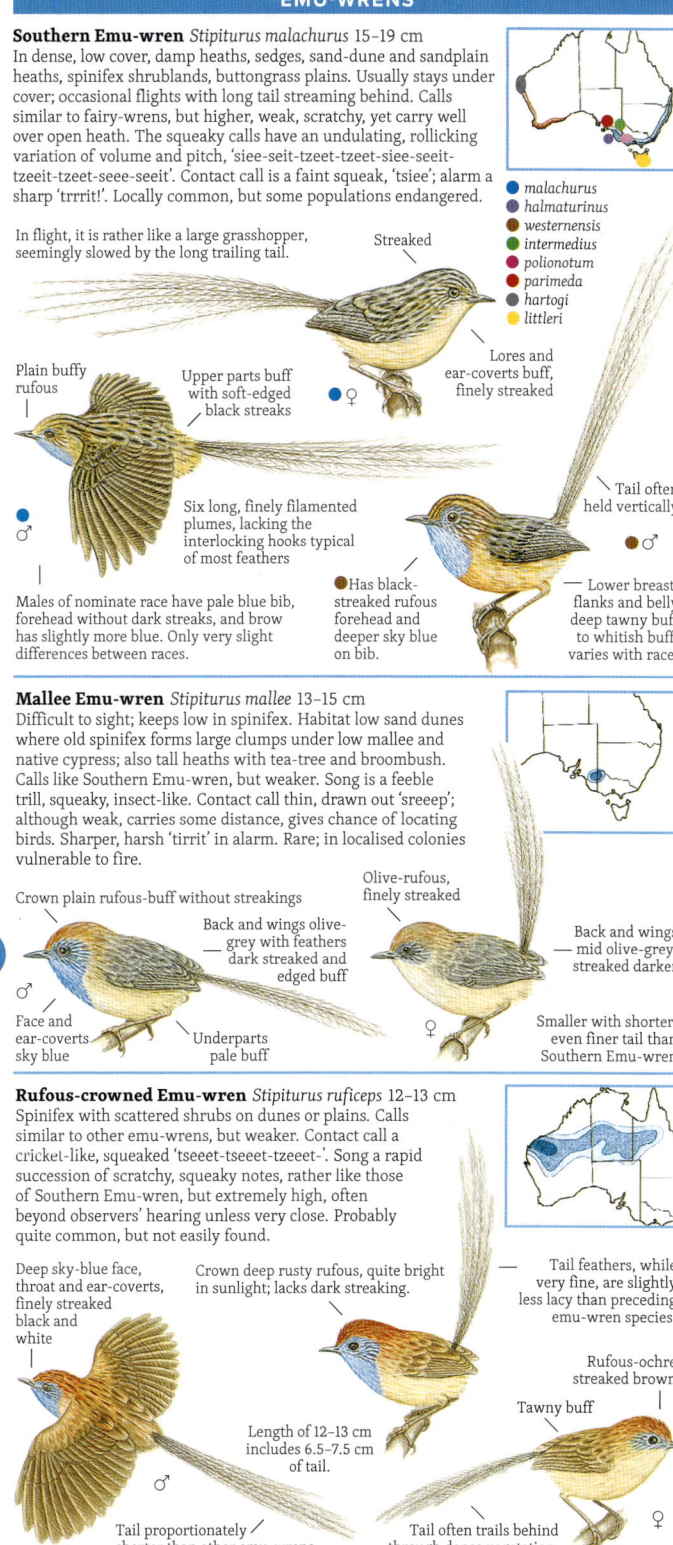

244

## CHESTNUT-SHOULDERED FAIRY-WRENS

**Lovely Fairy-wren** *Malurus amabilis* 13–14 cm
Distinctive, upright tail often briefly fanned, flicked, showing large white tips. Favours thickets of rainforest margins, monsoon, vine scrubs; forages short distances into closed forest and nearby woodland; also swamps, mangroves. Gives slightly descending, rippling trill. Contact call a high yet rather soft 'streee', given often to keep family party together; churring in alarm. Sedentary; common.

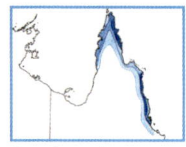

Several races in past, now considered but one, with slight, gradual northwards increase in brightness and purity of blues

**Blue-breasted Fairy-wren** *Malurus pulcherrimus* 14–15 cm
Dry and semi-arid woodlands, mallee, open sandplain heaths. Shy and secretive. Song begins with high, squeaky, thin notes leading in to several seconds of lower pitched, very rapid, hard, mechanical, vibrating trilling. More often heard are rattling, mechanical alarms, 'trrrt-trt-trtt-trt-trrrt'. Sparse; locally common in remnant habitat.

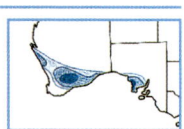

**Red-winged Fairy-wren** *Malurus elegans* 14–16 cm
Inhabits wetter SW of WA in undergrowth of wet karri forests, dense coastal heaths. Also in drier jarrah and wandoo forests in lush vegetation along creeks and around swamps. Call begins with several squeaky notes followed by rattling trill, quite unlike song of Splendid, the usual overlapping species, but not so different from call of Blue-breasted, which is usually in drier habitat.

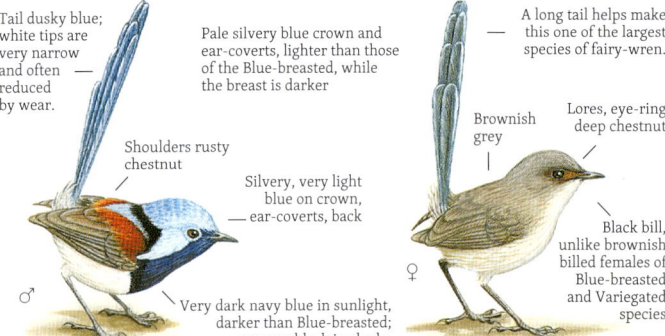

# VARIEGATED & PURPLE-BACKED FAIRY-WRENS

## Variegated Fairy-wren
*Malurus lamberti* 13–14 cm
Widespread chestnut-shouldered fairy-wren once considered a species with five subspecies until the Purple-backed (right) was split off with all other races. Along the east coast, *M. lamberti* inhabits forest undergrowth, coastal heath, dune vegetation, roadside and garden thickets. Often in pairs or families of up to five birds. Tends to be shy, keeping beneath undergrowth in presence of intruders. Song heard mainly in breeding season: a high, metallic, squeaky and rather clockwork or mechanical rattling trill at fairly uniform pitch, 'tririt-tirirrit-tirit-tirit-tirrririt-tirit-tirit'. A common call is a hard, staccato, erratic 'trrt-trtt, rrrrt-trrt-trt', strong rather than high or squeaky, and apparently as a warning. Varied contact calls during foraging; differ race to race – a rapid, sharp 'trrit' from nominate race; a thinner, squeakier, plaintive 'sreee'. Usually sedentary; moderately common.

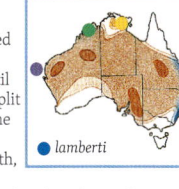

- ● *assimilis*
- ● *dulcis*
- ● *rogersi*
- ● *bernieri*
- ● *lamberti*

## Purple-backed Fairy-wren
*Malurus assimilis* 13–14 cm
A former race of the Variegated Fairy-wren but now considered a separate species under the nominate *assimilis* with four subspecies. Nominate and northern races use mulga, mallee and spinifex habitats. The northern, sometimes called 'Lavender-flanked', races *rogersi* and *dulcis* inhabit rugged country where spinifex and low shrubbery grows among boulders of ranges and gorges, and along pandanus and melaleuca thickets of watercourses. Race *bernieri* on Bernier Is., WA.

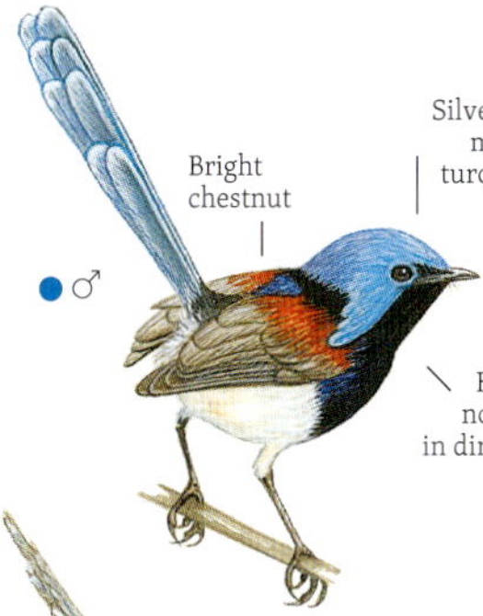

Bright chestnut
Silvery azure crown merges into light turquoise-blue ear-coverts.
Breast black, not dark blue, in direct sunlight
● ♂

Crown, mantle and back deep violet-cobalt; ear-coverts pale turquoise
Tail longer, fine white tips
Purple-backed race occurs in drier country with sheltering shrubbery.
● ♂

Tail grey-brown with faint blue tint
Grey-brown
Lores and eye-ring deep chestnut
Off-white
Females of Variegated and Purple-backed are similar.
● ♀

### MALE CROWN AND EAR-COVERT COMPARISONS

**Variegated Fairy-Wren & Purple-backed Fairy-Wren:** crown is blue-violet; ear-coverts are pale turquoise.

**Blue-breasted Fairy-wren:** ear-coverts are mid blue-violet like the crown.

**Red-winged Fairy-wren:** crown is paler silvery blue; ear-coverts are very light blue.

● Female **'Top End Lavender-flanked Fairy-wren'**
Upper parts dark blue-grey
Lores and eye-ring white
Tail has fine white tips.
● ♀

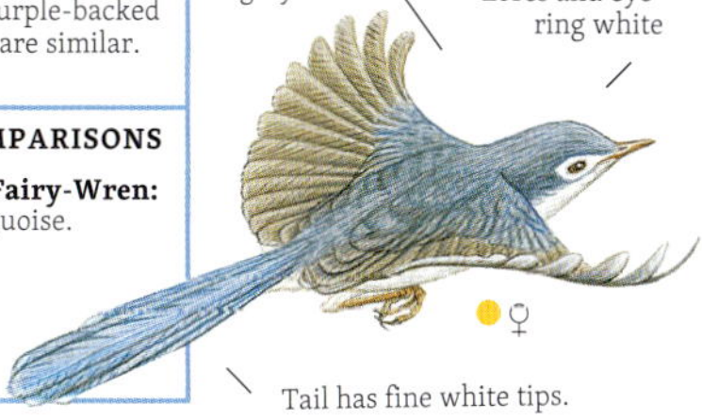

**'Kimberley Lavender-flanked Fairy-wren'** 
Wide white tips to tail
Chestnut scapulars
Crown deep purple-blue, ear-coverts lighter, brighter, blue
Breast black
Lavender tint along flanks
Males of the Top End Lavender-flanked, race *dulcis* have narrow white tips to tails, otherwise like race *rogersi*.
● ♂

Blue-grey; larger white tips
Lores and eye-ring light chestnut
Bill pinkish brown
● ♀

## BLUE-BLACK FAIRY-WRENS

**Superb Fairy-wren** *Malurus cyaneus* 13–14 cm
Prefers dense undergrowth of grass, bracken, shrubbery in forests, heaths, gardens, roadsides and inland watercourses. Often in family parties – a dominant male, one or several females, and brownish juveniles or immatures. Song is a vigorous trill, beginning squeakily, but quickly strengthening into a strong, downward cascade of louder, less sharp, musical notes. Foraging party maintains contact with weak, sharp 'trrit' or plaintive 'treee' from those falling behind; strident version in alarm, 'terrrrit!'. Common.

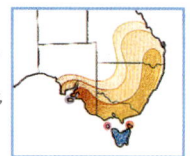

- cyaneus
- cyanochlamys
- samueli
- elizabethae
- ashbyi
- leggei

**Splendid Fairy-wren** *Malurus splendens* 13–14 cm
Varied habitats: race *splendens*, shrubby undergrowth, open forests, woodlands, mallee, mulga. Races *melanotus* and *musgravi*, both in semi-arid woodland, mallee, mulga, belar. Song a fast trill, the 'reeling' much less pronounced than in call of White-winged, but gives Splendid's song an undulating quality. Begins with very rapid, high, insect-like squeaks for a second or two, then lower, stronger or whistled notes, typically a reeling, rhythmic pattern: 'trit-triiit-trit-tirreet-tirreet-trit-tirreet-trit-tirreet'. Also high, vibrating trills, a loud, warning trill, frequent weak 'trrrip' contact squeaks and a plaintive, soft 'sreee' seeking contact. Sedentary; often locally common.

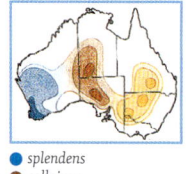

- splendens
- callainus
- melanotus
- emmottorum: extreme NW of range of melanotus

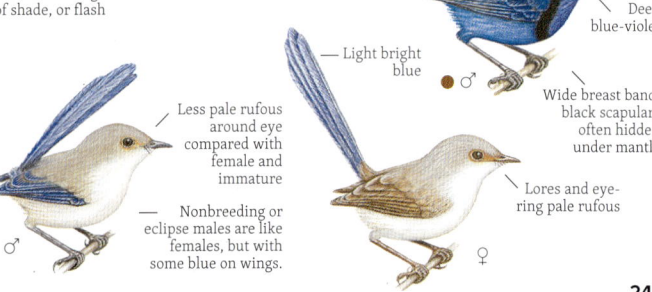

# PURPLE-CROWNED & WHITE-WINGED FAIRY-WRENS

**Purple-crowned Fairy-wren** *Malurus coronatus* 14–16 cm
Margins of rivers and creeks, pandanus and paperbark clumps with dense undercover of tall cane grass; occasionally in mangroves. Occur in pairs or small family parties; move quickly through their territory in short, low, direct flights; pause to hop about ground and low shrubbery after insects. Calls are distinctive, not the high and often weak trill of other fairy-wrens, but a strong, vigorous, reeling sequence with some almost harsh notes. Usually a vigorous, rollicking, scrub-wren-like 'zeepa-zeipa-zeipa' or 'cheipa-cheipa-'. Many variants: 'tcheip-tcheip-tcheip-' and 'tzwip-tzwip-tzwip-'. Moderately common in remnant suitable habitat; at many sites the essential water's-edge vegetation has been trampled, often destroyed, by cattle.

● *coronatus*
● *macgillivrayi*

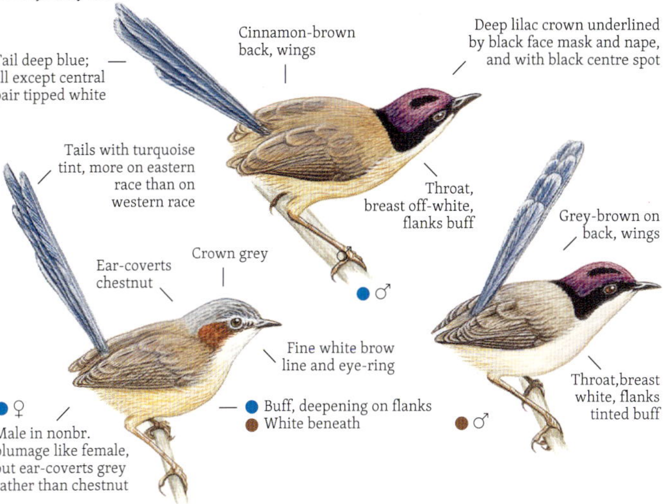

Tail deep blue; all except central pair tipped white

Cinnamon-brown back, wings

Deep lilac crown underlined by black face mask and nape, and with black centre spot

Tails with turquoise tint, more on eastern race than on western race

Throat, breast off-white, flanks buff

Grey-brown on back, wings

Ear-coverts chestnut

Crown grey

Fine white brow line and eye-ring

Throat, breast white, flanks tinted buff

● ♀
Male in nonbr. plumage like female, but ear-coverts grey rather than chestnut

● Buff, deepening on flanks
● White beneath

● ♂

---

**White-winged Fairy-wren** *Malurus leucopterus* 12–13 cm
On Dirk Hartog Is.; occurs in low scrub; on mainland, open plains of low saltbush, bluebush, spinifex, marshy margins of salt lakes, coastal heaths, cane grass. The black and white form was the first discovered, on Dirk Hartog Is. in 1818; mainland race *leuconotus* was described in 1865. Has an undulating, musical, reeling trill that carries some distance in open country and which has a metallic, mechanical quality to it, an uneven mechanical rotation, like a rapidly retrieving fishing reel or a sewing machine at fluctuating speed: 'chirr-IRR-irr-IRIRit-chirr-IRRirit-irirIRIRRIT-chirrir-IRRITirrit-', a pleasant, rollicking sound that is sustained and repeated. Also has an abrupt, high, weak, insect-like 'prrit!'. Locally nomadic; common.

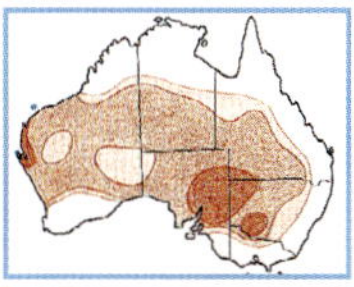

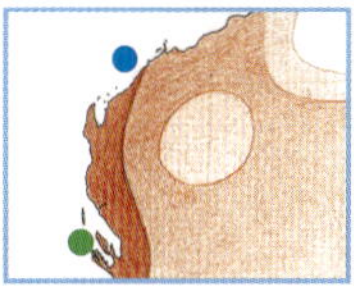

● Nominate: Dirk Hartog and Barrow Islands; glossy black in place of the blue of the mainland race
● Similar but tail shorter, female browner

● *leucopterus*
● *leuconotus*
● *edouardi*

Flickering wings of male, in distance, are like white butterflies.

Scapulars stark white set in cobalt blue

Plain faced, no obvious markings at eye or lores

Primaries brown; white spreads further out on secondaries of older males.

♀
Similar: female Red-backed Fairy-wren, but shorter brownish tail

Plumage of female almost plain light sandy brown; upper parts merge softly into off-white underparts.

♂

● Blue and white mainland race may be, or under some lighting conditions may appear to be, blackish or partly black.

240

# RED-BACKED FAIRY-WREN

**Red-backed Fairy-Wren** *Malurus melanocephalus* 10–13 cm
Habitat usually tall rank grasses with scattered trees and thickets of wattle or other shrubbery, coastal and inland, along watercourses; also in heaths, rainforest margins, swamp woodlands, spinifex. May be seen in pairs, small family groups or larger parties in which fully coloured males are greatly outnumbered by brownish birds, including females, immatures and males in eclipse plumage. In autumn, males moult to eclipse plumage similar to that of females; this is retained for five or six months. Older, dominant males moult direct to their new colourful breeding plumage, completely missing the dull eclipse stage. In common with males, females have no blue in plumage. Even the tail is without a trace of the blue that lightly tints the tails of most other fairy-wren females. Call begins with squeaks so high and weak that they would be heard only when very close, followed by a trill beginning with high squeaky notes, usually switching to lower, louder, rattling sounds: 'sreee-sreee-trichee-trichee-tchakka-tchakka-tchakka'. This is usually given by males in breeding season. Contact call within a foraging group is a soft 'tsiet'; the alarm call a louder, sharper 'trrrrit!'. Common; sedentary.

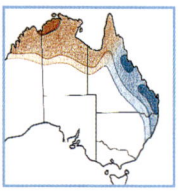

- 🔵 *melanocephalus*: nominate race, with orange-scarlet back
- 🔴 *cruentatus*: deeper crimson back

Tail square-tipped, relatively short, making this species the smallest fairy-wren in overall length

Back and mantle orange-scarlet

Patchy, scruffy male in transition to eclipse plumage

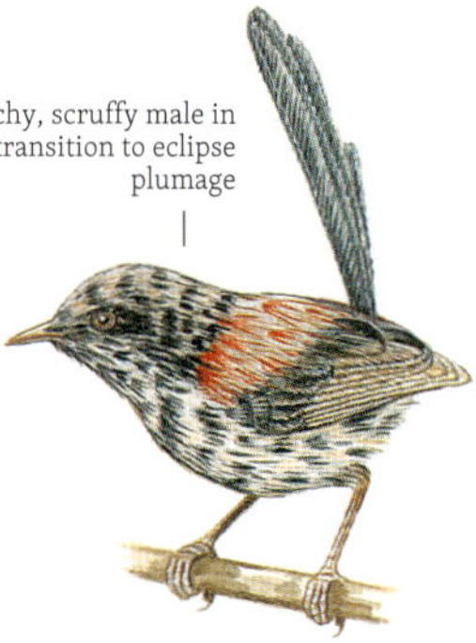

Predominantly black and red with some dark brown on wings

Back deeper crimson, tail even shorter

Crimson race occurs across tropical northern Australia.

In display, the male of this species fluffs out his crimson plumage in spectacular contrast against surrounding black, becoming a ball of fire when in flight or when bouncing between twigs and branches.

Tail pale brown like rest of upper parts, and relatively short. Females of otherwise similar White-winged Fairy-wren have longer, blue-grey tails.

Face plain; lacks identifying marks present on most other species.

Builds a nest of grass and bark strips hidden very low in a clump of rank grass or a dense low shrub; nests are placed higher, often in pandanus, in vicinity of watercourses.

# FAIRY-WRENS, EMU-WRENS, GRASSWRENS

The family Maluridae is divided into subfamilies: Malurinae (fairy-wrens and emu-wrens) and Amytornithinae (grasswrens). Most carry the long tail vertically, but the large grasswrens only rarely uplift it partly. Most flights are brief, low on relatively short wings, the long tail trailing. Members of this family occupy almost every habitat; most are insectivorous, but seeds are important for some grasswrens. The fairy-wrens are cooperative breeders, often with several nonbreeding adults or immatures helping the parent pair. Grasswrens, however, appear to have only the parents at the nest.

Scarlet Robin = 12 cm

## malurinae
### page 239

Red-backed Fairy-wren

Blue-breasted Fairy-wren

Purple-crowned Fairy-wren

Red-winged Fairy-wren

White-winged Fairy-wren

Superb Fairy-wren

Southern Emu-wren

Splendid Fairy-wren

Mallee Emu-wren

Variegated Fairy-wren

Rufous-crowned Emu-wren

Lovely Fairy-wren

## amytornithinae
### page 245

Black Grasswren

Short-tailed Grasswren

White-throated Grasswren

Dusky Grasswren

Carpentarian Grasswren

Kalkadoon Grasswren

Thick-billed Grasswren

Grey Grasswren

Western Grasswren

Eyrean Grasswren

Striated Grasswren

## TREECREEPERS

### Rufous Treecreeper *Climacteris rufus* 15–17 cm
Open woodlands and forest, including dense wet SW coastal karri, mid-west jarrah and wandoo, and inland through dry woodland to sparse mallee of semi-desert. Forages spiralling up tree trunks and limbs, using jerky two-footed jumps, then gliding down to base of next tree. Over greater distances uses low undulating flight; pale cinnamon wing bands conspicuous. Much time spent among fallen timber and often on open ground. Call is a sharp, ringing 'chip, chip, chip', widely and regularly spaced, 2–5 seconds apart. The trill is very high and sharp, descending and slowing, lasting 2 or 3 seconds. Moderately common.

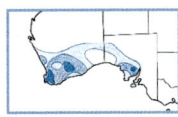

Colours on birds near W coast are more intense than on those of interior.

Most colourful of treecreepers, with much bright rufous in plumage

Rufous edges of primaries show.

Brow, face, underparts bright rufous

Male has black-edged pale streaks on breast.

Bright rufous-buff band across wing

♂

Barred coverts

Feet large, strong, enabling bird to hop up vertical surfaces

♀ Pale streaks

### Black-tailed Treecreeper *Climacteris melanurus* 17–20 cm
Tropical open forests, woodlands, spinifex with scattered shrubs, tree-lined rivers in arid regions. Through most of its range in the N, this is the only treecreeper. Spirals up trees; often feeds on the ground. Flight direct, only slight undulation; shows buff wing band. Calls with sharp, regular 'chip, chip, chip,' at intervals of about a second, similar to call of Rufous; also as a rapid descending trill, 'chip-chip-chipchipipipip'. Common; sedentary.

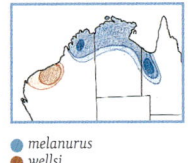

● *melanurus*
● *wellsi*

Largest treecreeper; darkness of upper parts rivalled only by a very dark race of the Brown Treecreeper, *melanota*

In flight, shows buffy white wing bar.

● Overall lighter in tone, and brighter rufous

♂

Under-tail-coverts barred

♀ White throat with rufous streaks
♂ Black throat with white streaks

More marked sexual dimorphism than in other treecreepers, with different colour and pattern at throat; both races

● ♂

● ♀

### Brown Treecreeper *Climacteris picumnus* 16–18 cm
Eucalypt forests and woodlands, scrubs of the drier areas, river-edge trees, timbered paddocks. Feeds spiralling up trees, and often on fallen timber or ground, where it tends to hop with head high. Flight swift, undulating, with glides that display colour of wing bar. Gregarious, often in family groups of 4 to 5, but a territory may be defended by just 1 or 2 birds. Call is a clear, high 'whit, whit, whit, whit'; each note has a strong, sharp finish. Common; sedentary.

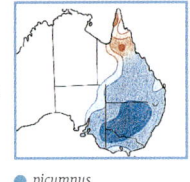

● *picumnus*
● *melanota*

Nominate race light to mid grey-brown. Buff edge of wing bands may be visible.

Long brow line, off-white, conspicuous

Ear coverts softly streaked

● Upper parts blackish-brown.

● ♂

Male: fine black lines

Female: fine rufous lines

Female has rufous streaks; male black.

Dark band across all except central pair

● ♀

Dark tail band not evident

● 'Black Treecreeper', NE Qld

# TREECREEPERS

**White-throated Treecreeper** *Cormobates leucophaea* 14 cm
Rainforest, eucalypt forest and woodland; river margins of drier regions. Also lower scrubs of banksia, mallee and bingalow. Searches bark vigorously, spiralling rapidly up trunks and branches, swooping down to begin low on next tree; rarely hunts on ground. Long flights are fast, direct, undulating. In pairs or family parties only in breeding season; else usually solitary. Has loud, ringing, clear, musical 'whit-whit-whit-whit-', rippling 'quit-quit-quit-' and more mellow, musical 'twiet-twiet-twiet-'. Also a loud, rippling trill. Sedentary, common.

- leucophaeus
- minor
- intermedius
- metastasis
- grisescens

**Red-browed Treecreeper** *Climacteris erythrops* 14–15 cm
Forested ranges with big trees; rainforest margins. Distinctive dark treecreeper with red brow, white throat. Favours tall, smooth-barked trees of wet eucalypt forests; forages among long strips of bark. Flight swift, undulating; shows grey-buff wing band. Gregarious, usually in small family parties that appear to contain only one female. Call is a series of squeaks at very high pitch: weak, thin, iep, iep, iep. Also a fast, scratchy trill; begins extremely high and squeaky and ends with several stronger, harsher squeaks: 'iep-ip-ip-ip-i-i-i-i-ip--chzip-chziep-chziep'. In the distance sounds more like the trill of a fairy-wren than of a treecreeper. Uncommon; sedentary.

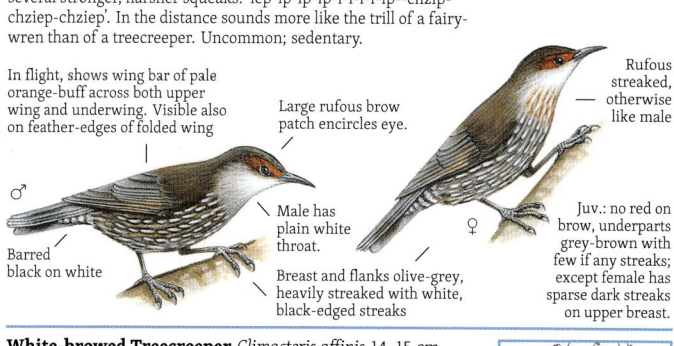

**White-browed Treecreeper** *Climacteris affinis* 14–15 cm
Semi-arid interior: mulga, mallee, desert oak, callitris; spinifex with scattered trees; trees along watercourses. Spirals upwards on tree trunks and limbs, searching crevices and hollows for insects; also searches bark and crevices of fallen logs, forages on ground. Usually in pairs or small family parties. Calls sharp, abrupt, very metallic and penetrating: 'chrip, chrip, chrip' and also a double noted 'chip-chip, chip-chip'. Has also a rapid trill, a ripple of sharp, metallic sound lasting several seconds. Sedentary; uncommon.

- affinis
- superciliosa

## BLUE-WINGED PITTA, SCRUB-BIRDS

**Blue-winged Pitta** *Pitta moluccensis* 18–21 cm
In SE Asia, inhabits rainforests, mangroves; in N Aust., recorded W Kimberley and drier Pilbara. Call similar to Noisy Pitta, but more husky, slightly lower, a scratchy 'quaraerk, quaraerk'. Calls have also been described as 'taew-laew', 'pu-wi-u, pu-wi-u' and 'pu-wiu'; also has a loud 'skieew!' alarm call. Rare vagrant or occasional visitor.

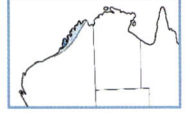

**Rufous Scrub-bird** *Atrichornis rufescens* 17–19 cm
Famed for powerful, varied calls. Secretive, using areas of dense groundcover where sighting is difficult, even when close. This species is confined to small parts of coastal ranges, in temperate rainforest and adjacent eucalypt forest, 600–1000 m altitude. Fossicks for insects and prey in forest litter. Powerful ringing calls carry far. At close range, extremely loud – 'cheip, cheip, cheip, cheip', the 'cheip' with the central 'ei' strongly emphasised. Also a more abrupt, equally loud, 'chep-chep-chep-', a sharper 'chwiep-chwiep-chwiep' and a whip-cracked 'chweeip-chweeip-'. Fragmented populations on high ranges; uncommon.

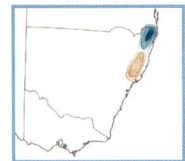

● *rufescens*
● *ferrieri*

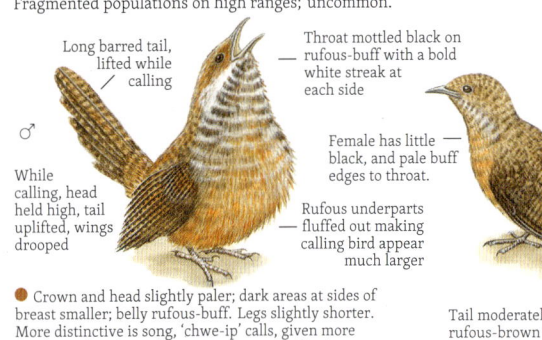

● Crown and head slightly paler; dark areas at sides of breast smaller; belly rufous-buff. Legs slightly shorter. More distinctive is song, 'chwe-ip' calls, given more slowly, and with much stronger upward inflection.

**Noisy Scrub-bird** *Atrichornis clamosus* 21–23 cm
One of Australia's rarest birds. In dense heath, rushes, tall sedges, under stunted scrubby trees of gullies in coastal hills. Other populations in similar dense thickets around coastal sandplain swamps. Powerful calls, but elusive, difficult to sight even when close. Gives a series of notes, initially slow, sharp, becoming lower, richer, faster, increasingly powerful, almost painfully loud if the bird comes close under low cover: 'chwiep, chwiep, chwip-CHWIP-CHWIP-CHIWIEP-chwp-chwp'. Rare, restricted habitat; reintroduced to other S-coast sites where new populations are establishing well.

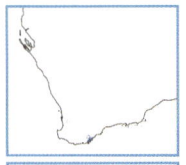

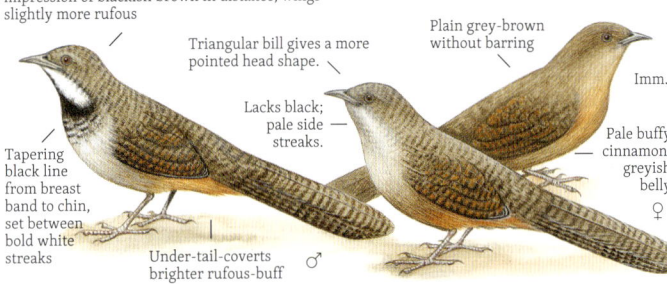

## PITTAS

**Papuan Pitta** *Erythropitta macklotii* 17 19 cm
Rainforest, monsoon forest. Rakes vigorously through forest floor litter for insects and other small prey. Presence usually first revealed by calls, beginning low, rising steadily; mellow and tremulous at first, finishing quite high and clear. The second note follows at slightly lower pitch and volume, softer and more mellow, 'gr-r-r-r-rork, groir'. Calls both from ground and higher perches. A breeding summer migrant from NG, Nov.–Apr.; probably uncommon.

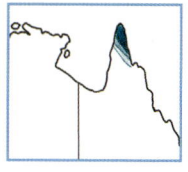

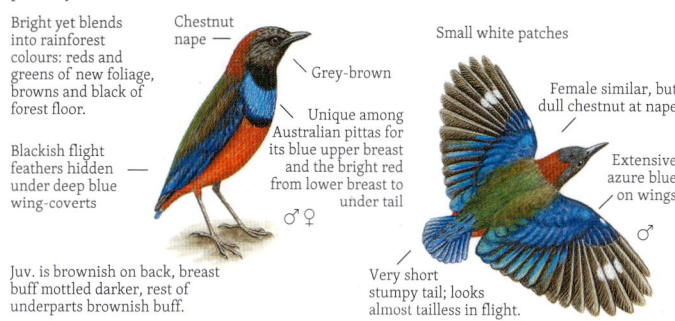

**Rainbow Pitta** *Pitta iris* 16–19 cm
An inhabitant of monsoon rainforests or scrubs, vine scrubs, thickets of cane-like native bamboo. The one Aust. pitta that does not also occur in NG or other islands to the N. Forages in rainforest leaf litter, scratching about with long legs and large, strong feet; takes invertebrates from ground debris, and snails that are broken against an 'anvil' rock. A soft 'kwe', followed by a call similar to that of the Noisy, but somewhat sharper: 'kwe, KWEEIK-a-kweek'. Sedentary; moderately common in suitable habitat.

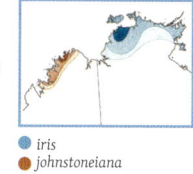

- *iris*
- *johnstoneiana*

**Noisy Pitta** *Pitta versicolor* 18–20 cm
Tropical and subtropical rainforests, monsoon and dense eucalypt forests. Forages in litter of forest floor; large snails commonly taken, these broken using a rock as anvil. Shy, usually silent except in breeding season, when calls almost incessantly. Has loud, far-carrying, double noted, mellow whistles, 'quarreek-quarrik', commonly described as 'walk-to-work'; calls from ground and from high in forest canopy. Also an occasional single, harsh, mournful 'kieow'. Common.

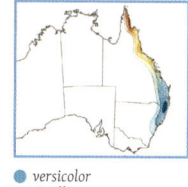

- *versicolor*
- *simillima*
- *intermedia*

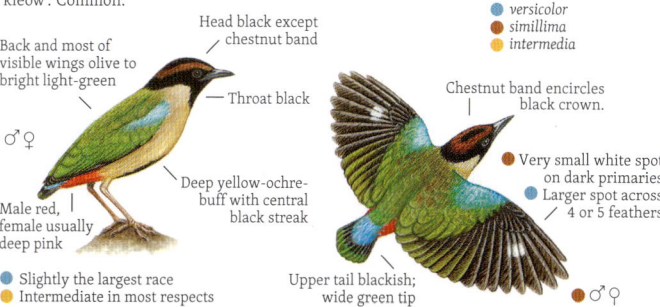

# LYREBIRDS

**Albert's Lyrebird** *Menura alberti* ♂ 90 cm; ♀ 75 cm
Rugged ranges with subtropical rainforest and dense undergrowth, where it forages by scratching about in leaf litter, rotting logs and loose topsoil, finding various small prey. Displays are similar to those of the Superb, but the shorter tail filaments do not envelop the bird so completely. Powerful voice, extremely varied calls and song, a mixture of own notes and mimicry, the latter restricted to fewer species. In alarm gives a shrill shriek. Even when calling strongly, this shy and elusive species is not easily sighted in the dense tangled vegetation of its habitat. Locally quite common, but only within parts of a very restricted range.

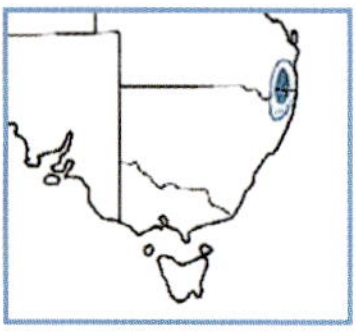

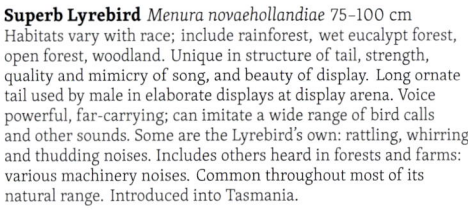

The veil of tail plumes is dense compared with the Superb Lyrebird's open tracery. The central long pair of ribbons, gracefully curving, are set among many bushy, filamentous plumes.

Straight or curved and twisted in cramped nest

Rufous throat and fore-neck

♀

The Albert's Lyrebird has a much warmer, overall rufous-brown colour than the grey-brown Superb Lyrebird.

The outermost pair, the lyrates, are broader, shorter, lack notched pattern.

Lower breast creamy buff, becoming deeper buff on belly, flanks and under-tail-coverts.

Rump rufous

Inner pair long, slender, ribbon-like; tend to curve in, cross over; filamentous plumes rather bushy

♂

---

**Superb Lyrebird** *Menura novaehollandiae* 75–100 cm
Habitats vary with race; include rainforest, wet eucalypt forest, open forest, woodland. Unique in structure of tail, strength, quality and mimicry of song, and beauty of display. Long ornate tail used by male in elaborate displays at display arena. Voice powerful, far-carrying; can imitate a wide range of bird calls and other sounds. Some are the Lyrebird's own: rattling, whirring and thudding noises. Includes others heard in forests and farms: various machinery noises. Common throughout most of its natural range. Introduced into Tasmania.

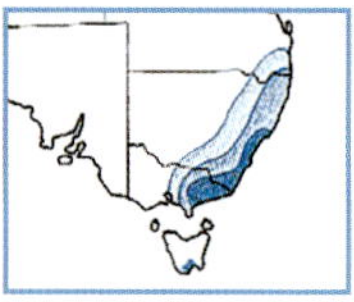

- ● *novaehollandiae*
- ● *edwardi*: in N
- ● *victoriae*: in S and Tas.

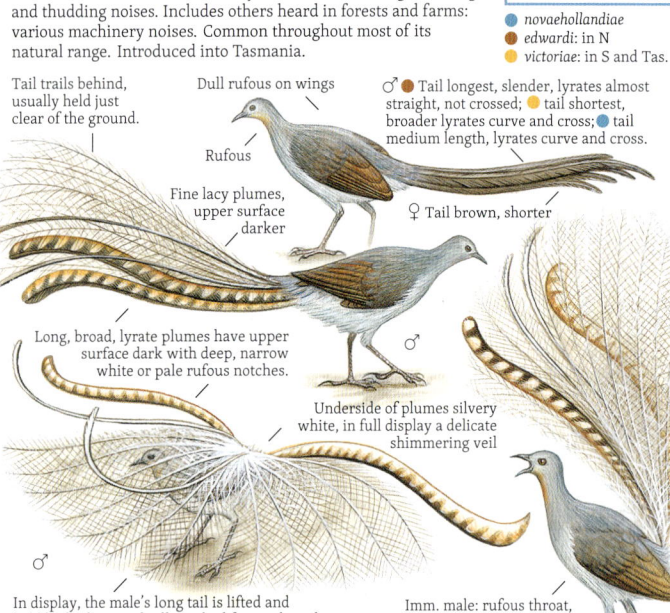

Tail trails behind, usually held just clear of the ground.

Dull rufous on wings

Rufous

Fine lacy plumes, upper surface darker

♂ ● Tail longest, slender, lyrates almost straight, not crossed; ● tail shortest, broader lyrates curve and cross; ● tail medium length, lyrates curve and cross.

♀ Tail brown, shorter

Long, broad, lyrate plumes have upper surface dark with deep, narrow white or pale rufous notches.

Underside of plumes silvery white, in full display a delicate shimmering veil

♂

In display, the male's long tail is lifted and held high, then gradually arched forward until, at the climax of the display, he is enveloped in a shimmering, silvery, lace-like veil, all the while pouring forth his loud and rich song.

Imm. male: rufous throat, usually lost by fourth year; the tail plumes take about six years to reach maximum length.

Imm. ♂

# PITTAS, LYREBIRDS

This family group begins the very large songbird order, the Passeriformes, whose families make up slightly more than half of Australia's bird species, and of its land birds a considerably greater proportion. Beginning the Passeriformes are the brilliantly coloured birds of the family Pittidae. These ground-feeding songbirds are widely distributed through tropical rainforests from Africa to China and through SE Asia to Australia, where four species occur. The two species of scrub-birds, family Atrichornithidae, are unique to Australia, as are the two species of lyrebirds, family Menuridae. Completing this family group are the treecreepers, Climacteridae, with six Australian species.

Scarlet Robin = 12 cm

Birds of Groups 15 to 26 belong to the 'songbird' order, the Passeriformes.

## pittidae
page 234

Papuan Pitta

Blue-winged Pitta

Noisy Pitta

Rainbow Pitta

## atrichornithidae
page 235

Rufous Scrub-bird

Noisy Scrub-bird

## menuridae
page 233

Albert's Lyrebird

Superb Lyrebird

## climacteridae
page 236

White-throated Treecreeper

Red-browed Treecreeper

White-browed Treecreeper

Rufous Treecreeper

Black-tailed Treecreeper

Brown Treecreeper

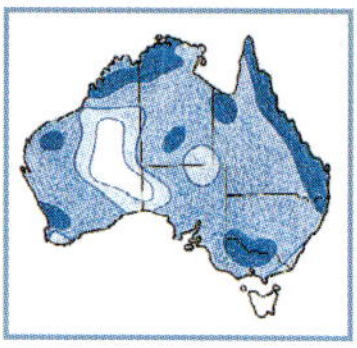

## BEE-EATER & ORIENTAL DOLLARBIRD

**Rainbow Bee-eater** *Merops ornatus* 23–27 cm
Orange wings flash against blue sky as a Rainbow Bee-eater twists and turns in pursuit of fast-flying insects, and then glides back to its perch to beat the captured bee or dragonfly against the wood. Finds open airspace in and over woodland, open forest, semi-arid scrub, grassland, clearings in heavier forest, farmland. Avoids heavy forest that would hinder its aerial pursuit of insects. In breeding season, requires also an open clearing or paddock with loamy soil soft enough for tunnelling, yet firm enough to support the net tunnel. The plumage combines delicate pastel tints with strong, bright hues; wings are long and pointed, tail usually has long central streamers. Calls are loud, clear, sharp, cheerily vibrating, often given in flight, carrying far and attracting attention: 'pirr-pirrrp-pirr-pirrr-pirrp-pirr-pirrr-pirr-pirrp-'. Perched birds exchange a similar, softer, slower call with less vibrancy, 'quiera-quiera-quirea-'. Common; regular summer migrant to southern Aust., Sep.–Apr. In the N, they are resident and remain to breed.

Orange underwings catch the sun; against cool background, like blue sky or green trees, the wings seem to give out a fiery inner glow.

Bright red eye set in long, wide, black eye stripe.

Juv.: crown greener
Eye brown
Bill long, evenly tapered, only slightly downcurved
Orange of crown blends softly downwards into yellow-green of body.
Black edge
Light bright green merges into bright pale blue of under-tail-coverts.
Black crescent
Male has long, fine tail streamers with slightly clubbed tips. ♂
Juv.
Juv. tail lacks tail streamers of adults.
Female has shorter, thicker streamers. ♀

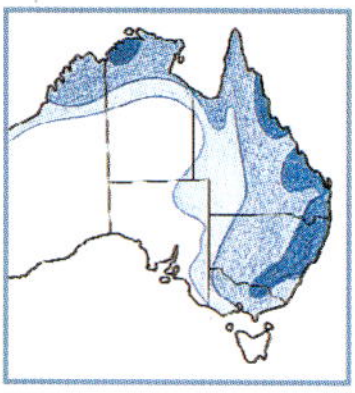

**Oriental Dollarbird** *Eurystomus orientalis* 25–29 cm
A regular breeding migrant to Australia. Conspicuous and noisy early in breeding season with its diving, rolling display flights. Hunts from high bare limbs; takes prey in flight in powerful pursuits and acrobatic aerial manoeuvring, snapping up large flying insects. Call is a rapid cackle – an abrupt, grinding, harsh, metallic, jarring, buzzing, rattle, 'kzak-kzak-kzak-kza-kzak-' or 'kak-kak-kak-'. Sometimes it gives similar, slightly slower, stronger 'kzaak!' calls at wide erratic intervals. Other calls include a deeper, slower, growling 'kzarrk-' and a higher, raucous 'kraak-kraak-'. Habitats diverse: woodland, edges of heavier forest, open country with widely scattered trees, inland watercourse trees, farmland, suburbs. Summer breeding migrant to Aust.; common in N, sparse in drier and SE regions.

Deep, relaxed, strong wing-beats and glides display bold rounded crescents of white near wing tips.

Dark brown; may appear almost black.
Throat blue-violet
Brown of back merges into deep viridian on wings.
Bill fleshy pink to red, tipped black
Flight feathers deep violet-blue
Colours of adult female are very slightly softer than those of male.
Large white wing patches give name 'dollar' bird.
Juv. has dull colours, brownish bill and feet, lacks blue at throat.
Dark tipped

231

# KOOKABURRAS

**Blue-winged Kookaburra** *Dacelo leachii* 38–42 cm
More colourful than Laughing Kookaburra but has rather unpleasant, staring white eye. Uses open forest, semi-arid woodland, tropical woodlands, tree-lined rivers. Small family parties, up to 10 or so birds. Calls described as maniacal, demonic. Call is a raucous cacophony, begins slowly, builds louder: 'kok-kok-kok-QUOR-R-RORK-QUOR-R-RORK-QUOR-R-RORK-car-r-aark-craark-crak-qwok-quok'. One bird starts calling, followed by others in unsynchronised fashion, building up, rising and falling in cacophony. Only similar bird is the Laughing Kookaburra, but less blue in wings, dark eye streak, pleasant laughter. Sedentary, or local movements; moderately common.

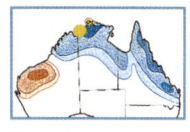

- *leachii*, including darker former race *kempi*
- *cervina*
- *occidentalis*

Eye white
Head off-white, streaked dark brown
Massive bill, dark above, pale creamy pink beneath
Off-white to buff, variable fine, broken, rufous barring or scalloping
Extensive bright blue along wings
Bright pale blue
Males, both races, have blue tails.
Females, both races, have red tails.

Lighter streaking, head almost white
Brown, pale-tipped
Paler buff, more lightly barred
Juv.: similar to adult female, but heavier streaking on crown
Similar, but head darker

**Laughing Kookaburra** *Dacelo novaeguineae* 40–47 cm
Occupies diverse habitats: open forest, woodland, partly cleared farmland, watercourse trees of semi-arid inland, parks and gardens. The call is a far more pleasant, jovial 'laughter' than that of the Blue-winged; merry chuckling rising to raucous laughter, fading away again to a slow throaty chuckle: 'chok, kok-kak-KAK-KAK--KAK-KOK-KAK-KOK-kook-kook-kok, kok, kok', often others joining in. Sedentary; common.

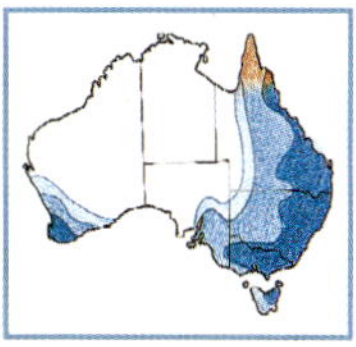

Both species have white bar at base of primaries.
Brown, feathers pale-edged; white margins more evident on juveniles

- *novaeguineae*
- *minor*: considerably smaller, darker crown

Massive deep bill with pale underside
Tips of wing-coverts very pale blue or often almost white
Feet relatively small, weak
Both sexes have rufous-barred black tail, tipped and edged white.
Blue in coverts
Flight feathers grey-brown

## YELLOW-BILLED & FOREST KINGFISHERS

### Yellow-billed Kingfisher *Syma torotoro* 18–20 cm

Although brightly coloured, can be difficult to sight; sits still in rainforest, merging into foliage where orange and red of deciduous leaves and new growth are spread through the greenery. Habitat is tropical rainforest, monsoon forest, mangroves. Favours rainforest margins; often hunts and nests in adjoining tropical woodland. The unique calls are most likely to reveal its presence, but seem to echo through the rainforest so that the direction of their origin can be hard to judge. Sits very still, often quite low; occasionally drops to ground. Call is a loud, ringing, musical trill, at first quite rapid, but rising steeply in pitch and tempo up to a shrill, undulating, vibrating whistle, finally fading away. Usually soon repeated, often with increasing intensity. Scolds and screeches when defending nest site. Sedentary; common within the very restricted range.

Nominate in NG; Aust. race is *flavirostris*.

Head rich rufous-buff; has black nape spots, from behind look like eyes.

Female has black centre to crown.

Wing-coverts blue-green to olive-green

Flight feathers grey-black to blackish olive-brown

Rump light bright blue

Black eye-ring

Top of outer half of bill is usually black. Cutting edge is finely serrated along outer part (both sexes).

Buffy white

♂

♀

Deep blue

Juv. has white patch on nape between the two black spots; bill black, becomes mostly yellow within first year.

Some birds have dark upper surface to outer bill, others have entirely yellow bills.

In flight, under-wing rufous-buff

### Forest Kingfisher *Todiramphus macleayii* 17–23 cm

Wide range of habitats: open forest, woodland; favours vicinity of rivers, margins of swamps, billabongs; also mangroves, farmlands. Hunts from exposed perches in fairly open country; dives to ground or shallow water to take its prey; conspicuous, noisy in breeding season. Nominate race *macleayii* found only in NT; race *incinctus* from Cape York to SE Aust., a third in NG. Call is a sharp, piercing, very rapid 'kik-kik-kik-kikkikikik-' in long sequences; much faster rate than quite similar call of the Sacred Kingfisher. Most alike are Little Kingfisher, but far smaller; Sacred and Torresian but greener; none has white wing spot. Nominate race *macleayii* sedentary; common. Eastern *incinctus* part migratory – some winter in NG, at which time absent from SE Aust. and less common in NE.

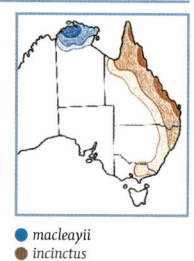

● *macleayii*
● *incinctus*

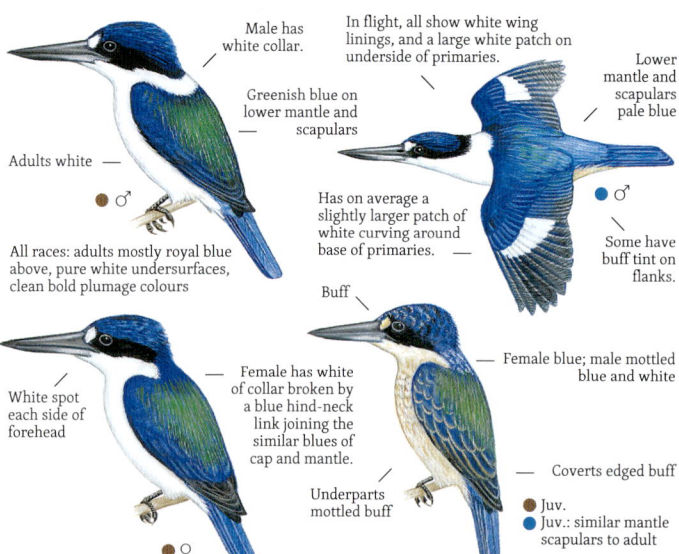

Male has white collar.

In flight, all show white wing linings, and a large white patch on underside of primaries.

Lower mantle and scapulars pale blue

Greenish blue on lower mantle and scapulars

Adults white

Has on average a slightly larger patch of white curving around base of primaries.

Some have buff tint on flanks.

All races: adults mostly royal blue above, pure white undersurfaces, clean bold plumage colours

Buff

White spot each side of forehead

Female has white of collar broken by a blue hind-neck link joining the similar blues of cap and mantle.

Female blue; male mottled blue and white

Underparts mottled buff

Coverts edged buff

Juv.

Juv.: similar mantle scapulars to adult

## RED-BACKED, SACRED & TORRESIAN KINGFISHERS

**Red-backed Kingfisher** *Todiramphus pyrrhopygius* 20–24 cm
Usually occurs throughout drier inland regions in semi-arid woodland, mulga and mallee scrubland, spinifex and other almost treeless country, often far from water. Hunts from perch in open; drops to ground to take small reptiles, occasionally mouse-sized mammals, large insects. Call is a mournful 'kee-ip', the last '-ip' ending abruptly, given in a long, slow series at intervals of several seconds: 'kweee-ip, kweee-ip, kweee-ip'. Also a slightly deeper, nasal 'quaeirr-ip, quaeirr-ip', which has an even more melancholic quality than the higher call. Moderately common; nomadic and migratory.

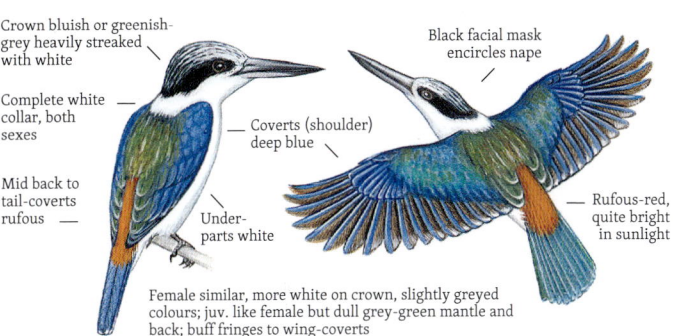

**Sacred Kingfisher** *Todiramphus sanctus* 20–23 cm
Well-known, widespread, common kingfisher; migrates into southern Aust. for summer breeding, attracting attention with persistent calling. Uses open forest, woodland, semi-arid scrubland, mangroves. Hunts mostly on dry land for small reptiles and large insects; occasionally uses wetlands. Early in breeding season, a loud, carrying 'kik-kik-kik-kik-', usually of 3 to 5 notes, occasionally 'kek-kek-kek-kek'; these calls repeated almost incessantly through day. Between a pair near the nest, a clear, shrill 'kieer-kieer-kieer-'; on arrival at nest tree, 'kek-krreee-krreee-kreee'. In defending nest site, dives with loud harsh scream. Some sedentary in N, others summer migrants to S.

Nominate race in Aust., other races Norfolk and Lord Howe Is.

**Torresian Kingfisher**
*Todiramphus sordidus* 24–28 cm
Large, stocky, heavy billed; hunts over low-tide mud and shallow pools. Calls are like those of the Sacred Kingfisher, but slower, stronger, at greater intervals: 'kik-, kik-, kik-', perhaps closer to 'kiek' than the quick 'kik' of the Sacred; also a rattling, piercing 'kr-r-riek-kr-r-riek-'. Inhabits coastal mangroves, favouring mangrove creeks and along seaward side of the mangrove belts, but also hunts out over nearby mudflats and beaches. Sedentary in N, migratory in S of range; quite common.

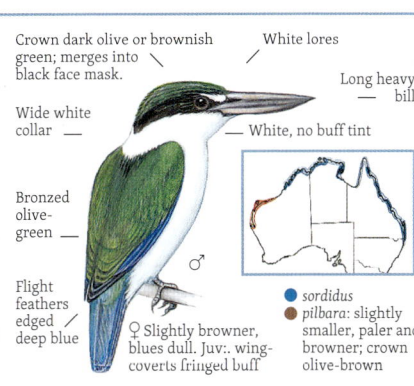

● *sordidus*
● *pilbara*: slightly smaller, paler and browner; crown olive-brown

## PARADISE-KINGFISHERS

**Buff-breasted Paradise-Kingfisher** *Tanysiptera sylvia* 33 cm
Keeps to the mid level and lower canopy of the rainforest, where it
tends to perch in dense foliage, remaining almost motionless and
silent. Unless it moves, dropping to the ground to take some small
lizard or frog, or calls, its presence might easily remain undetected.
Its habit of occasionally flicking the tail upwards while perched helps
to reveal its presence. It is best seen soon after its arrival from NG,
early in Nov., when it is in its brightest plumage, the long tail plumes
not yet stained or damaged by nest digging. Its frequent territorial
calls at that time also make it easier to find; later, with eggs or young
in the nest, the birds are more secretive. The call is a rapid succession of notes, not high
pitched, but rather metallic with a touch of harshness. The call carries far, resonating through
the rainforest: 'karow-karow-karow-karow-' or 'charow-tcharow-'; the 'ow' slightly uplifting,
stronger, higher and emphasised. Each call is rapid, the usual series of six to ten calls taking
only 3 or 4 seconds. It also gives a mellow, musical trill that begins strongly and continues for
3 or 4 seconds, fading away in pitch and volume. The alarm, if the nest is disturbed, is a high
screech. Habitat is rainforest of coastal lowland and lower slopes of ranges, riverine thickets
and tropical gardens. Prefers rainforest with open lower levels, being usually a terrestrial
hunter. Locally common in restricted range and habitat; breeding migrant Nov. to Mar.

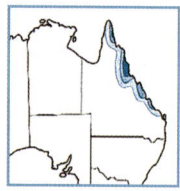

Deep blue crown, black through eye to upper back

Bill red, long and thick towards base

Juv.: underparts dull ochre-buff, shoulders mottled ochre

White back

Bill brown

Heavy red bill

Centre of the tail is white, joining white rump and back.

Feet red

Entirely white underparts make the **Common Paradise-Kingfisher** unlikely to be confused with the Buff-breasted Paradise Kingfisher.

Tail is initially very short, dull blue with only central shafts white.

All tail feathers are blue, tipped white.

The tail, with two greatly elongated central feathers, makes up about half the total length of the Buff-breasted Paradise-Kingfisher. The outer feathers of the tail are deep royal blue, and the central pair white.

Buff-breasted has only a slight enlargement of the tail tip; Common has large spatulate tips to the tail.

In flight through the rainforest, this bird is spectacular, its long, white tail plumes streaming behind, showing orange-buff under-surfaces of wings and body.

**Common Paradise-Kingfisher** *Tanysiptera galatea* 33–43 cm
Common in lowland rainforests of NG. Possible vagrant to NE Qld or Australian islands
of Torres Str.; one doubtful record. Calls are mournful notes, leading into a rapid, loud trill.

## AZURE & LITTLE KINGFISHERS

**Azure Kingfisher** *Ceyx azureus* 17–19 cm
Named for the azure blue that dominates the plumage – in sunlight a deep blue with violet sheen. Often overlooked unless a high squeak draws attention as the kingfisher darts across a river pool, skimming low across the water, a momentary glimpse of blue vanishing around a river bend or into shadowy over-hanging vegetation. When perched, can be hard to detect, but regularly lifts tail, occasionally bobs head. Most likely to draw attention by dropping to the water with a small splash to take a yabby or small fish. Often silent, but frequent sharp squeaking when breeding. The high, thin 'eeeet' given in flight is easily audible if bird darts past very close, but, at any distance, needs nearly perfect high-frequency hearing. Similar are the Little and Forest Kingfishers, but both have white underparts. Usual habitat is the well-vegetated banks of creeks, swamps, lakes and mangroves. Common where habitat remains suitable, mainly in far N, otherwise uncommon; most sedentary, some migratory.

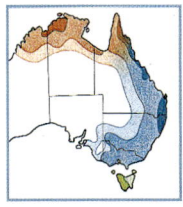

● *azurea*
● *ruficollaris*
● *diemenensis*

Small kingfisher with large, heavy bill; rufous-buff underparts

● The largest, heaviest race also has the darkest violet-blue crown.

Buff-white spot

Buffy white

Lighter colours than other races, blue slightly less violet, underparts paler rufous or orange-buff.

White streak

Deeper rufous-buff and with more extensive blue-violet along full length of sides of breast and flanks.

● Is smallest race, but has proportionately the largest bill. Blue of upper parts more violet tinted or azure.

Juv. has shorter, pale tipped bill, light scalloping on crown.

Feet red; only two forward toes

**Little Kingfisher** *Ceyx pusillus* 12–13 cm
Even smaller than Azure Kingfisher. Not often seen. Sits quietly, motionless except for an occasional bob of the head, on perches low over water. Dives to take small fish, crustaceans, water beetles and other aquatic creatures. The flight is fast and direct, skimming the surface, to a perch rarely more than several metres above the water. It is unlikely to be seen over land. Gives an extremely high-pitched squeak, 'tzeeet' or 'szeeit', often in flight; sometimes a rapid, pipping 'tseet-tseet-tseet-tseet'. These calls are even higher pitched and weaker than those of the Azure, and probably inaudible to many observers unless the bird is close. Inhabits densely vegetated channels of mangroves, swamps, creeks, rainforest streams; occurs on coastal lowlands and altitudes to 750 m in mountain rainforests. Generally uncommon, but probably more common in remote northern rainforests and mangroves; sedentary.

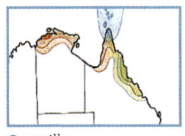

● *pusilla*
● *ramsayi*
● *halli*

● Nominate of this species occurs NG and S to islands of Torres Str.; also possibly on NE tip of Cape York. Its blue is more violet, quite bright. Breast band only narrowly broken at centre.

Neatly divided: deep royal or cerulean blue above, clean white beneath – quite unlike the much warmer violet-blue and rufous-buff combination of the Azure Kingfisher

Bill long, black, more slender than that of the Azure

Deep royal blue

White spot

Has only the very slight beginning of breast band

White streak

This race has blue intrusions onto each side of breast that are larger, closer to meeting in centre as a breast band; even more marked on the juveniles.

Slightly brighter cerulean blue

Juv.: blue is dusky, the breast is lightly scalloped with brown.

Feet dark brown

## LARGE SWIFTS

**Fork-tailed Swift** *Apus pacificus* 17–18 cm
A medium-large swift with white throat and rump, and deeply forked tail that forms a single fine point when fully closed, or spreads into a wide 'V' shape in slow aerial manoeuvres. Occupies low to very high airspace over varied habitat, rainforest to semi-desert; most active just ahead of summer storm fronts. In flocks, rarely one or two individuals. Hawks for insects, at times very high, and sometimes in mixed flocks with the larger White-throated Needletail. Often, when hawking for food, flocks make circling sweeps, gathering flying insects in wide-gaping bills; at times scoop water on the wing from inland lakes or small pools. Flight buoyant, more erratic and fluttering than the Needletail; stays on the wing day and night, sleeping in high, circling flocks. Voice is a high pitched buzzing, lorikeet-like, and shrill, excited twittering. Summer migrant, Oct.–Apr.

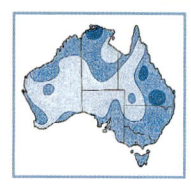

Long, sickle-shaped, pointed wings

Large swift with long, back-swept wings, forked or tapering tail

Long, tapered body

Black or brownish black in worn plumage

Dull white rump, quite bright under direct sunlight

White throat

Long tail is usually narrowly forked, but may be closed to a long fine point or spread to a wide, deep fork.

When moulting, loss of outer feathers gives square-cut tail shape; alters flight jiz.

Many variations in wing and tail positions and shapes as the swifts soar, glide and engage in twisting, sharply turning, pursuit of flying insects

Underbody scalloped but not visible unless close

---

**White-throated Needletail** *Hirundapus caudacutus* 19–21 cm
Australia's largest swift. A rather heavy looking body merges to a broad, short, square-cut tail; long winged, powerful. Uses high open air spaces above almost any habitat, including oceans; at times gathers over ranges, headlands, often in humid, unsettled weather preceding thunderstorms. Gives sharp, musical, uneven twittering with high to very high squeaks, very rapid, some clear, some scratchy: 'chi-chiet-chit-chi-chit-chiet-chi-'. Breeds in northern Asia, migrates S to reach Aust. early Oct., and the SE corner by Dec.; returns from Mar. with most gone by end of Apr.; moderately common, locally common E coast and ranges.

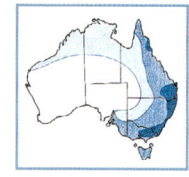

Upper parts greenish-black; centre of back paler brown

White

White throat

White inner edge to tertials

White under-tail-coverts extend forward onto the flanks.

Rear body is thickset rather than tapering towards tail.

Short, broad tail is tipped with protruding spines, but these are not usually visible at distance.

Tail usually with square-cut tip

One of the fastest birds, but also makes slower aerial manoeuvres, wings forward and broadly flared. In steep, fast dive, wings curve back.

## UNIFORM SWIFTLET & HOUSE SWIFT

**Uniform Swiftlet** *Aerodramus vanikorensis* 11–12 cm
As the name indicates, almost uniform in its dusky dark colour. Similar in shape and only slightly larger than the resident Australian Swiftlet and vagrant Glossy Swiftlet. The juvenile is probably indistinguishable in the field. When flying low and close, scratchy, high twittering may be heard. Habitat mostly over lowland rainforest in NG, but will use airspace over almost any terrain, including coastal and offshore islands. Common and widespread across N islands, many races; very rare vagrant with few Aust. sightings – only one confirmed record from Cape York.

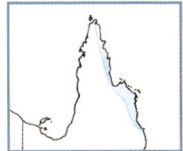

Numerous subspecies through NG and islands to NW of Aust.; race most likely to reach Aust. is *yorki* from lowlands of southern NG.

Slightly larger and more thickset than resident Glossy Swiftlet. Wing tip is very slightly more sharply pointed.

In open country, flocks will hawk for insects almost to ground level; at other times, just above forest canopy, or very much higher.

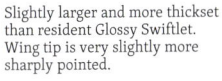

Upper parts dark grey-brown with greenish bronze sheen

Rump dark

Tail has slightly shallower fork than has tail of Glossy Swiftlet.

Typical erratic swiftlet flight, hawking for insects in large flocks

Upper parts uniformly dark like the Glossy, but unlike the Australian Swiftlet, which has a white to dull grey rump

Throat usually slightly paler than rest of the under-parts. (On the similar Glossy Swiftlet the throat is not paler.)

Underparts almost uniformly dark (or only very slightly paler at throat) compared with the pale-bellied undersurface of the Glossy Swiftlet

Wing position and shape change greatly between soaring, slow and fast glide, powered flight, and tight twists and turns.

When any all-dark swiftlet is sighted, especially in NW of WA, other potential visitors from islands to the N include Edible-nest Swiftlet, *Aerodramus fuciphagus*, and Black-nest Swiftlet, *A. maximus*.

**House Swift** *Apus nipalensis* 13–14 cm
A small, rather stocky swift. Large, clear-cut white rump patch wraps around, usually visible from below. Flight is not as fast, powerful or graceful as larger swifts; steady but rather fluttering, weaker, often gliding. Calls are a harsh, shrill, rippling or rattling trill; occasionally high, squeaky shrieks. Habitat is the airspace above most environs within its range, including coastal cliffs, caves and buildings. Common from Indonesia to Asia and Africa; possible rare vagrant across N Aust. Any swifts of this species reaching Aust. are most likely to be from the nearest breeding population on Indonesian islands. Also occurring in Java, Sumatra and further N is the Silver-rumped Spinetail – it has one possible sighting in company with other swifts in the Kimberley of WA.

Wing shape and position varies greatly.

Compared with swiftlets, House Swift is more compact, which, with white throat and rump, gives appearance and jiz similar to the White-breasted Woodswallow.

Wings rather blunt-tipped compared with other local swifts

White rump patch extends to sides; may be glimpsed from below.

White throat

Closed, has small notch

Square if partly spread, rounded if widely spread

As with other swifts, sunlight makes thin, translucent inner flight feathers brighter than opaque plumage.

Very shallow fork, becomes square tipped if only partially spread.

A possible vagrant, the slightly smaller Silver-rumped Spinetail, *Rhapidura leucopygialis*, has silvery white from rump over the tail-coverts almost to tip of tail; otherwise blackish, including throat.

## SWIFTLETS

### Glossy Swiftlet *Collocalia esculenta* 9–11.5 cm
Smallest Australian swift, no white on rump, underside dark merging softly to whitish belly. Flight erratic, on shallow fluttering, downcurved wings; abrupt turns, stalls, glides. Sweeps very low through forest spaces, often over watercourses or, in faster flight, high above canopy. Usually silent, but occasional weak, sharp twitter. Habitat is the airspace above and through the forests. Occasional vagrant to NE Qld.; one record from Broome, WA.

Race *esculens* breeds on Christmas Is. in the Indian Ocean. It has upper parts glossed greenish and more white showing in the fanned tail.

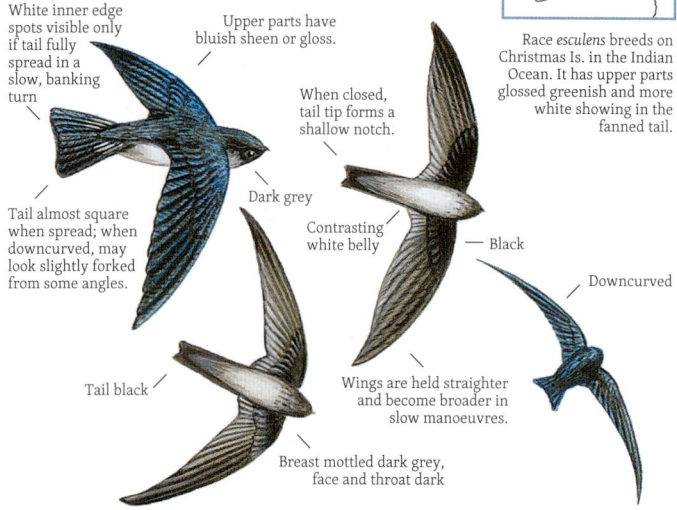

### Australian Swiftlet *Aerodramus terraereginae* 11–12 cm
Small, dark, grey-brown swiftlet with dull white rump. Coastal ranges, gorges, islands, woodlands to 1000 m altitude. Small parties to large flocks; flight erratic, twisting and turning in pursuit of flying insects, occasionally quite low, at other times high. Calls are high-pitched twitterings. Usually seen in vicinity of breeding caves. The race *terraereginae* occurs along the Qld coast and offshore islands, nesting in cave-like crevices among boulders; race *chillagoensis* is confined to a smaller range, using limestone caves near Chillagoe as nest sites. Sedentary; common within restricted range.

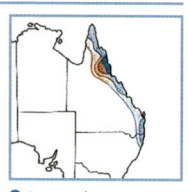

- *terraereginae*
- *chillagoensis*